# Methods in Enzymology

Volume 105
OXYGEN RADICALS IN
BIOLOGICAL SYSTEMS

# METHODS IN ENZYMOLOGY

EDITORS-IN-CHIEF

Sidney P. Colowick     Nathan O. Kaplan

*Methods in Enzymology*

*Volume 105*

# Oxygen Radicals in Biological Systems

EDITED BY

Lester Packer

MEMBRANE BIOENERGETICS GROUP
UNIVERSITY OF CALIFORNIA
BERKELEY, CALIFORNIA

## Advisory Board

Bruce Ames
Anthony Diplock
Lars Ernster
Irwin Fridovich

Rolf J. Mehlhorn
William A. Pryor
Trevor F. Slater

1984

ACADEMIC PRESS, INC.

*(Harcourt Brace Jovanovich, Publishers)*

Orlando   San Diego   San Francisco   New York   London
Toronto   Montreal   Sydney   Tokyo   São Paulo

QP 601
C71
v. 105
1984.

COPYRIGHT © 1984, BY ACADEMIC PRESS, INC.
ALL RIGHTS RESERVED.
NO PART OF THIS PUBLICATION MAY BE REPRODUCED OR
TRANSMITTED IN ANY FORM OR BY ANY MEANS, ELECTRONIC
OR MECHANICAL, INCLUDING PHOTOCOPY, RECORDING, OR ANY
INFORMATION STORAGE AND RETRIEVAL SYSTEM, WITHOUT
PERMISSION IN WRITING FROM THE PUBLISHER.

ACADEMIC PRESS, INC.
Orlando, Florida 32887


*United Kingdom Edition published by*
ACADEMIC PRESS, INC. (LONDON) LTD.
24/28 Oval Road, London NW1 7DX


LIBRARY OF CONGRESS CATALOG CARD NUMBER:     54-9110

ISBN 0-12-182005-X

PRINTED IN THE UNITED STATES OF AMERICA

84 85 86 87     9 8 7 6 5 4 3 2 1

# Table of Contents

## Section I. Chemistry and Biochemistry of Oxygen and Intermediate States of Its Reduction

## Section II. Isolation and Assays of Enzymes or Substances Resulting in Formation or Removal of Oxygen Radicals

### A. One-Electron Transfer Enzyme Reactions Resulting in $O_2^-$ Production

## B. Isolation, Purification, Characterization, and Assay of Antioxygenic Enzymes

## C. Antioxidants/Prooxidants

## D. Detection and Characterization of Oxygen Radicals

## B. Plasma Membrane

# Contributors to Volume 105

Article numbers are in parentheses following the names of contributors.
Affiliations listed are current.

R. B. ABAKERLI (3), *Industrias, Monsanto, Estrada Municipal, PLN 393 Paulinia, São Paulo, Brazil*

HUGO AEBI (13), *Medizinisch-chemisches Institut, Universität Bern, CH-3000 Bern 9, Switzerland*

THEO P. M. AKERBOOM (59), *Institut für Physiologische Chemie I, Universität Düsseldorf, D-4000 Düsseldorf, Federal Republic of Germany*

BRUCE N. AMES (29, 44), *Department of Biochemistry, University of California, Berkeley, California 94720*

PATRICIA C. ANDREWS (45), *Department of Medicine, Tufts-New England Medical Center, Boston, Massachusetts 02111*

KOZI ASADA (56), *The Research Institute for Food Science, Kyoto University, Uji, Kyoto 611, Japan*

KLAUS-DIETER ASMUS (20), *Hahn-Meitner-Institut Berlin, Bereich Strahlenchemie, D-1000 Berlin 39, Federal Republic of Germany*

BERNARD M. BABIOR (45), *Department of Medicine, Tufts-New England Medical Center, Boston, Massachusetts 02111*

J. V. BANNISTER (9), *Inorganic Chemistry Laboratory, University of Oxford, Oxford OX1 3QR, England*

W. H. BANNISTER (9), *Nuffield Department of Clinical Biochemistry, Radcliffe Infirmary, University of Oxford, Oxford OX2 6HE, England*

BENON H. J. BIELSKI (8), *Department of Chemistry, Brookhaven National Laboratory, Upton, New York 11973*

R. P. BIRD (35), *Ludwig Institute for Cancer Research, Toronto, Ontario M4Y 2L4, Canada*

CARMIA BOREK (62), *Radiological Research Laboratory, Cancer Center, Institute of Cancer Research, Columbia University, College of Physicians and Surgeons, New York, New York 10032*

ALBERTO BOVERIS (57), *Departamento de Química Biológica, Facultad de Farmacia y Bioquímica, Universidad de Buenos Aires, 1113 Buenos Aires, Argentina*

M. BRUNORI (2), *Department of Biochemistry, Biological Sciences, and Experimental Medicine, II University of Rome, Rome 00100, Italy*

JUDITH L. BUTTRISS (15), *Department of Biochemistry, Guy's Hospital Medical School, London SE1 9RT, England*

ENRIQUE CADENAS (26), *Institut für Physiologische Chemie I, Universität Düsseldorf, D-4000 Düsseldorf, Federal Republic of Germany*

LAWRENCE CASTLE (34), *Department of Chemistry, Louisiana State University, Baton Rouge, Louisiana 70803*

RICHARD CATHCART (44), *Department of Biochemistry, University of California, Berkeley, California 94720*

ARTHUR I. CEDERBAUM (68), *Department of Biochemistry, Mount Sinai School of Medicine, New York, New York 10029*

MICHAEL F. CHRISTMAN (29), *Department of Biochemistry, University of California, Berkeley, California 94720*

DENNIS P. CLIFFORD (51), *Webb-Waring Lung Institute, University of Colorado Medical Center, Denver, Colorado 80262*

GERALD COHEN (36, 67, 68), *Department of Neurology, Mount Sinai School of Medicine, New York, New York 10029*

GIDON CZAPSKI (24), *Department of Physical Chemistry, Hebrew University of Jerusalem, Jerusalem 91904, Israel*

ROLANDO DEL MAESTRO (49), *Brain Research Laboratory, Department of Clini-*

cal Neurological Sciences, Victoria Hospital, London, Ontario N6A 4G5, Canada

INDRAJIT D. DESAI (16), Division of Human Nutrition, University of British Columbia, Vancouver, British Columbia V6T 1W5, Canada

CORA J. DILLARD (41), Department of Food Science and Technology, University of California, Davis, California 95616

ANTHONY T. DIPLOCK (15), Department of Biochemistry, Guy's Hospital Medical School, London SE1 9RT, England

THOMAS A. DIX (43), Department of Chemistry, Pennsylvania State University, State College, Pennsylvania 16802

H. H. DRAPER (35), Department of Nutrition, University of Guelph, Guelph, Ontario N1G 2W1, Canada

HERMANN ESTERBAUER (38), Institute of Biochemistry, University of Graz, A-8010 Graz, Austria

L. FLOHÉ (10, 12), Grünenthal GmbH, Center of Research, D-5100 Aachen, Federal Republic of Germany

ROBERT A. FLOYD (27), Biomembrane Research Program, Oklahoma Medical Research Foundation, Oklahoma City, Oklahoma 73104

CHRISTOPHER S. FOOTE (3), Department of Chemistry, University of California, Los Angeles, California 90024

L. G. FORNI (21), Department of Biochemistry, Brunel University, Uxbridge UB8 3PH, England

IRWIN FRIDOVICH (5), Department of Biochemistry, Duke University Medical Center, Durham, North Carolina 27710

CHERYL A. FURTEK (31), Laboratories of Molecular Biology and Genetics, University of Wisconsin, Madison, Wisconsin 53706

BRUCE L. GELLER (11), Department of Biological Chemistry, UCLA School of Medicine, Los Angeles, California 90024

ERIC A. GLENDE, JR. (40), Department of Physiology, School of Medicine, Case

Western Reserve University, Cleveland, Ohio 44106

MONIKA J. GREEN (1), Inorganic Chemistry Laboratory, University of Oxford, Oxford OX1 3QR, England

WOLFGANG A. GÜNZLER (12), Grünenthal GmbH, Center of Research, D-5100 Aachen, Federal Republic of Germany

JOHN M. C. GUTTERIDGE (4), National Institute for Biological Standards, Hampstead, London HW3 6RB, England

BARRY HALLIWELL (4), Department of Biochemistry, King's College, University of London, London WC2R 2LS, England

HOSNI M. HASSAN (30, 53, 69), Departments of Food Science and Microbiology, North Carolina State University, Raleigh, North Carolina 27650

LINDA HATCH (19), Institute for Toxicology, University of Southern California, Los Angeles, California 90033

OSAMU HAYAISHI (6), Osaka Medical College, Takatsuki 569, Japan

RICHARD E. HEIKKILA (67), Department of Neurology, Rutgers Medical School, Piscataway, New Jersey 08854

H. ALLEN O. HILL (1), Inorganic Chemistry Laboratory, University of Oxford, Oxford OX1 3QR, England

PAUL HOCHSTEIN (19), Institute for Toxicology, University of Southern California, Los Angeles, California 90033

MONICA HOLLSTEIN (29), Department of Biochemistry, University of California, Berkeley, California 94720

SATORU IKENOYA (17), Development Division, Eisai Co., Ltd., 4-6-10 Koishikawa, Bunkyo-ku, Tokyo, Japan

KENDALL A. ITOKU (42), Medical School, Health Science Center, University of Missouri, Columbia, Missouri 65211

HARRY S. JACOB (48), Division of Hematology, Mayo Memorial Building, University of Minnesota Hospitals, Minneapolis, Minnesota 55455

EDWARD G. JANZEN (22), Guelph-Waterloo

Center for Graduate Work in Chemistry, Department of Chemistry and Biochemistry, University of Guelph, Guelph, Ontario N1G 2W1, Canada

RICHARD B. JOHNSTON, JR. (46), Department of Pediatrics, National Jewish Hospital and Research Center, and University of Colorado School of Medicine, Denver, Colorado 80206

HAROLD P. JONES (50), Department of Biochemistry, University of South Alabama, Mobile, Alabama 36688

KOUICHI KATAYAMA (17), Tsukuba Research Laboratories, Eisai Co., Ltd., Tokodai, Toyosato-machi, Tsukuba-gun, Ibaragi 300-26, Japan

SEYMOUR J. KLEBANOFF (52), Department of Medicine, University of Washington School of Medicine, Seattle, Washington 98195

NORMAN I. KRINSKY (18), Department of Biochemistry and Pharmacology, Tufts University School of Medicine, Boston, Massachusetts 02111

SIMO LAAKSO (14), Department of Biochemistry, University of Turku, SF-20500 Turku 50, Finland

JOHANNA LANG (38), Institute of Biochemistry, University of Graz, A-8010 Graz, Austria

GLEN D. LAWRENCE (36), Institute of Human Nutrition, Columbia University College of Physicians and Surgeons, New York, New York 10032

STEPHEN A. LESKO (71), Division of Biophysics, The Johns Hopkins University School of Hygiene and Public Health, Baltimore, Maryland 21205

DAVID E. LEVIN (29), Department of Biochemistry, University of California, Berkeley, California 94720

C. ANN LEWIS (27), Department of Biochemistry and Microbiology, University of St. Andrews, St. Andrews Fife Ky16 9AL, Scotland

ESA-MATTI LILIUS (14), Department of Bio-

chemistry, University of Turku, SF-20500 Turku 50, Finland

W. LOHMANN (60), Institut für Biophysik, Justus-Liebig-Universität, D-6300 Giessen, Federal Republic of Germany

J. WILLIAM LOWN (70), Department of Chemistry, University of Alberta, Edmonton, Alberta T6G 2G2, Canada

MICHÈLE MARKERT (45), Laboratoire Central de Chimie Clinique, Centre Hospitalier Universitaire Vaudois, 1011 Lausanne, Switzerland

LAWRENCE J. MARNETT (43, 54), Department of Chemistry, Wayne State University, Detroit, Michigan 48202

RONALD P. MASON (55), Laboratory of Molecular Biophysics, National Institute of Environmental Health Sciences, Research Triangle Park, North Carolina 27709

JOE M. McCORD (50), Department of Biochemistry, University of South Alabama, Mobile, Alabama 36688

ROLF J. MEHLHORN (25), Membrane Bioenergetics Group, Lawrence Berkeley Laboratory, University of California, Berkeley, California 94720

ANDREW R. MIKSZTAL (7), Department of Chemistry and Biochemistry and Molecular Biology Institute, University of California, Los Angeles, California 90024

GREGORY V. MILLER (42), Medical School, Health Science Center, University of Missouri, Columbia, Missouri 65211

CHARLES F. MOLDOW (48), Department of Medicine, Veterans Administration Medical Center, Minneapolis, Minnesota 55417

CARMELLA S. MOODY (30), Department of Microbiology, North Carolina State University, Raleigh, North Carolina 27650

ARMIN MÜLLER (37), Institut für Physiologische Chemie I, Universität Düsseldorf, D-4000 Düsseldorf, Federal Republic of Germany

KENNETH D. MUNKRES (31), Laboratories

of Molecular Biology and Genetics, University of Wisconsin, Madison, Wisconsin 53706

H. NEUBACHER (60), Institut für Biophysik, Justus-Liebig-Universität, D-6300 Giessen, Federal Republic of Germany

LARRY W. OBERLEY (61), Radiation Research Laboratory, Department of Radiology, The University of Iowa, Iowa City, Iowa 52242

PETER J. O'BRIEN (47), Department of Biochemistry, Memorial University of Newfoundland, St. John's, Newfoundland A1B 3X9, Canada

STEN ORRENIUS (66), Department of Forensic Medicine, Karolinska Institutet, S-104 01 Stockholm, Sweden

F. ÖTTING (10), Grünenthal GmbH, Center of Research, D-5100 Aachen, Federal Republic of Germany

LESTER PACKER (25), Membrane Bioenergetics Group, University of California, Berkeley, California 94720

RONALD PETHIG (28), School of Electronic Engineering Science, University College of North Wales, Bangor, Gwynedd LL57 1UT, Wales

NED A. PORTER (32), Department of Chemistry, Duke University, Durham, North Carolina 27706

WILLIAM A. PRYOR (34), Departments of Chemistry and Biochemistry, Louisiana State University, Baton Rouge, Louisiana 70803

ELMER J. RAUCKMAN (23), Department of Surgery, Duke University Medical Center, Durham, North Carolina 27710

RICHARD O. RECKNAGEL (40), Department of Physiology, School of Medicine, Case Western Reserve University, Cleveland, Ohio 44106

J. LESLIE REDPATH (65), Division of Radiation Oncology, Department of Radiological Sciences, California College of Medicine, University of California, Irvine, California 92717

JOHN E. REPINE (51), Webb-Waring Lung Institute, University of Colorado Medical Center, Denver, Colorado 80262

CHRISTOPH RICHTER (58), Laboratory of Biochemistry I, Swiss Federal Institute of Technology, CH-8092 Zürich, Switzerland

GERALD M. ROSEN (23), Department of Pharmacology, Duke University Medical Center, Durham, North Carolina 27710

HENRY ROSEN (52), Department of Medicine, University of Washington School of Medicine, Seattle, Washington 98195

G. ROTILIO (2), Department of Biology, Faculty of Sciences, II University of Rome, Rome 00100, Italy

DONALD T. SAWYER (7), Department of Chemistry, University of California, Riverside, Riverside, California 92521

ELIZABETH SCHWIERS (44), Department of Biochemistry, University of California, Berkeley, California 94720

ROGER C. SEALY (63), National Biomedical ESR Center, Medical College of Wisconsin, Milwaukee, Wisconsin 53226

ALEX SEVANIAN (19), Institute for Toxicology, University of Southern California, Los Angeles, California 90033

F. C. SHOOK (3), Pilot Chemical Company, Santa Fe Springs, California 90670

PAUL H. SIEDLIK (54), Pharmacology Department, Warner Lambert-Parke Davis, Ann Arbor, Michigan 48105

HELMUT SIES (26, 37, 59), Institut für Physiologische Chemie I, Universität Düsseldorf, D-4000 Düsseldorf, Federal Republic of Germany

T. F. SLATER (33, 38), Department of Biochemistry, Brunel University, Uxbridge, Middlesex UB8 3PH, England

MARTYN T. SMITH (66), Department of Biomedical and Environmental Health Sciences, University of California, Berkeley, California 94720

R. S. SOHAL (64), Department of Biology, Southern Methodist University, Dallas, Texas 75275

DOUGLAS R. SPITZ (61), Radiation Research Laboratory, Department of Radiology, The University of Iowa, Iowa City, Iowa 52242

WILLIAM S. STARK (42), Division of Biological Sciences, University of Missouri, Columbia, Missouri 65211

MASAHIRO TAKADA (17), Tsukuba Research Laboratories, Eisai Co., Ltd., Tokodai, Toyosato-machi, Tsukuba-gun, Ibaragi 300-26, Japan

AL L. TAPPEL (41), Department of Food Science and Technology, University of California, Davis, California 95616

HJÖRDIS THOR (66), Department of Forensic Medicine, Karolinska Institutet, S-104 01 Stockholm, Sweden

PEKKA TURUNEN (14), Wallac Biochemical Laboratory, SF-20101 Turku 10, Finland

JOAN S. VALENTINE (7), Department of Chemistry and Biochemistry and Molecular Biology Institute, University of California, Los Angeles, California 90024

ANN M. WALTERSDORPH (52), Department of Medicine, University of Washington School of Medicine, Seattle, Washington 98195

SUDHAKAR WELANKIWAR (18), Department of Biochemistry and Pharmacology, Tufts University School of Medicine, Boston, Massachusetts 02111

R. L. WILLSON (21), Department of Biochemistry, Brunel University, Uxbridge UB8 3PH, England

DENNIS R. WINGE (11), Departments of Biochemistry and Medicine, University of Utah, Salt Lake City, Utah 84132

PETER K. WONG (27), Biomembrane Research Program, Oklahoma Medical Research Foundation, Oklahoma City, Oklahoma 73104

KUNIO YAGI (39), Institute of Applied Biochemistry, Yagi Memorial Park, Mitake, Gifu 505-1, Japan

RYOTARO YOSHIDA (6), Department of Medical Chemistry, Kyoto University Faculty of Medicine, Kyoto 606, Japan

TERUAKI YUZURIHA (17), Tsukuba Research Laboratories, Eisai Co., Ltd., Tokodai, Toyosato-machi, Tsukuba-gun, Ibaragi 300-26, Japan

SYLVIA ZADRAVEC (38), Institute of Biochemistry, University of Graz, A-8010 Graz, Austria

# Preface

Increasingly, researchers and clinicians are recognizing that biological damage mediated by reactive species of oxygen is an important factor in disease and aging. Oxygen radical damage has been implicated in inflammation, arthritis, adult respiratory distress syndrome, myocardial infarction, pulmonary dysfunction in hemodialized patients, Purtscher's syndrome, Bloom's syndrome, systemic lupus erythematosus, mutagenicity and carcinogenicity, and other pathologies. While more experimental data to support the involvement of oxygen radicals in these conditions are needed, the explosive interest of many investigators in the field more than warrants a volume that covers the most current methods used in studying the role of oxygen radicals in biological systems.

Much of the recent progress in research concerning the role of oxygen radicals in disease has been due to the discovery and utilization of antioxidant substances. Since the discovery by McCord and Fridovich that the plasma protein, erythrocuprein, functions to dismutate the superoxide anion radical, this oxygen radical, as well as other oxygen radical species and products derived therefrom, have received a great deal of attention. These other compounds include hydrogen peroxide, hydroxyl radical $(OH \cdot)$, and singlet oxygen $(^1O_2)$. The study of oxygen radical reactions in simple chemical and biochemical systems has defined parameters governing the initiation, propagation, and termination of destructive free radical chain reactions, and this understanding has suggested oxygen radical involvement in a remarkably diverse and extensive array of pathological states.

Dramatic therapeutic effects of antioxidant substances, such as vitamin E and selenium, in clinical settings have provided further support for the involvement of oxygen radicals in certain disease conditions. Recent research results of chemists, biologists, clinicians, and epidemiologists have created a new multidisciplinary field considered by some to be among the most exciting and important areas of biological research.

As with most research, progress is frequently accelerated by the publication of a single volume that contains all of the state-of-the-art methodologies used in the field. The major emphasis of this volume is on the methods and on the basic chemistry and biochemistry of oxygen radicals and their effects in biological systems. Such a volume, it is hoped, will stimulate new research, heighten awareness of recent discoveries in basic research, and focus attention on their implications for human health.

In preparing this volume, I would be remiss in not pointing out the very great benefit derived from the active participation of the advisory board (Bruce Ames, Anthony Diplock, Lars Ernster, Irwin Fridovich, Rolf J. Mehlhorn, William A. Pryor, and Trevor Slater) and Al L. Tappel in the selection of the methods to be included and in identifying the leading investigators to make these contributions to the volume. Also I acknowledge the valuable editorial and administrative assistance provided by Mr. John Hazlett who worked closely with me in bringing this volume to fruition.

LESTER PACKER

# METHODS IN ENZYMOLOGY

## EDITED BY

### Sidney P. Colowick and Nathan O. Kaplan

VANDERBILT UNIVERSITY  
SCHOOL OF MEDICINE  
NASHVILLE, TENNESSEE

DEPARTMENT OF CHEMISTRY  
UNIVERSITY OF CALIFORNIA  
AT SAN DIEGO  
LA JOLLA, CALIFORNIA

# METHODS IN ENZYMOLOGY

### EDITORS-IN-CHIEF

## Sidney P. Colowick     Nathan O. Kaplan

VOLUME VIII. Complex Carbohydrates
*Edited by* ELIZABETH F. NEUFELD AND VICTOR GINSBURG

VOLUME IX. Carbohydrate Metabolism
*Edited by* WILLIS A. WOOD

VOLUME X. Oxidation and Phosphorylation
*Edited by* RONALD W. ESTABROOK AND MAYNARD E. PULLMAN

VOLUME XI. Enzyme Structure
*Edited by* C. H. W. HIRS

VOLUME XII. Nucleic Acids (Parts A and B)
*Edited by* LAWRENCE GROSSMAN AND KIVIE MOLDAVE

VOLUME XIII. Citric Acid Cycle
*Edited by* J. M. LOWENSTEIN

VOLUME XIV. Lipids
*Edited by* J. M. LOWENSTEIN

VOLUME XV. Steroids and Terpenoids
*Edited by* RAYMOND B. CLAYTON

VOLUME XVI. Fast Reactions
*Edited by* KENNETH KUSTIN

VOLUME XVII. Metabolism of Amino Acids and Amines (Parts A and B)
*Edited by* HERBERT TABOR AND CELIA WHITE TABOR

VOLUME XVIII. Vitamins and Coenzymes (Parts A, B, and C)
*Edited by* DONALD B. MCCORMICK AND LEMUEL D. WRIGHT

VOLUME XIX. Proteolytic Enzymes
*Edited by* GERTRUDE E. PERLMANN AND LASZLO LORAND

# Methods in Enzymology

Volume 105
OXYGEN RADICALS IN
BIOLOGICAL SYSTEMS

# Section I

## Chemistry and Biochemistry of Oxygen and Intermediate States of Its Reduction

## [1] Chemistry of Dioxygen

*By* Monika J. Green and H. Allen O. Hill

*To be tired of the chemistry of oxygen is to be tired of Life*

The chemistry of molecular oxygen, or dioxygen as it is increasingly coming to be called, is dominated by the relative reluctance with which the element reacts with most, *but not all,* compounds. This quality is rarely thermodynamic in origin; rather the slow rate of reaction is associated with either the strong oxygen–oxygen bond or the character of the ground state of dioxygen or both. As is well known, dioxygen is paramagnetic, a consequence of the two unpaired electrons which lead to its description as a triplet. (Three states are revealed when the molecule is placed in a magnetic field.) No orbital angular momentum is associated with the ground state; thus the molecule is well described by the following term: $^3\Sigma_g^-$. Most molecules are diamagnetic (with ground states described as singlets). If a reaction is to take place with dioxygen in its ground state, there must be a change of spin at some stage during the reaction. This is forbidden, within the limitations of the descriptions used; at the very least, the reactions are improbable. For dioxygen to react rapidly, this spin restriction should be removed or the oxygen–oxygen bond should be partially weakened, or both of these should occur simultaneously. The conversion of dioxygen to its excited state singlet forms, either photochemically or thermally, both weakens the O—O bond and removes the spin restriction. The reduction of dioxygen also serves to weaken the O—O bond and gives rise to species whose reactions are not subject to a kinetic barrier associated with a spin restriction. Of course, if the species with which dioxygen reacts also contains unpaired electrons or can provide a ready interconversion of spin states, then the spin restriction is removed or irrelevant. This does not obviate the problem associated with the strong O—O bond and indeed in such cases it is possible that at least the initial products will contain dioxygen with the O—O bond intact or nearly so. It is not surprising that much of the chemistry of dioxygen is concerned with reactions with paramagnetic species, with electron-donating species, with light, or with various combinations of these three factors. We will be concerned primarily with the chemistry of the reduction products of dioxygen but one should be aware of the marked influence of

Copyright © 1984 by Academic Press, Inc.
All rights of reproduction in any form reserved.
ISBN 0-12-182005-X

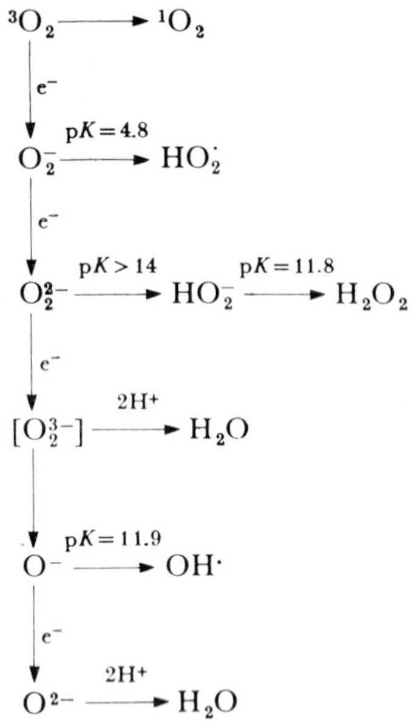

$$^3O_2 \longrightarrow {}^1O_2$$

$$\Big\downarrow e^-$$

$$\overset{pK=4.8}{O_2^- \longrightarrow HO_2^{\cdot}}$$

$$\Big\downarrow e^-$$

$$\overset{pK>14}{O_2^{2-} \longrightarrow} HO_2^- \overset{pK=11.8}{\longrightarrow} H_2O_2$$

$$\Big\downarrow e^-$$

$$[O_2^{3-}] \overset{2H^+}{\longrightarrow} H_2O$$

$$\Big\downarrow$$

$$O^- \overset{pK=11.9}{\longrightarrow} OH^{\cdot}$$

$$\Big\downarrow e^-$$

$$O^{2-} \overset{2H^+}{\longrightarrow} H_2O$$

FIG. 1. The products derived from the successive one-electron reductions of dioxygen.

all of these factors on most reactions of dioxygen. They are often, if not always, responsible for the promotion or catalysis of oxidation reactions.[1]

### The Reduction Products of Dioxygen

The sequence of one-electron reductions of dioxygen is shown in Fig. 1. The first reduction product, the superoxide ion, $O_2^-$, is a base with the equilibrium with its conjugate acid, the hydroperoxyl radical, $O_2H$, described by a $pK$ of 4.8. This is sufficiently close to biologically and physiologically relevant pH values to demand consideration of the participation of $HO_2$ in reactions ostensibly of the superoxide ion. The second reduction product, the peroxide ion, is a very strong base. The first $pK$ can only be estimated. The second, 11.8, is such that both hydrogen peroxide and the hydroperoxyl anion, $O_2H^-$, are possible reactants in systems involving the former. This is particularly true in the presence of multivalent cations which can lower the apparent $pK$. The hypothetical three-electron reduction product, $[O_2^{3-}]$, can be considered the precursor of oxide and oxene. The former is trivially the conjugate base of water; the latter of the hydroxyl radical, $pK = 11.9$. Thus in considering the highly reactive

[1] H. A. O. Hill, *Ciba Found. Symp. New Ser.* **65,** 5 (1979).

hydroxyl radical as the three-electron reduction product of dioxygen, it is possible to relate the chemistry of dioxygen, through superoxide and peroxide, with the chemistry of oxygen through oxene and oxide.

The relationship between dioxygen and its reduction products is further described in the oxidation state diagram (Fig. 2). This provides a convenient and concise representation of the thermodynamic properties of the species as contained in the electrode potentials. From a plot of the product of the oxidation state and the electrode potential against the oxidation state, the electrode potential relating any two species can be

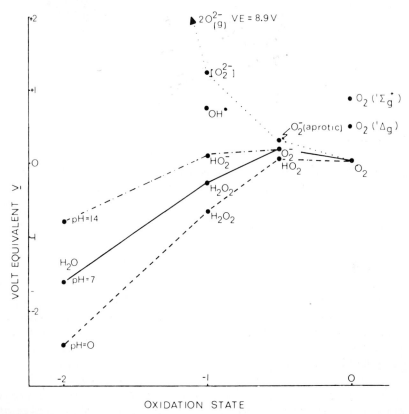

OXIDATION STATE

FIG. 2. The oxidation-state diagram for dioxygen: $V$, the volt equivalent ($VE$), is the standard electrode potential at 25° (versus the normal hydrogen electrode) multiplied by the oxidation state (the *formal* charge per atom of element in a *compound* or *ion*, assuming that each oxygen atom has a charge of $-2$ and each hydrogen atom a charge of $+1$; the oxidation state of the elemental form is zero). The potentials shown are, therefore those of the cell:

Pt, $H_2|a_{H^+} = 1\|$Standard couple $pO_2 = 1$ atm, pH solvent as indicated|Pt

derived simply by taking the slope of the line joining the two relevant points. As with all thermodynamic quantities, no account is taken of the rates of the reactions involved. Additionally, the electrode potentials are dependent on the experimental conditions and besides pH, are sensitive to solvent and the presence of metal ions, even those that are not redox active. Consequently, even assuming that conditions close to equilibrium prevail in the system under investigation, differences in the experimental conditions and the intrusion of factors that determine the rates of the reaction mean that extrapolation from the data contained in Fig. 2 must proceed with caution. Thus though it is obvious from Fig. 2 that hydrogen peroxide should disproportionate to dioxygen and water, it does not readily do so unless the reaction is catalyzed by redox-active metal complexes of which catalase is the supremely effective exponent. The predicted disproportionation of the superoxide ion is observed to proceed relatively rapidly though, here again, catalysis by metal complexes and the copper–zinc, manganese, and iron dismutases is efficient. However an examination of the pH dependence of the reaction shows that the uncatalyzed reaction proceeds via the intermediacy of the hydroperoxyl radical [Eqs. (1)–(3)]:

$$HO_2 + HO_2 \rightarrow O_2 + H_2O_2 \qquad k = 8.6 \times 10^5 \ M^{-1} \ s^{-1} \qquad (1)$$
$$HO_2 + O_2^- \rightarrow O_2 + HO_2^- \qquad k = 1.02 \times 10^8 \ M^{-1} \ s^{-1} \qquad (2)$$

but

$$O_2^- + O_2^- \rightarrow O_2 + O_2^{2-} \qquad k = 0.35 \ M^{-1} \ s^{-1} \qquad (3)$$

The reduction products of dioxygen, and the excited singlet forms of the element, are all predicted to be good oxidants with respect to reduction to water; this is indeed the case for the hydroxyl radical. The catalytic properties of metal complexes are of value to the synthetic chemist and reach a pinnacle of refinement in, e.g., the peroxidases. However, they introduce complications in the interpretation of the results and lead to variables often outside the control of the experimentalist. Thus the Haber–Weiss reaction [Eq. (4)] provides an example of a reaction that has "become slower with time"; as the catalytic properties of metal ions were recognized and it became possible to control the levels of "adventitious" metal ions, so the observed rate of the reaction decreased.

$$H_2O_2 + O_2^- \rightarrow O_2 + OH^- + OH\cdot \qquad k \le 0.13 \ M^{-1} \ s^{-1} \qquad (4)$$

The catalytic properties of redox-active metal ions, in particular those of iron and copper, are thought to involve the following reactions [Eq. (5)–(6)]:

$$H_2O_2 + M^{n+} \rightarrow M^{(n+1)+} + OH^- + OH\cdot \qquad (5)$$
$$O_2^- + M^{(n+1)+} \rightarrow O_2 + M^{n+} \qquad (6)$$

Obviously, reductants other than the superoxide ion can effect the "recycling" of the metal ion. Nevertheless, it is observed that dioxygen, in the presence of sources of the superoxide ion and redox-active metal ions, provides a ready source of the hydroxyl radical. The product of the Fenton reaction, i.e., Eq. (5) with $M^{n+} = Fe^{2+}$, is presumed to be the hydroxyl ion. (Though most interest centers on adventitious iron, it should be remembered that the rate of reaction of $Cu^+$ with $H_2O_2$ is many orders of magnitude faster than that of $Fe^{2+}$.) However, it is possible that its conjugate base, the oxene anion, $O^-$, can act as a Lewis base and thus as a ligand to metal ions [Eqs. (7) and (8)]:

$$H^+ + O^- \rightarrow HO\cdot \tag{7}$$
$$M^{n+} + O^- \rightarrow MO^{(n-1)+} \tag{8}$$

Ironically, the evidence for the participation of such metal complexes of the oxene anion is more convincing for reactions involving redox enzymes, such as the ferryl intermediates in the peroxidases and monooxygenases, than it is for deceptively simple chemical systems.

Metal ions therefore are a potent influence on the chemistry of dioxygen and its reduction products. In order to assess their role in any biochemical reaction, their concentrations must be controlled. There are two ways of approaching this problem: to remove them or sequester them with chelating agents.[2] The former route is not a trivial exercise. Indeed for many systems of biological interest it is impossible since "free" metal ions may be required for the integrity of the enzyme, organelle, or cell. If adopted, it will be necessary to remove the metal ions from glassware, equipment, and reagents including proteins. If such an approach is not feasible, it is usual to attempt to control the level of the "adventitious" metal ions by adding sequestering agents. For the most careful work, it should be remembered that commercially available sequestering agents often contain metal ions; addition of chelating agents may therefore *increase* the amount of metal ions in the system. It should not be presumed that, even if the metal ion is completely sequestered, i.e., the concentration of the "free" aquated metal ion is vanishingly low, the potential for catalysis has been removed. Many metal complexes are *better* catalysts than the free metal ion. (Obviously this is the case for most, if not all, metalloenzymes!) It should also be remembered that chelating agents may affect reactions in a manner not necessarily related to their metal ion sequestering properties. The recommended protocol is to scrupulously remove metal ions from those components of the experiment not dependent upon them. If it is necessary to add chelating agents, use the mini-

[2] B. Halliwell, *FEBS. Lett.* **92**, 321 (1978).

mum quantity of *purified* reagent and assess the specificity of the chelating agent by using more than one.

### The Analytical Chemistry of Dioxygen and Its Reduction Products

We will be principally concerned with the analysis of superoxide, hydrogen peroxide, and the hydroxyl radical. Procedures relevant to, e.g., the analysis of hydrogen peroxide in organelles or cells have been considered in other volumes. Likewise the chemistry of the superoxide ion in aprotic media is described in [10]. (At this stage we will revert to usage more common to analytical procedures and refer to oxygen rather than dioxygen when describing the elemental form.)

### Oxygen Electrode

Commercially available oxygen electrodes are normally of the Clark type. They consist of a two-electrode cell with a platinum cathode and a silver anode encased in an epoxy block. A thin, gas-permeable membrane, normally polyethylene or a fluorinated plastic, is stretched over the electrodes and kept in place by a rubber O ring. A layer of electrolyte solution (2.33 $M$ KCl) between electrodes and the membrane maintains electrical contact. A typical electrode is shown in Fig. 3. Oxygen can diffuse across the membrane from the test solution into the electrolyte

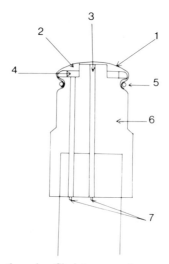

Fig. 3. A cross section through a Clark-type membrane-covered polarographic oxygen detector (adapted from Ref. 3); 1, membrane; 2, electrolyte reservoir; 3, cathode; 4, anode; 5, O-ring; 6, epoxy coating of detector; 7, electrode connections.

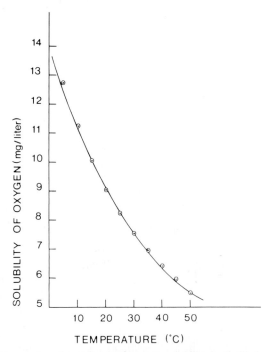

FIG. 4. The temperature dependence of oxygen solubility in distilled water (pressure = 760 mm Hg). (Data taken from Ref. 3.)

film and is then reduced at the cathode causing a current to flow. The cathode is held at a sufficiently negative potential with respect to the other electrode, which serves both as a counter electrode and a reference electrode, to reduce all the oxygen reaching it. The current is therefore diffusion controlled and thus proportional to the oxygen concentration or activity.

The solubility of oxygen in aqueous solution is[3] critically dependent on the temperature (Fig. 4), pressure, and ionic strength. Errors in measurement may be very high if care is not taken to consider these factors. Nomograms are available[3] to determine oxygen solubility in air-saturated fresh water and air-saturated saline water.

*Reagents*

1. 5% (w/v) solution of dimethyldichlorosilane in chloroform
2. 2% solution of sodium dithionite
3. Doubly distilled water

[3] M. L. Hitchman, "Measurement of Dissolved Oxygen." Wiley, New York, 1978.

*Procedures*. When using biological samples it is necessary to siliconize the glass of the electrode chamber prior to use to prevent cells from sticking to the surface. This is achieved by immersing the *dry* chambers into reagent 1. Care should be taken as addition of water to reagent 1 produces HCl. The chamber is then thoroughly washed with reagent 3.

*Calibration*. The oxygen electrode is immersed in a glass chamber in a thermostated bath. The solutions within the chambers are constantly stirred. The current output is monitored using a y-t recorder. There are a number of methods of calibrating oxygen electrodes including the use of mixtures of gases and the Winkler method. The easiest way of calibrating the electrode is to use air-saturated water and oxygen-free water, the upper limit being set using water that has been doubly distilled and aerated at a constant temperature while being well stirred. The response at zero oxygen tension is checked using reagent 2. It is often advisable to check the calibration during the course of an experiment. The electrode is now ready to use. The sample (3 ml) is allowed to warm up to constant temperature. The electrode is inserted and the system is checked to ensure that it is free of all air bubbles and that it is airtight. Care must be taken that the magnetic stirrer does not touch the electrode and that it is stirring efficiently. A steady current is obtained and then the perturbation, e.g., stimulant, enzyme, or substrate, is added. Misleading results may arise if efficient stirring is not maintained since depletion of oxygen close to the electrode may occur. After each experiment, the electrode is carefully washed.

### Oxygen Analysis

*Principle*. The method, due to Winkler,[4] is based on the quantitative oxidation of Mn(II) to Mn(III) in alkaline solution with the subsequent oxidation of $I^-$ by the Mn(III) in acid solution [Eqs. (9)–(13)]. The free iodine is then titrated with thiosulfate.

$$Mn^{2+} + 2OH^- \rightarrow Mn(OH)_2 \qquad (9)$$
$$2Mn(OH)_2 + \tfrac{1}{2}O_2 + H_2O \rightarrow 2Mn(OH)_3 \qquad (10)$$
$$2Mn(OH)_3 + 6H^+ + 3I^- \rightarrow 2Mn^{2+} + I_3^- + 6H_2O \qquad (11)$$
$$I_3^- \rightarrow I_2 + I^- \qquad (12)$$
$$I_2 + 2S_2O_3^{2-} \rightarrow 2I^- + S_4O_6^{2-} \qquad (13)$$

Great care must be taken with the control of the pH and the concentration of iodide. The method is[5] accurate to ±0.1 mg/liter of dissolved oxygen.

---

[4] L. W. Winkler, *Ber. Dtsch. Chem. Ges.* **21**, 2843 (1888).
[5] G. Alsterberg, *Biochem. Z.* **159**, 36 (1925).

*Reagents*

1. 2.15 $M$ $MnSO_4$
2. A solution of NaI (0.9 $M$) and $NaN_3$ (0.15 $M$) in NaOH (12.5 $M$)
3. Concentrated $H_2SO_4$
4. 0.5% starch solution
5. $Na_2S_2O_3 \cdot 5H_2O$ (0.025 $M$)

*Method.* To a 300-ml sample of the solution to be analyzed for dissolved oxygen content, add 2 ml of reagent 1 and 2 ml of reagent 2. Allow the precipitate of $Mn(OH)_3$ to settle and then add 2 ml of reagent 3. Shake well. Remove 203 ml of solution and titrate with reagent 5 using reagent 4 as an indicator: 1 ml $Na_2S_2O_4$ ≡ 1 mg/liter of dissolved $O_2$ in solution.

This method works well for solutions with $Fe^{2+}$ contents less than 1 mg/liter. Several modifications of this method exist and are well described[3] elsewhere.

### Superoxide Salts

*Preparation.* The preparation of other salts of the superoxide ion and its preparation in aprotic media, where it is stable, are considered elsewhere in this volume [7].

*Determination of Superoxide.* Steady-state levels of superoxide can be detected by ultraviolet absorption spectroscopy[6] or by electron paramagnetic resonance spectroscopy (EPR)[7] of frozen samples at 77 K (Fig. 5). These methods are not readily applicable to biological samples and those employing EPR spectroscopy are of qualitative use only. Chemically based methods measure rates of superoxide generation in aqueous solutions by trapping it with suitable indicators and observing the effect of exogenous superoxide dismutase added as an inhibitor. These methods work well for systems which allow access of superoxide dismutase to the site of generation. This may not be the case when, for example, the superoxide is generated at a membrane or within a vesicle which excludes the enzyme. Those assays based on the production of luminescent products are not given as they are far from specific for the superoxide ion. The recommended method is that based on the reduction of acetylated cytochrome *c*.

### Assays for Superoxide

### 1. Reaction with Tetranitromethane

*Principle.* A spectrophotometric assay that can be used to determine the concentration of the superoxide anion in $N,N$-dimethylformamide

---

[6] W. Slough, *J. Chem. Soc. Chem. Commun.*, 184 (1965).
[7] M. R. Green, H. A. O. Hill, and D. R. Turner, *FEBS. Lett.* **103**, 176 (1979).

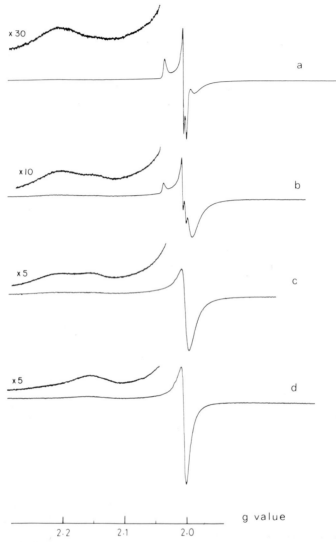

FIG. 5. EPR spectra of solutions of the superoxide ion in DMF at 77 K frozen as glasses. (a) 5 m$M$ tetraethylammonium superoxide; (b) +18 m$M$ $H_2O$; (c) +55 m$M$ $H_2O$; (d) +220 m$M$ $H_2O$. Power 12.5 mW; modulation 1 G; frequency 9.268 GHz; scan rate 1 G s$^{-1}$; time constant 0.25 sec. Gain: (a) 3.2 × $10^2$; (b) 5 × $10^2$; (c,d) 6.3 × $10^2$. (From Ref. 7.)

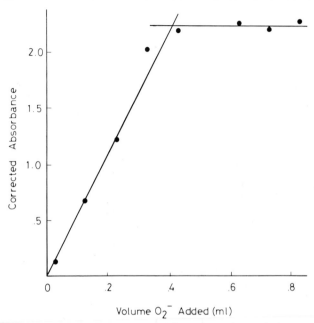

Fig. 6. Addition of a solution of electrochemically generated tetraethylammonium super-oxide in DMF to 2.4 ml of tetranitromethane ($1.5 \times 10^{-4}$ $M$) in ethanol in a cuvette of 1 cm pathlength. The $A_{350}$, due to the formation of the nitroform anion, has been corrected for volume changes. (From Ref. 7.)

(DMF) is based on its reaction with tetranitromethane to form[7] [Eq. (14)] the nitroform anion ($\varepsilon_{350} = 14.8 \times 10^3$ cm$^{-1}$ $M^{-1}$):

$$O_2^- + C(NO_2)_4 \rightarrow O_2 + C(NO_2)_3^- + NO_2 \tag{14}$$

### Reagents

1. Tetranitromethane ($1.5 \times 10^{-4}$ $M$) in freshly distilled ethanol
2. A solution of a superoxide salt in DMF
3. Linde 4A molecular seive

*Assay Procedure.* The tetranitromethane should first be washed with water and then dried with a Linde 4A molecular seive. The solution in ethanol should be prepared immediately prior to use as it deteriorates on standing.

The end point of the titration is determined spectrophotometrically at 350 nm. The cuvette contains 2.4 ml of reagent 1. Aliquots of reagent 2 are added, the cuvette is vigorously shaken, and the absorbance increase is immediately noted. Figure 6 shows such a titration.

## 2. Reduction of Ferricytochrome c and Acetylated Ferricytochrome c

*Principle.* The reduction of ferricytochrome $c$ is one of the most frequently used assays.[8] Superoxide-independent reactions can occur in samples containing cytochrome oxidase or cytochrome reductase activity. The use of acetylated cytochrome $c$[9] will minimize these potentially misleading reactions.

### Reagents

1. Potassium phosphate buffer (50 m$M$), pH 7.5
2. Horse heart cytochrome $c$ (300 $\mu M$), grade III (Sigma)
3. Saturated sodium acetate solution
4. Acetic anhydride
5. Ammonium phosphate (0.01 $M$)
6. Bovine superoxide dismutase (1 mg/ml)

### Preparation of Acetylated Cytochrome c

METHOD. To a 2 m$M$ solution of reagent 2 in water, add an equal volume of reagent 3 at 0°. A 10% excess of reagent 4 is added and the mixture stirred for 30 min. The mixture is then dialyzed, with several changes, against reagent 5 for 24 hr. Removal of native cytochrome $c$ is achieved by passage of the mixture down a Whatman CM 23 column (1.6 cm diameter, 50 cm long) equilibrated with reagent 5. The native cytochrome will bind. The acetylated cytochrome should be dialyzed against several changes of distilled water and then freeze dried and stored at −20°.

Superoxide-dependent reduction of ferricytochrome $c$ or acetylated cytochrome $c$ [Eq. (15)]:

$$\text{Fe}^{\text{III}}\text{cyt } c + O_2^- \rightarrow \text{Fe}^{\text{II}} \text{ cyt } c + O_2 \tag{15}$$

can be monitored at 550 nm at a constant temperature, e.g., 25 or 37° using a split beam spectrophotometer with an extinction coefficient of 18,500 $M^{-1}$ cm$^{-1}$. Two 3-ml cuvettes can be used, and both sample and reference cuvettes should be thermostatted and contain 300 $\mu$l of reagent 2. The reference cuvette should contain 75 $\mu$l of reagent 6. The system to be analyzed is placed in both cuvettes and made up to 3 ml with reagent 1. The reaction is initiated by, e.g., addition of enzyme or substrate. It has been found that more reproducible results are obtained if wet air is bubbled through all solutions prior to use. The superoxide-dependent reduc-

[8] J. M. McCord and I. Fridovich, *J. Biol. Chem.* **244**, 6049 (1969).
[9] A. Azzi, C. Montecucco, and C. Richter, *Biochem. Biophys. Res. Commun.* **65**, 597 (1975).

tion of ferricytochrome $c$ is monitored for several minutes and the rate of reaction in the linear phase calculated. It is recommended that the assay be repeated at least three times. Acetylated cytochrome $c$ should replace the native form when conditions demand.

## 3. Superoxide-Dependent Reduction of Nitro Blue Tetrazolium Salts

*Principle.* The superoxide anion can[10] reduce nitro blue tetrazolium ion (yellow) (NBT) to the insoluble blue formazan. This assay is particularly applicable to gels. Care should be taken when using this assay as reduction of nitro blue tetrazolium is not specific to the superoxide anion. It should not be used to measure superoxide concentrations in dimethyl sulfoxide solutions as interference in the assay arises from several species, including[11] the dimysl ion ($CH_3SOCH_2^-$). It is, however, a reasonably sensitive assay as the extinction coefficient is high ($\varepsilon_{490} = 100$ m$M^{-1}$ cm$^{-1}$).

### Reagents

1. A solution of nitro blue tetrazolium chloride, (2,2'-di-$p$-nitrophenyl-5,5'-diphenyl-3,3'[3,3'-dimethoxyl-4,4'-diphenylene] ditetrazolium chloride) (0.4 m$M$), in reagent 2
2. Potassium phosphate (50 m$M$), pH 7.5
3. Bovine superoxide dismutase (1 mg/ml)

*Procedures.* The reduction of the nitro blue tetrazolium ion is monitored as an increase in absorbance at 490 nm. The use of plastic disposable cuvettes is recommended as formazan is insoluble in water and tends to adhere to glass. The final concentration of NBT in the cuvettes should be 40 $\mu M$. NBT should be present in both sample and reference cuvettes which should be thermostated. The reference cuvette should also contain superoxide dismustase at a final concentration of 25 $\mu$g/ml. The reactants should then be added and the change in absorbance due to the superoxide-dependent reduction of NBT to formazan followed. Assays should be repeated at least three times.

### Hydrogen Peroxide Analysis

A number of analytical methods have been devised for the identification and quantification of hydrogen peroxide in biological systems. Some are well described in Volume LII. Many of these methods are sensitive to concentrations of peroxide as low as $10^{-6}$ mol/dm$^3$, but suffer from the

[10] C. Auclair, M. Torres, and J. Hakim, *FEBS. Lett.* **89**, 26 (1978).
[11] R. L. Arudi, A. O. Allen, and B. H. J. Bielski, *FEBS. Lett.* **135**, 265 (1981).

disadvantage that they detect total peroxide concentration not just hydrogen peroxide. However methods which detect only hydrogen peroxide depend on the peroxidase (Px) catalyzed oxidation of a suitable substrate to yield a colored product which can then be detected spectrophotometrically.

$$H_2O_2 + \text{substrate} \xrightarrow{\text{peroxidase}} \text{colored product} \tag{16}$$

A method that is sensitive to micromolar concentrations of hydrogen peroxide is a modified colorimetric assay using 4-aminoantipyrine and phenol as donor substrates.[12]

*Principle.* Hydrogen peroxide is a potent oxidizing agent which will couple oxidatively with 4-aminoantipyrine and phenol to yield a quinone-imine dye with a maximum absorbtion at 505 nm ($\varepsilon = 6.4 \times 10^3$ mol$^{-1}$ cm$^{-1}$)

$$2H_2O_2 + \text{4-aminoantipyrine} + \text{phenol} \xrightarrow{Px} \text{Chromagen} + 4H_2O \tag{17}$$

A small quantity of hydrogen peroxide in the stock reagent solution eliminates any interference in the assay by trace impurities in the 4-aminoantipyrine.

*Reagents*

1. Phenol
2. 4-Aminoantipyrine
3. Potassium phosphate buffer pH 6.9 (0.1 $M$)
4. Horseradish peroxidase
5. Hydrogen peroxide

*Procedure.* The reagent solution (100 ml) is prepared using 0.234 g reagent 1, 0.10 g reagent 2, and 1 ml of reagent 3 and contains $2 \times 10^{-8}$ $M$ reagent 4 and $2.5 \times 10^{-6}$ $M$ reagent 5. The reagent solution (4 ml) is mixed with the peroxide sample and made up to 10 ml with doubly distilled water. The change in the absorbance at 505 nm is measured until a constant reading is obtained (approximately 5 min at ambient temperature). A 4-ml aliquot of the reagent solution made upto 10 ml with doubly distilled water which also serves as the reference. Changes in pH or buffer concentration have a insignificant affect on the extinction coefficient but may affect the stability of the chromagen formed.

## 3. Hydroxyl Radical Detection

There are many assays for the detection of hydroxyl radicals. The three methods described are more direct than many. The most commonly

[12] J. E. Frew, P. Jones, and G. Scholes, *Anal. Chim. Acta* (1983), in press.

used assay is based on the gas chromatographic detection of ethene formed from methional [Eq. (18)], $\beta$-methylmercaptopropanal, methionine, or 4-methylthio-2-oxobutanoic acid.

$$CH_3SCH_2CH_2CHO + HO\cdot \rightarrow CH_2{=}CH_2 + HCOOH + \tfrac{1}{2}CH_3SSCH_3 \qquad (18)$$

Pryor and Tang[13] have shown that a variety of organic radicals are also capable of producing ethene from methional with different yields of ethene, the difference presumably reflecting the relative ability of the radicals to abstract the aldehydic hydrogen in competition with other available reactions. It is a convenient exploratory method. Halliwell has developed a number of chromatographically based methods and one is described. A useful method of detecting hydroxyl radicals, and many others, is[14] by the technique of "spin trapping."

### a. Ethene from 4-Methylthio-2-oxobutanoic Acid

*Principle.* As previously stated, the method must be used with caution. It is not a specific method for the detection of the hydroxyl radical. Other radicals, and one-electron oxidants, can form ethene from 4-methylthio-2-oxobutanoic acid (KMBA) which is used in preference to methional because of the offensive odor of the latter.

### Reagents

1. 10 mM KMBA
2. 20 mM diethylenetriaminepentaacetic acid (DTPA)
3. 50 mM potassium phosphate buffer pH 7.5
4. Ethene standard, 5% ethene in 95% nitrogen (Phase Separation Inc.)

*Procedure.* The reactions leading to the generation of ethene are performed in 25-ml Erlenmeyer flasks sealed with septum caps and incubated, with shaking, in a water bath at 30 or 37°, depending on the reaction of interest. The final solution in the flask should be 1 mM in KMBA, 1 mM in DTPA in a final volume, including the material to be assayed, of 2 ml. If cellular matter is to be analyzed, the flask should be siliconized prior to use (see the section on the oxygen electrode). The flasks are shaken and the gas above the flask is sampled using a 500-$\mu$l gas-tight microsyringe. The gases are then injected into the heated injection port of a gas chromatograph. A typical column use is 3 mm × 2 m, packed with phase separation Q-S Porapack adsorbant or Chromosorb 102 and maintained at 140°. The ethene can be detected using a flame ionization detec-

[13] W. A. Pryor and R. H. Tang, *Biochem. Biophys. Res. Commun.* **81**, 498 (1978).
[14] E. G. Janzen, *in* "Free Radicals in Biology" (W. A. Pryor, ed.), Vol. 4, p. 116. Academic Press, New York, 1980.

tor. The area of the peak associated with ethene is compared to that of the standard.

### b. Hydroxylation of Phenol

*Principle.* Hydroxyl radicals attack aromatic rings at diffusion-controlled rates to form hydroxylated products which are colored and absorb at 510 nm. A spectrophotometric assay based on this method is not, however, very sensitive. It is preferable to employ[15] a gas chromatographic analysis of derivatized phenol and its products.

*Reagents*

1. Triply distilled phenol in water (25 m$M$)
2. DTPA (10 m$M$)
3. Potassium phosphate buffer (50 m$M$)
4. HCl (11.6 $M$)
5. *o*-Hydroxybenzyl alcohol (200 n$M$)
6. Triply distilled trifluoroacetic anhydride
7. Diethyl ether
8. Toluene
9. Anhydrous $Na_2SO_4$

*Method.* The sample to be analyzed is added to 200 $\mu$l reagent 1 and 20 $\mu$l reagent 2. Reagent 3 is added to give a final volume of 2 ml contained in a stoppered tube. The tube is kept at a constant temperature, e.g., 25° for 60 min, after which reagent 4 (0.1 ml) is added to stop any further reaction. The internal standard, reagent 5 (0.1 ml), is added. The tube is cooled in ice and 2 ml of reagent 7 is added. The tube is well shaken to achieve good mixing of the two phases and allowed to settle, and 200 $\mu$l of the upper phase is removed and added to another tube to which 0.1 ml of reagent 6 is added. The tube is then sealed and incubated at 65° for 1 hr. The contents are then poured into 1 ml of 1 $M$ $KH_2PO_4$. Reagent 8 (1 ml) is then added and the mixture is well shaken for a few seconds. The toluene layer is separated and dried with 100 mg of reagent 9 and then analyzed by gas–liquid chromatography. The column used in the original work was a 25-m OV101-WCOT column. Using a $^{63}Ni$ electron capture detector it was possible to analyze solutions containing picomoles of product.

### c. Spin Trapping of Hydroxyl Radicals

*Principle.* The superoxide anion, its conjugate acid, the hydroperoxyl radical, and the hydroxyl radical are paramagnetic species and can be

---

[15] R. Richmond, B. Halliwell, J. Chauhan, and A. Darbre, *Anal. Biochem.* **118**, 328 (1981).

detected directly by electron paramagnetic resonance spectroscopy. *Direct* detection is only possible at low temperatures and thus EPR spectroscopy has only occasionally been used to monitor the continuous formation of these radical species. Recently, spin trapping has been developed to detect and *identify* low concentrations of free radicals in reacting systems. This involves the reaction of a radical X· with a diamagnetic compound to form a more stable radical product which has a readily observable EPR spectrum and one, ideally, that is characteristic *of the radical trapped.* Nitrones or nitroso compounds have been found to act as suitable traps [Eqs. (19) and (20)].

$$X\cdot + -CH=\overset{\longrightarrow}{\underset{+}{N}}-\overset{O^-}{|} \longrightarrow -\overset{O\cdot}{\underset{\underset{X}{|}}{CH}}-N- \tag{19}$$

$$X\cdot + -N=O \longrightarrow X-\overset{}{\underset{|}{N}}-O\cdot \tag{20}$$

Nitrones can trap a large number of different radicals including carbon-, hydrogen-, oxygen-, nitrogen-, and halogen-centered radicals, whereas nitroso spin traps react mainly with carbon-centered radicals. The hydroxyl radical forms an unstable adduct with the nitroso compounds which disproportionates at diffusion-controlled rates to yield diamagnetic and, thus, EPR-silent species.

Nitrones can trap hydroxyl radicals and the spin adducts are relatively long lived. Examples of commercially available spin traps are *N*-tert-butyl-α-phenylnitrone (PBN) and 5,5'-dimethylpyroline-*N*-oxide (DMPO):

PBN                         DMPO

(I)                          (II)

It should be noted that DMPO will form adducts with *both* $HO_2$ and $O_2^-$. Their rates of formation and the stability of the adducts formed, together with some of their properties, are given in the table with the simulated spectra in Fig. 7. The adduct, Fig. 7b, is detectable only when $O_2^-$ or $HO_2$ is being continuously formed. The adduct is thought to decompose into that derived from the hydroxyl radical (Fig. 7a) and a non-radical species. Care should therefore be taken in interpreting experiments which show the formation of the hydroxyl radical.

SOME PHYSICAL PROPERTIES OF THE HYDROPEROXYL AND HYDROXYL RADICAL
ADDUCTS OF DMPO

| Property | DMPO–O$_2$H | DMPO–OH | Reference |
|---|---|---|---|
| Rate of formation | $6.6 \times 10^3 \ M^{-1} \ sec^{-1}$ (pH 7.8, 25°) | $2.1–3.4 \times 10^9 \ M^{-1} \ sec^{-1}$ | a |
| Half-life of adduct | 35 sec (pH 7.8) | 2 hr | b |
| Hyperfine splitting | $A_N = 14.2$ G $A_H = 11.6$ G, 1.2 G | $A_N = 14.8$ G $A_H = 14.8$ G | c |
| g-value | 2.006 | 2.005 | |

[a] E. Finkelstein, G. M. Rosen, and E. J. Rauckman, *J. Am. Chem. Soc.* **102,** 4992 (1980).
[b] G. R. Buettner and L. W. Oberley, *Biochem. Biophys. Res. Commun.* **83,** 69 (1978).
[c] E. G. Janzen, *in* "Free Radicals in Biology" (W. A. Pryor, ed.), Vol. 4, p. 15. Academic Press, New York, 1980.

### Reagents

1. Activated charcoal
2. DMPO (1 *M*)
3. DTPA (25 m*M*)

*Purification of DMPO.* DMPO obtained commercially contains impurities and must be purified prior to use. This is readily achieved by diluting the DMPO with 7 to 8 parts of distilled water (v/v) and stirring with 1 part of neutral, activated charcoal for 1 hr. The suspension is then filtered, checked for any contaminating nitroxides by EPR spectroscopy (see below), and, if uncontaminated, stored in small aliquots, frozen at $-20°$, in the dark. The concentration of the stock solution is determined spectrophotometrically, $\varepsilon_{234} = 7700 \ M^{-1} \ cm^{-1}$ in ethanol.[16]

*Procedures.* Typically, DMPO is added to the sample to give a final concentration of 10–100 m*M*. Obviously it is important to check that at the concentration used, DMPO does not adversely affect the system to be investigated. To minimize the adventitious formation of radical species through Fenton-type chemistry, it is normal to add the sequestering agent, DTPA, to give a final concentration of 1 m*M*. Care should be taken that *all buffers used are metal free.* Aqueous samples can be measured at ambient temperatures using, in the usual rectangular cavity, a flat quartz cell. There are many methods whereby the sample can be introduced to the quartz cell. One method, which has been found convenient, is as follows. A 1-ml syringe is fitted with a Luer needle, inserted through a

[16] R. Bonnett, R. F. C. Brown, V. M. Clark, I. O. Sutherland, and A. Todd, *J. Chem. Soc.* 2094 (1959).

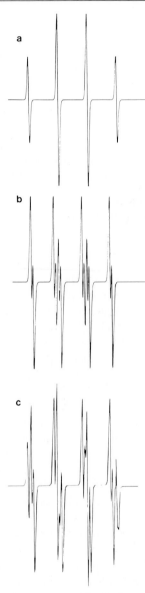

FIG. 7. Simulated EPR spectra of free radical adducts of DMPO: (a) spectrum of the DMPO–OH adduct, (b) spectrum of the DMPO–O$_2$H adduct, (c) spectrum of an equimolar mixture of DMPO–OH and DMPO–O$_2$H (see also the table).

subseal at the top of the flat cell, and the sample drawn in through the bottom of the flat cell (N.B.: *no power should be incident on the cavity at this stage*) through a Teflon stopper which has a short piece of narrow gauge Teflon tubing passed through it. EPR scans can be commenced some 20 sec after introduction of the sample, i.e., after retuning of the cavity.

# [2] Biochemistry of Oxygen Radical Species

*By* M. BRUNORI and G. ROTILIO

"Oxygen radical" is a term currently used in the biochemical literature to indicate reactive oxygen species which are generated as intermediates in redox processes leading from oxygen to water. The five possible species are indicated in Table I (see also [1]); in the present chapter, enzyme-related aspects of oxygen radical biochemistry are briefly introduced, with a special focus on processes in which these species are formed catalytically, and the reactions which they undergo in the presence of enzymes either in a catalytic or in a noncatalytic mode.

## Formation

A schematic view of the reactions leading to or dealing with the various possible oxygen species is provided in Table I, together with the indication of the localization of the enzymes in the animal cell. Besides these catalytic processes, it should be recalled that the other major potential source of oxygen radicals is the oxidation of oxyhemoglobin in red blood cells.[1]

Oxidases are enzymes that reduce oxygen to water without producing detectable "free" oxygen radicals under normal, or "coupled," conditions. They contain iron and/or copper in a multicenter pattern of sites. For this group of enzymes, oxygen radicals bound at the active site(s) of the enzyme have been detected and carefully investigated in some cases. Other oxidases containing flavin (e.g., amino acid oxidase) or a single copper (e.g., amine oxidase, galactose oxidase) produce free $H_2O_2$.

A single-electron reduction of oxygen is established for only one soluble enzyme, i.e., xanthine oxidase,[2] but it is likely to be an unnatural

---

[1] H. P. Mirsa and I. Fridovich, *J. Biol. Chem.* **247**, 6960 (1972).

[2] P. F. Knowles, J. F. Gibson, F. M. Pick, and R. C. Bray, *Biochem. J.* **111**, 53 (1969).

Copyright © 1984 by Academic Press, Inc.
All rights of reproduction in any form reserved.
ISBN 0-12-182005-X

TABLE I

REDOX REACTIONS OF BIOLOGICAL RELEVANCE INVOLVING OXYGEN RADICALS

| Reaction type | $1e^-$ oxidation (e.g., xanthine oxidase, NAD(P)H oxidase) | $1e^-$ dismutation (e.g., superoxide dismutase) | $2e^-$-$H_2O_2$ forming oxidation (e.g., uricase, amino acid oxidase) | $2e^-$-$H_2O$ forming oxidation (e.g., peroxidase) | $2e^-$ dismutation (e.g., catalase) | $3e^-$ oxidation dismutation (metal catalyzed Haber–Weiss reaction) | $4e^-$ oxidation (e.g., cytochrome oxidase) |
|---|---|---|---|---|---|---|---|
| $1e^-$   $O_2$ | | | | | | | |
|   $HO_2$ | | | | | | | |
| $2e^-$   $H_2O_2$ | | | | | | | |
| $3e^-$   $OH\cdot + OH^-$ | | | | | | | |
| $4e^-$   $2H_2O$ | | | | | | | |
| Examples of localization in animal cells | Granulocyte membrane | Cytoplasm, mitochondrion | Peroxisome | Cytoplasm granules of granulocytes | Peroxisome | Granulocytes | Mitochondrion |

situation (the enzyme being a dehydrogenase *in vivo*). Membrane-bound NAD(P)H oxidases (for example from granulocytes and macrophages), however, are well-documented cases of $O_2^-$ production (see other chapters in this volume).

OH· is produced in the reactions of many reduced metalloproteins with $H_2O_2$ (Fenton's reaction); if the reductant is $O_2^-$ this is a protein-driven Haber–Weiss reaction. It has been suggested that such a cycle may be catalytic with iron-containing proteins of neurophils, such as lactoferrin and myeloperoxidase.[3]

*Reactions*

Reactions of enzymes with oxygen and oxygen radicals may occur in either a noncatalytic or a catalytic mode.

*Noncatalytic Reactions.* $O_2$, $O_2^-$, and $H_2O_2$ are effective redox reagents for many enzyme-active sites, depending on redox potentials and kinetic accessibility of the site; these reactions may lead, in some cases, to inactivation. Autoxidation of oxyhemoglobin occurs continuously under physiological conditions, with formation of $O_2^-$ and inactivation of the oxygen carrier function of hemoglobin.[1,4] The mechanism of inactivation is often complex, the various oxygen species being involved at different stages of multistep processes. On the other hand, the OH· radical is generally believed[5] to be the ultimate inactivating agent in these processes, because of its redox potential and kinetic reactivity (see [1]). For example, some metalloenzymes, in particular those containing iron–sulfur centers and lacking a site for the complete reduction of oxygen (see the discussion on cytochrome oxidase below), are very sensitive to exposure to air (like nitrogenase or succinate dehydrogenase). This aerobic inactivation is likely to be started by the primary formation of $O_2^-$ on oxidation of the reduced metal centers by $O_2$,[6] with subsequent accumulation of $H_2O_2$ by spontaneous dismutation of $O_2^-$, and final attack of the enzyme, either at the active site or at crucial amino acids, by OH· produced by a Fenton reaction of the reduced metal(s) with $H_2O_2$.

$H_2O_2$ can be either reducing or oxidizing depending on conditions. The copper of superoxide dismutase[7] and the iron of cytochrome $c$ (cyt $c$)[8] are

[3] J. V. Bannister, W. M. Bannister, H. A. O. Hill, and P. J. Thornally, *Biochim. Biophys. Acta* **715**, 116 (1982).

[4] M. Brunori, G. Falcioni, E. Fioretti, B. Giardina, and G. Rotilio, *Eur. J. Biochem.* **53**, 99 (1975).

[5] I. Fridovich, *in* "Free Radicals in Biology" (W. Pryor, ed.), Vol. I, pp. 239–272. Academic Press, New York, 1976.

[6] H. P. Misra and I. Fridovich, *J. Biol. Chem.* **246**, 6686 (1971).

[7] R. C. Bray, S. H. Cockle, E. M. Fielden, P. B. Roberts, G. Rotilio, and L. Calabrese, *Biochem. J.* **139**, 43 (1974).

[8] C. Barnofsky and S.-V. C. Wanda, *Biochem. Biophys. Res. Commun.* **111**, 231 (1983).

reduced by a slight excess of peroxide. At neutral pH reduced cyto-chrome $c$ is oxidized by $H_2O_2$ with consequent production of OH$\cdot$ radicals; a similar reaction occurs with reduced superoxide dismutase and leads to inactivation of the enzyme.[7]

$O_2^-$ is a reducing agent for many metalloenzymes and this reaction has been extensively used to "scavenge" $O_2^-$ and thus to detect its forma-tion. The reaction with ferricytochrome $c$[9] is a current probe system in assays of superoxide dismutase. The reaction with ceruloplasmin[10] may serve a physiological role, also because, at variance with cyt $c$ lactoferrin or ceruloplasmin is oxidized by $O_2$ much faster than by $H_2O_2$, and there-fore it seems not to support an $O_2^-$-driven Haber–Weiss cycle.

The reactions of OH$\cdot$ with enzymes have been studied in two ways. (1) *The indirect approach* is based on measuring the enzymatic activity in the presence of potential sources of OH$\cdot$ and the protective effects of proper OH$\cdot$ scavengers (such as primary alcohols, in particular mannitol).[5] An increased effect upon addition of both superoxide dismutase and catalase represents supporting evidence for OH$\cdot$ involvement, since this radical is usually formed by a cycle involving both $O_2^-$ and $H_2O_2$. (2) *Direct studies* have made use of pulse radiolysis, in which the radical can be quantita-tively obtained by radiolysis of water under appropriate experimental conditions (see [11]).

It has been established that OH$\cdot$ reacts with several amino acids[11] at nearly diffusion-controlled rates. Hydrogen atom abstraction has been observed with aliphatic amino acids and cysteine, while ring addition occurs for the aromatic side chains and the trisulfide RSSSR is formed in a complex reaction with cystine. Well-documented pulse radiolysis studies of inactivation by OH$\cdot$ are papain[11] (where an essential single free cys-teine SH group undergoes H atom abstraction), carboxypeptidase A, al-cohol dehydrogenase, ribonuclease, and superoxide dismutase.[12]

*Catalytic Reactions.* $O_2$, $H_2O_2$, and $O_2^-$ are the substrates of enzy-matic reactions which are extensively discussed in several chapters of this volume and in other volumes of this series; therefore specific references are not given in the following, which is a general comment to Table I.

$O_2$ is the substrate of oxidases and oxygenases depending on whether its reduction occurs by electron abstraction from a substrate (oxidase), or, at least partially, by electron sharing with a substrate (oxygen incorpora-

[9] J. Butter, G. G. Jaysore, and A. J. Swallow, *Biochim. Biophys. Acta* **126,** 269 (1975).

[10] I. M. Goldstein, H. B. Kaplan, M. S. Edelson, and G. Weissmann, *J. Biol. Chem.* **254,** 4040 (1979).

[11] G. E. Adams and P. Werdman, *in* "Free Radicals in Biology" (W. Pryor, ed.), Vol. III, pp. 53–95. Academic Press, New York, 1977.

[12] P. B. Roberts, E. M. Fielden, G. Rotilio, L. Calabrese, J. V. Bannister, and W. M. Bannister, *Radiat. Res.* **60,** 441 (1974).

tion by oxygenases). In general, the oxygen reduction step distinctly follows substrate oxidation in the oxidases, while ternary complexes are the rule for oxygenases. $H_2O_2$ is the substrate for catalase and peroxidase, the difference between the two enzymes being the specificity for electron donor molecules used to oxidize $H_2O_2$ to water; this is $H_2O_2$ itself in the case of catalase, while it is a reduced compound (like GSH or a halide ion) in the case of peroxidases.

$O_2^-$ is the substrate in the superoxide dismutation reaction, which redistributes electrons between two $O_2^-$ via a ping-pong redox cycle (see below). In indoleamine dioxygenase (see [6]), $O_2^-$ acts as a real substrate, forming with the organic substrate a ternary complex in which electrons and oxygen atoms are incorporated into indoleamine (such as tryptophan and its hydroxy derivatives) from a $(Fe^{3+}-O_2^-)$ complex of the enzyme. While this mechanism is likely to apply also to nitropropane-dioxy-genase,[13] other oxygenase reactions seem to involve $O_2^-$ as an enzyme-bound intermediate (see below).

## Classification

The enzymology of oxygen and oxygen radicals is very complex, which makes it difficult to settle on a single classification criterion and often several criteria have been used. We shall introduce two of these.

### Oxygen Radical Involvement

From the viewpoint of the type of oxygen radical involvement, a possible classification includes (1) enzymes which *utilize bound oxygen radicals:* these are the $H_2O$-forming oxidases and oxygenases, which never release in the bulk significant amounts of oxygen intermediate reduction products; (2) enzymes which *utilize free oxygen radicals,* which include catalases and peroxidases, utilizing free hydrogen peroxide, and indole-amine oxygenase and superoxide dismutases, utilizing free superoxide; (3) enzymes which *release free oxygen* radicals, such as $O_2^-$ and $H_2O_2$-forming oxidases.

### Oxygen Involvement

From the standpoint of oxygen itself, a possible classification may include (1) enzymes which have *$O_2$ as a substrate:* oxidases and oxy-genases; and (2) enzymes which have *$O_2$ as a product:* hydrogen perox-idase dismutase (catalase) and superoxide dismutase. Obviously, the latter classification excludes peroxidases. Cytochrome $c$ oxidase and super-

---

[13] K. Soda, T. Kido, and K. Asada, "Biochemical and Medical Aspects of Active Oxygen" (O. Hayaishi and R. Asada, eds.), pp. 119–133. Univ. of Tokyo Press, Tokyo, 1977.

oxide dismutase are representative of two distinct types of redox mechanisms, i.e., a sequential 4 $e^-$ transfer in the case of cytochrome $c$ oxidase and a ping-pong 1 $e^-$ transfer in the case of superoxide dismutase. They will be described as two examples of methodological approaches to the enzymology of oxygen radical reactions, also being closest to the experience of the authors.

Superoxide Dismutase

*Generalities*

Superoxide dismutases are enzymes that undergo cycles of reduction and reoxidation by $O_2^-$ at rates much faster than with any other redox molecule, including $O_2$ or $H_2O_2$. The net result of such reactions at steady state is dismutation of $O_2^-$ into $H_2O_2$ and $O_2$ according to the following:

$$\text{Enzyme}_{ox} + O_2^- \rightarrow \text{Enzyme}_{red} + O_2 \tag{1}$$
$$\text{Enzyme}_{red} + O_2^- + 2H^+ \rightarrow \text{Enzyme}_{ox} + H_2O_2 \tag{2}$$

These enzymes[5] have a transition metal at the active center, which is copper ($Cu^{2+}, Cu^+$), nonheme iron ($Fe^{3+}, Fe^{2+}$), or manganese ($Mn^{3+}, Mn^{2+}$).

The best characterized superoxide dismutase is the enzyme isolated from bovine erythrocytes, for which a highly refined X-ray structure analysis is available.[14] It contains a binuclear cluster, with the active copper and zinc bridged by a common ligand, the imidazole nucleus of histidine 63. The following discussion relates primarily to studies of the Cu–Zn superoxide dismutase.

*Methodological Approach to the Mechanism of Action*

$O_2^-$ fluxes high enough to put micromolar enzyme concentrations into a steady state within a few microseconds can be obtained by (1) chemical and enzymatic methods and (2) physicochemical methods. For (1) see other chapters in this volume. These methods, however, are not appropriate for mechanistic studies since rates of superoxide dismutase reaction are obtained only indirectly from the extent of inhibition of another reaction competing for $O_2^-$. Moreover these methods consist of several chemical reactions subject to interference in particular when enzyme activity is studied as a function of pH, ionic strength and the presence of potentially inactivating molecules. (3) Physicochemical methods are pulse radiolysis and polarography. Pulse radiolysis (see [20]) is suitable for either steady-

[14] J. A. Tainer, E. D. Getroff, K. M. Beem, J. S. Richardson, and D. C. Richardson, *J. Mol. Biol.* **162**, 181 (1982).

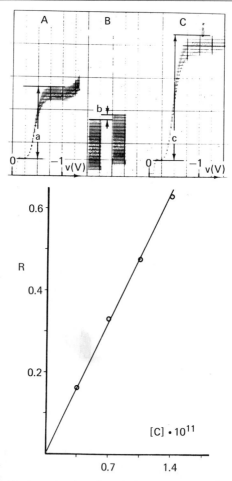

FIG. 1. Polarographic determination of catalytic rate constant of superoxide dismutase. Upper panel: (A) Polarographic wave of oxygen monoreduction in 0.1 $M$ sodium borate, pH 9.8, equilibrated with air and containing $5 \times 10^{-4}$ $M$ triphenylphosphine oxide. The $v$ scale is the polarographic voltage in volts (V). (B) To the solution as in (A), $3.5 \times 10^{-11}$ $M$ superoxide dismutase (Cu,Zn) was added and the current value recorded at $-1$ V (at increased amplification). The same aliquot of enzyme is added 3–4 times and the current value recorded after each addition. (C) Polarographic wave of oxygen bireduction, as observed after addition of $1 \times 10^{-7}$ $M$ superoxide dismutase (Cu,Zn) to set the full scale of current. For the symbols see legend of lower panel. It should be noticed that, in the absence of spontaneous dismutation, either spontaneous or enzymatic, $c = 2a$ and thus $I_d = a$. Lower panel: Plot of the experimental data of upper panel according (19) to the equation

$$R = [(7.42 I_l/I_d) - 7.42]/[2.25 - (I_l/I_d)] = k[C]t_g$$

where $I_l$ and $I_d$ are the mean limiting currents in the presence and in the absence of enzyme, state measurements, by following the decay of $O_2^-$ at 260 nm, or direct measurement of the individual rate constants, by following valence-related optical density changes of the metal centers. It has also been used in connection with rapid-freezing EPR to measure the steady-state valence state of the copper in the bovine enzyme[15] since the intensity of the EPR signal gives the concentration of the Cu(II) state of the enzyme.

Much lower concentrations of enzyme, which may be easily brought to the steady state by very small fluxes of $O_2^-$ such as those expected in physiological conditions, can be used by applying another spectroscopic method, that is, $^{19}F$ nuclear magnetic relaxation.[16] The enhancement of the rate of this process is a specific property of the cupric state of copper–zinc superoxide dismutase and can detect the steady-state fraction of Cu(II) enzyme at micromolar concentrations (i.e., the concentration of the enzyme in most cells). Moreover, this method, in the kinetic mode, can measure very low concentrations of $O_2^-$ in a most direct and nonperturbing way.[17]

In the polarographic method of catalytic currents[18] $O_2^-$ is produced by reduction of $O_2$ at the dropping mercury electrode in the presence of a surfactant (e.g., triphenylphosphine oxide), which inhibits proton transfer to $O_2^-$ at the electrode surface. At pH values as high as 9–10 very high fluxes of $O_2^-$ (up to millimolar concentrations) are produced. The rate constant of dismutation may be calculated by measuring the increment of the polarographic wave as a function of enzyme concentration, according to the equation of Koutecky et al.[19] (see Fig. 1). In a more recent version utilizing a rotating disk electrode coated with mercury, significant $O_2^-$ fluxes can be obtained even at physiological pH values.[20]

[15] E. M. Fielden, P. B. Roberts, R. C. Bray, D. J. Lowe, G. N. Mautner, G. Rotilio, and L. Calabrese, *Biochem. J.* **139**, 49 (1974).
[16] P. Viglino, A. Rigo, E. Argese, L. Calabrese, D. Cocco, and G. Rotilio, *Biochem. Biophys. Res. Commun.* **100**, 125 (1981).
[17] A. Rigo, P. Ugo, P. Viglino, and G. Rotilio, *FEBS Lett.* **132**, 78 (1981).
[18] A. Rigo, P. Viglino, and G. Rotilio, *Anal. Biochem.* **68**, 1 (1975).
[19] J. Koutecky, R. Bridcka, and V. Hames, *Collect. Czech. Chem. Commun.* **18**, 611 (1953).
[20] E. Argese, B. De Carli, E. Orsega, A. Rigo, and G. Rotilio, *Anal. Biochem.* **132**, 110, (1983).

respectively, measured as indicated in the upper panel; $t_g$ is the drop time of mercury electrode (sec); $k$ is the rate constant of enzymatic dismutation ($M^{-1}$ $sec^{-1}$); and $[C]$ is the enzyme concentration (moles per liter) reported on the abscissa. From the slope of the plot and the drop time, $k$ may be calculated.

TABLE II

SECOND-ORDER RATE CONSTANTS OF SUPEROXIDE DISMUTATION BY
SUPEROXIDE DISMUTASES AS MEASURED BY DIRECT METHODS

| Type | Source | Rate constant $(M^{-1} sec^{-1})$ | pH | Reference |
|---|---|---|---|---|
| Cu, Zn | Bovine erythrocytes | $2.3 \times 10^9$ | 9.8 | 18,21 |
| Mn | *Bacillus stearo-thermophilus* | $6.0 \times 10^8$ | 8.0 | 26 |
| Fe | *Photobacterium ceioghathi* | $5.5 \times 10^8$ | 8.0 | 25 |

## Brief Description of Results

*General Features of the Mechanism of Dismutation.* In all types of superoxide dismutases, the rates for reactions (1) and (2) are the same and equal to the turnover rate. Typical values from pulse radiolysis and polarographic measurements are given in Table II.

In the case of the bovine erythrocyte enzyme, the rate is independent of pH between pH 5 and 10 and then drops between pH 10 and 12,[18] this effect being reversible.[18,21]

Iron and manganese dismutases are much more pH dependent, usually with maximum rates between pH 6 and 8 and a steep inactivation curve above neutrality.[20] For all enzymes, activity drops below pH 5, where metal removal and competition of spontaneous dismutation become dominant processes. No substrate saturation is seen in pulse radiolysis experiments for the copper–zinc enzyme.[21] Polarographic measurements can detect a definite $K_m = 3.55 \times 10^{-4} M$ at pH 9.9, and a competitive pattern of inhibition by single charged anions.[22] In line with spectroscopic measurements[23] and the known X-ray structure,[14] this indicates that anions displace an $H_2O$ molecule from the first copper coordinating sphere. $O_2^-$ seems to interact via an inner-sphere mechanism at this water coordination position. Less clear is the inhibition of the iron enzyme by $N_3^-$.[24,25] No anion effect is apparent with the Mn enzyme.[26]

*Fine Details of the Mechanism of Bovine Erythrocyte Cu,Zn Dismutase.* The expected steady-state level of Cu(II) is 50% of the total copper.

[21] G. Rotilio, R. C. Bray, and E. M. Fielden, *Biochim. Biophys. Acta* **268,** 605 (1972).

[22] A. Rigo, R. Stevanato, P. Viglino, and G. Rotilio, *Biochem. Biophys. Res. Commun.* **79,** 776 (1977).

[23] G. Rotilio, L. Morpurgo, C. Giovagnoli, L. Calabrese, and B. Mondovì, *Biochemistry* **11,** 2187 (1972).

[24] T. O. Slykhouse and J. A. Fee, *J. Biol. Chem.* **251,** 5472 (1976).

[25] F. Lavelle, M. E. McAdam, E. M. Fielden, P. B. Roberts, K. Puget, and A. M. Michelson, *Biochem. J.* **161,** 3 (1977).

[26] M. F. McAdam, F. Lavelle, R. A. Fox, and E. M. Fielden, *Biochem. J.* **165,** 81 (1977).

However, both pulse radiolysis and $^{19}F$ nuclear magnetic relaxation have measured higher steady-state fractions, approaching 75%, in freeze-dried samples with no substantial loss of activity.[15,16] Apparently, under some circumstances, the enzyme appears to dismute $O_2^-$ with a half-the-site mechanism, in which the working copper seems to prevent the site on the other subunit from functioning in catalysis. Positively charged amino acids side chains help to funnel $O_2^-$ to the active metal. Neutralization of Arg-143, which is positioned at the entrance of a channel connecting the solvent with the copper site, leads to alteration of the copper geometry and inhibition of the enzyme activity.[27] Neutralization of lysines has no effect on copper geometry but leads to decrease of enzyme activity. The lysine-modified protein is less sensitive to ionic strength, which strongly inhibits the enzyme activity of the native protein.[28]

Removal of the zinc decreases the redox potential of the copper, but does not alter the rates of its reaction with $O_2^-$ at pH 6.[29,30] However at higher pH values the zinc seems to participate in the catalytic activity because it stabilizes the p$K$ of the imidazole nucleus of His-63 bridging the two metals. Studies with the enzyme in which Co(II) replaces Zn and builds in the protein an additional spectroscopic probe near the copper site have been extremely illuminating in this respect. In fact it was demonstrated that the bridging imidazole is protonated, on the copper-facing side, in the reductive step of the mechanism, and releases a proton in the oxidative step.[31] This bridging imidazole can thus be a nearby well-oriented proton source in order to protonate the highly unstable peroxide dianion ($O_2^{2-}$) product at a rate compatible with that of oxidation of the copper.[15]

Altogether, these studies on the role of the zinc show how valuable a proper methodology of selective removal and replacement of metals in enzymes may be.

### Cytochrome c Oxidase

*Generalities*

Cytochrome $c$ oxidase is the terminal enzyme of the mitochondrial respiratory chain, whose main function is to transfer electrons flowing

[27] D. P. Malinowski and I. Fridovich, *Biochemistry* **18,** 5909 (1979).
[28] D. Cocco, L. Rossi, D. Barra, F. Bossa, and G. Rotilio, *FEBS Lett.* **150,** 303 (1982).
[29] P. O'Neil, E. M. Fielden, D. Cocco, L. Calabrese, and G. Rotilio, *in* "Oxy Radicals and Their Scavenger Systems: Molecular Aspects" (G. Cohen and R. A. Greenwood, eds.), p. 316. Elsevier, Amsterdam, 1983.
[30] M. W. Pantoliano, J. S. Valentine, A. R. Burger, and S. J. Lippard, *J. Inorg. Biochem.* **17,** 325 (1982).
[31] M. E. McAdam, E. M. Fielden, P. Lavelle, L. Calabrese, D. Cocco, and G. Rotilio, *Biochem. J.* **167,** 271 (1977).

down the chain to dioxygen. The "$O_2$ activation" process occurring on the enzyme overcomes the intrinsic kinetic inertness of the terminal electron acceptor.[32,33]

The complete transfer of 4 electrons to $O_2$ with formation of $2H_2O$ occurs "on-the-enzyme" without release of "oxygen radicals" into the bulk. To achieve such a goal, oxidase has four electron-accepting sites per functional unit, a structural feature common to other oxidases (e.g., laccase) leading to the formation of water as the terminal product. Thus the decay of $O_2$ radicals has to be investigated while bound to the enzyme, and to achieve this goal rapid reaction methods have been successfully employed.

In the overall "$O_2$ activation" process, a crucial step is the binding of the gaseous ligand with formation of an irreversible complex. This is achieved through the action of a binuclear metal site, which in cytochrome oxidases is a heme $a$-copper center (i.e., that of cyte $a_3$ and Cu $a_3$). An interesting question from the standpoint of the chemistry occurring at this binuclear cluster is the understanding of the role played by the protein in determining the macromolecule toward being either an "oxygen carrier" (with reversible $O_2$ binding) or an "oxidase" (with essentially irreversible $O_2$ binding). A comparison between hemocyanin and laccase will be extremely informative, since both contain a binuclear copper binding site, but the latter is an oxidase while the former is an oxygen carrier.

Several recent reviews are available on the structure and function of cytochrome $c$ oxidase.[32–35]

### Methodological Approach to Detect $O_2$ Intermediates

The reaction of reduced cytochrome oxidase with $O_2$ is very fast indeed,[36] a feature common to all natural oxygen binding proteins, whether $O_2$ carriers or oxidases. Typical second-order rate constants are in the range $10^7$–$10^8$ $M^{-1}$ sec$^{-1}$ at 20°.[37] Therefore the experimental approach to the study of intermediates is based on the principles of rapid reaction techniques, with detection of species by different types of spectroscopy (typically electronic absorption at room temperature).

[32] M. Wikström, K. Krab, and M. Saraste, *Annu. Rev. Biochem.* **50,** 623 (1981).

[33] B. G. Malmstrom, *Annu. Rev. Biochem.* **51,** 21 (1982).

[34] M. Brunori, E. Antonini, and M. T. Wilson, *in* "Metal Ions in Biological Systems" (H. Sigel, ed.), Vol. 13, p. 187. Dekker, New York, 1981.

[35] R. A. Capaldi, *in* "Membrane Proteins in Energy Transduction" (R. A. Capaldi, ed.), Vol. 2, p. 201. Dekker, New York, 1979.

[36] C. Greenwood and Q. H. Gibson, *J. Biol. Chem.* **242,** 1782 (1967).

[37] E. Antonini and M. Brunori, "Hemoglobin and Myoglobin in their Reactions with Ligands," Vol. 21. North-Holland Publ., Amsterdam, 1971.

In view of (1) the very rapid reaction with $O_2$ (microsecond to millisecond time range) and (2) the irreversibility of the process, the methodological approach employed for detection of $O_2$ intermediates in cytochrome oxidase is based on the photochemical initiation of the reaction, starting with the reduced CO derivative of the enzyme in the presence of excess $O_2$. This *flow-flash method* is depicted below:

$$a^{2+}\ a_3^{2+}\!-\!CO \searrow \longrightarrow a^{2+}\ a_3^{2+}\!-\!CO \xrightarrow{\ flash\ } a^{2+}\ a_3^{2+} \overset{CO}{\underset{O_2}{\diagdown}} \longrightarrow products$$
$$+$$
$$O_2 \qquad\qquad O_2$$

This approach, therefore, takes advantage of (1) the slow rate constant for the dissociation of CO from reduced cytochrome oxidase, even in the presence of excess $O_2$ ($k$, 0.02–0.03 $sec^{-1}$ at 20°); (2) the high quantum yield for the photodissociation of CO ($\varphi \sim 1$), and (3) the very fast combination of $O_2$ with the photochemically stripped liganded binding site (cyt $a_3^{2+}$–Cu $a_3^+$ center). This experimental setup was used initially by Greenwood and Gibson[36] (see also Refs. 38 and 39) to follow the time course of the $O_2$ reaction of cytochrome oxidase between 2 and 20° by optical spectroscopy. The time course is complex (see below), and is complete within a few milliseconds above freezing.

In view of the rather limited structural information content of optical spectroscopy, investigation of the oxygen intermediates of cytochrome oxidase has been extended to use EPR and EXAFS.[40–42] This has been made possible by the introduction of a "freeze trapping" technique, which takes advantage of low-temperature stabilization of intermediates.

The method introduced by Chance and collaborators[40] and used by others,[41] is based on a series of experimental steps which may be summarized as follows. (1) A solution of cytochrome oxidase, or a suspension of mitochondria, in the presence of 1 atm of pure CO and reductants (e.g. ascorbate) is cooled down to $-15/22°$. (2) After addition of $O_2$ in water–ethylene glycol ($O_2$ conc = 2 m$M$ at 1 atm and $-23°$) to the precooled solution (1), the sample is stirred and cooled down to $-78°$; at this temperature the complex with CO is stable for hours. (3) The $O_2$ reaction is activated photochemically (see above) and the time course of the events followed at different temperatures by different spectroscopic means.

[38] C. Greenwood, M. T. Wilson, and M. Brunori, *Biochem. J.* **137**, 205 (1974).
[39] M. Brunori and Q. H. Gibson, *EMBO J.* **2** (1983).
[40] B. Chance, C. Saronio, J. S. Leigh, Jr., M. J. Ingledew, and T. E. King, *Biochem. J.* **171**, 787 (1978).
[41] G. M. Clore, L. E. Andréasson, B. Karlsson, R. Aasa, and B. G. Malmstrom, *Biochem. J.* **185**, 139 (1980).
[42] L. Powers, B. Chance, Y. Ching, and P. Angiolillo, *Biophys. J.* **34**, 465 (1981).

This approach has provided very important information on the nature of some of the $O_2$ intermediates in the overall reaction leading to water (see below); it should be pointed out, however, that the very nature of the experimental approach (freeze-trapping photochemical activation starting from the CO derivative) leads to the following problems. (1) At low temperatures (e.g., below $-100°$), the photochemically dissociated CO binds to Cu $a_3$ in the active-site pocket, without diffusing out in the bulk.[43] This should be taken into proper consideration in the interpretation of the spectroscopic properties of the species following photochemical activation. (2) The CO derivative of reduced cytochrome oxidase is in the "resting" conformational state, as indicated by the analysis of room temperature kinetic data on the mixed-valence state of the enzyme.[44] Thus it seems likely that also at low temperatures the events following the $O_2$ reaction will involve exclusively the "resting" state of the enzyme (and thus no information on the more active "pulsed" state[45,46] may be obtained by this approach).

*Intermediates in the Oxygen Reaction*

Room-temperature flow-flash experiments[36,38] have shown that the time course of the reaction with oxygen is characterized by three kinetic phases. (1) The faster kinetic component ($t_{1/2} \sim 10$–$100 \mu sec$) is associated to oxidation of cyt $a_3$, on the basis of the spectral distribution in the Soret and $\alpha$-band regions. The measured rate constant is $O_2$ concentration dependent only in the lower concentration range, yielding an apparent second-order combination rate constant of $\sim 10^8 M^{-1} sec^{-1}$. (2) The second phase, associated with oxidation of Cu $a$ (as indicated by observation in the near infrared), is observed also in the Soret region,[39] and has a rate constant of $3 \times 10^3 sec^{-1}$. (3) The last phase, with a rate constant of $7 \times 10^2 sec^{-1}$, is associated with oxidation of cyt $a$ ($t_{1/2} \sim 1$–$5 msec$). This experimental information led to the formulation of a kinetic scheme in which electrons are transferred to bound dioxygen in a sequential pathway.

Low-temperature freeze-trapping experiments have confirmed a sequential scheme and have led to the identification of several intermediates, some of which have been well characterized. Although there are

[43] J. O. Alben, P. P. Moh, F. G. Fiammingo, and R. A. Altschuld, *Proc. Natl. Acad. Sci. U.S.A.* **78**, 234 (1981).

[44] M. Brunori, A. Colosimo, P. Sarti, E. Antonini, and M. T. Wilson, *FEBS Lett.* **152**, 75 (1983).

[45] E. Antonini, M. Brunori, A. Colosimo, C. Greenwood, and M. T. Wilson, *Proc. Natl. Acad. Sci. U.S.A.* **74**, 3128 (1977).

[46] R. W. Shaw, R. E. Hansen, and H. Beinert, *Biochim. Biophys. Acta* **548**, 386 (1979).

differences of interpretation as to the electron distribution in the various species and to side paths which may be artificially populated at low temperatures, some clear-cut species have been identified.

Intermediate I (or A) is the first product of $O_2$ reaction after photolysis, and is a complex of $O_2$ with the binuclear center, having the overall charge [cyt $a_3$–Cu $a_3$–$O_2$]$^{3+}$. The extent of electron transfer from the two metals to the bound dioxygen is a matter of debate, but the view that bound peroxide may be the first observed intermediate has gained some support[32,33]; thus the binuclear center may transfer a net two electrons to dioxygen. This species is the only one significantly populated when starting with the mixed-valence state of cytochrome oxidase (in which cyt $a$ and Cu $a$ are both oxidized). Thus a peroxo complex, bridging the two metals, is a likely interpretation of the structure of intermediate I.

Identification of intermediate II, with addition of one more electron to intermediate I, has been predicted and recently supported by the discovery at very low temperature of an unusual copper EPR signal.[47] The localization of the unpaired electron on the copper nucleus and the unusual relaxation properties of the EPR signal have led to the proposal of a possible structure for intermediate II, i.e., $a_3^{4+} = O\ Cu_{a_3}^{2+}$. Thus addition of a third electron leads to rupture of the O—O bond on the enzyme, liberation of the first water molecule and formation of a cupric $Cu_{a_3}^{2+}$ and ferryl cyt $a_3^{4+}$ bound to the second oxygen atom.

Formation of subsequent intermediates is less clearly established, and will not be dealt with in this context. The illustration of these two species, however, is representative of the enzyme-bound $O_2$ radical species which are formed in the sequential electron transfer in cytochrome oxidase.

Finally, two points deserve comment. The first is related to the features of these intermediates when the $O_2$ reaction is started with the "pulsed" state of the enzyme. It is well accepted that cytochrome oxidase exists in two distinct conformational states with different catalytic and spectroscopic features.[34,45,46] In view of a number of reports, data on oxygen intermediates for both states are necessary. The second point deals with the significance of these intermediates at room temperature (or at physiological temperatures), since a recent study carried out with a specially built steady-state rapid-freeze instrument has been unable to detect significant amounts of these species above 0°.[48] Additional experimental advances seem necessary to answer these and other unsettled questions.

[47] B. Karlsson, R. Aasa, T. Vanngård, and B. G. Malmström, *FEBS Lett.* **313,** 186 (1981).
[48] M. T. Wilson, P. Jensen, R. Aasa, B. G. Malmström, and T. Vanngård, *Biochem. J.* **203,** 483 (1982).

## [3] Characterization of Singlet Oxygen

By Christopher S. Foote, F. C. Shook, and R. B. Abakerli

Considerable recent interest has focused on the question of the intermediacy of singlet molecular oxygen ($^1O_2$) in biological processes, formed either as a result of natural processes or under the effect of exogenous agents such as photosensitizers.[1,2] In well-characterized chemical systems, singlet oxygen can be produced by both photochemical and nonphotochemical processes, and its intermediacy can be quantitatively determined.[3] In biological systems (which are often inhomogeneous) demonstration of its intermediacy in the oxidation of target species is not easy. The main problem is that singlet oxygen is likely to be accompanied by other reactive oxygen species such as superoxide ion ($O_2^-$), hydroxyl radical (OH·), and alkoxy and peroxy radicals. The reactions of these species may compete with or be confused with those of singlet oxygen.[4]

The four main methods for detection of singlet oxygen in a reacting system are to compare observed products with those known to be produced by singlet oxygen, to determine the change in the amount of reaction or effect produced on adding substances that modify the lifetime of singlet oxygen, to measure the luminescence of singlet oxygen, or to produce singlet oxygen independently in the system and compare its reactions with those observed under the test conditions. All of these methods have difficulties; the greatest degree of certainty is obtained by using as many independent techniques as possible in combination. At the present state of development, no one method used alone is likely to provide definitive evidence for the intermediacy of singlet oxygen in a complex system, nor is it likely that a single technique will even be usable in all systems. Thus this chapter provides a brief overview of a few of the available methods, and specific directions for only one. It is hoped that the technique described will be used with these warnings in mind.

[1] N. I. Krinsky, in "Singlet Oxygen" (H. H. Wasserman and R. W. Murray, eds.), p. 597. Academic Press, New York, 1979.
[2] C. S. Foote, Free Radicals Biol. 2, 85–133 (1976).
[3] R. W. Murray and H. H. Wasserman, eds., "Singlet Oxygen." Academic Press, New York, 1979.
[4] C. S. Foote, in "Biochemical and Clinical Aspects of Oxygen" (W. S. Caughey, ed.), p. 603. Academic Press, New York, 1979.

METHODS IN ENZYMOLOGY, VOL. 105

Copyright © 1984 by Academic Press, Inc.
All rights of reproduction in any form reserved.
ISBN 0-12-182005-X

Sources of Singlet Oxygen

*Nonphotochemical Routes*

There are a number of nonphotochemical reactions that have been shown to produce singlet oxygen in good yield. This subject has been reviewed.[5] Most of these reactions use strong oxidizing species (e.g., sodium hypochlorite, triphenyl phosphite ozonide) that could react directly with oxidizable species present in a biological system. Probably the best choice for a nonchemical source of singlet oxygen is to use an aromatic endoperoxide that liberates singlet oxygen on warming[6,7]; in principle, these derivatives can be individually designed to reflect desired solubility characteristics and the temperature at which singlet oxygen is liberated, although not much has yet been done along these lines.

*Photosensitized Reactions*

There are two fundamental types of sensitized photooxygenation.[8] They differ in that the triplet sensitizer reacts directly with the substrate in the first (the Type I reaction), while in the second (Type II), it reacts first with oxygen to produce singlet oxygen. Thus even when photooxygenation (the most studied source of singlet oxygen) is used, the reactive intermediate is not certain to be singlet oxygen, but must be demonstrated in each case.

$$\text{Sens} \rightarrow {}^1\text{Sens} \rightarrow {}^3\text{Sens}$$

$$\text{Type I} \xleftarrow[\text{Subst.}]{} {}^3\text{Sens} \xrightarrow[\text{O}_2]{} \text{Type II}$$

$$\downarrow \qquad\qquad\qquad\qquad \downarrow$$

$$\text{Radicals} \qquad\qquad\qquad\quad {}^1\text{O}_2$$

An important characteristic of these reactions is that the process observed is often very concentration dependent. The reason is that there is always a competition between substrate and oxygen for the sensitizer triplet.[8] If the substrate concentration is too low, or the oxygen concen-

[5] R. W. Murray, *in* "Singlet Oxygen" (H. H. Wasserman and R. W. Murray, eds.), p. 59. Academic Press, New York, 1979.

[6] N. J. Turro, M. F. Chow, and J. Rigaudy, *J. Am. Chem. Soc.* **103,** 7218 (1981).

[7] H. H. Wasserman and J. R. Scheffer, *J. Am. Chem. Soc.* **89,** 3073 (1967).

[8] C. S. Foote, *in* "Pathology of Oxygen" (A. P. Autor, ed.), p. 21. Academic Press, New York, 1982.

tration is too high, the Type I reaction may become very inefficient, and the Type II reaction is favored. Of course, if there is no good substrate for singlet oxygen available at reasonable concentration, the Type II process will result in quenching of the excited sensitizer with no reaction observed. In systems where the sensitizer is bound to an easily oxidized biomolecule, the local concentration of substrate becomes very high and Type I reactions become highly favored.

In the Type I reaction, the sensitizer interacts directly with the substrate (for example, a hydrogen or electron donor) with a resulting hydrogen atom or electron transfer to produce radicals. These radicals can subsequently react with oxygen to produce oxidized products or other reactive species. The products are often peroxides, which can in turn break down to induce free radical chain autoxidation, leading to further oxidation in a nonphotochemical step.

$$ROOR' \rightarrow RO\cdot + R'O\cdot$$

Sensitizers can produce superoxide ion by undergoing electron transfer processes with the substrate or oxygen, as shown below.

$$^3Sens + Subs \rightarrow Sens^- + Subs_{ox}$$
$$Sens^- + O_2 \rightarrow Sens + O_2^-$$

or

$$^3Sens + O_2 \rightarrow Sens^+ + O_2^-$$

These reactions produce $O_2^-$, which can subsequently give the very reactive hydroxyl radical (OH·) by several pathways. These radicals can react with organic molecules in a variety of ways, or can initiate radical chain autoxidation.[9] The sorts of compounds that react in the Type I reaction are those that are electron rich or have easily abstractable hydrogens. Particularly reactive, for example, are aromatic amines, phenols, and sulfhydryl compounds.

In the Type II reaction, the sensitizer can transfer its excitation energy to a ground state oxygen molecule.[8] Singlet molecular oxygen is produced. Singlet oxygen is a metastable species with a lifetime varying from about 4 $\mu$sec in water to 25 to 100 $\mu$sec in nonpolar organic media that are reasonable models for lipid regions of the cell.[10] It is quite reactive, and reacts with many organic molecules to give peroxides or other oxidized products; however, it is also quite selective and fails to react with molecules that are not electron rich enough and simply returns to the ground state. Probably the best photochemical source of singlet oxygen for use in

[9] A. Singh, *Can. J. Physiol. Pharmacol.* **60**, 1330 (1982).
[10] F. Wilkinson and J. G. Brummer, *J. Phys. Chem. Ref. Dat.* **10**, 809 (1981).

complex systems is one of the polymer-bound rose bengal derivatives[11,12]; these must be carefully extracted to remove unbound rose bengal, but they are less likely than free dyes to give Type I reactions in biological systems.

$$^3O_2 \xleftarrow[\text{Decay}]{} {}^1O_2 \xrightarrow[\text{Acceptor}]{} AO_2$$

A number of compounds have been found to quench (i.e., deactivate without reaction) singlet oxygen efficiently.[13] For example, $\beta$-carotene inhibits photooxidation of 2-methyl-2-pentene efficiently at $10^{-4}$ $M$ without itself being appreciably oxidized. Certain amines are quenchers, e.g., DABCO (1,4-diazabicyclooctane). Other amines both quench singlet oxygen and react with it, depending on conditions. Azide ion is a somewhat better quencher; phenols also quench singlet oxygen; some also react chemically with it.

$$^1O_2 + \text{Quencher} \rightarrow {}^3O_2 + \text{Quencher}$$

The major biological targets for singlet oxygen are now well known.[14,15] Membranes are peroxidized, leading to fragility and easy lysis. The initial hydroperoxides appear to break down in a subsequent slower step, probably involving formation of free radicals and subsequent radical chain autoxidation to cause increased degradation of unsaturated molecules in the membrane.[16]

Nucleic acids are also an important target for singlet oxygen; guanine is the major target of this reaction. Many proteins and enzymes are damaged by singlet oxygen; the major targets are histidine, methionine, and tryptophan (although tryptophan can probably react by a Type I mechanism also). Tyrosine and cysteine are also photooxidation targets, although these appear more likely to be Type I reactions.

### Determination of Mechanism

In homogeneous solution, the problem of mechanism determination is at its easiest and a variety of kinetic and trapping techniques can be used.

[11] A. P. Schaap, A. L. Thayer, E. C. Blossey, and D. C. Neckers, *J. Am. Chem. Soc.* **97**, 3741 (1975).
[12] A. P. Schaap, A. L. Thayer, K. A. Zaklika, and P. C. Valenti, *J. Am. Chem. Soc.* **101**, 4016 (1979).
[13] C. S. Foote, in "Singlet Oxygen" (H. H. Wasserman and R. W. Murray, eds.), p. 139. Academic Press, New York, 1979.
[14] C. S. Foote, in "Oxygen and Oxy-Radicals in Chemistry and Biology" (M. A. J. Rodgers and E. L. Powers, eds.), p. 425. Academic Press, New York, 1981.
[15] J. D. Spikes, in "Oxygen and Oxy-Radicals in Chemistry and Biology" (M. A. J. Rodgers and E. L. Powers, eds.), p. 421. Academic Press, New York, 1981.
[16] A. A. Lamola, T. Yamane, and A. M. Trozzolo, *Science* **179**, 1131 (1973).

It is important to recognize that the simple detection of a reactive species such as singlet oxygen in a system does not necessarily mean that it is the reactive species in the oxidation of the target molecules; its presence is a necessary but not sufficient condition for its intermediacy. In other words, we need to know not only how much of the intermediate is produced, but what fraction of the target molecules are actually oxidized via that intermediate. In this case, a *negative* result is easier to make definitive: if singlet oxygen is absent, it cannot be the reactive intermediate. It is useful in this case to set upper limits for the fraction of the reaction proceeding via singlet oxygen if possible.

It is of the utmost importance that the technique used for the detection of singlet oxygen be specific for that intermediate, since many chemical reagents can react with different oxidizing species to give similar products.[4]

There are a number of standard tests for the intermediacy of reactive oxygen species other than singlet oxygen; these have been reviewed,[9] and will not be discussed here, but their use should be considered in conjunction with tests for singlet oxygen. Specificity of the tests for other species is also a matter of concern. Radical chain processes also need to be ruled out; these may occur following the production of an initial peroxide molecule, and can lead to the oxidation of many more molecules than formed in the initial step. Radical chain oxidations are usually detected by their inhibition by a chain terminator such as a phenol, but this is often difficult to do with specificity in the presence of other oxidizing agents.

$$RO\cdot + PhOH \rightarrow ROH + PhO\cdot \quad \text{(Chain Termination)}$$

## Singlet Oxygen Tests

### Traps

A large number of traps for singlet oxygen have been developed that give isolable products. The obvious hope in using these traps is that they will give these products *only* with singlet oxygen, but this is often not the case. The first class of traps to be used in biological systems was the furans; unfortunately, these are the least specific since they are oxidized by almost all strong oxidants to diketones.[4]

A more specific class of traps is provided by substituted anthracene derivatives; these can be made water soluble with suitable substituents (S) and an isolable endoperoxide is formed.[17,18] Unfortunately, these compounds sensitize their own photooxidation, so that the production of small amounts of product is difficult to avoid except in the complete absence of light. Also, their usefulness in complex systems may be limited because of adsorption and compartmentalization problems.

One of the more specific traps that has been developed is cholesterol; this gives a single product (the 5$\alpha$-hydroperoxide) on reaction with singlet oxygen.[19,20] Radical oxidation gives a very complex mixture that contains none of the 5$\alpha$ product. Cholesterol is useful as a marker for the presence of singlet oxygen in biological systems; however, a major drawback is its low reactivity, which results in poor sensitivity. It is also nearly completely insoluble in aqueous systems. The system using radiolabeled cholesterol bound to dispersible polymer beadlets,[21,22] described below, is quantitatible and overcomes the solubility and sensitivity limitations.

### Kinetic Methods

The second method of detecting singlet oxygen is to inhibit its reactions by adding something that changes its lifetime by reacting with it, quenching it, or modifying its decay rate. This method can be quite effective in homogeneous solution, since the rate constants for a large number of compounds are well known.[10] To be most effective, inhibition studies should be carried out quantitatively, so that all rate constants in the system are determined. It is much less effective simply to carry out one

[17] A. P. Schaap, A. L. Thayer, G. R. Faler, K. Goda, and T. Kimura, *J. Am. Chem. Soc.* **96,** 4025 (1974).

[18] J. M. Aubry, J. Rigaudy, C. Ferradini, and J. Pucheault, *J. Am. Chem. Soc.* **103,** 4965 (1981).

[19] L. L. Smith, J. I. Teng, M. J. Kulig, and F. L. Hill, *J. Org. Chem.* **38,** 1763 (1973).

[20] L. L. Smith, W. S. Matthews, J. C. Price, R. C. Bachmann, and B. Reynolds, *J. Cromatogr.* **27,** 182 (1967).

[21] C. S. Foote, R. B. Abakerli, R. L. Clough, and R. I. Lehrer, *in* "Bioluminescence and Chemiluminescence" (M. A. DeLuca and W. D. McElroy, eds.), p. 81. Academic Press, New York, 1980.

[22] C. S. Foote, F. C. Shook, and R. B. Abakerli, *J. Am. Chem. Soc.* **102,** 2503 (1980).

point inhibition experiments, because all singlet oxygen quenchers and reagents are compounds of low oxidation potential and will also react with other strong oxidants. The kinetic parameters provide a fingerprint for the reaction species being trapped by the inhibitor. Of course, it is much more difficult to use quantitative techniques in inhomogeneous systems, since targets and quenchers are localized, and their local concentrations are usually not known. The inhibition method, if carefully quantitated, allows the determination of the fraction of target molecules reacting with singlet oxygen. The methodology of the kinetic techniques has been reviewed.[13]

Another way of testing for the intermediacy of singlet oxygen is to determine the effect on the rate of the observed reaction on substituting $D_2O$ for water.[23] This technique is based on the fact that the lifetime of singlet oxygen in $D_2O$ is longer than in $H_2O$, so that reactions of singlet oxygen with substrates may be more efficient, since more singlet oxygen survives to react. This method requires that the reaction be carried out in such a way that only a small fraction of the singlet oxygen reacts with the substrate or with other quenchers and solvent deactivation determines the lifetime. This technique can and should be quantitatively compared with the effect calculated, based on known rate constants. Another problem is that other reactive species may show solvent deuterium isotope effects on their lifetimes. For example, $O_2^-$ is known to live substantially longer in $D_2O$ than in water.[24]

### Luminescence

There are two types of singlet oxygen luminescence, the direct emission from a single molecule at 1.27 $\mu$m, and "dimol" luminescence at 634 and 704 nm.[25] Both are extremely inefficient in solution.

$$^1O_2 \rightarrow h\nu \quad (1.27 \ \mu m)$$
$$2(^1O_2) \rightarrow h\nu \quad (634,704 \ nm)$$

Dimol luminescence has often been used for the identification of singlet oxygen in biological systems. Unfortunately, this weak luminescence is usually accompanied by light of other wavelengths in complex systems. Although these emissions have been assigned to higher vibrational states of singlet oxygen,[25] it seems more likely that the extraneous emission comes from excited carbonyls or other species in the system. Precise wavelength determination is essential if this technique is to be used for the

[23] R. Nilsson and D. R. Kearns, *Photochem. Photobiol.* **17**, 65 (1973).
[24] B. H. J. Bielski and E. Saito, *J. Phys. Chem.* **75**, 2263 (1971).
[25] M. Kasha and A. U. Khan, *Ann. N.Y. Acad. Sci.* **171**, 5 (1970).

detection of singlet oxygen; it appears to be of qualitative usefulness at best, since the luminescence depends on a second order process.

More hope exists for quantitation of the 1.27 $\mu$m emission. This luminescence, though weak, can be detected, both in steady-state[26,27] and time-resolved systems,[28–31] and the wavelength is specific. We have used this system to detect singlet oxygen produced by the fungal photosensitizer cercosporin.[32] This method may be useful for the detection of singlet oxygen in biological systems, although sensitivity requirements will be extreme. Changes in the lifetime of singlet oxygen measured using this technique could be interpreted in terms of the degree to which it is reacting with host molecules.

### Sample Procedure

The procedure given below is a sample of the technique using the cholesterol beadlet preparation. The formation of the 5$\alpha$-cholesterol oxidation product, determined by counting after thin-layer chromatography (TLC) separation, is a measure of the amount of singlet oxygen production; the presence of the isomeric 7$\alpha$ and 7$\beta$ products are indications of free radical processes.[19,20] Included is a method for determining the efficiency with which singlet oxygen is trapped, which depends on the method of preparation of the beadlets and the concentration of the dispersion. Knowing the efficiency of singlet oxygen trapping and the amount of 5$\alpha$ product formed allows calculation of the absolute amount of singlet oxygen formed. However, this determination must be supplemented by an inhibition study to allow determination of the quantitative involvement of singlet oxygen in the oxidation of target species.

### Methods

#### Materials

*Purification of Radiolabeled Cholesterol.* Labeled [4-$^{14}$C]cholesterol (New England Nuclear, 50 $\mu$Ci in 2.5 ml benzene) was applied to a 10 × 20 cm × 0.25 mm preparative thin layer (TLC) plate (silica gel HF-254,

[26] A. U. Khan and M. Kasha, *Proc. Natl. Acad. Sci. U.S.A.* **76,** 6047 (1979).
[27] A. A. Krasnovskii, Jr., *Photochem. Photobiol.* **36,** 733 (1982).
[28] K. I. Salokhiddinov, I. M. Byteva, and B. M. Dzhagarov, *Opt. Spektrosk.* **47,** 487 (1979).
[29] P. R. Ogilby and C. S. Foote, *J. Am. Chem. Soc.* **104,** 2069 (1982).
[30] J. R. Hurst, J. D. MacDonald, and G. B. Schuster, *J. Am. Chem. Soc.* **104,** 2065 (1982).
[31] J. G. Parker and W. D. Stanbro, *J. Am. Chem. Soc.* **104,** 2067 (1982).
[32] D. C. Dobrowolsky and C. S. Foote, *Angew. Chem.* in press (1983).

Merck). Unlabeled cholesterol was applied to the same plate, in a different spot, as a standard. The plate was irrigated twice with benzene/ethyl acetate (17:8). The portion of the plate containing the unlabeled cholesterol was sprayed with 50% $H_2SO_4$ and heated gently with a heat gun to develop the bright red color characteristic of cholesterol. The corresponding unsprayed region was excised from the plate; the labeled cholesterol was then extracted with dry diethyl ether. The ether was removed and the cholesterol redissolved in 2.5 ml benzene. This solution was stored under argon in a tightly capped vial at $-28°$. Aliquots were analyzed frequently; when the presence of significant activity outside the cholesterol band was seen, the sample was rechromatographed.

*Cholesterol Beadlets.* A benzene solution (0.740 ml) of TLC-purified [$^{14}$C]cholesterol (17.6 $\mu$Ci/ml) was dried under a nitrogen stream and redissolved in 1 ml methanol. To this solution, 0.5 ml of a 10% polystyrene latex beadlet suspension (Dow Diagnostics) was added and dried under nitrogen. The residue was resuspended in 1 ml of doubly distilled water and diluted to 2 ml with phosphate-buffered saline (PBS). The suspension was filtered through a capillary, homogenized, and 10 $\mu$l was checked for radioactivity.

*Standards.* The 3$\beta$,5$\alpha$-diol was synthesized by the procedure of Kulig and Smith[33]; the 7-ketone by that of Nickon and Mendelson[34]; the 3$\beta$-, 7$\alpha$-, $\beta$-diols through lithium aluminum hydride (LAH) reduction of the 7-ketone, and the 5$\alpha$,6$\alpha$-epoxide by reacting cholesterol and *m*-chloroperoxybenzoic acid at room temperature. All standards were purified by column chromatography. A mixture containing the 5$\alpha$-, the 7$\alpha$-, and the 7$\beta$,3$\beta$-diols was conveniently prepared as follows.

Cholesterol (0.10 g) (MCB, recrystallized twice from methanol) was dissolved in 20 ml dry benzene; 5.0 mg of zinc tetraphenylporphine was added as sensitizer. The solution was photooxidized for 4 hr. To facilitate conversion of the 5$\alpha$-hydroperoxide to the 7-hydroperoxides, the reaction mixture was transferred to a stoppered flask and allowed to stand at ambient temperature for 48 hr. The hydroperoxides were reduced by addition of excess $NaBH_4$ and stirring for 1 hr. The solution was filtered, the benzene removed *in vacuo,* and the residue taken up in ether. An aliquot was removed and analyzed by TLC. Development showed the presence of cholesterol and the 5$\alpha$-, 7$\alpha$-, and 7$\beta$-diols. The mixture was stored under $N_2$ at $-28°$.

[33] M. J. Kulig and L. L. Smith, *J. Org. Chem.* **38,** 3639 (1973).
[34] A. Nickon and W. L. Mendelson, *J. Am. Chem. Soc.* **87,** 3921 (1965).

*Chromatography*

Cholesterol and oxidation products were analyzed by thin-layer chromatography on 20 × 20 cm × 0.25 mm preparative TLC plates (silica gel HF-254, Merck). The sample was applied in a thin streak using a commercial applicator. The plate was irrigated twice with benzene/ethyl acetate (17 : 8). Development of a narrow strip with a 50% $H_2SO_4$ spray and gentle heating typically gave four bands: ($R_f$, color with $H_2SO_4$, composition) Band 1: 0.70, red-violet, cholesterol; Band 2: 0.24, blue, cholesterol 5α-diol; Band 3: 0.19, blue, cholesterol 7β-diol; Band 4: 0.13, blue, cholesterol 7α-diol. The assignments were made by comparison with $R_f$ values of known samples.

For scintillation counting, the appropriate bands were excised from the TLC plate. The silica gel was transferred to a glass scintillation vial. BioFluor (10 ml), a high efficiency cocktail (New England Nuclear), was added. Total $^{14}C$ activity was recorded; typical counting times were 5 or 10 min. Quenching was determined by the internal standards method.[35]

*Photooxidation*

All photooxidations were performed using a Sylvania DWY tungsten halogen lamp (60–80 V). A 1% $K_2Cr_2O_7$ solution was used as a filter. Oxygen was passed through the solution, but $O_2$ uptake was not measured. Unless otherwise stated, photooxidations were performed at ambient temperature.

*Assay*

This procedure will need to be modified somewhat, depending on the system to be studied. This version was developed for use with polymorphonuclear leukocytes. The samples were reduced with 1 ml of a methanol solution of triphenylphosphine ($1.0 \times 10^{-3}$ M) for at least 1 hr, then extracted with methanol three times, followed by centrifugation. The combined extracts were dried under nitrogen and the residue redissolved in 0.2 ml of toluene and checked for recovered radioactivity. The samples were applied to a silica gel plate. The unlabeled diols and cholesterol were applied as standards, and the plate was developed and assayed as above. After the identification of the standards, autoradiography was done by allowing the TLC plate to stay in contact with the film (Kodak SB-5 with Dupont Cronex Quanta II intensifying screens) for approximately 1 week at −70°.

[35] A. Dyer, "An Introduction to Liquid Scintillation Counting," p. 57. Heyden, New York, 1974.

*Quantitation of $^1O_2$ Trapping Efficiency*

Fifty microliters of a solution of [4-$^{14}$C]cholesterol in benzene (purified by TLC) was transferred by syringe to a clean, dry Pyrex test tube, and 0.2 ml of a 10% aqueous suspension of polystyrene latex beads was added via pipet. Approximately 0.5 ml ether and 5.0 ml of $5 \times 10^{-4}$ $M$ histidine in pH 8 phosphate buffer (0.1 $M$) was added. The solution was stirred for 0.5 hr while $N_2$ was passed over the surface to remove the ether. Then 10 $\mu$l of $1.0 \times 10^{-3}$ $M$ methylene blue was added to give a final concentration of methylene blue of $2 \times 10^{-6}$ $M$.

The test tube was irradiated for 10 min, then transferred to a separatory funnel and extracted with four 10-ml portions of $CH_2Cl_2$. The organic layers were combined and excess triphenylphosphine added. The solution was stirred for 1 hr.

The organic layer was dried over anhydrous $Na_2SO_4$ and the solvent removed *in vacuo*. The residue was taken up in ether and chromatographed. The appropriate bands were excised and counted to determine the amount of cholesterol oxidation product.

The aqueous layer was transferred to a centrifuge tube and centrifuged for 10 min. At the end of that time, there had collected at the bottom of the tube a small amount of suspended $CH_2Cl_2$ and beads, leaving the aqueous layer only very faintly turbid. This solution was treated with the Pauly reagent[36] as follows: to 5 ml of solution was added 1 ml 1% sulfanilic acid in 1.0 $N$ HCl and 1 ml of 5% $NaNO_2$. The solutions were mixed and allowed to stand for 5 min, then 3 ml of 20% $Na_2CO_3$ was added. The absorption spectra of the photooxidized sample, as well as that of a similarly treated unreacted sample of histidine, was obtained and compared to a reagent blank. This measurement allowed the amount of histidine photooxidized to be determined. A small correction was required since the amount of light absorbed by the sensitizer was decreased by light scattering in the turbid beadlet solution. The amount of correction is determined by photooxidizing a beadlet-free sample of histidine in parallel. The trapping efficiency assay should be done on each beadlet–cholesterol preparation, and under the same concentration conditions used in the test system, since it has been found that the trapping efficiency depends on the concentration.

*Calculation of Trapping Efficiency.* The trapping efficiency (*TE*) is obtained from the formula

$$TE = PH/\Delta H(\beta + H)$$

where $P$ is the amount of $5\alpha$ product, $H$ the histidine concentration, $\Delta H$

---

[36] H. T. MacPherson, *Biochem. J.* **40**, 470 (1946).

its change, $\beta$ the concentration of histidine in $H_2O$ at which half the singlet oxygen is trapped ($3.0 \times 10^{-3} M^{37,38}$); $CF$ is the correction factor for light scattering. A typical value for $CF$ is 1.34 and for $TE$ is $2.5 \times 10^{-5}$. The amount of $^1O_2$ formed in the reaction is then

$$\text{Amt. } ^1O_2 = (P/TE)CF$$

Units of all the concentrations are molar.

Acknowledgment

This work was supported by NIH Grant GM-20080.

[37] L. A. A. Sluyterman, *Rec. Trav. Chim.* **80**, 989 (1961).
[38] L. Weil, *Arch. Biochem. Biophys.* **110**, 57 (1965).

# [4] Role of Iron in Oxygen Radical Reactions

*By* BARRY HALLIWELL and JOHN M. C. GUTTERIDGE

The discovery of the enzymatic production of the superoxide ($O_2^-$) radical and of the presence of superoxide dismutase (SOD) enzymes in aerobic cells led directly to the proposal that $O_2^-$ is a major factor in oxygen toxicity and that SOD constitutes an important defense against it.[1] Systems generating the $O_2^-$ radical have been shown to have a number of damaging effects, some of which are summarized in Table I. The superoxide radical itself in organic solvents is a powerful base and nucleophile,[2] which may have relevance to reactions taking place within the interior of cell membranes. In aqueous solution it is far less reactive, acting mainly as a reducing agent and undergoing the dismutation reaction,

$$O_2^- + O_2^- + 2H^+ \rightarrow H_2O_2 + O_2 \tag{1}$$

Yet all of the damage observed in Table I refers to reactions carried out in aqueous solution. In many cases damage is decreased by addition not only of SOD but also of catalase, and it was proposed[3] that $O_2^-$ and $H_2O_2$ can combine together directly to generate the highly reactive hydroxyl radical, OH·

$$H_2O_2 + O_2^- \rightarrow OH \cdot + OH^- + O_2 \tag{2}$$

Indeed, damage is often decreased by scavengers of this radical such as mannitol, sodium formate, and thiourea. It must be emphasized that in

[1] I. Fridovich, *Annu. Rev. Biochem.* **44**, 147 (1975).
[2] D. T. Sawyer and M. T. Gibian, *Tetrahedron* **35**, 1471 (1979).
[3] C. Beauchamp and I. Fridovich, *J. Biol. Chem.* **245**, 4641 (1970).

Copyright © 1984 by Academic Press, Inc.
All rights of reproduction in any form reserved.
ISBN 0-12-182005-X

TABLE I

SOME DELETERIOUS EFFECTS OF SYSTEMS GENERATING THE SUPEROXIDE RADICAL[a]

| Source of $O_2^-$ | System studied | Damage | Comments |
|---|---|---|---|
| Heart muscle submitochondrial particles | Activity of NADH-CoQ reductase complex | Activity lost | Damage prevented by SOD; the $O_2^-$ generated by the complex inactivates it unless SOD is present; catalase not protective |
| Illuminated FMN[b] | Bacteria | Loss of viability | Protection by SOD |
| Xanthine + xanthine oxidase | Human synovial fluid | Degradation—loss of viscosity and lubricating power | Both SOD and catalase protect |
| Xanthine + xanthine oxidase | Bacteriophage R17 | Inactivation | SOD protects partially |
| Illuminated FMN | Ribonuclease | Loss of activity | SOD protects partially |
| Illuminated FMN | Calf myoblast cells | Growth abnormality, some cell death | SOD protects partially |
| Hypoxanthine + xanthine oxidase | Rat brain membrane $Na^+,K^+$-ATPase | Inactivation | SOD protects partially |

| Acetaldehyde + xanthine oxidase | Erythrocyte membranes | Lysis | SOD protects |
|---|---|---|---|
| Acetaldehyde + xanthine oxidase | Arachidonic acid | Oxidation | Both SOD and catalase protect |
| Hypoxanthine + xanthine oxidase | DNA | Degradation, single-strand breaks, attack on sugar moiety | Both SOD and catalase protect |
| Autoxidation of dihydroxy-fumarate | Rat thymocytes | Inhibition of $Na^+$-dependent amino acid uptake | SOD protects, but not catalase |
| Hypoxanthine + xanthine oxidase | Cheek pouch of living hamster (perfused with $O_2^-$ generating system). | Increased permeability of blood vessels; leakage of contents | SOD protects |
| Autoxidation of dialuric acid | Escherichia coli | Loss of viability | Both SOD and catalase protect |
| Acetaldehyde + xanthine oxidase | Staphylococcus aureus | Loss of viability | Both SOD and catalase protect; traces of iron chelates needed for killing |
| Xanthine + xanthine oxidase | Rat lung in vivo (instilled into lungs) | Acute lung injury, edema | SOD protects but not catalase |

[a] For sources of information see Refs. 1, 7, and 34.
[b] Flavin mononucleotide.

| System used to produce OH· | Reaction mixture | Nanomoles of hydroxylated product typically produced | Comments |
|---|---|---|---|
| Hypoxanthine + xanthine oxidase[14] | Final volume 2 ml containing final concentrations 2.5 m$M$ salicylate, 0.3 m$M$ EDTA, 0.1 m$M$ FeSO$_4$ or FeCl$_3$ (fresh solutions just before use), 0.2 m$M$ hypoxanthine, 150 m$M$ KH$_2$PO$_4$–KOH buffer, pH 7.4; add 40 $\mu$l xanthine oxidase in buffer (0.4 enzyme units/ml); incubate at 25° for 90 min (treat as above) | 150–200 | OH· formation inhibited by SOD, catalase |
| GSH + H$_2$O$_2$[15] | Final volume 2 ml containing final concentrations 0.1 m$M$ FeCl$_2$ (fresh solution just before use), 0.1 m$M$ GSH (fresh solution every day), 0.1 m$M$ H$_2$O$_2$, 150 m$M$ KH$_2$PO$_4$–KOH buffer, pH 7.4; incubate at 25° for 90 min (treat as above) | 30–40 | OH· formation inhibited by SOD, catalase |
| NAD(P)H + H$_2$O$_2$[16] | As for GSH but using NADH or NADPH | 20–30 (NADH) 30–40 (NADPH) | OH· formation inhibited by SOD, catalase |
| Ascorbic acid + H$_2$O$_2$[17] | Final volume 2 ml containing final concentrations 150 m$M$ KH$_2$PO$_4$–KOH buffer, pH 7.4, 0.1 m$M$ FeCl$_3$ (fresh solution just before use), 0.1 m$M$ H$_2$O$_2$, 0.1 m$M$ ascorbate (fresh solution every day); incubate at 25° for 90 min (treat as above) | 100 | OH· formation inhibited by catalase but not by SOD[17,18] |

(*continued*)

such experiments a range of scavengers should be used and it ought to be possible to correlate the degree of protection that they offer with the known rate constants for reaction of the scavengers with OH·.

Unfortunately, the rate constant for reaction (2) is very small[4-6] and it certainly could not occur at the low steady-state concentrations of $O_2^-$ and $H_2O_2$ likely to be found in biological systems. Several authors proposed that salts of transition metals could catalyze reaction (2) (for a review see Ref. 7) although most direct evidence for this has come from work with iron salts.

### Detection of Hydroxyl Radicals

Formation of a radical identified as OH· in $O_2^-$-generating systems has been identified by its ability to oxidize methional into ethene[3] and to degrade tryptophan,[8] but neither of these methods is specific since other oxidizing species can bring the reactions about.[9-11] Technically, the easiest method to use in detecting OH· is spin trapping (see [22] and [23]) although the results are sometimes difficult to interpret and the technique requires access to ESR facilities.[12] Alternatively, the ability of OH· to attack aromatic compounds with the formation of hydroxylated products may be used. For example, if salicylate (2-hydroxybenzoate) at a concentration of 2 or 2.5 mM is included in the reaction mixture the hydroxylated products may be extracted into ether and assayed by a colorimetric method.[13] Table II indicates how this is done and gives some examples.

[4] G. J. McClune and J. A. Fee, *FEBS Lett.* **67**, 294 (1976).

[5] B. Halliwell, *FEBS Lett.* **72**, 8 (1976).

[6] A. Rigo, R. Stevanato, A. Finnazzi-Agro, and G. Rotilio, *FEBS Lett.* **80**, 130 (1977).

[7] B. Halliwell, *in* "Age Pigments" (R. S. Sohal, ed.), p. 1. Elsevier, Amsterdam, 1981.

[8] J. M. McCord and E. D. Day, *FEBS Lett.* **86**, 139 (1978).

[9] W. A. Pryor and R. H. Tang, *Biochem. Biophys. Res. Commun.* **81**, 498 (1978).

[10] A. Singh, H. Singh, W. Kremers, and G. W. Koroll, *Bull. Eur. Physiopathol. Resp.* **17**, 31 (1981).

[11] R. J. Youngman and E. F. Elstner, *FEBS Lett.* **129**, 265 (1981).

[12] E. Finkelstein, G. M. Rosen, and E. J. Rauckman, *Mol. Pharmacol.* **16**, 676 (1979).

[13] B. Halliwell, *FEBS Lett.* **92**, 321 (1978).

TABLE III (*footnote*)

   [a] Include salicylate at a final concentration of 2 or 2.5 mM in the reaction mixture. At the end of the reaction add 80 μl 11.6 M HCl, 0.5 g NaCl, and 4 ml chilled diethyl ether. Mix on a Whirlimix for 30 sec. Remove 3 ml of upper ether layer, place in a boiling tube, and evaporate to dryness at 40°. Dissolve residue in 0.25 ml cold double-distilled water. Add in order 0.125 ml 10% (w/v) trichloroacetic acid dissolved in 0.5 M HCl, 0.25 ml 10% (w/v) sodium tungstate, and 0.25 ml 0.5% (w/v) sodium nitrite (make sodium nitrite solutions fresh every day). Stand for 5 min and then add 0.5 ml 0.5 M KOH and read absorbance at 510 nm. Construct a standard curve using 2.3-dihydroxybenzoate solutions carried through the same extraction and assay procedure; 200 nmol gives an absorbance of approximately 0.651.

The method appears to pick up about 70% of the OH· radicals formed.[14] For increased sensitivity the hydroxylated products may be separated by gas-liquid chromatography (GLC) after conversion to volatile derivatives using trifluoroacetic anhydride.[14,19] Figure 1 shows the separation that can be achieved. This GLC technique has been used to detect formation of OH· radicals in the hypoxanthine/xanthine oxidase system[14] and from reduced paraquat.[19] We prefer to use phenol as a detector molecule in GLC experiments since its hydroxylated products can be quantitatively and simultaneously derivatized much more easily than those from salicylate. On the other hand, salicylate derivatives give a more intense color in the colorimetric method (Table II). The characteristic end products of OH· radical attack on phenol are catechol and hydroquinone in approximately a 3 : 1 ratio, very little resorcinol being observed.

The GLC technique is much more tedious than the colorimetric method but also much more sensitive, detecting hydroxylated products in picomole amounts.[19]

### The Role of Iron Salts in Hydroxyl Radical Formation: Chelation Experiments

Formation of hydroxyl radicals in the systems listed in Table II is dependent on the presence of iron salts and appears to occur by the following reactions[1,3,8,20]:

$$Fe^{3+} + O_2^- \rightarrow Fe^{2+} + O_2 \tag{3}$$
$$Fe^{2+} + H_2O_2 \rightarrow Fe^{3+} + OH\cdot + OH^- \tag{4}$$
$$\text{Net } O_2^- + H_2O_2 \rightarrow O_2 + OH\cdot + OH^- \tag{2}$$

The physiological relevance of these reactions has been discussed.[1,7,21] Many biochemical reagents are contaminated with iron salts in amounts sufficient for some OH· formation to occur even without iron salt addition. Several laboratories have sought for chelators of iron salts which prevent generation of OH· from $O_2^-$ and $H_2O_2$. Three have so far been discovered, namely, diethylenetriaminepentaacetic acid,[13,22]

[14] R. Richmond, B. Halliwell, J. Chauhan, and A. Darbre, *Anal. Biochem.* **118,** 328 (1981).
[15] D. A. Rowley and B. Halliwell, *FEBS Lett.* **138,** 33 (1982).
[16] D. A. Rowley and B. Halliwell, *FEBS Lett.* **142,** 39 (1982).
[17] D. A. Rowley and B. Halliwell, *Clin. Sci.* **64,** 649 (1983).
[18] C. C. Winterbourn, *Biochem. J.* **182,** 625 (1979).
[19] R. Richmond and B. Halliwell, *J. Inorg. Biochem.* **17,** 95 (1982).
[20] B. Halliwell, *Bull. Eur. Physiopathol. Resp.* **17,** 21 (1981).
[21] B. Halliwell, *in* "Copper Proteins" (R. Lontie, ed.), Vol. 2. CRC Press, Boca Raton, Florida, in press, 1983.
[22] G. R. Buettner, L. W. Oberley, and S. W. H. C. Leuthauser, *Photochem. Photobiol.* **21,** 693 (1978).

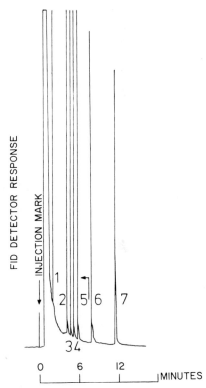

FIG. 1. Formation of hydroxyl radicals from reduced paraquat demonstrated by gas–liquid chromatography of trifluoroacetyl (TFA) derivatives of phenol. To the paraquat reaction mixture (see Ref. 19 for details) was added 6 g Analar NaCl followed by 0.5 ml 11.6 $M$ Analar HCl and 0.1 ml of an ethyl acetate standard solution containing 0.575 $\mu$mol each of 3,5-xylenol, $p$-methoxyphenol, and naphthalene. To a control reaction mixture containing only buffer were added NaCl and HCl as above followed by 0.66 $\mu$mol each of phenol, catechol, resorcinol, hydroquinone, 3,5-xylenol, $p$-methoxyphenol, and naphthalene. Samples were placed in an ice bath for 10 min and 3.5 ml chilled triple-distilled diethyl ether was added, the tubes gently inverted by hand 10 times, and left to stand in an ice bath for 10 min. This process was repeated three times and then 2.6–2.7 ml of the top ether layer transferred into stoppered tubes containing 0.1 g anhydrous $Na_2SO_4$. These were shaken and left to stand at 6° for 30 min. Into 2-ml white neutral colorbreak ampules (Baird and Tatlock Ltd.) was put 2.4 ml of the water-free ether and evaporated to 150–200 $\mu$l on ice by a gentle $N_2$ stream. Acetonitrile (100 $\mu$l) and 100 $\mu$l of triple-distilled trifluoroacetic anhydride were then added using a glass constriction pipet. The tubes were liquid $N_2$ cooled and flame torch sealed and then incubated at 70° for 1 hr. Gas chromatography was carried out using a flame ionization detector.[19] The figure shows the separation of TFA derivatives achieved: 1, TFA-phenol; 2, TFA-catechol; 3, TFA-resorcinol; 4, TFA-hydroquinone; 5, TFA-3,5-xylenol; 6, TFA-$p$-methoxyphenol; 7, naphthalene. Naphthalene is not derivatized by TFA and is more volatile than any of the phenolic compounds, so it controls for loss of material during extraction and derivatization.

TABLE III

USE OF DESFERRIOXAMINE TO DETECT THE ROLE OF
IRON SALTS IN SUPEROXIDE-DEPENDENT FORMATION
OF HYDROXYL RADICALS[a]

| [Desferrioxamine] present (mmol/liter) | [$Fe^{3+}$ complex] present (mmol/liter) | Percentage inhibition of salicylate hydroxylation |
|---|---|---|
| 0 | 0 | 0 |
| 0.05 | 0 | 26 |
| 0.1 | 0 | 47 |
| 0.2 | 0 | 56 |
| 0.4 | 0 | 71 |
| 0.5 | 0 | 78 |
| 0.6 | 0 | 81 |
| 0.8 | 0 | 86 |
| 1.0 | 0 | 89 |
| 0 | 0.05 | 0 |
| 0 | 0.1 | 1 |
| 0 | 0.25 | 7 |
| 0 | 0.5 | 11 |
| 0 | 1.0 | 16 |

[a] Both desferrioxamine and its $Fe^{3+}$ complex[28] react with the hydroxyl radical with a second-order rate constant equal to $1.3 \times 10^{10}$ $M^{-1}$ sec$^{-1}$. The effect of the chelator on formation of OH· radicals from a hypoxanthine–xanthine oxidase system in the presence of 0.1 m$M$ iron salt (Table I) is shown. At a concentration of 1 m$M$ the $Fe^{3+}$ complex is beginning to scavenge OH· directly and so concentrations of chelator greater than this should not be used in experiments. The inhibition by desferrioxamine at lower concentrations must be due to its ability to bind $Fe^{3+}$, since the $Fe^{3+}$ chelate has no effect.

bathophenanthroline sulfonate,[23] and desferrioxamine.[24] The last of these three is by far the most effective inhibitor; it is a specific chelator of $Fe^{3+}$, whereas the others also chelate other metal ions and can be injected into humans and experimental animals.[25] Iron(III) bound to the above chelating agents cannot easily be reduced by $O_2^-$, whereas iron bound to EDTA

[23] B. Halliwell, *FEBS Lett.* **96**, 238 (1978).
[24] J. M. C. Gutteridge, R. Richmond, and B. Halliwell, *Biochem. J.* **184**, 469 (1979).
[25] D. R. Blake, N. D. Hall, P. A. Bacon, P. A. Dieppe, B. Halliwell, and J. M. C. Gutteridge, *Ann. Rheum. Dis.* **42**, 89 (1983).

TABLE IV
DETECTION OF NON-PROTEIN-BOUND IRON SALTS IN
EXTRACELLULAR FLUIDS BY THE BLEOMYCIN ASSAY[a]

| Human body fluid studied | Non-protein-bound iron concentration ($\mu$mol/liter) (mean $\pm$ SD) |
|---|---|
| Plasma or serum | 0 |
| Synovial fluid (rheumatoid patients) | 2.8 $\pm$ 1.2 |
| Cerebrospinal fluid | |
|   Normal patients | 2.2 $\pm$ 1.3 |
|   Multiple sclerosis | 1.8 $\pm$ 0.8 |
|   Epilepsy | 2.5 $\pm$ 1.9 |
|   Infantile NCL[b] | 3.9 $\pm$ 2.5 |
|   Juvenile NCL | 6.2 $\pm$ 1.9[c] |

[a] All reagents except for bleomycin are made up in pyrogen-free water and treated with Chelex 100 resin to remove contaminating iron. Ascorbic acid solution is freshly prepared by dissolving 0.7 g Analar solid in 10 ml water, shaking with 0.4 g of Chelex, centrifuging at 2000 $g$ to remove the Chelex, and diluting 1 : 50 with water. Calf thymus DNA is dissolved in water at 1 mg/ml by gentle mixing and allowing to stand at 4° for at least 12 hr before use. Bleomycin sulfate (1 mg/ml) is dissolved in pyrogen-free water. Reactions are carried out in new plastic tubes to which is added in the order stated 0.5 ml DNA solution, 0.05 ml bleomycin solution, 0.1 ml 50 m$M$ MgCl$_2$, 0.05 ml 10 m$M$ HCl, 0.1 ml H$_2$O, 0.1 ml sample, 0.1 ml ascorbate solution. Tubes are incubated at 37° for 2 hr in a shaking water bath. Controls are run in which pyrogen-free water replaces the sample and also in which sample is added but no bleomycin (this corrects for thiobarbituric acid-reactive material present in the sample or formed in it during incubation). The reaction is stopped by adding 0.1 ml of 0.1 $M$ EDTA, then 1 ml of 1% (w/v) thiobarbituric acid dissolved in 0.05 $M$ NaOH is added, plus 1 ml of 25% (v/v) HCl. The mixtures are transferred to glass tubes and heated at 100° for 10 min. After cooling the absorbance at 532 nm is read. A standard curve is prepared using FeCl$_3$ in pyrogen-free water. The pH optimum for the reaction is 7.3. Addition of 0.05 ml 10 m$M$ HCl normally brings the pH into this range. The assay is linear in the range 0–50 $\mu$mol/liter. Iron salt (10 $\mu$mol/liter) should give $A_{532}$ of approximately 0.5.
[b] NCL, Neuronal ceroid lipofuscinosis.
[c] Results significantly higher than normal.[33]

is reduced more readily than "unchelated" $Fe^{3+}$ salts.[26] Desferrioxamine also prevents reduction of $Fe^{3+}$ salts by ascorbic acid,[27] another potential source of OH· radicals *in vivo*.[17,18] All these chelating agents, like most other compounds, are scavengers of the OH· radical if added to a reaction mixture at high enough concentrations and so, if they are to be used as "probes" for iron-dependent hydroxyl radical generation the concentrations must be carefully controlled. Table III illustrates this point for desferrioxamine.

### Iron Salts *in Vivo*

If reactions (3) and (4) are relevant *in vivo* then iron salts must be present. The iron-saturated forms of transferrin and lactoferrin may well participate in OH· formation[28,29,30] and there is considerable evidence for a small pool of intracellular iron salts, probably chelated to phosphate esters such as ADP, ATP, and GTP (see Refs. 21 and 31 for reviews). An assay has been developed to measure the content of "catalytic" iron salts in extracellular fluids, based on the ability of the antibiotic bleomycin to degrade DNA in a reaction absolutely dependent on traces of iron salts.[32] DNA degradation is followed by measuring the release of products that react on heating with thiobarbituric acid under acidic conditions to form a chromogen. Table IV explains how the assay is performed and gives some typical results. The reaction is not interfered with by normal constituents of body fluids,[33] unlike conventional assays for iron-dependent formation of OH· radicals and lipid peroxidation.[34] There is evidence for increased iron salt concentrations in certain disease states.[34] No catalytic iron salts can be detected in normal blood serum or plasma but they are present in other body fluids. The bleomycin method does not detect iron bound to proteins such as transferrin, ferritin, catalase, or hemoglobin.[32]

[26] J. Butler and B. Halliwell, *Arch. Biochem. Biophys.* **218,** 174 (1982).
[27] S. F. Wong, B. Halliwell, R. Richmond, and W. R. Skowroneck, *J. Inorg. Biochem.* **14,** 127 (1981).
[28] S. Hoe, D. A. Rowley, and B. Halliwell, *Chem. Biol. Interact.* **41,** 75 (1982).
[29] D. R. Ambruso and R. B. Johnston, *J. Clin. Invest.* **67,** 352 (1981).
[30] J. V. Bannister, W. H. Bannister, H. A. O. Hill, and P. J. Thornalley, *Biochim. Biophys. Acta* **715,** 116 (1982).
[31] B. Halliwell and J. M. C. Gutteridge, "Free Radical Aspects of Biology and Medicine" Clarendon, Oxford, in press, 1984.
[32] J. M. C. Gutteridge, D. A. Rowley, and B. Halliwell, *Biochem. J.* **199,** 263 (1981).
[33] J. M. C. Gutteridge, D. A. Rowley, and B. Halliwell, *Biochem. J.* **206,** 605 (1982).
[34] J. M. C. Gutteridge, D. A. Rowley, B. Halliwell, and T. Westermarck, *Lancet* **1,** 459 (1982).

# Section II

## Isolation and Assays of Enzymes or Substances Resulting in Formation or Removal of Oxygen Radicals

### A. One-Electron Transfer Enzyme Reactions Resulting in $O_2^-$ Production
*Articles 5 through 8*

### B. Isolation, Purification, Characterization, and Assay of Antioxygenic Enzymes
*Articles 9 through 14*

### C. Antioxidants/Prooxidants
*Articles 15 through 19*

### D. Detection and Characterization of Oxygen Radicals
*Articles 20 through 28*

### E. Genetic Methods for Detection/Assay of Oxygen Radical Species
*Articles 29 through 31*

# [5] Overview: Biological Sources of $O_2^-$

## By IRWIN FRIDOVICH

The univalent reduction of dioxygen to $O_2^-$ is a commonplace event in both abiotic and biological systems. Indeed, because of the spin restriction,[1-5] the univalent route of dioxygen reduction is its most facile pathway. Numerous biologically relevant sources of $O_2^-$ have been described. These run the gamut from autoxidizable small molecules such as leucoflavins,[6] catecholamines,[7] and tetrahydropterins[8] to whole cells such as neutrophils,[9] monocytes,[10] and macrophages.[11] In between these two extremes we find autoxidizable proteins such as reduced ferredoxins[12] and hemoproteins,[13,14] oxidative enzymes such as xanthine oxidase,[15] and subcellular organelles such as mitochondria,[16] chloroplasts,[17] microsomes,[18] and nuclei.[19]

Although $O_2^-$ is undoubtedly produced in aerobic cells the quantitative aspects of its production are hard to arrive at both because of its intrinsic instability and because of the ubiquity of the superoxide dismutases, which catalyze its dismutation to $O_2$ plus $H_2O_2$. In respiring cells most of the dioxygen consumed is reduced to water, by the cytochrome oxidase and the "blue" copper oxidases, without the release of intermedi-

---

[1] H. Taube, *J. Gen. Physiol.* **49**, 29 (1965).

[2] J. P. Collman, *Acc. Chem. Res.* **1**, 136 (1968).

[3] G. A. Hamilton, *Adv. Enzymol.* **32**, 55 (1969).

[4] G. A. Hamilton, *Prog. Bioorg. Chem.* **1**, 88 (1971).

[5] G. A. Hamilton, *in* "Molecular Mechanisms of Oxygen Activation" (O. Hayaishi, ed.), p. 405. Academic Press, New York, 1974.

[6] D. Ballou, G. Palmer, and V. Massey, *Biochem. Biophys. Res. Commun.* **36**, 898 (1969).

[7] H. P. Misra and I. Fridovich, *J. Biol. Chem.* **247**, 3170 (1972).

[8] D. B. Fisher and S. Kaufman, *J. Biol. Chem.* **248**, 4300 (1973).

[9] B. M. Babior, *New Engl. J. Med.* **298**, 659, 721 (1978).

[10] R. B. Johnston, Jr., J. E. Lehmeyer, and L. A. Guthrie, *J. Exp. Med.* **143**, 1551 (1976).

[11] R. B. Johnston, Jr., C. A. Godzik, and Z. A. Cohn, *J. Exp. Med.* **148**, 115 (1978).

[12] H. P. Misra and I. Fridovich, *J. Biol. Chem.* **246**, 6886 (1971).

[13] H. P. Misra and I. Fridovich, *J. Biol. Chem.* **247**, 6960 (1972).

[14] R. H. Cassell and I. Fridovich, *Biochemistry* **14**, 1866 (1975).

[15] J. M. McCord and I. Fridovich, *J. Biol. Chem.* **243**, 5753 (1968).

[16] A. Boveris, N. Oshino, and B. Chance, *Biochem. J.* **128**, 617 (1972).

[17] K. Asada and K. Kiso, *Agric. Biol. Chem.* **27**, 453 (1973).

[18] S. D. Aust, D. L. Roerig, and T. C. Pederson, *Biochem. Biophys. Res. Commun.* **47**, 1133 (1972).

[19] G. M. Bartoli, T. Galeotti, and A. Azzi, *Biochim. Biophys. Acta* **497**, 622 (1977).

Copyright © 1984 by Academic Press, Inc.
All rights of reproduction in any form reserved.
ISBN 0-12-182005-X

ates such as $O_2^-$.[20] Since cytochrome oxidase is inhibited by $CN^-$, one can use the $CN^-$-resistant respiration as an upper limit measure of $O_2^-$ production. In *Escherichia coli* this amounts to 3% of total respiration.[21] Another approach was applied to extracts of *Streptococcus faecalis* supplied with NADH. In this case an inhibitory antibody was used to specifically suppress the superoxide dismutase (SOD) and $O_2^-$ production was found to account for 17% of total $O_2$ uptake.[22] Mammalian liver has been estimated to generate 24 nmol of $O_2^-$ min$^{-1}$ g$^{-1}$ and the steady-state level of $O_2^-$ in mitochondria is estimated at only $1 \times 10^{-11}$ $M$, due to the action of SOD.[23]

There is a copious body of compelling evidence that $O_2^-$ is an agent of oxygen toxicity and that superoxide dismutases, which catalyze reaction (1), provide an essential defense. Much of this has been reviewed recently.[24] This being the case, it makes perfect sense that, to the maximum extent possible, the production of $O_2^-$ should be avoided in living systems. Yet, there are specialized cells which, when activated, generate $O_2^-$ as the major reduction product of dioxygen. These are the phagocytes which, during the respiratory burst, produce $O_2^-$ as part of the devices applied to the killing of microorganisms.[9] The importance of the $O_2^-$, and of the $H_2O_2$ derived therefrom, is underscored by the genetic disorder *chronic granulomatous disease,* in which inability to produce $O_2^-$ and $H_2O_2$ is associated with a serious impairment of the microbicidal action of the phagocytes and a consequent susceptibility to infection.[9]

$$O_2^- + O_2^- + 2H^+ \rightarrow H_2O_2 + O_2 \tag{1}$$

$O_2^-$ produced by activated phagocytes also seems to play a role in the inflammatory process, by converting a serum component into a powerful neutrophil chemoattractant.[25] The precursor of this chemoattractant may be arachidonic acid bound to serum albumin, since arachidonate has itself been reported to generate a chemoattractant upon exposure to a flux of $O_2^-$.[26,27] Since activated neutrophils can thus generate a chemoattractant, which is not also an activator, the local congregation of neutrophils

[20] E. Antonini, M. Brunori, C. Greenwood, and B. G. Malmström, *Nature (London)* **228,** 936 (1970).
[21] H. M. Hassan and I. Fridovich, *J. Biol. Chem.* **254,** 10846 (1979).
[22] L. Britton, D. P. Malinowski, and I. Fridovich, *J. Bacteriol.* **134,** 229 (1978).
[23] B. Chance, H. Sies, and A. Boveris, *Physiol. Rev.* **59,** 527 (1979).
[24] I. Fridovich, *in* "Oxygen and Living Processes" (D. L. Gilbert, ed.), p. 250. Springer-Verlag, Berlin and New York, 1981.
[25] W. F. Petrone, D. K. English, K. Wong, and J. M. McCord, *Proc. Natl. Acad. Sci. U.S.A.* **77,** 1159 (1980).
[26] H. D. Perez, B. B. Weksler, and I. M. Goldstein, *Inflammation* **4,** 313 (1980).
[27] E. J. Goetzl, H. R. Hill, and R. R. Gorman, *Prostaglandins* **19,** 71 (1980).

would continue to be augmented until the activator was exhausted. When that happened, newly arrived neutrophils would fail to be activated and the gradient of chemoattractant would dissipate. Such a mechanism would work very well when the local activator was a population of opsonized microorganisms, but could get out of control when the activator was endogenously generated, as in autoimmunity; or was not susceptible to removal by the activated neutrophils, as would be the case with urate crystals or asbestos fibers. SOD might be expected to exert an antiinflammatory effect, which it does.[28]

$O_2^-$ should be seen as an intermediate of dioxygen reduction which is an important cause of oxygen toxicity and of the oxygen-dependent toxicities of redox-active compounds and which also plays a key role in the microbicidal action of neutrophils, in the inflammatory process, and perhaps in other physiological processes yet to be elucidated.

---

[28] J. M. McCord, and K. Wong, *Ciba Found. Symp. (N.S.)* **65**, 343 (1979).

# [6] Overview: Superoxygenase

By RYOTARO YOSHIDA and OSAMU HAYAISHI

Oxygenases catalyze the incorporation of either one or two atoms of molecular oxygen into their substrates, and are classified into two major groups, monooxygenases and dioxygenases. In the metabolism of tryptophan in mammals, two most important pathways, leading to the formation of pyridine nucleotide coenzymes and indoleamines, are initiated by the two well-known oxygenases, hepatic tryptophan 2,3-dioxygenase and tryptophan 5-hydroxylase (monooxygenase), respectively.

In 1937, Kotake and Ito found that rabbits fed D-tryptophan excreted D-kynurenine in the urine.[1] Since the hepatic tryptophan 2,3-dioxygenase was assumed to be specific for L-tryptophan,[2] they suggested the occurrence of an enzyme in rabbit tissues that catalyzed the oxidative ring cleavage of D-tryptophan. About 30 years later, Higuchi, Kuno, and Hayaishi discovered an enzyme activity capable of oxidizing D-tryptophan to N-formyl-D-kynurenine in a homogenate of rabbit small intestine.[3] Since the purified enzyme from rabbit intestine had a broad substrate specificity for various indoleamine derivatives[4] and utilized the superoxide anion as

---

[1] Y. Kotake and N. Ito, *J. Biochem.* **25**, 71 (1937).
[2] M. Civen and W. E. Knox, *J. Biol. Chem.* **235**, 1716 (1960).
[3] K. Higuchi, S. Kuno, and O. Hayaishi, *Arch. Biochem. Biophys.* **120**, 397 (1967).
[4] F. Hirata and O. Hayaishi, *Biochem. Biophys. Res. Commun.* **47**, 1112 (1972).

METHODS IN ENZYMOLOGY, VOL. 105

Copyright © 1984 by Academic Press, Inc.
All rights of reproduction in any form reserved.
ISBN 0-12-182005-X

an oxygen source,[5–11] the name "indoleamine 2,3-dioxygenase (superoxygenase)" was proposed to designate this enzyme.[12]

## Purification and Properties of Indoleamine 2,3-Dioxygenase

The tissue distribution of the enzyme activity was determined with the high-speed supernatants of homogenates of various organs of rabbit using D-tryptophan as a substrate. The enzyme activity was found to be ubiquitously distributed. The highest enzyme activity was observed in the small intestine, lung, and colon in that order. In mice, however, the tissue distribution of the enzyme activity was quite different from that in rabbit; the highest enzyme activity was observed in the epididymis, colon, and small intestine in that order.[13] Since the antibody for the enzyme from rabbit small intestine cross-reacted with the enzymes distributed in other organs of rabbit but not with the mouse enzymes, indoleamine 2,3-dioxygenase has been isolated from the rabbit small intestine[14] and the mouse epididymis.[15] In this chapter, we describe the purification method of rabbit enzyme.

Starting from crude extracts of the rabbit small intestine, the enzyme was purified about 500-fold with an overall yield of about 11% by the conventional procedures; streptomycin treatment, ammonium sulfate fractionation, P-cellulose chromatography, hydroxyapatite chromatography, gel filtration on Sephadex G-100, isoelectric focusing, and gel filtration on Sephadex G-100. The most highly purified enzyme preparation is essentially homogeneous upon polyacrylamide gel electrophoresis and analytical ultracentrifugation. The native enzyme is a monomeric protein with a molecular weight of about 41,000 and an $s_{20,w}$ value of 3.45 S. It has a relative abundance of hydrophobic amino acids, and contains approxi-

[5] F. Hirata and O. Hayaishi, *J. Biol. Chem.* **246,** 7825 (1971).

[6] F. Hirata and O. Hayaishi, *J. Biol. Chem.* **250,** 5960 (1975).

[7] T. Taniguchi, F. Hirata, and O. Hayaishi, *J. Biol. Chem.* **252,** 2774 (1977).

[8] O. Hayaishi, F. Hirata, T. Ohnishi, J. P. Henry, I. Rothenthal, and A. Katoh, *J. Biol. Chem.* **252,** 3548 (1977).

[9] F. Hirata, T. Ohnishi, and O. Hayaishi, *J. Biol. Chem.* **252,** 4637 (1977).

[10] T. Ohnishi, F. Hirata, and O. Hayaishi, *J. Biol. Chem.* **252,** 4643 (1977).

[11] T. Taniguchi, M. Sono, F. Hirata, O. Hayaishi, M. Tamura, K. Hayashi, T. Iizuka, and Y. Ishimura, *J. Biol. Chem.* **254,** 3288 (1979).

[12] O. Hayaishi and R. Yoshida, *in* "Oxygen Free Radicals and Tissue Damage" (Ciba Foundation Symp. 65), 199. Elsevier, Amsterdam, 1979.

[13] R. Yoshida, T. Nukiwa, Y. Watanabe, M. Fujiwara, F. Hirata, and O. Hayaishi, *Arch. Biochem. Biophys.* **212,** 629 (1981).

[14] T. Shimizu, S. Nomiyama, F. Hirata, and O. Hayaishi, *J. Biol. Chem.* **253,** 4700 (1979).

[15] K. Nakata, Y. Watanabe, R. Yoshida, and O. Hayaishi, Manuscript in preparation, 1983.

mately 5% carbohydrate by weight. One mole of enzyme has 0.8 mol of protoheme IX as a sole prosthetic group. Brady suggested from the experiments with copper chelators that indoleamine 2,3-dioxygenase was a copper-containing hemoprotein.[16] However, only a trace amount of copper was found in the highly purified enzyme preparation and the ratio of copper to heme was less than 0.03.

### Assay Procedures of Indoleamine 2,3-Dioxygenase

In our laboratory, three assay procedures have been employed. The first and the second methods were used mostly for the assay of crude systems such as homogenates, slices, and *in vivo* experiments. For the assay of purified enzyme preparations the third procedure was employed.

### Assay with L-[ring-2-*14*C]Tryptophan

The most convenient method is the use of [ring-2-14C]tryptophan as originally described by Peterkofsky in 1968 for the tryptophan 2,3-dioxygenase assay.[17] The standard assay mixture (0.2 ml) in a Thunberg-type tube contained 50 m$M$ potassium phosphate buffer, pH 6.6, 0.2 m$M$ L-[ring-2-14C]tryptophan (500 cpm/nmol), 50 $\mu$g of catalase, 0.4 unit of formamidase, 25 m$M$ sodium formate, 10 m$M$ ascorbic acid, 25 m$M$ methylene blue, and enzyme. Formamidase (purified from the rat liver as originally described by Knox[18] with some modifications[10] was added to complete the hydrolysis of formylkynurenine to form kynurenine and formic acid. Sodium formate was added as an internal carrier. The reaction (37°, 30 min) was started by the addition of substrate and terminated by the addition of 0.2 ml of 10% trichloroacetic acid. [14C]Formic acid produced from the radioactive product, formylkynurenine, was sublimed into a sidearm immersed in a dry ice/ethyl alcohol bath under reduced pressure and the radioactivity of aliquots (0.4 ml) was determined with a liquid scintillation counter. The reaction rate was expressed as the amount of [14C]formic acid formed during 60 min.

### Assay with L-[methylene-*14*C]Indoleamines

[ring-2-14C]Tryptamine, 5-hydroxy-D- and L-tryptophan, or serotonin is commercially unavailable. With L-[methylene-14C]tryptophan or other indoleamines as substrate in the presence of formamidase, L-[methylene-

[16] F. O. Brady, *FEBS Lett.* **57**, 237 (1975).
[17] B. Peterkofsky, *Arch. Biochem. Biophys.* **128**, 637 (1968).
[18] This series, Vol. 2, p. 242.

[14C]kynurenine or the corresponding anthraniloylalkylamines was identified as the major product by cellulose thin-layer chromatography with a solvent system of 20% KCl or other solvent mixtures.[19] The reaction rate was expressed as the amount of [14C]kynurenine or the corresponding anthraniloylalkylamines formed during 60 min.

### Spectrophotometric Method

The standard assay medium contained 100 m$M$ potassium phosphate buffer, pH 6.6 (for L-tryptophan and 5-hydroxy-L-tryptophan), pH 7.4 (for tryptamine and serotonin) or pH 7.8 (for the D-isomers of tryptophan and 5-hydroxy-tryptophan), 25 $\mu M$ methylene blue, 10 m$M$ ascorbic acid, 50 $\mu$g of catalase, various indoleamines as substrate and enzyme in a total volume of 0.2 ml. Incubation was carried out at 37° and the reaction was terminated by the addition of 5% zinc acetate (0.2 ml) and 0.18 $N$ sodium hydroxide (0.2 ml). After centrifugation at 1000 $g$ for 5 min, an aliquot (0.4 ml) of the solution was added to 1 ml of 1 $M$ Tris–HCl buffer, pH 7.0. When the indole ring of the substrate (e.g., tryptophan) was cleaved, the absorption maximum (e.g., 280 nm for tryptophan) shifted by about 40 nm to a longer wavelength (e.g., 321 nm for formylkynurenine). In the presence of formamidase in the reaction mixture, the formyl group was hydrolyzed and the absorption maximum shifted by another 40 nm (e.g., 360 nm for kynurenine). Authentic compounds of the reaction products such as $N$-formylkynurenamine oxalate ($\varepsilon$ at 324 nm = 3700 $M^{-1}$ cm$^{-1}$), $N$-formyl-5-hydroxy-L-kynurenine ($\varepsilon$ at 341 nm = 3800 $M^{-1}$ cm$^{-1}$), and $N$-formyl-5-hydroxy-kynurenamine oxalate ($\varepsilon$ at 345 nm = 3500 $M^{-1}$ cm$^{-1}$) were chemically synthesized from tryptamine, 5-hydroxy-L-tryptophan, and serotonin, respectively.[14] The reaction rate was expressed as the amount of formyl compounds/minute at 37°.

Substrate specificity of the purified indoleamine 2,3-dioxygenase was examined spectrophotometrically at 24°. The spectra of the reaction products in the absence and presence of formamidase were compared with those of authentic compounds. A single enzyme protein catalyzed the oxygenative ring cleavage of D- and L-tryptophan, 5-hydroxy-D- and L-tryptophan, and serotonin.[14] $N,N'$-Dimethyltryptamine, a psychogenic substance which is produced in lung or brain, was also metabolized at much lower rate.[20] The maximal turnover number (99 mol/min/mol of enzyme at 24°) and the lowest $K_m$ value (20 $\mu M$) were obtained with L-

[19] M. Fujiwara, M. Shibata, Y. Watanabe, T. Nukiwa, F. Hirata, N. Mizuno, and O. Hayaishi, J. Biol. Chem. 253, 6081 (1978).
[20] O. Hayaishi, F. Hirata, M. Fujiwara, T. Ohnishi, and T. Nukiwa, Proc. Fed. Eur. Biochem. Soc. Meet., 10th 131 (1975).

tryptophan. A marked substrate inhibition was observed with the L-isomers of tryptophan and 5-hydroxy-tryptophan above 0.2 and 0.06 m$M$, at pH 6.6, respectively. The compounds such as skatole, indole, indoleacetic acid, 5-hydroxy-indoleacetic acid, $N$-acetyltryptophan, melatonin, and $\alpha$-methyl-DL-tryptophan were all inert as substrate.

Participation of $O_2^-$ in the Reaction of Indoleamine 2,3-Dioxygenase

One of the most interesting features of indoleamine 2,3-dioxygenase is that the purified enzyme is inactive unless the superoxide anion is present in the incubation mixture.

*$O_2^-$ Generation and Indoleamine 2,3-Dioxygenase Activity*

When the purified enzyme was incubated with the substrate under appropriate conditions, there was no activity unless either ascorbic acid or xanthine oxidase or glutathione reductase with their substrates was added to the reaction mixture. Ascorbic acid, the xanthine oxidase system, and the glutathione reductase system are all known to generate the superoxide anion as well as hydrogen peroxide. However, neither hydrogen peroxide as such, nor the glucose oxidase and amino acid oxidase systems, both of which are known to generate hydrogen peroxide, was able to support the enzyme activity.[6] By directly infusing $KO_2$, potassium superoxide, into the reaction mxiture, we were able to demonstrate that superoxide per se could support the enzyme activity.[10]

*Inhibition of Indoleamine 2,3-Dioxygenase Reaction by Superoxide Dismutase*

Superoxide dismutase from the bovine erythrocytes or *Escherichia coli*, which catalyzes the dismutation of $O_2^-$ to molecular oxygen and hydrogen peroxide, inhibited the reaction catalyzed by the enzyme when the inhibitors were added to the reaction mixture either during the steady state of the reaction or prior to the incubation.[5,6,10]

*Reaction of Indoleamine 2,3-Dioxygenase with $O_2^-$*

The native enzyme purified from the rabbit small intestine showed an absorption spectrum typical of a high-spin ferric ($Fe^{3+}$) hemoprotein having the absorption peaks at 406 nm in the Soret region and 499 and 630 nm in the visible regions at pH 8.0.[9] Upon infusion of $O_2^-$ (e.g., $KO_2$ dissolved in an aprotic solvent such as dimethyl sulfoxide) into the reaction mixture in the presence of catalase, the spectrum of the ferric enzyme changed to that of the oxygenated form of the enzyme which had absorption peaks at 415, 542, and 576 nm.[9] The formation of this new spectral species was completely abolished in the presence of superoxide dismutase, indicating that this process involved the binding of superoxide anion to the ferric form of the enzyme. This spectrum is similar to that of oxygenated hemoglobin,[21] oxygenated myoglobin,[22] peroxidase compound III,[23] or the enzyme·substrate·$O_2$ ternary complex of tryptophan 2,3-dioxygenase.[24] This apparent enzyme·$O_2$ complex could be reversibly converted to the ferrous enzyme and oxygen by degassing. When an excess amount of substrate, such as D-tryptophan, was introduced into a reaction mixture containing the enzyme·$O_2$ complex, free enzyme and the product were immediately released. These results are consistent with the interpretation that the native, ferric form of the enzyme binds to the superoxide anion to form the oxygenated enzyme ($Fe^{3+}O_2^-$).

### Enhancement of Intracellular Indoleamine 2,3-Dioxygenase Activity by Diethyldithiocarbamate Treatment

The intracellular indoleamine 2,3-dioxygenase activity was markedly enhanced by the addition of 5 m$M$ diethyldithiocarbamate, an inhibitor of superoxide dismutase.[7] This agent, however, did not activate the purified indoleamine 2,3-dioxygenase *in vitro*, indicating that the intracellular accumulation of $O_2^-$ resulted in acceleration of the *in situ* dioxygenase activity.

These results taken together support a conclusion that indoleamine 2,3-dioxygenase is a unique enzyme which utilizes the superoxide anion rather than molecular oxygen as the oxidizing agent both *in vitro* and *in vivo*.

### Incorporation of $^{18}O_2^-$ into the Reaction Product

$^{18}O$-Labeled oxygen atoms were incorporated into the reaction product when $K^{18}O_2$ was used as the oxidizing agent. However, when the reaction was carried out in the presence of heavy oxygen-containing air, oxygen-18 was also incorporated into the reaction product, albeit to a lesser extent,[8] indicating that both $O_2^-$ and $O_2$ were utilized as the source

[21] A. E. Sidwell, Jr., R. H. Munch, E. S. G. Barron, and T. R. Hogness, *J. Biol. Chem.* **123**, 335 (1938).

[22] I. Yamazaki, K. Yokota, and K. Shikama, *J. Biol. Chem.* **239**, 4151 (1964).

[23] K. Yokota and I. Yamazaki, *Biochem. Biophys. Res. Commun.* **18**, 48 (1965).

[24] Y. Ishimura, M. Nozaki, O. Hayaishi, T. Nakamura, M. Tamura, and I. Yamazaki, *J. Biol. Chem.* **245**, 3593 (1976).

of oxygen incorporated into the product. These results are consistent with the kinetic experiments as will be described below.

## Kinetical Properties of Indoleamine 2,3-Dioxygenase

In order to further understand the reaction mechanism of indoleamine 2,3-dioxygenase, Taniguchi et al.[11] in our laboratory determined the following rate constants: $k_1$ [in the presence ($2.3 \times 10^6 M^{-1} sec^{-1}$) or absence ($3.3 \times 10^6 M^{-1} sec^{-1}$) of L-tryptophan (0.2 m$M$)] for the binding of $O_2^-$ to the ferric enzyme, $k_2$ [in the presence ($6.3 \times 10^6 M^{-1} sec^{-1}$) or absence ($7.4 \times 10^6 M^{-1} sec^{-1}$) of L-tryptophan (0.2 m$M$)] for the binding of $O_2$ to the ferrous enzyme, $k_3$ [in the presence ($0.028 sec^{-1}$) or absence ($4.7 \times 10^{-4} sec^{-1}$) of L-tryptophan (0.2 m$M$)] for the conversion of the ternary complex to the ferric enzyme, $k_4$ ($2.0 sec^{-1}$) for the product formation, and $k_5$ ($22 sec^{-1}$) for the direct oxidation of the ferrous enzyme to the ferric form. These results suggest that in the principal pathway, the ferrous enzyme binds to the organic substrate and then with molecular oxygen to yield a ternary complex, and that the ternary complex thus formed decomposes to the ferrous enzyme with concomitant product formation. However, the ternary complex of indoleamine 2,3-dioxygenase was unstable and decomposed rapidly to the ferric enzyme. The ferric enzyme thus produced bound to $O_2^-$. Therefore, indoleamine 2,3-dioxygenase requires and utilizes the superoxide anion.

## Physiological Significance of Indoleamine 2,3-Dioxygenase

### Induction of Indoleamine 2,3-Dioxygenase during Virus Infection

Because the lung is an aerobic organ, presumably produces the superoxide anion, and contains high levels of indoleamine 2,3-dioxygenase, the lung of specific pathogen-free mice was selected as an enzyme source for the studies of various changes of this enzyme activity under a variety of physiological and pathological conditions. The indoleamine 2,3-dioxygenase activity in various tissues of mice is relatively stable under various physiological conditions, although it exhibits a daily rhythmic cycle and age-dependent changes.[13] Recently, we found that the enzyme was remarkably (approximately 100-fold) induced particularly in the lung during in vivo influenza virus,[25] HVJ (hemagglutination virus of Japan) and MHV (mouse hepatitis virus) infection. The pulmonary indoleamine 2,3-dioxygenase activity increased linearly from the fifth day after influenza virus

[25] R. Yoshida, Y. Urade, M. Tokuda, and O. Hayaishi, Proc. Natl. Acad. Sci. U.S.A. **76**, 4084 (1979).

infection, reached the highest level (approximately 120-fold) around the eleventh day, and then gradually decreased to normal values in about 3 weeks. The time course of the increase in the enzyme activity was quite different from that (a peak obtained by the third day and persisted until the ninth day) of virus replication in the lung or that (started to rise on the ninth day) of the serum antibody content. Rather, it is closely related to the perivascular and peribronchial infiltration of mononuclear and lymphocytic cells.

### Induction of Indoleamine 2,3-Dioxygenase by Lipopolysaccharide

To simplify the experimental conditions, we then used bacterial lipopolysaccharide (LPS) instead of influenza virus. The LPS, an outer membrane component of gram-negative bacteria, is an inflammatory agent and causes nonspecific immune responses. The enzyme induction was also observed in the lung after an intraperitoneal administration of LPS (20 $\mu$g/ mouse).[26,27] The effect appeared to be specific for indoleamine 2,3-dioxygenase because other enzyme activities in the mouse lung such as superoxide dismutase, monoamine oxidase, acid phosphatase, prostaglandin synthase, and lipoxygenase did not change significantly with this treatment.

The cells, in which the enzyme was enriched after LPS treatment, were totally dispersed with a protease solution, fractionated on Percoll density gradient, and the type of cells was identified with electron microscopy or using various marker enzyme activities of main types of lung cells.[28] The enzyme induced by LPS treatment was found to be exclusively localized in the alveolar interstitial cells of the mouse lung.

The enzyme induction was observed in various strains of mice including C3H/He, BALB/c, and A/J. This LPS-mediated indoleamine 2,3-dioxygenase induction was not observed in C3H/HeJ mice possessing abnormal bone marrow-derived lymphocytes. However, in this strain of mice, the enzyme induction could almost normally occur as a response to LPS, if the animals were irradiated and then reconstituted with exogenous spleen lymphocytes isolated from C3H/He mice.[29] Based on these experiments and other immunological studies using various mutant mice, it was

[26] R. Yoshida and O. Hayaishi, *Proc. Natl. Acad. Sci. U.S.A.* **75**, 3998 (1978).
[27] R. Yoshida, Y. Urade, K. Nakata, Y. Watanabe, and O. Hayaishi, *Arch. Biochem. Biophys.* **212**, 629 (1981).
[28] Y. Urade, R. Yoshida, H. Kitamura, and O. Hayaishi, *J. Biol. Chem.* **258**, 6621 (1983).
[29] R. Yoshida, Y. Urade, S. Sayama, O. Takikawa, Y. Ozaki, and O. Hayaishi, *in* "Oxygenases and Oxygen Metabolism" (M. Nozaki, S. Yamamoto, Y. Ishimura, M. J. Coon, L. Ernster, and R. W. Estabrook, eds.), p. 569. Academic Press, New York, 1982.

deduced that the enzyme induction in the alveolar interstitial cells was triggered by the interaction of lymphocytes with viruses and other inducers such as LPS and poly(I) · poly(C).

### Induction of Indoleamine 2,3-Dioxygenase by Interferon

In an attempt to clarify the mechanism of the enzyme induction by LPS or during virus infection *in vivo,* we developed an *in vitro* system using mouse lung slices. When lung slices were incubated with LPS (5 μg/ml), indoleamine 2,3-dioxygenase activity in the high-speed supernatant fraction increased (approximately 10-fold) for at least 48 hr and gradually decreased. A similar increase in the enzyme activity was observed by the addition of poly(I) · poly(C) (0.5 mg/4 ml), a synthetic double-stranded RNA, to the lung slices. It is well known that virus, LPS and poly(I)· poly(C) are potent inducers of interferon. When LPS was substituted by interferon (10⁴ units/ml, isolated from mouse L cell or mouse brain) in these *in vitro* systems using cultured mouse lung slices, the enzyme was induced 10- to 15-fold by this treatment.[30] However, human leukocyte interferon showed little effect on the enzyme activity, since interferon is species-specific. When an interferon preparation was treated either by heat, α-chymotrypsin, or anti-interferon serum, such increase in the enzyme activity was diminished essentially to the same extent as seen in the antiviral activity. However, the effects of LPS or poly(I) · poly(C) on the enzyme activity were retained intact by these treatments. These results suggest that the increase in the indoleamine 2,3-dioxygenase activity is caused by interferon itself and not by possible contaminants such as LPS or double-stranded RNA derived from viruses.

### Interferon-Mediated Indoleamine 2,3-Dioxygenase Induction and the Unsaturated Fatty Acid Metabolism

The enzyme induction by interferon was completely suppressed by the addition of glucocorticoids (e.g., dexamethasone, betamethasone, and cortisone) or nonsteroidal antiinflammatory agents possessing the prostaglandin synthase inhibitory activity (e.g., indomethacin, phenylbutazone, and aspirin).[31] Similar effects of these various agents were also observed with the enzyme induction by LPS, suggesting that the enzyme induction by virus, LPS, and interferon was mediated by a metabolite(s)

[30] R. Yoshida, J. Imanishi, T. Oku, T. Kishida, and O. Hayaishi, *Proc. Natl. Acad. Sci. U.S.A.* **78,** 129 (1981).
[31] S. Sayama, R. Yoshida, T. Oku, J. Imanishi, T. Kishida, and O. Hayaishi, *Proc. Natl. Acad. Sci. U.S.A.* **78,** 7327 (1981).

of unsaturated fatty acid. Although interferon is reported to induce several enzymes such as 2′,5′-oligoadenylate synthetase and protein kinase and the interferon-mediated antiviral activity has been assumed to be mediated by these enzymes,[32] these enzyme inductions by interferon were not suppressed by nonsteroidal antiinflammatory agents.[31]

### Conclusion

It has been almost a quarter of a century since the discovery of oxygenases, and extensive studies have been carried out on the properties and functions of oxygenases. Indoleamine 2,3-dioxygenase (a superoxygenase) is a unique heme-containing oxygenase, because it requires the superoxide anion for the initiation of the reaction and for maintenance of the catalytic cycle during the steady state. A dramatic and specific induction of pulmonary indoleamine 2,3-dioxygenase by virus and LPS was mediated by interferon, and the interferon-mediated enzyme induction was completely suppressed by inhibitors of prostaglandin biosynthesis. Recently, it was also reported that the antiviral activity of interferon was inhibited by nonsteroidal antiinflammatory agents such as indomethacin and aspirin. However, oligoadenylate synthetase or protein kinase induction by interferon, both of which are candidates of antiviral protein(s), was not inhibited by these agents. Therefore, indoleamine 2,3-dioxygenase may play an important role in the inflammatory processes, immune responses, and/or the mode of action of interferon.

### Acknowledgments

This work was supported in part by research grants from the Sakamoto Foundation, Nippon Shinyaku Co., Ltd., and the Intractable Diseases Division, Public Health Bureau, Ministry of Health and Welfare, Japan and by Grants-in-Aid for Scientific Research and Cancer Research from the Ministry of Education, Science and Culture, Japan.

[32] A. Kimchi, L. Schulman, A. Schmidt, Y. Clernajovsky, A. Fradin, and M. Revel, *Proc. Natl. Acad. Sci. U.S.A.* **76**, 3208 (1979).

## [7] Methods for the Study of Superoxide Chemistry in Nonaqueous Solutions

By JOAN S. VALENTINE, ANDREW R. MIKSZTAL, and DONALD T. SAWYER

Our current knowledge of the chemical properties of superoxide anion, $O_2^-$, is based on studies of superoxide reactions in aqueous and nonaqueous solution.[1] With some few exceptions,[2] these studies have been carried out since 1969, much of the interest in superoxide chemistry having been inspired by the discovery of the superoxide dismutase enzymes at that time.[3] The superoxide dismutases were postulated to play a protective role within the cell by reducing the steady-state levels of superoxide, a species which is inevitably produced in some unknown amount as a by-product of aerobic metabolism.[4,5] It is clearly necessary to have information concerning the basic principles of superoxide reactivity if we are to understand completely the possible roles that superoxide might play in aerobic metabolism and the possibility that it plays a key role in the mechanism of dioxygen toxicity.[6,7] In aqueous solution, studies of superoxide chemistry require specialized rapid reaction techniques because of the rapid, spontaneous disproportionation of superoxide in protic media.[1,2]

$$2O_2^- + 2H^+ \rightarrow O_2 + H_2O_2 \tag{1}$$

These techniques and the results of such studies are reviewed elsewhere

[1] Recent reviews of superoxide chemistry are: D. T. Sawyer and J. S. Valentine, *Acc. Chem. Res.* **14**, 393 (1981); D. T. Sawyer, D. T. Richens, E. J. Nanni, Jr., and M. D. Stallings, *Dev. Biochem.* **11A**, 1 (1980); H. A. O. Hill, *in* "Oxygen Free Radicals and Tissue Damage" (Ciba Foundation Symposium 65), p. 5. Excerpta Medica, Amsterdam, 1979; J. S. Valentine, *in* "Biochemical and Clinical Aspects of Oxygen" (W. S. Caughey, ed.), p. 659. Academic Press, New York, 1979; D. T. Sawyer and M. J. Gibian, *Tetrahedron* **35**, 1471 (1979); E. Lee-Ruff, *Chem. Soc. Rev.* **6**, 195 (1977); J. A. Fee and J. S. Valentine, *in* "Superoxide and Superoxide Dismutases" (A. M. Michelson, J. M. McCord, and I. Fridovich, eds.), p. 19. Academic Press, New York, 1977.

[2] For references to studies in aqueous solution carried out before 1969, see B. H. J. Bielski and J. M. Gebicki, *Adv. Radiat. Chem.* **2**, 177 (1970); G. Czapski, *Annu. Rev. Phys. Chem.* **22**, 171 (1971).

[3] J. M. McCord and I. Fridovich, *J. Biol. Chem.* **244**, 6049 (1969).

[4] I. F. Fridovich, *Adv. Inorg. Biochem.* **1**, 67 (1979).

[5] J. A. Fee, *Trends Biochem. Sci.* **7**, 84 (1982).

[6] D. T. Sawyer and J. S. Valentine, *Acc. Chem. Res.* **15**, 200 (1982).

[7] Irwin Fridovich, *Acc. Chem. Res.* **15**, 200 (1982).

METHODS IN ENZYMOLOGY, VOL. 105

Copyright © 1984 by Academic Press, Inc.
All rights of reproduction in any form reserved.
ISBN 0-12-182005-X

within this volume. Our interest in the reactivity of superoxide in non-aqueous media began with our discoveries that electrochemical reduction of dioxygen in dimethyl sulfoxide yielded stable solutions of superoxide,[8] and that crown ethers would solubilize $KO_2$ in aprotic solvents and could be used to prepare stable solutions of superoxide for various uses.[9] These two methods and a third method, using the soluble superoxide salt, tetramethylammonium superoxide, $(Me_4N)O_2$, developed by Foote and co-workers,[10] are the methods currently in common use for the study of superoxide reactions in nonaqueous solutions. The results of studies of superoxide chemistry in aqueous and nonaqueous solution are the subject of numerous reviews.[1,2] In this chapter, we review the methods for preparation of solutions of superoxide in nonaqueous, aprotic solvents and the uses of these methods for the study of superoxide chemistry in such media.

## Electrochemical Methods

Solutions of superoxide can be conveniently prepared by controlled potential reduction of dioxygen in the presence of a supporting electrolyte.[8,11] Solvents commonly used in this method are dimethyl sulfoxide (Me$_2$SO), dimethylformamide (DMF), acetonitrile (MeCN), or pyridine (py) and concentrations as high as 10 m$M$ can be achieved. The electrochemical method for preparation of solutions of superoxide has the advantage that $O_2^-$ can be generated *in situ* in the presence of a substrate, as long as that substrate is not excessively reactive with $O_2$ itself. In addition, the ability to monitor electrochemically the superoxide and electroactive reactants, intermediates, and products during the course of the reaction and after the reaction is complete can be useful. Such an approach has been used successfully in several studies of superoxide reactivity.[12,13] The disadvantages of the electrochemical method are (1) the relatively low concentrations of superoxide achievable by this method, which makes it unsuitable for most synthetic uses; (2) the possibility of interference from reactions of electrolytes or electrodes; (3) the relatively short half-life of superoxide solutions prepared by this method, which is generally less than 7 hr; and (4) the requirement for specialized electrochemical equipment.

[8] D. T. Sawyer and J. L. Roberts, Jr., *Electroanal. Chem.* **12,** 90 (1966).

[9] J. S. Valentine and A. B. Curtis, *J. Am. Chem. Soc.* **97,** 224 (1975).

[10] J. W. Peters and C. S. Foote, *J. Am. Chem. Soc.* **98,** 873 (1976).

[11] A. D. Goolsby and D. T. Sawyer, *Anal. Chem.* **40,** 83 (1968); D. T. Sawyer, G. Chiericato, Jr., C. T. Angelis, E. J. Nanni, Jr., and T. Tsuchiya, *Anal. Chem.* **54,** 1720 (1982).

[12] M. V. Merritt and D. T. Sawyer, *J. Org. Chem.* **35,** 2157 (1970).

[13] J. L. Roberts, Jr., and D. T. Sawyer, *J. Am. Chem. Soc.* **103,** 712 (1981).

*Experimental Procedure.* Figure 1 illustrates the reversible cyclic voltammogram for the reduction of $O_2$ in pyridine at a platinum electrode. The $O_2^-$ is reoxidized to $O_2$ when the scan is reversed and reduced to $O_2^{2-}$ if the negative scan is continued to more negative potentials.

*Reagents.* The solvents that are used for electrochemical generation of superoxide must be as pure and dry as possible; the "distilled-in-glass" grade from Burdick & Jackson of dimethyl sulfoxide, dimethylformamide, and pyridine can be used as supplied. Acetonitrile solvent from this source can be further dried by passing the solvent through a column of Woelm N Super I activated alumina prior to its use (resulting in a reduction of the water level to $< 1$ m$M$). Tetraethylammonium perchlorate (TEAP), from Southwestern Analytical, is normally used as the supporting electrolyte (0.1 $M$) and is dried on a vacuum line prior to use.

*Instrumentation.* A three-electrode potentiostat such as the Princeton Applied Research Model 173 potentiostat/galvanostat with a Model 175 universal programmer and Model 179 digital coulometer can be used for cyclic voltammetric and controlled-potential electrolysis experiments. Cyclic voltammograms are recorded with an X-Y recorder such as the Series 2000 recorder; the scans should be initiated at the rest potential of the solution. The Bioanalytical Systems Model SP-2 potentiostat is another useful system for electrosynthesis of superoxide.

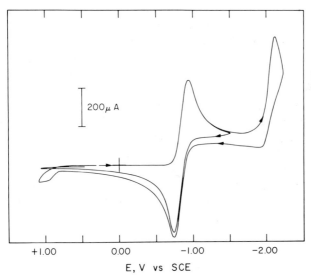

FIG. 1. Cyclic voltammogram for 5 m$M$ $O_2$ (1 atm) in pyridine (0.1 $M$ tetraethylammonium perchlorate) at a platinum electrode (area 0.23 cm$^2$). Scan rate, 0.1 V sec$^{-1}$. Saturated calomel electrode (SCE) vs normal hydrogen electrode (NHE), +0.244 V.

A Brinkmann electrochemical cell equipped with an O-ring seal is a convenient container for the electrochemical generation of superoxide. For product analysis, a Beckman platinum-inlay electrode (area 0.23 cm²) or a Bioanalytical Systems glassy carbon electrode (area 0.11 cm²) is an appropriate sensor with linear-sweep voltammetry. A Ag–AgCl reference electrode in a Pyrex tube that is closed with a soft-glass cracked tip is convenient; the inner compartment is filled with an aqueous tetramethylammonium chloride solution and adjusted to 0.00 V vs SCE. The reference electrode is placed in a Luggin capillary. The controlled-potential electrolysis experiments make use of a platinum-mesh working electrode or sheets of graphitized carbon. The platinum-mesh auxiliary electrode is placed inside a glass compartment that is closed with a medium-porosity glass frit. Electrochemical generation of superoxide in acetonitrile or pyridine requires the use of a double-bridged H cell with a center compartment that can be flushed with the supporting electrolyte and/or TEAOH–TEAP solutions. The concentration of electrochemically generated superoxide in acetonitrile can also be determined from its absorption spectrum. Superoxide ion in acetonitrile has an absorption band centered at 252 nm ($\varepsilon = 6.5 \pm 1.0$ m$M^{-1}$ cm$^{-1}$). The UV cutoff of Me$_2$SO, DMF, and pyridine precludes accurate absorption measurements of O$_2^-$.

*Electrogeneration.* Solutions of superoxide are prepared by use of a O$_2$-presaturated (1 atm O$_2$) aprotic solvent. A cyclic voltammogram is obtained by scanning negatively from the rest potential (see Fig. 1). The peak potential for the reduction of dioxygen plus $-0.14$ V is used to set the control voltage for the potentiostat. During the electrolysis, dioxygen is continuously bubbled through a stirred solution. Upon complection of the electrolysis, argon is passed through the solution to purge the dissolved dioxygen. Argon is continuously passed over the solution during the cyclic voltammetry experiment. The potential is scanned positively from the rest potential of the deaerated superoxide solution. The anodic peak current of the superoxide voltammogram provides a measure of the superoxide concentration. These values, along with the integrated coulometric currents, are used to calculate the efficiency of the electrosynthesis of superoxide in the various aprotic solvents. The electrochemical parameters for O$_2$ and O$_2^-$ are summarized in the table.

The electrosynthesis of O$_2^-$ in MeCN by controlled-potential electrolysis of dioxygen is 27% efficient at a platinum gauze electrode and 67% efficient at a graphitized carbon electrode. The percentage efficiency of superoxide generation varies from 50 to 75% in DMF and from 50 to 80% in Me$_2$SO using a platinum gauze electrode. The efficiency is limited by the presence of protic sources and Lewis acids in the solvent and on the surfaces of the electrodes and the interior walls of the electrolysis cells.

ELECTROCHEMICAL PARAMETERS FOR $O_2$ AND $O_2^-$ IN SEVERAL APROTIC
SOLVENTS (0.1 $M$ TETRAETHYLAMMONIUM PERCHLORATE)

A. Redox potentials for $O_2/O_2^-$ at various electrodes[11][a]

$E_{p,c}/E_{p,a}$ (V vs SCE)

| Electrode | MeCN | Me$_2$SO | DMF | Pyridine |
|---|---|---|---|---|
| Platinum | −1.23/−0.58 | −0.92/−0.71 | −1.11/−0.75 | −1.27/−0.72 |
| Gold | −1.12/−0.63 | −0.93/−0.71 | −1.03/−0.78 | −1.24/−0.74 |
| Graphitic carbon | −0.95/−0.78 | −0.92/−0.72 | −1.02/−0.78 | −1.23/−0.73 |
| Glassy carbon | −0.83/−0.73 | −0.80/−0.72 | −0.87/−0.79 | −1.07/−0.68 |

B. Normalized voltammetric peak currents and diffusion coefficients
($D_{O_2^-}$) for $O_2^-$ in several aprotic solvents that contain 0.1 $M$
tetraethylammonium perchlorate[a]

$i_{p,a}$ ($\mu$A − m$M^{-1}$)

| Solvent | 0.23 cm$^2$ pt, 0.1 V sec$^{-1}$ | 0.11 cm$^2$ C, 0.1 V sec$^{-1}$ | $D_{O_2^-}$ (cm$^2$ sec$^{-1}$) ($\times$ 10$^{-5}$) |
|---|---|---|---|
| MeCN | 75 ± 4 | 35 ± 4 | 1.49 |
| DMF | 49 ± 2 | 23 ± 2 | 0.63 |
| Me$_2$SO | 37 ± 2 | 17 ± 2 | 0.36 |

$$I = i_{p,a}/AC_{O_2} \cdot D_{O_2^-}^{-1/2} = 84 \times 10^3 \ \mu A \ cm^{-2} \ mM^{-1}$$

[a] Scan rate, 0.1 V sec$^{-1}$.

Both $H_2O$ and trace levels of transition metals are the major cause of
decomposition of $O_2^-$ solutions via disproportionation[1]

$$O_2^- + HA \rightarrow A^- + \tfrac{1}{2}H_2O_2 + \tfrac{1}{2}O_2 \qquad (2)$$

Fairly stable solutions of superoxide can be generated only with the
utmost care. All glassware used in the electrosynthesis of superoxide
must be scrupulously cleaned with an ammoniacal–EDTA solution.

## Solubilization Using Crown Ethers or Cryptates

Solutions of superoxide can also be conveniently prepared by dissolution of the ionic salt potassium superoxide in a variety of aprotic solvents
containing either crown ethers[14] or cryptates.[15] Each of these reagents is

[14] C. J. Pederson and H. K. Frensdorff, *Angew. Chem. Int. Ed.* **11**, 16 (1972).
[15] B. Dietrich, J. M. Lehn, J. P. Sauvage, and J. Blanzat, *Tetrahedron* **29**, 1629 (1973).

commercially available. Typical solvents used are dimethyl sulfoxide, dimethylacetamide, tetrahydrofuran, and pyridine. This method is suitable for synthetic procedures since relatively concentrated solutions of superoxide can be prepared. Solutions of superoxide thus prepared in acetonitrile or dimethylformamide appear to be relatively unstable, however. Alternatively, reactions of solid $KO_2$ can be carried out in a two-phase system using the crown ether or cryptate as a phase-transfer catalyst.[16] Suitable solvents include those listed above as well as benzene, toluene, or acetonitrile. This latter method has the advantage that superoxide concentrations in the solution are never high and side reactions with solvent are thereby minimized. Reactions of solid $KO_2$ with substrates in dimethyl sulfoxide generally do not require phase transfer catalysts since the salt has appreciable solubility in that solvent. In fact, it is frequently not necessary to use crown ethers and cryptates to solubilize $KO_2$, which is significantly soluble in $Me_2SO$ (1.3 m$M$) and DMF (0.3 m$M$). Such concentrations are frequently adequate for studies of superoxide reactivity in those solvents. [With 0.1 $M$ tetraethylammonium perchlorate present, the solubility of $KO_2$ in $Me_2SO$ increases to 5.4 m$M$; probably by metathesis, $KO_{2(s)} + (Et_4N)ClO_4 \rightarrow (Et_4N)O_2 + KClO_{4(s)}$.] Reactions can also be carried out in mixed solvent systems by addition of a superoxide solution in one solvent to a solution of a reagent in another solvent.

The principal advantages of this method for preparation of superoxide solutions are that the materials needed are commercially available, relatively high concentrations are achievable, and no specialized equipment is required. The chief disadvantage of the method is the dependence on the purity of the $KO_2$ as purchased, since there are no methods developed for its purification and the commercial product is only 96.5% pure *at best*.

*Experimental Procedure*

*Materials.* Potassium superoxide (Alfa-Ventron, 96.5% purity) is ground to a fine powder with a mortar and pestle in a dry, $CO_2$-free atmosphere, preferably in an inert atmosphere chamber. The purified crown ether and cryptate are both very hygroscopic and are therefore purified and stored under an inert atmosphere using the following procedures. The crown ether, 18-crown-6 (1,4,7,10,13,16-hexaoxacyclooctadecane; Aldrich or PCR), is purified by precipitation from dry dimethoxyethane.[17] The cryptate, Kryptofix-222 (4,7,13,16,21,24-hexaoxa-1,10-diazabicyclo[8.8.8]hexacosane), purchased from MCB as a yellowish,

[16] C. M. Starks and C. Liotta, "Phase Transfer Catalysis." Academic Press, New York, 1978.
[17] M. Nappa and J. S. Valentine, *J. Am. Chem. Soc.* **100**, 5075 (1978).

granular solid, is dried in a vacuum desiccator over $P_2O_5$ for 3 days and then recrystallized from hot heptane with addition of activated charcoal. Final purification is achieved by vacuum distillation at 170°. The melting point of the snow white crystalline solid is 71.5° (lit. 68–69°).[18] Dimethyl sulfoxide (Burdick & Jackson) is vacuum distilled from $CaH_2$ (40°, 0.2 Torr) and stored over 4A molecular sieves. Dimethylacetamide (Burdick & Jackson) is vacuum distilled from BaO (53°, 10.0 Torr), transferred into an inert atmosphere chamber, and then passed through a column of activated alumina to remove residual traces of water. It is stored over 4A molecular sieves. Pyridine (Mallinckrodt) is distilled from $CaH_2$ under argon and stored over 4A molecular sieves. All other solvents are distilled from drying agents under an inert atmosphere. All solvents are stored over drying agents under an inert atmosphere and are transferred by syringe.

*Preparation of a 0.1 M $KO_2$ Solution in Dimethyl Sulfoxide Using Crown Ether.* Finely powdered $KO_2$ (71.1 mg, 1.00 mmol) and 18-crown-6 (approx. 500–800 mg. 2–3 mmol) are placed in a dry 50-ml Schlenk tube[19] containing a Teflon-coated magnetic stirring bar. The flask is evacuated and flushed with dry argon. Dimethyl sulfoxide (10.0 ml) is introduced into the flask by syringe. Stirring the solution for 0.5–3 hr gives a clear pale yellow solution. The color of the solution is variable and is apparently due to impurities. Concentrated solutions of $KO_2$ prepared in this manner are occasionally colorless.

The stability of solutions prepared in this manner is variable. Under optimal conditions of solvent and reagent purity and dryness, such solutions are occasionally observed to decompose less than 2% in 24 hr. More typically, 10–30% decomposition is observed in that time. However, such solutions can be stored at −20° for weeks with no apparent decomposition.

Solutions of superoxide prepared in this manner have ESR (electron spin resonance) and UV spectra typical of superoxide.[1] The ESR spectrum has $g_{\parallel} = 2.11$ and $g_{\perp} = 2.01$.[20] The UV spectrum is obtained using matched 0.1-mm pathlength cells to minimize solvent absorption. It shows a maximum at 251 nm with $\varepsilon = 2686 \pm 29\ M^{-1}\ cm^{-1}$.[21] We find the UV spectrum to be the most convenient method to determine superoxide concentrations in dimethyl sulfoxide. A recent report by Arudi *et al.*[22]

[18] B. Dietrich, J. M. Lehn, and J. P. Sauvage, *Tetrahedron Lett.* 2885 (1969).
[19] D. F. Shriver, "The Manipulation of Air-Sensitive Compounds," p. 141. McGraw-Hill, New York, 1969.
[20] J. S. Valentine, Y. Tatsuno, and M. Nappa, *J. Am. Chem. Soc.* **99**, 3522 (1977).
[21] S. Kim, R. DiCosimo, and J. San Filippo, Jr., *Anal. Chem.* **51**, 679 (1979).
[22] R. L. Arudi, A. O. Allen, and B. H. J. Bielski, *FEBS Lett.* **135**, 265 (1981).

that such a method is not satisfactory because of interference by UV absorption of $O_2$, dimsyl ion ($CH_3SOCH_2^-$), and $K^+$-18-crown-6 is not in agreement with our observations and has recently been contradicted by Gampp and Lippard.[23] The latter authors[23] also report that the slow decomposition of such superoxide solutions does eventually give dimsyl ion in addition to dimethyl sulfone according to reaction (3):

$$2O_2^- + 2Me_2SO \rightarrow O_2 + Me_2SO_2 + OH^- + MeSO(CH_2)^- \tag{3}$$

*Preparation of KO₂ Solutions in Dimethylacetamide and Pyridine by Use of Cryptate.* A similar procedure is used to prepare potassium superoxide solutions in these solvents, with the substitution of K-222 for 18-crown-6. Thus a 0.005 $M$ superoxide solution can be prepared in dimethylacetamide and a 0.010 $M$ solution in pyridine, in each case using 2–3 equivalents of K-222 per equivalent of $KO_2$. A cryptate is required in place of a crown ether in these solvents because of the increased complexing ability of the former for potassium ion. The resulting solutions are usually yellow, as in dimethyl sulfoxide. The resulting solutions are used immediately as they are considerably less stable than solutions in dimethyl sulfoxide.

The strong UV absorbances of these solvents preclude measurement of the UV maximum of superoxide thus dissolved. The ESR spectral parameters are similar to those for DMSO solutions.

*Hazards.* Crown ethers, cryptates, pyridine, and dimethyl sulfoxide are potentially toxic materials. Potassium superoxide is itself a potentially strong oxidant and gives $O_2$ and $H_2O_2$ upon exposure to water. Contact with paper has led to fires. All solution preparation and reactions should be carried out with protection against possible explosion. The procedure of Rosenthal[24] for preparation of $K^{18}O_2$ in one instance in our laboratory led to a violent explosion.

### Tetramethylammonium Superoxide

From many points of view, use of tetramethylammonium superoxide is the preferred method to prepare solutions of superoxide in aprotic solvents. The salt itself can be prepared in such a fashion that its purity is 99%, it is highly soluble in such solvents, and the purity of the resulting solutions is remarkably high. The principal disadvantage of the method is that the method of preparation of the salt is cumbersome and potentially dangerous. The preparation of tetramethylammonium superoxide ($Me_4N)O_2$ from potassium superoxide and ($Me_4N)OH \cdot 5H_2O$ via a

[23] H. Gampp and S. J. Lippard, *Inorg. Chem.* **22**, 357 (1983).
[24] I. Rosenthal, *J. Labelled Comp. Radiopharm.* **12**, 317 (1976).

solid-phase metathesis reaction and subsequent extraction into liquid ammonia was first described by McElroy and Hashman in 1964.[25] Recently, Sawyer et al.[26] have optimized the synthetic procedure and characterized the isolated salt in terms of its electrochemical, spectroscopic, and magnetic properties. They[26] obtained solutions of tetramethylammonium superoxide in pure anhydrous solvents such as acetonitrile, dimethyl sulfoxide, dimethylformamide, and pyridine that were stable for days.

*Synthetic Procedure.* In an inert atmosphere chamber, $KO_2$ (70 g, 1 mol) and $(Me_4N)OH \cdot 5H_2O$ (10 g, 0.055 mol) are ground separately in mortars to fine powders prior to their addition to a 500-ml round-bottom flask. Approximately 40 ml of 3-mm-diameter glass beads is added to the flask as a mixing and grinding agent. The flask is then sealed with a roto-evaporator top and moved from the chamber to a roto-evaporator. The dry mixture is rotated under vacuum for at least 3 days and then returned to the chamber. The contents are then transferred to the glass filtration cup of a specially designed Soxhlet extractor (Fig. 2). [Caution: *Paper collection cups will lead to fire and should not be used.*] A cold finger condenser is mounted in the top of the extractor. The stopcock of the condenser is closed and the bottom opening of the Soxhlet extractor is sealed with a glass stopper prior to transfer of the assembled apparatus to a fume hood behind safety shields. The cold finger condenser is connected through a two-way stopcock to compressed gas cylinders of ammonia and ultrapure nitrogen. The stopcock allows either a mixture of both gases or pure nitrogen to pass to the cold finger. A trap that contains 4A molecular sieves is placed in the gas line to reduce the level of residual moisture. This trap is immersed in an ice-water bath to cool the gases prior to passage to the cold finger. With a slight flow of nitrogen through the system to minimize the opportunity for atmospheric water contamination, the glass stopper is replaced with the 500-ml round-bottom collection flask. The sidearm of the collection flask is vented through a mineral oil bubbler, which prevents back-diffusion of water into the system. Next, the reservoir of the cold finger is filled with a dry ice/acetone slurry, and the two-way stopcock is switched to allow passage of both nitrogen and ammonia into the system. (Condensed ammonia extracts the tetramethylammonium superoxide from the glass extraction cup into the collection flask.) A slight flow of nitrogen is necessary to prevent the development of subatmospheric pressures within the system as the ammonia condenses to a liquid. After the introduction of sufficient ammonia for the equivalent of three extractions, the flow of ammonia is stopped and the system allowed to stand with nitrogen flowing through it. The collection

[25] A. D. McElroy and J. S. Hashman, *Inorg. Chem.* **3**, 1798 (1964).
[26] D. T. Sawyer, T. S. Calderwood, K. Yamaguchi, and C. T. Angelis, *Inorg. Chem.* **22**, 2577 (1983).

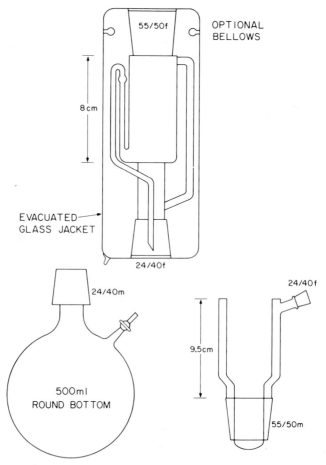

FIG. 2. Special Soxhlet apparatus for the extraction with liquid ammonia of $(Me_4N)O_2$ from the reaction mass of its solid-state metathesis [20:1 mol ratio of $KO_2$ and $(Me_4N)OH \cdot 5H_2O$]. A glass cup with a fritted-glass bottom is used to hold the solid material and is placed inside the upper vacuum-jacketed section.

flask is removed and closed with a glass stopper. The residual liquid ammonia is allowed to evaporate through the sidearm via the mineral oil bubbler. After the ammonia is completely removed, the collection flask is transferred to the inert atmosphere chamber. The tetramethylammonium superoxide in the flask (about 4.6 g) is transferred to a preweighed glass container and sealed.

Separate lots of the salt prepared in this fashion have been found to have a purity ranging from 85 to 99%. Analysis was carried out by reaction with water followed by titration of the resulting hydroxide with acid and of the resulting hydrogen peroxide by standard techniques.[26] Dissolution of $(Me_4N)O_2$ in *rigorously* dried metal-free acetonitrile gives $\lambda_{max} = 289$ nm attributed to the presence of the dimer, $[(Me_4N)O_2]_2$.[26] The spectrum of this solution slowly reverts to that characteristic of monomeric $(Me_4N)O_2$ with $\lambda_{max} = 257$ nm. Solutions of $(Me_4N)O_2$ prepared in acetonitrile purified by conventional techniques give the latter spectrum.[26]

*Base-Induced Generation of Superoxide from Hydrogen Peroxide in Pyridine.* Substantial yields of superoxide are obtained when tetramethylammonium hydroxide is added to hydrogen peroxide in pyridine.[27]

$$2H_2O_2 + OH^- \xrightarrow{py} O_2^- + H_2O + 1/n[py(\cdot OH)]_n \tag{4}$$

Such a procedure with 30% $H_2O_2$ is essentially stoichiometric and the other products are inert.

### Acknowledgments

This work was supported by the National Science Foundation under Grants CHE82-04859 (J.S.V.) and CHE79-22040 (D.T.S.) and by the National Institutes of Health-USPHS under Grant GM-22761 (D.T.S.). We are grateful to Professor C. S. Foote and Dr. H. J. Guiraud (Department of Chemistry and Biochemistry, UCLA) for helpful advice and assistance with the optimization of the synthetic procedures for $(Me_4N)O_2$ and to Mr. Kenneth Yamaguchi (Department of Chemistry, University of California, Riverside) for assistance with the electrosynthesis procedures.

[27] J. L. Roberts, Jr., M. M. Morrison, and D. T. Sawyer, *J. Am. Chem. Soc.* **100**, 329 (1978).

## [8] Generation of Superoxide Radicals in Aqueous and Ethanolic Solutions by Vacuum-UV Photolysis

### By BENON H. J. BIELSKI

While generation of the perhydroxyl $(HO_2)$ and superoxide $(O_2^-)$ radicals by UV photolysis (380–200 nm) of hydrogen peroxide has been the subject of numerous studies,[1-5] the generation of these species by vacuum-UV (VUV, $\lambda < 180$ nm) photolysis of water in presence of appropri-

[1] C. J. Hochanadel, *Radiat. Res.* **17**, 286 (1962).

[2] J. H. Baxendale, *Radiat. Res.* **17**, 312 (1962).

[3] C. H. Bamford and R. P. Wayne, *in* "Photochemistry and Reaction Kinetics" (P. G. Ashmore, F. S. Dainton, and T. M. Sugden, eds.), Chap. 3. Cambridge Univ. Press, London and New York, 1967.

[4] M. T. Downes and H. C. Sutton, *J. Chem. Soc. Faraday Trans. I* **69**, 263 (1973).

[5] A. Nadezhdin and H. B. Dunford, *J. Phys. Chem.* **83**, 1957 (1979).

Copyright © 1984 by Academic Press, Inc.
All rights of reproduction in any form reserved.
ISBN 0-12-182005-X

ate scavengers has not been fully explored for practical uses until recently.[6-10] Although kinetic and spectral studies of $HO_2/O_2^-$ by flash-photolysis of peroxide solutions proved quite successful at pH below 8,[5] the presence of high concentrations of hydrogen peroxide in such studies may not always yield unequivocal results because of possible side reactions. Therefore, overall studies on the properties of $HO_2/O_2^-$ are best carried out at as low hydrogen peroxide as is experimentally feasible.

A recently developed technique for the preparation of alkaline (pH 11–13) superoxide solutions which is both simple and versatile utilizes primary radicals H and OH produced by decomposition of water by VUV light[7,10]:

$$H_2O + h\nu \rightarrow H + OH \tag{1}$$

In the presence of oxygen and appropriate scavengers such as formate or ethanol, the primary radicals are converted to superoxide radicals, the quantum yield of which is twice the quantum yield of reaction (1). The reaction mechanisms and experimental conditions under which these reactions occur have been studied extensively by radiation chemists and are well established.[11-23]

In aqueous formate solutions, the mechanism by which $HO_2/O_2^-$ is generated is given by the following equations[22]:

$$H + O_2 \rightarrow HO_2 \tag{2}$$
$$HO_2 \rightleftharpoons H^+ + O_2^- \qquad K_{HO_2} = 2 \times 10^{-5}\ M^{24,25} \tag{3}$$
$$HCOO^- + H \rightarrow H_2 + CO_2^- \tag{4}$$
$$HCOO^- + OH \rightarrow H_2O + CO_2^- \tag{5}$$
$$COO^- + O_2 \rightarrow CO_2 + O_2^- \tag{6}$$

[6] J. M. McCord and I. Fridovich, *Photochem. Photobiol.* **17**, 115 (1973).

[7] R. A. Holroyd and B. H. J. Bielski, *J. Am. Chem. Soc.* **100**, 5796 (1978).

[8] B. H. J. Bielski and J. M. Gebicki, *J. Am. Chem. Soc.* **104**, 796 (1982).

[9] B. H. J. Bielski and R. L. Arudi, *Anal. Biochem.* **133**, 170 (1983).

[10] R. A. Holroyd and B. H. J. Bielski, "Photochemical Method for Generating Superoxide Radicals ($O_2^-$) In Aqueous Solutions." United States Patent 4,199,419 (1980).

[11] G. G. Jayson, G. Scholes, and J. Weiss, *J. Chem. Soc.* **250**, 1358 (1957).

[12] A. Hummel and A. O. Allen, *Radiat. Res.* **17**, 302 (1962).

[13] G. E. Adams, R. L. Willson, *Trans. Faraday Soc.* **65**, 2981 (1969).

[14] J. Rabani, D. Klug-Roth, and A. Henglein, *J. Phys. Chem.* **78**, 2089 (1974).

[15] J. H. Baxendale and G. Z. Hughes, *Z. Phys. Chem. (Frankfurt/Main)* **14**, 323 (1958).

[16] I. A. Taub and L. M. Dorfman, *J. Am. Chem. Soc.* **84**, 4053 (1962).

[17] M. Simic, P. Neta, and E. Hayon, *J. Phys. Chem.* **73**, 3794 (1969).

[18] D. W. Johnson and G. A. Salmon, *J. Chem. Soc. Faraday Trans I* **71**, 583 (1975).

[19] E. Bothe and D. Schulte-Frohlinde, *Z. Naturforsch. Abt. B* **33**, 786 (1978).

[20] M. N. Schuchmann and C. von Sonntag, *J. Phys. Chem.* **83**, 780 (1979).

[21] M. N. Schuchmann and C. von Sonntag, *J. Phys. Chem.* **86**, 1995 (1982).

[22] I. G. Draganic and Z. D. Draganic, "The Radiation Chemistry of Water," p. 130. Academic Press, New York, 1971.

[23] C. von Sonntag and H. P. Schuchmann, *Adv. Photochem.* **10**, 59 (1977).

Although this mechanism is applicable over the entire pH range, here it is used only above pH 11.

Photolysis of ethanolic solutions in presence of oxygen and KOH yields superoxide radical over a wide concentration range of ethanol [e.g., 5 mM to 16.2 M (92% v/v)]. In dilute solutions $O_2^-$ is formed according to reactions (2) and (3) and (7) to (9).[11-23]

$$CH_3CH_2OH + OH \rightarrow CH_3\dot{C}HOH + H_2O \qquad (7)$$

$$CH_3\dot{C}HOH + O_2 \rightarrow CH_3\overset{\cdot O_2}{\underset{|}{C}}HOH \qquad (8)$$

$$CH_3\overset{\cdot O_2}{\underset{|}{C}}HOH + OH^- \rightarrow CH_3CHO + H_2O + O_2^- \qquad (9)$$

In concentrated ethanol solutions containing oxygen and KOH, it is the ethanol that is being photolyzed:

$$CH_3CH_2OH + h\nu \rightarrow CH_3\dot{C}HOH + H \qquad (10)$$
$$CH_3CH_2OH + h\nu \rightarrow CH_3CH_2O\cdot + H \qquad (11)$$
$$CH_3CH_2OH + H \rightarrow CH_3\dot{C}HOH + H_2 \qquad (12)$$
$$CH_3CH_2O\cdot + CH_3CH_2OH \rightarrow CH_3CH_2OH + CH_3\dot{C}HOH \qquad (13)$$

Any ethoxy radical formed in reaction (11) is rapidly converted to the $CH_3\dot{C}HOH$ radical by the excess of ethanol in reaction (13). For an excellent review of photolysis of alcohols the reader is referred to Ref. 23. Overall the photogeneration of superoxide radical in concentrated ethanol solutions is initiated by reactions (10) to (13) and followed by reactions (8) and (9). This protic solvent system has the advantage that it can be used in studies of certain water-insoluble compounds; also the chemical behavior of $HO_2/O_2^-$ is similar to that in pure water.[8]

In aqueous solutions the superoxide radical is always in equilibrium $(3, -3)$ with its conjugate acid the perhydroxyl radical and disproportionates spontaneously in the absence of other reactants to oxygen and hydrogen peroxide by the following second-order reactions[24,25]:

$$HO_2 + HO_2 \rightarrow H_2O_2 + O_2 \qquad (14)$$
$$HO_2 + O_2^- + H_2O \rightarrow H_2O_2 + O_2 + OH^- \qquad (15)$$

The corresponding rate constants are $k_{14} = 8.6 \times 10^5\ M^{-1}\ \text{sec}^{-1}$ and $k_{15} = 1.0 \times 10^8\ M^{-1}\ \text{sec}^{-1}$, respectively.[25] The overall disproportionation of $HO_2/O_2^-$ as a function of pH, which is described by Eq. (16),

$$k_{obs} = \left[\frac{k_{14} + k_{15}(K_{HO_2}/[H^+])}{(1 + K_{HO_2}/[H^+])^2}\right] \qquad (16)$$

[24] D. Behar, G. Czapski, J. Rabani, L. M. Dorfman, and H. A. Schwarz, *J. Phys. Chem.* **74**, 3209 (1970).
[25] B. H. J. Bielski, *Photochem. Photobiol.* **28**, 645 (1978).

indicates that the radicals disappear most rapidly at their respective $pK_{HO_2} = 4.7$ where reaction (15) is rate controlling. Above pH 6 Eq. (16) reduces to $k_{obs} = k_{15}[H^+]/K_{HO_2} = 5 \times 10^{12} [H^+] M^{-1} sec^{-1}$, that is the rate of disproportionation decreases by one order of magnitude per unit increase in pH. Earlier results suggested that the reaction $O_2^- + O_2^-$ may not occur at all ($k_{obs} = 0.3 M^{-1} sec^{-1}$ at pH 13),[25] and is certainly negligible. It is because of this relatively high stability of superoxide radical in alkaline solutions (free of catalytic amounts of metal impurities) that $O_2^-$ solutions prepared by VUV photolysis can be used directly in titration experiments or kinetic rate studies in a stopped-flow spectrophotometer.[9]

## Methods

Since decomposition of $HO_2/O_2^-$ is catalyzed by trace amounts of metallic impurities, it is essential that all chemicals be highly purified and that the glassware (preferentially quartz) be scrupulously clean. Also it is advisable that all preparative work be carried out in a laminar flow hood as airborne particulate matter was found to significantly affect the stability of $O_2^-$ solutions. All solutions were prepared from distilled water that had been further purified by passing it through a commercial Milli-Q purification system (Millipore Corp.).

### Cleaning of Glassware

All glassware was soaked for at least 1 hr in a hot mixture of concentrated $H_2SO_4/HNO_3$ (80 : 20 v/v). After thorough rinsing with ordinary distilled water the glassware was rinsed (for 1 min only) with a solution consisting of 375 ml $H_2O$, 10 g Alconox, 100 ml of 70% $HNO_3$, and 25 ml HF. This was followed by at least 10 rinses with ordinary distilled water and 10 rinses with MQ-purified distilled water.

### Chemicals

The following commercially available chemicals were used without further purification: KOH (Apache Chemicals Inc., 99.999% grade); ethanol (US Industrial Chemical Co., punctillious grade); $HClO_4$ (G. Frederick Smith Chemical Co., Vycor distilled grade); $H_2SO_4$ (BDH Chemicals Ltd., distilled grade "Aristar"). Oxygen or nitrogen (Matheson Co., brand UHP 99.999%) was used to oxygenate or deoxygenate solutions, respectively.

Commercially available chemicals that needed purification were recrystallized from the MQ-treated distilled water. When purifying chemicals it is advantageous to start with the purification of ethylenediaminetetraacetic acid (EDTA) and/or diethylenetriaminepentaacetic acid

(DTPA) (Sigma Chemical Co.) as these chemicals are subsequently used as sequestering agents in the purification of other chemicals. EDTA/DTPA as obtained from the manufacturer contain significant amounts of metallic impurities. They can be easily purified by recrystallization from water and usually three to four such recrystallizations yield very pure samples. Sodium formate was recrystallized three times from MQ water in presence of 0.1 m$M$ EDTA (recrystallized) and three times from MQ water to remove traces of EDTA.

*Apparatus*

A convenient bench-top apparatus for generating superoxide in aqueous or ethanolic solutions is shown in Fig. 1. It is a simple and inexpensive setup since it consists of only a 100-W microwave generator [i.e., KIVA model MPG-4M (D) with an Evenson-type cavity (F)] that powers a Xe-plasma lamp, and an assortment of glassware. A solution to be photolyzed flows from reservoir (A), where it is saturated with oxygen, through a cooling coil (B) to the plasma lamp (C) where superoxide radicals are generated. Thence the solution flows into a reaction vessel (E).

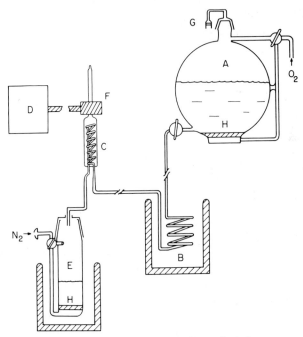

FIG. 1. Apparatus for the generation of superoxide radicals in aqueous and ethanolic solutions. (A) Reservoir; (B) cooling coil; (C) Xe-plasma lamp; (D) microwave generator; (E) reaction vessel; (F) microwave cavity; (G) glass frit (medium) to keep out dust; (H) glass frits (medium) for purging with $O_2$ or $N_2$.

The rate of flow is controlled by raising or lowering the reservoir which is connected to the cooling coil by a length of polethylene tubing (not shown in the figure). The reservoir has two coarse-sized frits, one for oxygenation (H) and another for eliminating air-borne dust (G). Precooling the solution improves the yield of $O_2^-$ as the plasma lamp generates heat, raising the temperature of the flowing solution. Aqueous solutions are conveniently cooled with ice, while alcoholic solutions are cooled with dry ice/solvent mixtures to lower temperatures ($-20$ to $-40°$).

The plasma lamp (C) consists of an outer jacket which is made either from quartz or Pyrex and an internal coil or loop made of Suprasil quartz tubing through which the aqueous/ethanolic solution flows. The outer envelope of the plasma lamp is filled with Xe (10–40 Torr), which forms a plasma that surrounds the Suprasil coil when excited by microwaves. The plasma emits intense far-ultraviolet light in the region between 155 and 195 nm. The formation of the plasma is induced by a spark from a Tesla coil after the microwave power has been turned on. Details for the construction of such lamps can be found in Refs. 7 and 10. A practical lamp has an inner Suprasil coil about 60 cm long with an i.d. of 0.1 cm. Overall the superoxide yield is controlled by the following:

1. Design and flow rate; since the light is absorbed in a very shallow layer of solution the total area of exposure (wall of inner coil) to total volume in the coil is critical.

2. Tuning and power; the extent to which the coil is surrounded by the plasma discharge depends upon the tuning of the cavity and the microwave power.

3. Photolysis time; the yield of $O_2^-$ as a function of photolysis time typically increases at first, passes through a maximum and then decreases. Similarly the ratio of superoxide radical/hydrogen peroxide formed is large initially and steadily decreases with photolysis time. Since each lamp has a slightly different profile of $O_2^-$ yield vs photolysis time, this characteristic has to be determined for each lamp and checked periodically as it changes with time.

4. Chemical composition of photolysis solution; see next section.

*Attention:* The plasma lamp and microwave cavity can constitute a serious health hazard if not handled properly. The plasma lamp should be shielded with UV-absorbing plastic or Pyrex glass (not shown in figure) and safety glasses should be worn at all times. Also the microwave cavity must be checked for leakage and should be adjusted with a proper tool and not with bare fingers.

*Preparation and Stabilization of Superoxide Solutions*

Depending upon the specific requirements of a given experiment, aqueous or ethanolic superoxide solutions can be prepared in a setup as shown in Fig. 1. The arrangement is such that $O_2^-$ solutions can be run directly from the lamp into a vigorously stirred and/or bubbled solution in reaction vessel (E) containing the substrate and a buffer or acid to give the desired final pH.

A typical photolysis solution contains from 0.25 to 1.25 m$M$ $O_2$, 1–100 m$M$ KOH (pH 11–13), and either 5–50 m$M$ sodium formate or 5 m$M$ to 16.2 $M$ (95%) ethanol. By varying either the flow-rate or microwave power the superoxide radical concentration can be varied over a wide range (1–2 $\mu M$ to 0.3 m$M$) and appropriate calibration curves can be obtained e.g. $[O_2^-]$ vs flow rate at constant microwave power or $[O_2^-]$ vs microwave power at constant flow rate. As these solutions are relatively stable (between pH 12 and 13, measured $k_{obs}$ vary on the average from 1 to 3 $M^{-1}$ sec$^{-1}$ depending upon how clean the system is) the superoxide concentrations can be monitored spectrophotometrically at 260 nm ($\varepsilon_{O_2^-}$ = 1925 $M^{-1}$ cm$^{-1}$ at 24°).[25]

It is advantageous to design experiments which do not require superoxide concentrations in excess of 10–20 $\mu M$ $O_2^-$ since such solutions are not only more stable (the lower the concentration the longer the half-life) but also contain less hdyrogen peroxide. Under ideal conditions, i.e., very short photolysis exposure and high concentrations of primary radical scavengers (HCOO⁻/EtOH and $O_2$), the only source of hydrogen peroxide is reaction (15), and high ratios of $[O_2^-]/[H_2O_2] \sim 100$ are attained. In this respect photolysis has an advantage over ionizing radiation where $H_2O_2$ is always formed as a molecular product and the best ratio for $[O_2^-]/[H_2O_2]$ to be expected is 8.

In anaerobic experiments, the superoxide solution can be added dropwise (approximately 0.05 ml/drop) to a large volume (50 ml) of a nitrogen bubbled solution containing the compound under study (H,E). The resulting oxygen concentration in such a reaction mixture would be $[O_2] < 0.25$ $\mu M$, if the superoxide solution was prepared from an air-saturated solution (250 $\mu M$ $O_2$). For more stringent anaerobic conditions the superoxide solution can be rendered virtually oxygen free by first collecting it in a bubbler and purging it with ultra pure nitrogen for 2 min before reacting it in vessel E. Molecular oxygen can also be removed from such solutions by freeze-thawing on a high vacuum line. Both procedures cause an approximate loss in $[O_2^-]$ by about 10–15%.[9]

In general aqueous or ethanolic superoxide solutions can be kept for prolonged time periods (10% loss in 1 month) if they are frozen to $-196°$

(liquid nitrogen). The resulting $O_2^-$ ices (and/or glasses in the case of ethanolic solutions) are much more stable if they are prepared in quartz vessels that have been rinsed several times with the $O_2^-$ solution before freezing. When such samples are prepared, care should be taken that $O_2$ from the atmosphere does not condense into the tubes since upon warming the rapidly expanding oxygen may cause the tubes to explode. Hence samples should be opened while the lower half is immersed in liquid nitrogen. For quantitative work, ices with known amounts of superoxide radical are prepared by monitoring the $O_2^-$ decay in an aliquot of the sample. When the absorbance at 260 nm has dropped to a desired value ($O_2^-$ concentration) the sample in the quartz tube is rapidly frozen to $-196°$, thus maintaining the desired concentration. This technique allows the preparation of large volume samples that do not differ in superoxide radical concentration by more than 5%. Samples containing initially 50 $\mu M$ $O_2^-$ lose approximately 7% $O_2^-$ per freeze-thaw cycle. Overall such ices/glasses can be used directly, either by adding a lump to a rapidly stirring solution of a reactant or, once brought to room temperature, by the various methods described in the earlier sections.

### Acknowledgments

The author wishes to thank Drs. D. E. Cabelli and R. L. Arudi for their constructive criticism of this manuscript.

This research was carried out at Brookhaven National Laboratory under contract with the U.S. Department of Energy and supported by its Office of Basic Energy Science.

## [9] Isolation and Characterization of Superoxide Dismutase

### By J. V. BANNISTER and W. H. BANNISTER

Superoxide dismutase (SOD) is firmly established as an important tool for the study of reactions involving the superoxide anion radical. For this reason we give here methods for the preparation of highly purified copper–zinc (Cu,Zn) SOD. We also describe the purification of manganese (Mn) and iron (Fe) SOD which, like Cu,Zn-SOD, are of considerable interest in their own right as metalloenzymes evolved to detoxicate $O_2$ at its first level of toxicity. The isolation of CuZn SOD can be started with organic solvents in the classical manner[1] or organic solvents can be eschewed.[2] There is consensus that organic solvent procedures do not alter

[1] T. Mann and D. Keilin, *Proc. R. Soc. London Ser. B* **126,** 303 (1939).
[2] J. W. Hartz and H. F. Deutsch, *J. Biol. Chem.* **247,** 7043 (1972).

Copyright © 1984 by Academic Press, Inc.
All rights of reproduction in any form reserved.
ISBN 0-12-182005-X

essential properties of CuZn SOD and we consider these procedures ideal for the isolation of this SOD from red blood cells because of their efficiency in removing hemoglobin. We give here an organic solvent procedure for red blood cells. For solid tissues we give a procedure without use of organic solvents which yields both Cu,Zn-SOD and Mn-SOD. The latter, like Fe-SOD, is not resistant to organic solvents.

## CuZn SOD from Red Blood Cells

Red blood cells are collected by centrifuging. Washing with isotonic saline is not necessary. The cells are lysed by addition of an equal volume of distilled water, containing 0.1% Triton X-100, and left overnight at 4°. One-half volume of ethanol–chloroform mixture (3 : 1, v/v) previously chilled to $-20°$, is added very slowly to the lysate, with stirring. The mixture is left standing for 30 min at 4°. One-fifth volume of 0.15 $M$ NaCl is then added, with stirring, and the mixture is centrifuged at 2000 $g$ for 60 min at 4°. The volume of the supernatant is measured for reference and protein is precipitated with 1/20 volume of saturated lead acetate or saturated zinc acetate solution (in this case maintaining neutral pH with solid Tris) and collected by centrifuging. The lead acetate precipitate is extracted twice with 0.3 $M$ phosphate buffer, pH 6 (1/20 volume in the first and 1/40 volume in the second extraction). The zinc acetate precipitate is extracted twice with 0.3 $M$ pyrophosphate–acetic acid buffer, pH 7 (1/20 volume in the first extraction keeping neutral pH with 1 $M$ acetic acid and 1/40 volume in the second extraction keeping neutral pH with solid Tris). The purpose of zinc acetate precipitation–pyrophosphate buffer extraction is to obviate thiol oxidation, human Cu,Zn-SOD having a free cysteine per subunit.[3] An alternative to lead acetate or zinc acetate precipitation is phase separation with solid $K_2HPO_4$ and precipitation of protein from the lighter phase with acetone.[4] The crude SOD extract is dialyzed against 0.02 $M$ Tris buffer (starting buffer, see below) at 4° prior to chromatography.

Chromatography is carried out on QAE-Sephadex A-50 at room temperature (15–20°). The column size is 5 × 40 cm for medium-scale preparations (10 to 20 liters of red blood cells). The column is loaded to one-fourth capacity as judged by the brown band formed at the top. A good elution strategy is to run dilute buffer of decreasing pH until the SOD is in the botton third of the column. Current buffer is then stirred into the column above the SOD band. The (top) part of the column is poured off

[3] D. Barra, F. Martini, J. V. Bannister, M. E. Schinina, G. Rotilio, W. H. Bannister, and F. Bossa, *FEBS Lett.* **120,** 53 (1980).
[4] J. M. McCord and I. Fridovich, *J. Biol. Chem.* **244,** 6049 (1969).

and the SOD is eluted off the remaining part. We give here well-tested elution procedures for the commonly prepared bovine and human SOD. For bovine SOD we recommend 0.02 $M$ Tris–HCl, pH 7.8, as starting buffer followed by 0.03 $M$ Tris–HCl, pH 7.0, 0.03 $M$ cacodylate–HCl, pH 7.0 (this elutes an impurity from the SOD band untouched by the preceding buffer), and 0.03 $M$ cacodylate–HCl, pH 7.0, containing 0.1 $M$ NaCl, for final elution. For human SOD we recommed 0.02 $M$ Tris–acetic acid, pH 7.8, as application buffer followed by 0.03 $M$ Tris–acetic acid buffers of pH 7.0, 6.0, 5.75, 5.5, 5.25, and 5.0, and 0.05 $M$ Tris–acetic acid buffer, pH 5.0, for final elution. The eluted SOD is neutralized with solid Tris. Human Cu,Zn-SOD starts to run appreciably (on QAE-Sephadex) at pH 5.5 with 0.03 $M$ Tris–acetic acid buffer. A fraction which elutes at higher pH, previously called SOD I,[5] is not seen with fairly fresh blood (or autospy liver) as starting material. To obtain highly purified SOD, whether human or bovine, chromatography on QAE-Sephadex is repeated twice. (Gel filtration will then show no other protein and is not required.) The final SOD is dialyzed against distilled water at 4°, then lyophilized and stored at $-20°$. The type of chromatography on QAE-Sephadex described here should be started at a pH as high as 8.9 with unfamiliar SODs since some Cu,Zn-SODs, for instance, swordfish,[6] and goat,[7] are not adsorbed at pH 7.8.

## Cu,Zn-SOD and Mn-SOD from Solid Tissues

We give here a procedure, of general nature, developed for human liver. Autopsy liver is stored at $-20°$ and used in 2-kg batches. The liver is thawed overnight at 4°. The thawed liver is cut into thin slices, washed exhaustively with cold 0.15 $M$ NaCl, and homogenized with 2 volumes of chilled 0.02 $M$ Tris–acetic acid buffer, pH 7.8, containing 0.2% Triton X-100 and 1 m$M$ phenylmethyl sulfonyl fluoride. A prechilled Waring blender is used at high speed for 3 min. The homogenate is stood for 30 min at 4° before centrifugation at 13,700 $g$ for 30 min at 4°. The supernatant is placed in a stainless-steel beaker and heated rapidly to 60°, with stirring, in an almost boiling water bath. It is then cooled rapidly to room temperature, with stirring, in an ice bath, adjusted to pH 5.5 with 1 $M$ acetic acid and centrifuged at 13,700 $g$ for 30 min at 4°. The pH of the supernatant is adjusted to 7.8 with solid Tris and the protein solution is

[5] W. H. Bannister, A. Anastasi, and J. V. Bannister, in "Superoxide and Superoxide Dismutases" (A. M. Michelson, J. M. McCord, and I. Fridovich, eds.), p. 107. Academic Press, New York, 1977.

[6] J. V. Bannister, A. Anastasi, and W. H. Bannister, Comp. Biochem. Physiol. 56B, 235 (1977).

[7] W. H. Bannister, unpublished observations.

dialyzed against 0.02 $M$ Tris–acetic acid buffer (pH 7.8) at 4°. When the protein solution has approximately the conductivity of 0.02 $M$ Tris–acetic acid buffer, pH 7.8, it is run at room temperature into a column of QAE-Sephadex A-50 equilibrated with this buffer. The column should have a bed volume of about 2 liters (6.4 × 60 cm dimensions). The protein solution is washed in with one bed volume of starting buffer. The column is then eluted with 0.02 $M$ Tris–acetic acid buffer, pH 7.0. The eluate is collected in 1-liter fractions. Those with pH 7.5 are kept, to a total of 5 bed volumes, for isolation of Mn-SOD. The column is then washed virtually free of hemoglobin by means of 0.03 $M$ Tris–acetic acid buffer, pH 5.5, and finally eluted with 0.3 $M$ sodium acetate monitoring copper in the eluate with the biquinoline spot test. The copper-positive fraction is kept for isolation of Cu,Zn-SOD by chromatography on QAE-Sephadex as already described.

To isolate Mn-SOD, the Mn-SOD pool is brought close to the conductivity of 0.06 $M$ Tris–acetic acid buffer, pH 5.5, by adding 1/23.5 volume of 1 $M$ Tris–acetic acid, pH 5.5. The pH is adjusted to 5.5 with 0.06 $M$ acetic acid and the protein solution is run at room temperature into a 5 × 40-cm column of CM-Sephadex C-50 equilibrated with 0.06 $M$ Tris–acetic acid, pH 5.5. The column is washed with 1 to 2 bed volumes of 0.06 $M$ Tris–acetic acid, pH 5.5, and eluted first with a gradient obtained by continuously running 0.12 $M$ sodium acetate (1 liter) into 0.06 $M$ Tris–acetic acid, pH 5.5 (1 liter), then with 0.12 $M$ sodium acetate. Mn-SOD appears in the eluate toward the end of the gradient. The SOD-active fraction is stored at 4° after adjusting the pH to 7.8 with solid Tris. The fractions from six batches of liver (12 kg) are dialyzed against 0.02 $M$ Tris–acetic acid buffer, pH 7.8, at 4° and rechromatographed on CM-Sephadex as already described. The SOD obtained is concentrated by ultrafiltration at 4°, using an Amicon YM10 membrane, and finally purified by two cycles of gel filtration on a 4.7 × 115-cm column of Ultrogel AcA 44 run with 0.1 $M$ phosphate buffer, pH 7.8, at room temperature. The SOD preparation is concentrated by ultrafiltration at 4°, mixed with an equal volume of glycerol and stored at −20°.

## Fe-SOD from Bacteria

We give here a general procedure developed on *Pseudomonas putida*. Frozen bacterial paste (2 kg) is thawed and stirred into 4 liters of 10 m$M$ phosphate buffer, pH 7.8, containing 1 m$M$ phenylmethylsulfonyl fluoride, at 4°. The bacteria are sonicated for 10 min in batches of 500 ml in an ice bath. The sonicate is centrifuged at 13,700 $g$ for 30 min at 4°. The supernatant is kept as first extract. A second extract is obtained by ho-

mogenizing the precipitate with 4 liters of 10 m$M$ phosphate buffer, pH
7.8, containing 1 m$M$ phenylmethyl sulfonyl fluoride, in a Waring blender
and centrifuging again. The extracts are pooled and fractionated with
$(NH_4)_2SO_4$ at room temperature maintaining pH at 7.8 with solid Tris. The
50–85% $(NH_4)_2SO_4$ precipitate is collected by centrifuging at 13,700 $g$ for
60 min at 4°. It is dissolved in 0.1 $M$ phosphate buffer, pH 7.8. The
solution is adjusted to pH 7.8 with 1 $M$ NaOH and dialyzed against 5 m$M$
phosphate buffer, pH 7.8, at 4°. After dialysis it is clarified by centrifuging
and reduced to a volume of 1 liter by ultrafiltration at 4° using an Amicon
YM10 membrane. It is possible to cut off at this stage a large quantity of
yellow material by a few cycles of dilution with 5 m$M$ phosphate buffer,
pH 7.8, and concentration in a 2-liter ultrafiltration cell fitted with a YM10
membrane. The final solution is run at room temperature into a 6.4 × 60-
cm column of DEAE-Sephadex A-50 equilibrated with 5 m$M$ phosphate
buffer, pH 7.8, and washed in with 1 to 2 bed volumes of starting buffer.
The protein fraction which passes through the column can be kept for
preparation of Mn-SOD (see above for general procedure) in the case of
bacteria which also possess this SOD and for preparation of bacteriocu-
prein[8] in the case of *Photobacterium leiognathi*. Elution of the column is
started with 0.015 $M$ phosphate buffer, pH 7.8, and continued with
stepwise increase of the buffer concentration. *Pseudomonas putida* Fe-
SOD is eluted with 0.045 $M$ phosphate buffer, pH 7.8, and buffer steps of
0.015, 0.03, 0.045, and 0.06 $M$ are made. Larger steps may be needed for
the Fe-SODs of other bacteria. For instance, *P. leiognathi* Fe-SOD is
eluted with 0.18 $M$ phosphate buffer, pH 7.8, and buffer steps of 0.015,
0.03, 0.06, 0.12, 0.18, and 0.24 $M$ are most expedient with this SOD. The
SOD-active fraction of the eluate is dialyzed against 5 m$M$ phosphate
buffer, pH 7.8, at 4°, concentrated by ultrafiltration at 4°, and rechromatographed on a 5 × 40-cm column of DEAE-Sephadex at room temperature. The SOD fraction from this column is concentrated by ultrafiltration
at 4° and submitted to gel filtration at room temperature on a 4.7 × 115-cm
column of Sephadex G-75 run with 0.1 $M$ phosphate buffer, pH 7.8. It is
possible to obtain pure SOD by repeating this gel filtration step but sometimes it is necessary to rechromatograph the SOD on DEAE-Sephadex
and then do a second gel filtration. The final product is concentrated by
ultrafiltration at 4° and kept in 0.1 $M$ phosphate buffer, pH 7.8, at 4°.

## Characterization of SOD

Characterization of SOD as a highly purified preparation is most conveniently done by demonstrating coincidence of protein staining and

[8] K. Puget and A. M. Michelson, *Biochem. Biophys. Res. Commun.* **58,** 830 (1960).

enzyme activity after polyacrylamide gel electrophoresis.[9] Of the many assays available for SOD we find the xanthine–xanthine oxidase–cytochrome $c$ system of McCord and Fridovich[4] to be the most reliable for purposes of detection or measurement of specific activity. Activity of purified SOD is conveniently related to protein concentration measured spectrophotometrically according to Murphy and Kies[10] with the correction factor of Weisiger and Fridovich[11] for Mn-SOD.

[9] C. O. Beauchamp and I. Fridovich, *Anal. Biochem.* **44**, 276 (1971).
[10] J. B. Murphy and M. W. Kies, *Biochim. Biophys. Acta* **45**, 382 (1960).
[11] R. A. Weisiger and I. Fridovich, *J. Biol. Chem.* **248**, 3582 (1973).

# [10] Superoxide Dismutase Assays

*By* L. Flohé and F. Ötting

The term superoxide dismutase (SOD, EC 1.15.1.1) shall be used for a variety of metalloproteins catalyzing the reaction

$$2O_2^- + 2H^+ \rightarrow H_2O_2 + O_2$$

The primary difficulty in assaying SOD for its enzymatic activity consists in the free radical nature of its substrate $O_2^-$ which can only be supplied by generation within the assay medium. In addition, the substrate $O_2^-$ cannot easily be detected directly by conventional analytical tools. Routine testing of SOD therefore is performed according to the following general principle.

$O_2^-$ is generated enzymically or nonenzymically in the test medium which also contains an easily measurable indicator reacting with $O_2^-$. The SOD content of the sample is then calculated from the change of the indicator reaction. Numerous systems have been employed of which an overview is given in Table I. It is quite obvious from this list that any entity which scavenges $O_2^-$ or reacts with the indicator or changes the rate of $O_2^-$ formation will lead to erroneous determinations of the SOD activity. To overcome this problem different treatments of the assay sample have been proposed: extraction of the SOD with organic solvents,[1] acetylation of cytochrome $c$ (cyt $c$),[2] addition of cyanide,[3] and dialysis. Endogenous interfering substances can also be overcome by applying the

[1] J. M. McCord and I. Fridovich, *J. Biol. Chem.* **244**, 6049 (1969).
[2] A. Azzi, C. Montecucco, and C. Richter, *Biochem. Biophys. Res. Commun.* **65**, 597 (1975).
[3] C. Beauchamp and I. Fridovich, *Anal. Biochem.* **44**, 276 (1971).

METHODS IN ENZYMOLOGY, VOL. 105
Copyright © 1984 by Academic Press, Inc.
All rights of reproduction in any form reserved.
ISBN 0-12-182005-X

TABLE I

COMPONENTS OF INDIRECT ASSAYS FOR SOD

| Sources of $O_2^-$ | Indicators for $O_2^-$ |
|---|---|
| Xanthine + xanthine oxidase[a–e] | Reduction of cytochrome $c$[a,b,f] |
| Photoreduced flavins[c] | Reduction of nitroblue tetrazolium[b,c] |
| Autoxidation of epinephrine[g,h] | Reduction of tetranitromethane[a] |
| Autoxidation of pyrogallol[i,j] | Autoxidation of epinephrine[g,h] |
| Autoxidation of 6-hydroxydopamine[k] | Autoxidation of pyrogallol[i,j] |
| NADH oxidation by phenazine methosulfate[l] | Autoxidation of 6-hydroxydopamine[k] |
| Potassium superoxide[m] | Oxidation of 2-ethyl-1-hydroxy-2,5,5-trimethyl-3-oxazolidine[o] |
| Electrochemical reduction of $O_2$[a,n] | Chemiluminescence of luminol[p] |
| | Nitrite formation from hydroxylammonium chloride[e] |

[a] J. McCord and I. Fridovich, *J. Biol. Chem.* **244**, 6049 (1969). [b] M. L. Salin and J. M. McCord, *J. Clin. Invest.* **54**, 1005 (1974). [c] C. Beauchamp and I. Fridovich, *Anal. Biochem.* **44**, 276 (1971). [d] D. D. Tyler, *Biochem. J.* **147**, 493 (1975). [e] E. Elstner and A. Heupel, *Anal. Biochem.* **70**, 616 (1976). [f] V. Massey, S. Strickland, S. Mayhew, L. Howell, P. Engel, R. Matthews, M. Schuman, and P. Sullivan, *Biochem. Biophys. Res. Commun.* **36**, 891 (1969). [g] H. P. Misra and I. Fridovich, *J. Biol. Chem.* **247**, 3170 (1972). [h] M. Sun and S. Zigman, *Anal. Biochem.* **90**, 81 (1978). [i] K. Puget and A. M. Michelson, *Biochimie* **56**, 1255 (1974). [j] S. Marklund and G. Markland, *Eur. J. Biochem.* **47**, 469 (1974). [k] R. E. Heikkila and F. Cabbat, *Anal. Biochem.* **75**, 356 (1976). [l] R. Fried, *Biochimie* **57**, 657 (1975). [m] S. Marklund, *J. Biol. Chem.* **251**, 7504 (1976). [n] J. McCord and I. Fridovich, *J. Biol. Chem.* **243**, 5753 (1968). [o] G. M. Rosen, E. Finkelstein, and E. J. Rauckman, *Arch. Biochem. Biophys.* **215**, 367 (1982). [p] R. E. Bensinger and Ch. M. Johnson, *Anal. Biochem.* **116**, 142 (1981). [q] H. P. Misra and I. Fridovich, *Anal. Biochem.* **79**, 553 (1977). [r] Y. Kono, *Arch. Biochem. Biophys.* **186**, 189 (1978). [s] Y. Kobayashi, S. Okahata, K. Tanabe, and T. Usui, *J. Immunol. Methods* **24**, 75 (1978).

technique of parallel line analysis of variance.[4] As a note of general precaution it should therefore be stated that all these indirect procedures provide little more than a possibility of estimating relative SOD concentrations in samples of comparable composition. SOD assays based on direct monitoring of $O_2^-$ by sophisticated techniques (Table II) require an even higher degree of sample purity and shall not be dealt with here further. If an absolute measure of physiological levels of SOD is intended, we suggest a direct immunochemical method in addition to activity measurements. If the ratio of enzymatic activity to concentration of SOD antigen is unchanged independently of the composition of the test medium

[4] G. E. Eldred and J. R. Hoffert, *Anal. Biochem.* **110**, 137 (1981).

TABLE II
DIRECT ASSAYS OF SOD

Pulse radiolysis[a,b]
Rapid-freeze EPR[c,d]
Stopped-flow spectroscopy[e,f]
Polarographic techniques[g]
$^{19}$F NMR spectroscopy[h]

[a] D. Klug, J. Rabani, and I. Fridovich, *J. Biol. Chem.* **247,** 4839 (1972). [b] G. Rotilio, R. C. Bray, and E. M. Fielden, *Biochim. Biophys. Acta* **268,** 605 (1972). [c] D. Ballou, G. Palmer, and V. Massey, *Biochem. Biophys. Res. Commun.* **36,** 898 (1969). [d] W. H. Orme-Johnson and H. Beinert, *Biochem. Biophys. Res. Commun.* **36,** 905 (1969). [e] G. J. McClune and J. A. Fee, *FEBS Lett.* **67,** 294 (1976). [f] S. Marklund, *J. Biol. Chem.* **251,** 7504 (1976). [g] A. Rigo, P. Viglino, and G. Rotilio, *Anal. Biochem.* **68,** 1 (1975). [h] A. Rigo, P. Viglino, E. Argese, M. Terenzi, and G. Rotilio, *J. Biol. Chem.* **254,** 1759 (1979).

(for example during purification) and further, if recovery rates of known amounts of enzyme added to the test medium are invariably 100%, we consider the description of SOD activity properly done. Following this reasoning we therefore will focus on measuring SOD in biological media. We present (1) a qualitative test for SOD activity based on the reduction of nitro blue tetrazolium (NBT) by $O_2^-$; (2) a simple immunological determination of the SOD molecule of sufficient sensitivity and avoiding labeled reagents; and (3) an indirect measure of SOD activity to be used in purified samples.

## NBT Test

This assay was originally reported by Beauchamp and Fridovich,[3] The color reaction is ideally suited to serve as a fast and sensitive test for monitoring SOD in polyacrylamide or agarose gels.

### Principle

Flavins, e.g., riboflavin, can be photochemically reduced in the presence of an oxidizable substance, e.g., TEMED. Upon reoxidation in air reduced flavins will generate $O_2^-$,[5] and the superoxide radical in turn will reduce the colorless nitro blue tetrazolium to a blue formazan[6,7] which is practically insoluble. SOD by scavenging $O_2^-$ will inhibit color formation and appear as a colorless spot.

### Reagents

4-Nitro blue tetrazolium chloride (NBT, Serva)
Riboflavin (Sigma)
$N,N,N',N'$-Tetramethylethylenediamine (TEMED, Merck)
5,5-Diethylbarbituric acid sodium salt (barbital, Merck)
Agarose Type A (Pharmacia)
Gel bond film (Marine Colloids)
Solution A: 25 mg NBT and 10 mg riboflavin are dissolved in 100 ml doubly distilled water and kept absolutely dark in the cold. Under these conditions the reagent is stable at least for 6 weeks
Solution B: 1 g TEMED in 100 ml doubly distilled water

### Procedure

Immediately after termination of the electrophoresis or electrofocusing, the gel is soaked in solution A by pipetting ~1 ml onto a 100-cm$^2$ gel surface. After 20 min solution B is allowed to soak into the gel by applying the same procedure. Then, the gel is exposed to an appropriate light source (sun, lightbox etc.). SOD bands will appear colorless against a blue background. As the formazan dye is insoluble and stays in the gel matrix, excess reagents can be removed by soaking the gel in water. After drying the original can be kept as a record. An example is shown in Fig. 1.

### Comments

The detection limit of this method is around 2 ng. Though quantitative densitometry has been described,[8] we consider the test as a qualitative one: a fast identification of the enzyme. Complete inhibition of cuprozinc superoxide dismutases by 2 m$M$ cyanide may be utilized to distinguish these from mangano- and iron-type enzymes.[9]

[5] V. Massey, S. Strickland, S. G. Mayhew, L. G. Howell, P. C. Engel, R. G. Matthews, M. Schuman, and P. A. Sullivan, *Biochem. Biophys. Res. Commun.* **36**, 891 (1969).
[6] K. V. Rajagopalan and P. Handler, *J. Biol. Chem.* **239**, 2022 (1964).
[7] R. W. Miller, *Can. J. Biochem.* **48**, 935 (1970).
[8] W. Bohnenkamp and U. Weser, *Hoppe Seylers Z. Physiol. Chem.* **356**, 747 (1975).
[9] C. Beauchamp and I. Fridovich, *Biochim. Biophys. Acta* **317**, 50 (1973).

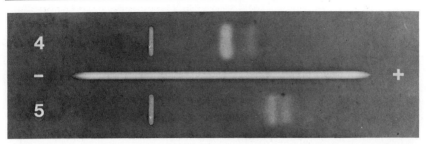

Fig. 1. NBT staining of bovine (lane 4, 60 ng) and human (lane 5, 80 ng) Cu,Zn-SOD after electrophoresis in agarose. Electrophoresis conditions: 250 V, 20 min in 150 m$M$ glycine, 20 m$M$ TRIS, pH 8.5. The agarose film was from Corning, as was the uncooled electrophoresis cell.

The NBT stain described above has been widely used for screening of SOD patterns after electrophoresis of crude tissue homogenates. Myeloperoxidase, like other peroxidases, will also mimic SOD activity in this test.[10] We can conclude that the NBT staining technique in order to be specific has to be combined with a SOD-specific means of recognition and/or separation, e.g., mobility in an electric field or binding to antibodies.

### Electroimmunoassay for Cu,Zn-SOD

Numerous immunological SOD assays have been used in the past to monitor SOD levels in tissue samples.[11-20] Because of regulatory restrictions to many laboratories, work with radioactive tracers, however, is a

[10] P. Patriarca, P. Dri, and M. Snidero, *J. Lab. Clin. Med.* **90,** 289 (1977).

[11] J. W. Harts, S. Funakoshi, and H. F. Deutsch, *Clin. Chim. Acta* **46,** 125 (1973).

[12] U. Reiss and D. Gershon, *Biochem. Biophys. Res. Commun.* **73,** 255 (1976).

[13] A. W. Eriksson, R. R. Frants, P. H. Jongbloet, and J. B. Bijlsma, *Clin. Genet.* **10,** 355 (1976).

[14] J. D. Crapo and J. M. McCord, *Am. J. Physiol.* **231,** 1196 (1976).

[15] K. Kelly, C. Barefoot, A. Sehon, and A. Petkau, *Arch. Biochem. Biophys.* **190,** 531 (1978).

[16] B. C. Del Villano and J. A. Tischfield, *J. Immunol. Methods* **29,** 253 (1979).

[17] A. Baret, P. Michel, M. R. Imbert, J. L. Morcellet, and A. M. Michelson, *Biochem. Biophys. Res. Commun.* **88,** 337 (1979).

[18] H. Joenje, R. R. Frants, F. Arwert, G. J. de Bruin, P. J. Kostense, J. J. van den Kamp, J. de Koning, and A. W. Eriksson, *Scand. J. Clin. Lab. Invest.* **39,** 759 (1979).

[19] A. Petkau, T. P. Copps, and K. Kelly, *Biochim. Biophys. Acta* **645,** 71 (1981).

[20] G. Bartosz, M. Soszynski, and W. Retelewska, *Mech. Ageing Dev.* **17,** 237 (1981).

burden and costly. Enzyme-labeled immunoassays for SOD would probably circumvent these restrictions, though the preparation of appropriate reagents again needs experience. In case a picogram-detection limit for SOD is not essential one might therefore consider one of the classical immunochemical determinations, e.g., an electroimmunoassay (EIA) according to Laurell and McKay[21] as described below.

*Principle*

In an electric field SOD as a charged antigen moves in an agarose matrix which contains the respective antibody. Under appropriate conditions the antibody–antigen complex precipitates and forms a rocket-shaped loop, the area of which is directly proportional to the amount of antigen present.

*Reagents*

Agarose Type A (Pharmacia)
Gel bond film (Marine Colloids)
5,5-Diethylbarbituric acid sodium salt (Barbital, Merck)
Staining solution (filter before use): 2.5 g Coomassie Brillant Blue G-250 (Serva), in 225 ml ethanol (96%), 225 ml water, 50 ml acetic acid
Destaining solution: as above but without dye
Antisera: the antisera were raised in rabbits or rats according to standard protocols. From the rabbit antisera the immunoglobulin fraction was isolated by threefold precipitation with ammonium sulfate (33% saturation)

*Cu,Zn-SOD*

Pure preparations ($\geq 98\%$ as judged by SDS-gradient gel electrophoresis) which were derived from liver (bovine) or placenta (human) were used as antigens or standards. Solutions of SOD below 10 $\mu g/ml$ were protected against adsorption losses by addition of 0.01% ovalbumin (w/v).

*Equipment*

A horizontal electrophoresis chamber Desaga Multiphor equipped with a water-cooled plate was used. The power supply was from LKB, type 2197.

---

[21] C. B. Laurell and E. J. McKay, this series, Vol. 73, p. 339.

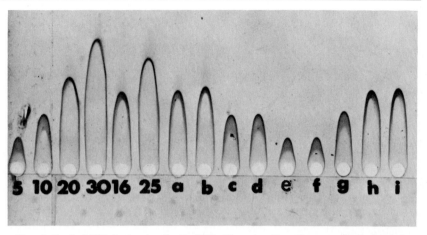

FIG. 2. Typical EIA for bovine Cu,Zn-SOD. Plate size 12 × 10 cm; antibody concentration: 4 mg IgG in 24 ml of 1.2% agarose in 17 mM sodium barbital buffer, pH 8.6. Duration of electrophoresis 16 hr, voltage 2.5 V/cm; loops: 5–30 standards in μg/ml; 16 and 25: controls; a–i: unknowns.

## Procedure

Twenty-four milliliters of 1.2% agarose in 17 mM barbital buffer, pH 8.6, is molten in a boiling waterbath. The solution is allowed to cool down to 55–57°. At that temperature the antibody solution (100–500 μl, depending on the titer) is carefully admixed and then quickly poured onto a cut polyester plastic support matirx (10 × 12 cm) which is held on a horizontal leveling tray by four rectangular aluminum bars, thus forming an edge for the liquid agarose–antibody mixture. After cooling a row of 16 holes of diameter 3 mm is punched into the agar 3 cm off the rim and 3 mm apart (Fig. 2). The wells are filled with 10 μl each of the standards and unknowns. For electrophoresis we use 17 mM Na barbital buffer pH 8.6 and 8.5 V/cm (3 hr) or 2.5 V/cm (16 hr). Voltage is measured directly in the gel with two electrodes through holes in the cover lid of the electrophoresis cell. The temperature of the cooling system is 5°.

After termination of the electrophoresis the gel is dried: first, several layers of filter paper are kept under pressure on the gel (10–15 min) and the remaining moisture is evaporated in a hot air oven (50°) or with a hair drier. Staining is performed for 15 min, followed by several changes of the destaining solution. Drying yields a film for permanent record.

The calibration curve is produced by plotting the peak heights of the standards, which is correct as long as the morphology of the "rockets" is

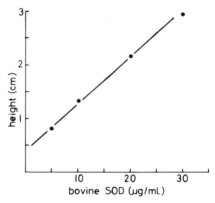

FIG. 3. Calibration curve of the EIA shown in Fig. 2.

identical in all of the samples. A typical experiment is presented in Fig. 2, the calibration curve of which is shown in Fig. 3.

Quality control data of the EIA for bovine Cu/Zn-SOD: The lower limit of sensitivity is ~10 ng, i.e., 1 $\mu$g/ml if 10-$\mu$l samples are used; the coefficients of variation for three independent quality control standards (80, 160, and 250 ng) used in the system were 5.1, 4.9, and 3.4%, respectively, during 85 assays. No change of the recovery rate of 100%—within the variation of the assay—could be observed with 1000-fold excess bovine serum albumin or ovalbumin, 20-fold excess human Cu,Zn-SOD.

*Comments*

In a program devoted to screening for extraction procedures from cell lysates this assay for bovine Cu,Zn-SOD proved very reliable and convenient. As at least some of the enzymatic activity of the SOD molecule is still present when the enzyme is bound to its antibody, the NBT-staining technique is also applicable in the EIA for SOD. This leads to an increase in the lower detection limit of ~1 ng (Fig. 4b). In Fig. 4a an example is given for the EIA for human SOD utilizing antisera from rats.

A principle drawback is that one measures the concentration of an antigen, not the catalytic activity of the SOD. We therefore suggest correlating the antigen concentration with data on the enzymatic activity. This is supported by an observation of Reiss and Gershon.[12] They described an age-related reduction of the enzymatic activity of the cytoplasmic SOD as compared to several structural parameters such as antigenicity and molecular weight which were unchanged. Similarly, Glass and Gershon[22]

[22] G. A. Glass and D. Gershon, *Biochem. Biophys. Res. Commun.* **103,** 1245 (1981).

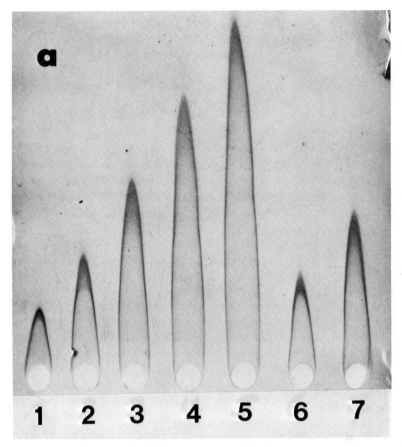

FIG. 4. (a) Example of an EIA for human Cu,Zn-SOD. The antiserum was developed in rats, with 200 μl used per 10 ml of 1.2% agarose; voltage, 2.5 V/cm for 16 hr. Loops 1–5, standards; 6 and 7, unknowns. (b) Example for the NBT staining of precipitin loops of bovine Cu,Zn-SOD. Loops 1–6, 25 ng, 12.5, 6.25, 3.1, 1.6, and 0.8 ng, respectively. The amount of antibody was reduced to 640 μg IgG/24 ml agarose. Electrophoresis was performed for 3 hr at 10 V/cm at pH 8.6.

reported a decrease of the antigen-related specific enzymatic activity of SOD during aging of rats and aging of erythrocytes.

### Ferricytochrome c Reduction Assay

*Principle*

The reduction rate of cytochrome $c$ by superoxide radicals is monitored at 550 nm utilizing the xanthine–xanthine oxidase system as the

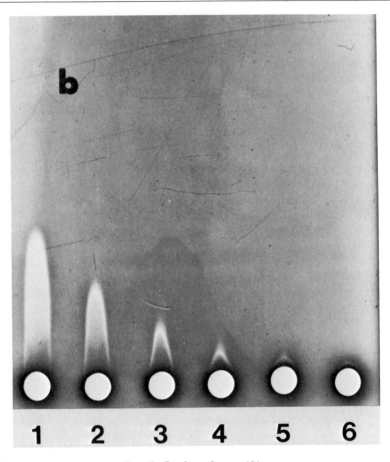

FIG. 4b. See legend on p. 101.

source for $O_2^-$. SOD will compete for superoxide and decrease the reduction rate of cytochrome $c$.[1]

*Reagents*

    Potassium dihydrogen phosphate p.a.
    Disodium hydrogen phosphate p.a.
    Ethylenediaminetetraacetic acid disodium salt p.a. (EDTA)
    Xanthine, crystalline (99–100%) (Merck)
    Xanthine oxidase from buttermilk grade I (Sigma)
    Cytochrome $c$ from horse heart, research grade (Serva, Kat.-No. 18020)
*Solution A.* 0.76 mg (5 $\mu$mol) xanthine in 10 ml 0.001 $N$ sodium hy-

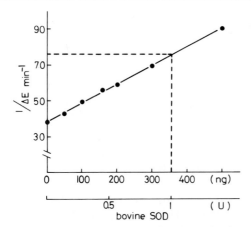

FIG. 5. Typical calibration curve of the cytochrome $c$ reduction assay. The rate of reduction of cytochrome $c$, inhibited by SOD, is plotted as the reciprocal absorbance change per minute versus concentration of SOD standards. Each individual point represents the mean value of a triplicate determination; 355 ng SOD per assay volume (3 ml) thus meets the definition of 1 unit of enzyme activity. This particular sample of bovine SOD therefore has a specific activity of 2817 U/mg.

droxide and 24.8 mg (2 $\mu$mol) cytochrome $c$ are admixed with 100 ml 50 m$M$ phosphate buffer pH 7.8 containing 0.1 m$M$ EDTA. The solution is stable for 3 days at 4°.

*Solution B*. Freshly prepared solution of xanthine oxidase in 0.1 m$M$ EDTA, ~0.2 U/ml. As the activity of the xanthine oxidase may vary, one should use sufficient enzyme to produce a rate of cytochrome $c$ reduction of 0.025 absorbance units/min in the assay without SOD.

*Procedure*

One unit of SOD is defined as that amount of enzyme which inhibits the rate of cytochrome $c$ reduction, under the conditions specified, by 50%[1] (see Fig. 5). To be able to extrapolate accurately to this value of 50% inhibition one should use several dilutions of one enzyme solution. For relative activity measurements we relate the data to a standard preparation utilizing a plot 1/$\Delta E$ min$^{-1}$ versus standards. Solution B is kept on ice; solution A is thermostated at 25° as is the cell compartment of the spectrophotometer. To adjust the proper wavelength at 550 nm cytochrome $c$ is reduced with some crystals of sodium dithionite and the maximum (550 nm) used as wavelength calibration. (1) Pipet 2.9 ml of solution A into a 3 ml cuvette; (2) add 50 $\mu$l of sample (water, SOD-standards or unknowns); (3) start the reaction with 50 $\mu$l solution B; (4) after mixing record the absorbance change at 550 nm; (5) plot $1/\Delta E$ min$^{-1}$—derived from the linear part of the reaction—versus concentration of SOD standards.

*Comments*

The assay works well in pure systems, the within-assay variation being below 10% (triplicates). The sensitivity of the assay defined as the lowest amount of SOD that can be significantly distinguished ($p < 0.05$) from the blank is ~15 ng/ml. A typical calibration curve is presented in Fig. 5.

It should be noted that the concentration of ferricytochrome $c$ is crucial. The apparent SOD activity is inversely related to the cytochrome $c$ concentration. Commercial lots may be quite variable in this respect. This difficulty can be overcome by adequate standardization of the cytochrome $c$ solution or by calibration of the test system with a standard SOD preparation.

If one intends to use the assay in nonpure systems the following controls and additions should be considered. Interfering reactants in the assay medium have to be checked in recovery experiments by addition of known amounts of a pure SOD preparation to the test sample.

Each sample has to be dialyzed. This will eliminate smaller, free molecules like ascorbate, reduced glutathione, catecholamines, etc. To block peroxidases the addition of 2 $\mu M$ potassium cyanide into the assay medium has been recommended. This concentration is claimed not to affect bovine Cu,Zn-SOD.[3] However, more recently Rigo *et al.*[23] had determined polarographically the $K_i$ values for cyanide and azide to be $1.77 \times 10^{-6}$ and $1.43 \times 10^{-2}$ $M$, respectively (pH 9.8 and 25°). It seems evident from these data that cyanide will not only affect peroxidases but also affect SOD. Azide, $10^{-5}$ $M$, is obviously more suitable for blocking peroxidases specifically.[24]

A different approach takes advantage of the fact that acetylated cytochrome $c$ though still reducible by $O_2^-$ is not susceptible to oxidases or reductases which use cytochrome $c$ as a substrate.[2,25] As the reduction rate of acetylated cytochrome $c$ is decreased, the sensitivity of the assay utilizing acetylated cytochrome $c$ is increased twofold. To distinguish mangano- or iron-type enzymes from the cuprozinc type, one makes use of the inhibition of the latter by 2 m$M$ cyanide in the assay medium.[9]

[23] A. Rigo, P. Viglino, and G. Rotilio, *Anal. Biochem.* **68,** 1 (1975).

[24] H. Theorell, *in* "The Enzymes" (J. B. Sumner and K. Myrback, eds.), Vol. II. Pt. 1, p. 397. Academic Press, New York, 1951.

[25] E. Finkelstein, G. M. Rosen, S. E. Patton, M. S. Cohen, and E. J. Rauckman, *Biochem. Biophys. Res. Commun.* **102,** 1008 (1981).

## [11] Subcellular Distribution of Superoxide Dismutases in Rat Liver

*By* BRUCE L. GELLER and DENNIS R. WINGE

Investigation of the distribution of superoxide dismutases (SOD) among various cells and tissues has inevitably led to probing the intracellular location of these enzymes. Eukaryotic cells contain two superoxide dismutases which have no apparent biochemical similarities except their abilities to catalyze the dismutation of the superoxide anion.[1] The manganese-containing enzyme (Mn-SOD) appears to be exclusively located in the soluble matrix of the mitochondria,[1-4] whereas the copper, zinc-containing enzyme (Cu,Zn-SOD) is primarily located in the cytosol. Mn-SOD has been reported to be located in the cytosol of baboon liver cells together with the Cu,Zn-enzyme,[5] although rigorous confirmation of this observation is lacking. Cu,Zn-SOD activity has also been suggested to be present in nuclei,[6] chloroplasts,[7-11] and mitochondria.[2-4,12-14] However, the mitochondrial location of Cu,Zn-SOD in rat liver was subsequently found to be attributable to contaminating lysosomes.[15] The latter finding not only clarified a misconception of the subcellular location of Cu,Zn-SOD, but also pointed out the need for a thorough reevaluation of the

[1] R. A. Weisiger and I. Fridovich, *J. Biol. Chem.* **248**, 3582 (1973).
[2] R. A. Weisiger and I. Fridovich, *J. Biol. Chem.* **248**, 4793 (1973).
[3] D. D. Tyler, *Biochem. J.* **147**, 493 (1975).
[4] C. Peeters-Joris, A. Vandevoorde, and P. Baudhuin, *Biochem. J.* **150**, 31 (1975).
[5] J. M. McCord, J. A. Boyle, E. D. Day, Jr., L. G. Rizzolo, and M. L. Salin, *in* "Proceedings of the First EMBO Workshop on Superoxide and Superoxide Dismutases" (A. M. Michelson, J. M. McCord, and I. Fridovich, eds.), p. 129. Academic Press, New York, 1978.
[6] A. Petkau, W. S. Chelack, K. Kelly, and H. G. Friesen, *in* "Oxygen Induced Pathology" (A. Autor, ed.), p. 45. Academic Press, New York, 1979.
[7] K. Asada, M. Urano, and M. Takahashi, *Eur. J. Biochem.* **36**, 257 (1973).
[8] J. F. Allen and D. O. Hall, *Biochem. Biophys. Res. Commun.* **52**, 856 (1973).
[9] B. Halliwell, *Eur. J. Biochem.* **55**, 355 (1975).
[10] C. Jackson, J. Dench, A. L. Moore, B. Halliwell, C. H. Foyer, and D. O. Hall, *Eur. J. Biochem.* **91**, 339 (1978).
[11] J. A. Baum and J. G. Scandalios, *Differentiation* **13**, 133 (1979).
[12] L. F. Panchenko, O. S. Brusov, A. M. Gerasimov, and T. D. Loktaeva, *FEBS Lett.* **55**, 84 (1975).
[13] G. P. Arron, L. Henry, J. M. Palmer, and D. O. Hall, *Biochem. Soc. Trans.* **4**, 618 (1976).
[14] L. E. A. Henry, R. Cammack, J. Schwitzguebel, J. M. Palmer, and D. O. Hall, *Biochem. J.* **187**, 321 (1980).
[15] B. L. Geller and D. R. Winge, *J. Biol. Chem.* **257**, 8945 (1982).

Copyright © 1984 by Academic Press, Inc.
All rights of reproduction in any form reserved.
ISBN 0-12-182005-X

putative subcellular distribution of Cu,Zn-SOD in nuclei and chloroplasts and Mn-SOD in the cytosol of primate liver cells.

The problems encountered in subcellular fractionation are not new to cell biology, and still pose a major hurdle in practical experimental procedures. Differential centrifugation has been the classic approach to subcellular fractionation, but this procedure is not adequate in resolving certain organelles such as mitochondria, peroxisomes, and lysosomes. Isopycnic gradient centrifugation has been combined with differential centrifugation to successfully resolve certain organelles.[16,17] We have used a variety of these methods to fractionate rat liver cells to distinguish lysosomal and mitochondrial superoxide dismutases.

### Subcellular Fractionation

*Materials*

> Homogenizing buffer: 0.25 $M$ sucrose (ultra pure quality), 10 m$M$ Tris–Cl, 1 m$M$ EDTA, pH 7.4
> Percoll (Pharmacia)
> 2.5 $M$ Sucrose (ultra pure) in water
> 25% Sucrose (w/w), 5% Dextran T-10 (w/v), 20 m$M$ Tris–Cl, 1 m$M$ EDTA, pH 7.4
> 45% Sucrose (w/w), 5% Dextran T-10 (w/v), 10 m$M$ Tris–Cl, 1 m$M$ EDTA pH 7.4
> Triton WR-1339, 200 mg/ml in water

*Procedure*

Rats are fasted overnight and then decapitated. The liver is perfused *in situ* with 75 ml homogenizing buffer at 4° by single pass injection through the portal vein and/or vena cava. This removes most of the red blood cells, and the resulting liver appears light tan in color. All further steps were done on ice. The liver is rinsed in homogenizing buffer, pat-dried with a paper towel, weighed, minced with scissors in 3 volumes (w/v) homogenizing buffer, and homogenized in a Potter–Elvehjem tissue grinder according to the following procedure: four strokes with a "loose" pestle (0.026-in. clearance between the pestle and mortar) followed by four strokes with a "tight" pestle (0.012 in. clearance), where each stroke is an up and down motion of 10 sec duration with the pestle rotating at 450

---

[16] F. Leighton, B. Poole, H. Beaufay, P. Baudhuin, J. W. Coffey, S. Fowler, and C. deDuve, *J. Cell Biol.* **37**, 482 (1968).

[17] S. Fleischer and M. Kervina, this series, Vol. 31, p. 6.

DISTRIBUTION OF MARKER ENZYMES IN THE SUBCELLULAR FRACTIONS[a]

| Fraction | Sulfite oxidase (mitochondrial intermembrane space) | Glutamate dehydrogenase (mitochondrial matrix) | Catalase (peroxisomes) | NADPH–cytochrome c reductase (microsomes) | Galactosyl-transferase (Golgi) | Acid phosphatase (lysosomes) | β-Glucos-aminidase (lysosomes) | Cu,Zn-superoxide dismutase | Mn-superoxide dismutase |
|---|---|---|---|---|---|---|---|---|---|
| Homogenate | 1.00 | 1.00 | 1.00 | 1.00 | — | 1.00 | — | 1.00 | 1.00 |
| 300 g supernatant | 0.64 | 0.74 | 0.64 | 0.85 | 1.00* | 0.90 | 1.00* | 0.79 | 0.81 |
| 10,000 g supernatant | 0.13 | 0.00 | 0.32 | 0.59 | 0.46 | 0.36 | 0.21 | 0.79 | 0.05 |
| 10,000 g pellet | 0.39 | 0.76 | 0.48 | 0.27 | 0.52 | 0.34 | 0.66 | 0.03 | 0.86 |

[a] The subcellular fractions were produced by differential centrifugation. All values are normalized to the homogenate value = 1.00, except *, which is normalized to the 300 g supernatant.

rpm.[17] The homogenate is diluted with another 6 volumes of homogenizing buffer and centrifuged at 300 $g$ for 15 min. The supernatant is carefully removed by pipet and saved, whereas the pellet is washed twice by resuspending it in 5 volumes of homogenizing buffer using four strokes of the loose pestle and centrifuging at 300 $g$ for 15 min. The percentage of cells ruptured based on marker enzyme activity in the crude homogenate and the 300 $g$ supernatant averaged 75%. The combined supernatants are centrifuged at 10,000 $g$ for 15 min. The supernatant is decanted and the pellet resuspended by gentle stirring with a "cold finger" (a test tube filled with ice water) to five times the pellet volume with homogenizing buffer. The homogenizing buffer must be added in small aliquots (0.5 ml) to obtain a homogeneous suspension. The resuspended pellet is centrifuged at 10,000 $g$ for 15 min, and the last washing step repeated. The extent of rupture of the fragile outer mitochondrial membrane based on sulfite oxidase activity was about 20%, whereas the inner mitochondrial membrane was less than 1% disrupted based on glutamate dehydrogenase activity. The resulting pellet, referred to in this text as the "10k pellet," contains nearly all the mitochondria, as well as a portion of the lysosomes, peroxisomes, microsomes, and Golgi, which are listed in the table.[18] Centrifugation of the 10k supernatant at 100,000 $g$ yields a supernatant containing the cytosolic Cu,Zn-SOD.

*Percoll Gradient*

Percoll was diluted to a final density of 1.10 g/ml in an isotonic solution of 0.25 $M$ sucrose by mixing 52.6 ml of Percoll (stock density = 1.13 g/ml), 10.0 ml of stock 2.5 $M$ sucrose, 1 ml of 2 $M$ Tris base, and 1 ml of 0.1 $M$ EDTA. The mixture was adjusted to pH 7.4 by dropwise addition of 0.3 $N$ HCl and then diluted to a final volume of 100 ml. A volume of 30 ml per tube was centrifuged at 39,000 $g$ for 20 min to form a nonlinear, sigmoidal gradient. Up to 1.5 ml of the resuspended 10k pellet is layered on top of the gradient and centrifuged again at 48,000 $g$ for 30 min. The gradient is

---

[18] All of the enzyme assays listed in this text are described in Ref. 15.

---

FIG. 1. Isopycnic gradient of 10k pellet. The 10k pellet from rat liver, prepared by differential centrifugation, was centrifuged on a Percoll density gradient. The fractions were assayed for ○, sulfite-cytochrome $c$ oxidoreductase (mitochondrial intermembrane space); ▲, Mn-SOD; ×, acid phosphatase (lysosomes); ●, Cu,Zn-SOD; ■, catalase (peroxisomes); ▽, NADPH–cytochrome $c$ oxidoreductase (microsomes); □, galactosyltransferase (Golgi); △, 5'-nucleotidase (plasma membrane); ▶, hemoglobin (erythrocytes). Enzyme values are expressed as percentage total activity on the entire gradient, except Cu,Zn-SOD and Mn-SOD which are in units/milliliter.

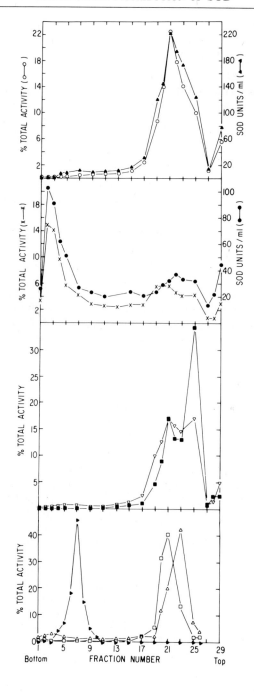

fractionated by inserting a tube through the gradient and pumping out the material from the bottom. The fractions are assayed for various subcellular marker enzymes, and a typical result is shown in Fig. 1. The Cu,Zn-SOD appeared to parallel the two lysosomal fractions. The bimodal distribution of lysosomes on colloidal silica density gradients has been reported by others.[19,20] Because the more buoyant lysosomes segregated coincidently with mitochondria, microsomes, peroxisomes, and Golgi, it remained unresolved whether the Cu,Zn-SOD associated with this peak was lysosomal.

### Sucrose Gradient

In an attempt to separate the lysosomes from the other particles, rats were intraperitoneally injected once with 85 mg Triton WR-1339 per 100 g body weight, 3–4 days prior to sacrifice, according to the method of Wattiaux et al.,[21] as described by Leighton et al.[16] The livers were excised and the 10k pellet was obtained by the procedure already described above. A 0.5-ml aliquot of the 10k pellet was then layered on top of a 10.8 ml linear gradient of 25 to 45% sucrose, as described by Fleischer and Kervina[17] (alternatively, a Percoll gradient of starting density 1.07 may be used, formed as described above, except 29.5 ml of Percoll per 100 ml is used). The sucrose solutions also contained 5% Dextran T-10, 20 m$M$ Tris–HCl, 1 m$M$ EDTA, pH 7.4. The gradient and 10k pellet were centrifuged in a swinging bucket-type rotor with average radius of 110 mm for a total of $9.73 \times 10^{10}$ radians$^2$/sec (for instance, Beckman SW-41 Ti at 36,000 rpm for 112 min). The gradient was fractionated as described for the Percoll gradient. The result, shown in Fig. 2, is a nearly total separation of mitochondria from the lysosomes which contain the Cu,Zn-SOD. The latter particles are now much less dense than all other subcellular particles due to the sequestration of the Triton WR-1339. Note the slight, but reproducible, partial resolution of mitochondria and peroxisomes. The pattern of the Mn-SOD is consistent with prior evidence[1-4] that this enzyme is located in mitochondria, and further eliminates the possibility that Mn-SOD was actually located in contaminating peroxisomes or lysosomes in the crude mitochondria derived from differential centrifugation alone.[1,2]

[19] L. H. Rome, A. J. Garvin, M. M. Allietta, and E. F. Neufeld, *Cell* **17**, 143 (1979).
[20] C. A. Surmacz, J. J. Wert, Jr., and G. E. Mortimore, *Am. J. Physiol.,* in press (1983).
[21] R. Wattiaux, M. Wibo, and P. Baudhuin, *in* "Ciba Foundation Symp; Lysosomes" (A. V. S. de Reuck and M. P. Cameron, eds.), p. 176. Little, Brown, Boston, Massachusetts, 1963.

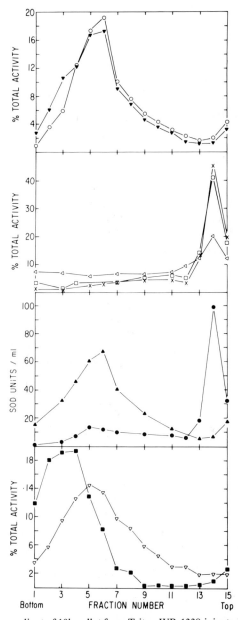

Fig. 2. Isopycnic gradient of 10k pellet from Triton WR-1339-injected rats. The 10k pellet from rats previously injected with Triton WR-1339 was centrifuged on a sucrose density gradient. The fractions were assayed for ○, sulfite–cytochrome $c$ oxidoreductase; ▼, succinate–cytochrome $c$ oxidoreductase (mitochondrial inner membrane); ×, acid phosphatase; □, $\beta$-galactosidase (lysosomes); ◁, $\beta$-glucuronidase (lysosomes); ▲, Mn-SOD; ●, Cu,Zn-SOD; ■, catalase; ▽, NADPH–cytochrome $c$ oxidoreductase. All values are expressed as in Fig. 1.

Digitonin Fractionation

The data from the above gradients are the basis for the localization of Cu,Zn-SOD in lysosomes. Other techniques such as digitonin fractionation,[22] which can distinguish between lysosomal enzymes, and other subcellular particle enzymes have been used to corroborate this line of evidence.[15]

*Materials*

Commercial digitonin is recrystallized by dissolving 1 g digitonin in 50 ml of boiling ethanol, cooling, filtering, and drying the precipitate. A 10% solution of digitonin in homogenizing buffer is solubilized by heating in a boiling water bath. The solution will remain stable for hours when cooled to room temperature.

*Procedure*[23]

Aliquots are diluted with homogenizing buffer and varying volumes of the 10% digitonin such that the final protein concentration is 50 mg/ml. The mixtures are incubated on ice for 15 min with gentle mixing every 2 min. Three volumes of homogenizing buffer are then added to each mixture and gently mixed. These solutions are centrifuged at 10,000 g for 10 min followed by removal of the supernatant. Both the supernatant and the resuspended pellet are used for marker enzyme assays.

The results of a typical digitonin titration experiment are shown in Fig. 3. The lysosomal enzyme, acid phosphatase, is released into the supernatant at a much lower digitonin to protein concentration ratio than that for enzymes from the mitochondrial intermembrane space and matrix. Cu,Zn-SOD is released at the same digitonin-to-protein concentration ratio, and to the same extent as lysosomal acid phosphatase. In similar experiments, the peroxisomal enzyme, catalase, was found to be released into the supernatant at about the same digitonin-to-protein concentration ratio as the mitochondrial intermembrane space enzyme, sulfite oxidase. These data further substantiate the lysosomal location of the Cu,Zn-SOD and the mitochondrial matrix location of the Mn-SOD.

Quantitation of Superoxide Dismutases

During these studies, we encountered a technical problem in quantitating Cu,Zn-SOD activity in the 10k pellet which contains Mn-SOD as the

[22] M. Levy, R. Toury, and J. Andre, *C. R. Soc. Biol., Ser.* D **262,** 1593 (1966).
[23] C. Schnaitman, V. G. Erwin, and J. W. Greenawalt, *J. Cell Biol.* **32,** 719 (1967).

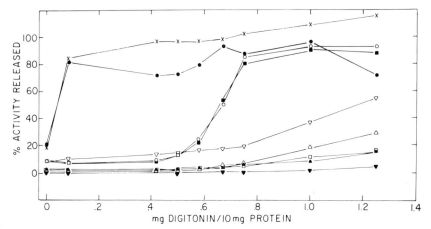

FIG. 3. Digitonin treatment of the crude mitochondria–lysosome fraction. The large granule fraction from rat liver was incubated with increasing concentrations of digitonin. The supernate and granules were then separated by centrifugation and assayed for the following: ×, acid phosphatase; ○, sulfite–cytochrome $c$ oxidoreductase; ■, adenylate kinase (mitochondrial intermembrane space); △, glutamate dehydrogenase (mitochondrial matrix); □, malate dehydrogenase (mitochondrial matrix); ▼, succinate–cytochrome $c$ oxidoreductase; ▲, Mn-SOD; ▽, rotenone-insensitive NADH–cytochrome $c$ oxidoreductase (mitochondrial outer membrane); ●, Cu,Zn-SOD. The values are expressed as percentage of total activity per mixture released into the supernate.

predominant superoxide dismutase. We developed a procedure to precisely assay the amount of Cu,Zn-SOD in such samples.[24] The method involves the inactivation of Mn-SOD by sodium dodecyl sulfate, a condition which does not affect Cu,Zn-SOD. The sodium dodecyl sulfate is then removed and the sample is assayed for activity with the xanthine oxidase–cytochrome $c$ method described by McCord and Fridovich.[25] The precision of the assay is superior to that of other methods, such as the cyanide subtraction method,[26] when the dismutase activity is predominantly Mn-SOD. This makes the method particularly useful for assaying subcellular fractions which contain the bulk of the mitochondria.

*Procedure*

The sample to be assayed is diluted to less than 9 mg protein/ml with 10–100 m$M$ Tris–Cl, pH 7.4, and 0.8 ml of this is mixed with 0.2 ml of 10% (w/v) sodium dodecyl sulfate in water. The mixture is incubated at 37° for

[24] B. L. Geller and D. R. Winge, *Anal. Biochem.* **128,** 86 (1983).
[25] J. M. McCord and I. Fridovich, *J. Biol. Chem.* **244,** 6049 (1969).
[26] J. D. Crapo, J. M. McCord, and I. Fridovich, this series, Vol. 53, p. 382.

30 min or longer, then cooled to 4°. Following this, 0.1 ml of 3 $M$ KCl in water is added and mixed vigorously. After 30 min at 4°, the mixture is centrifuged at 20,000 $g$ for 10 min. The supernatant is carefully removed and assayed for superoxide dismutase activity using the xanthine oxidase–cytochrome $c$ method, but including a final concentration of 50 $\mu M$ cyanide in the cuvette to inhibit any cytochrome oxidase activity.[26] At 50 $\mu M$ cyanide the Cu,Zn-enzyme is inhibited by 13%. The Mn-SOD activity is measured by assaying[25,26] the sample before sodium dodecyl sulfate treatment, in the presence of a final cyanide concentration of 2 m$M$, which inhibits the Cu,Zn-enzyme by 94%, but does not affect the Mn-enzyme.

Using these two measurements, the concentration of each superoxide dismutase activity is calculated using Eqs. (1) and (2):

$$x = \frac{\text{units Cu,Zn-SOD}}{\text{ml}} = \frac{(a)(0.8 \text{ ml} + 0.2 \text{ ml} + 0.1 \text{ ml})}{(1.00 - 0.13)(0.8 \text{ ml})(\text{dilution of sample})} \quad (1)$$

where $a$ is the measured activity of the SDS-treated sample at 50 $\mu M$ CN$^-$, 1.00 − 0.13 compensates for the partial inhibition of activity using 50 $\mu M$ CN$^-$, and the other factors adjust for the dilutions made;

$$y = \frac{\text{units Mn-SOD}}{\text{ml}} = b - (0.06)(x) \quad (2)$$

where $b$ is the measured activity of the untreated sample at 2 m$M$ CN$^-$, and $(0.06)(x)$ is the contribution of Cu,Zn-SOD activity at 2 m$M$ CN$^-$.

# [12] Assays of Glutathione Peroxidase

*By* LEOPOLD FLOHÉ and WOLFGANG A. GÜNZLER

The term glutathione peroxidase (glutathione : $H_2O_2$ oxidoreductase, EC 1.11.1.9) is reserved for the selenoprotein catalyzing the reaction:

$$\text{ROOH} + 2 \text{ GSH} \rightarrow \text{GSSG} + \text{ROH} + H_2O$$

It should not be applied to other proteins (e.g., glutathione $S$-transferases) exhibiting some overlap in activity.[1,2]

[1] L. Flohé, *in* "Free Radicals in Biology" (W. A. Pryor, ed.), Vol. V, p. 223. Academic Press, New York, 1982.
[2] L. Flohé, W. A. Günzler, and G. Loschen, *in* "Trace Metals in Health and Disease" (N. Kharasch, ed.), p. 263. Raven, New York, 1979.

Copyright © 1984 by Academic Press, Inc.
All rights of reproduction in any form reserved.
ISBN 0-12-182005-X

In order to determine glutathione peroxidase reliably, some factors of potential pitfall have to be considered, e.g., enzymatic side reactions of substrates (especially when crude tissue samples are assayed), high and variable spontaneous reaction rates of substrates, and the peculiar kinetics of the enzyme itself. With the best documented example, the enzyme of bovine red blood cells, ping-pong kinetics with infinite limiting maximum velocities and Michaelis constants have been established.[3,4] This means that the generally recommended conditions for determination of enzyme activity, i.e., "saturating" concentrations of all substrates, cannot possibly be fulfilled. In consequence, compromises are inevitable in the choice of substrate concentration for the assay and in the definition of the unit of activity.

### Fixed-Time Assay Measuring GSH Consumption

*Principle*

Enzyme-catalyzed or spontaneous reactions of GSH with $H_2O_2$ are stopped at fixed time ($t$) by addition of strong acid and the GSH content [GSH] is measured subsequently by polarography.

*Equipment*

> Waterbath 37°
> Recording polarograph; electrode: dropping mercury; reference electrode: $Hg/HgSO_4/0.1\ M\ K_2SO_4$.

*Reagents*

> 0.1 $M$ potassium phosphate, pH 7.0, 1 m$M$ in EDTA
> 4 m$M$ GSH in water; store refrigerated
> 5 m$M$ $H_2O_2$, prepared from a commercial solution (30%)
> 1.18 $M$ $HClO_4$, prepared from a commercial solution (70%)
> Enzyme sample, usually 0.5–3 $U_k$/ml buffer

*Procedure*

Assay A: A mixture containing 500 $\mu$l of each, 0.1 $M$ potassium phosphate buffer, buffered enzyme sample, and 4 m$M$ GSH, is preincubated for 10 min at 37°. The reaction is initiated by adding 500 $\mu$l of 5 m$M$ $H_2O_2$

[3] L. Flohé, G. Loschen, W. A. Günzler, and E. Eichele, *Hoppe-Seylers Z. Physiol. Chem.* **353**, 987 (1972).
[4] W. A. Günzler, H. Vergin, I. Müller, and L. Flohé, *Hoppe-Seylers Z. Physiol. Chem.* **353**, 1001 (1972).

prewarmed to 37°. Precisely after the incubation time (*t*), usually 60 sec, the reaction is stopped by vigorously injecting a total of 2.0 ml $HClO_4$. Enzymatic and nonenzymic GSH oxidation of enzyme samples is measured by this assay. The following assays are prepared for each set of experiments.

Assay B: As compared to assay A, buffer is used instead of enzyme sample and $HClO_4$ is added to the mixture prior to incubation and addition of $H_2O_2$. The GSH content at zero time is assessed by this assay.

Assay C: As compared to assay A, buffer is used instead of enzyme sample and $H_2O$ instead of GSH. This assay provides a reference for zero GSH content.

Assay D: As compared to assay A, only the enzyme sample is replaced by buffer. The spontaneous oxidation of GSH by $H_2O_2$ is assessed. Under the conditions given here the nonenzymic reaction rate is about 8% decrease in [GSH] per minute.

### Determination of GSH

The acidified assay mixtures are cooled to room temperature and deaerated by bubbling with nitrogen for 3 min. Polarograms are recorded between 0 and $-0.4$ V versus the mercuric sulfate reference electrode. The GSH contents of the assays A, B, and D, $[GSH]_T$, $[GSH]_0$, and $[GSH]_s$, respectively, are determined as functions of diffusion-limited currents at about $-0.3$ V using assay C for generation of a baseline.

### Definition of Units and Calculation of Activity

Under the assay conditions described, the reaction rate is independent of the concentration of $H_2O_2$, but directly proportional to the concentration of GSH. This means that GSH decreases according to first-order kinetics, and expressing the results as $\Delta[GSH]$ per time would be grossly misleading. Instead, $\Delta\log[GSH]$ per time properly reflects the enzymatic activity and, correspondingly, the amount of glutathione peroxidase inducing a net GSH decrease to 10% of the initial concentration in 1 min at 37° and pH 7.0 is defined as 1 unit ($U_k$):

$$U_k = \log([GSH]_0/[GSH])/t \qquad (1)$$

Accordingly, Eq. (2) is used for calculating the activity (*A*) of glutathione peroxidase in the enzyme sample:

$$A = \{\log([GSH]_0/[GSH]_T) - \log([GSH]_0/[GSH]_S)\}(V_i/V_s)/t \qquad (2)$$

The incubation time (*t*) is given in minutes; $[GSH]_0$, $[GSH]_T$, and $[GSH]_s$ are the respective GSH concentrations as determined polarographically and $V_i/V_s$ stands for the dilution of enzyme sample ($V_s$ = 0.5 ml) in the incubation mixture ($V_i$ = 2 ml).

### Comments

Due to interference of many inorganic (e.g., halides) and organic compounds with the polarographic determination of GSH the method is almost inapplicable to crude biological samples. However, it is a fast, reliable, and inexpensive procedure particularly suited for rapid screening of large numbers of samples during purification processes. The results are not seriously affected by a minor or even moderate degree of decomposition of the unstable substrates. The method is equally applicable to thiols other than GSH and therefore is useful, if the specificity of glutathione peroxidase is to be investigated.[5]

### Fixed-Time Assay Measuring $H_2O_2$ Consumption

### Principle

At fixed-time intervals the glutathione peroxidase reaction is stopped by acidification, unmetabolized GSH scavenged by mercuric salts, the solution neutralized, and the remaining $H_2O_2$ reacted with the fluorescent dye scopoletin in the presence of horseradish peroxidase.

### Comments

This procedure has been successfully used for the evaluation of the kinetics of glutathione peroxidase in combination with rapid flow techniques,[3] but is far too complicated for routine measurements and, therefore, shall not be described here in detail. However, this or similar principles have to be considered, if $H_2O_2$ at concentration below 1 $\mu M$ is to be determined at 10- to 100-fold excess GSH.

### Continous Monitoring of GSSG Formation

### Principle

GSSG formed during glutathione peroxidase reaction is instantly and continuously reduced by an excess of glutathione reductase activity providing for a constant level of GSH. The concomitant oxidation of NADPH is monitored photometrically.

[5] L. Flohé, W. A. Günzler, G. Jung, E. Schaich, and F. Schneider, *Hoppe-Seylers Z. Physiol. Chem.* **353**, 159 (1971).

*Equipment*

Photometer equipped with recorder and thermostated semi-micro cuvettes (1 ml).

*Reagents*

0.1 $M$ potassium phosphate buffer, pH 7.0; 1 m$M$ in EDTA

2.4 U/ml glutathione reductase (from yeast), prepared from a commercial crystal suspension (5 mg/ml; 120 U/mg): 20 $\mu$l suspension added to 5 ml 0.1 $M$ phosphate buffer pH 7. Prepare daily and keep cool

10 m$M$ GSH in water. Since the concentration is crucial, prepare carefully. If in doubt, check content using Ellman's reagent[6]

1.5 m$M$ NADPH in 0.1% NaHCO$_3$

12 m$M$ $t$-butyl hydroperoxide, prepared from a commercial solution (80%)

1.5 m$M$ H$_2$O$_2$, prepared from a commercial solution (30%)

Enzyme sample, approximately 0.05–1 U$_k$/ml buffer

*Procedure*

The following solutions are pipetted into a semi-micro cuvette: 500 ml 0.1 $M$ phosphate buffer (pH 7.0), 100 $\mu$l enzyme sample, 100 $\mu$l glutathione reductase (0.24 U), and exactly 100 $\mu$l of 10 m$M$ GSH. The mixture is preincubated for 10 min at 37°. Thereafter, 100 $\mu$l NADPH solution is added and the hydroperoxide-independent consumption of NADPH is monitored for about 3 min. The overall reaction is started by adding 100 $\mu$l of prewarmed hydroperoxide solution and the decrease in absorption at 340 (or 365) nm is monitored for about 5 min. The nonenzymic reaction rate is correspondingly assessed by replacing the enzyme sample by buffer.

*Calculation of Activity*

The decrease in NADPH concentration, $\Delta$[NADPH]/min, is calculated from the linear slopes of decreasing absorption using the appropriate extinction coefficient. The glutathione peroxidase-dependent reaction rate is obtained, when the nonenzymic and (in some cases) hydroperoxide-independent effects are subtracted from the overall reaction rate.

As GSH is regenerated continuously by glutathione reductase the concentration of GSH in the assay is kept at the initial level [GSH]$_0$. Therefore, the reaction of glutathione peroxidase proceeds according to kinet-

[6] H. Wenck, E. Schwabe, F. Schneider, and L. Flohé, *Z. Anal. Chem.* **258,** 267 (1972).

ics of pseudo-zero order. Despite the linear decrease in NADPH, however, results must not be expressed as substrate consumed per time without quoting the GSH level at which the reaction had proceeded. When, instead, the rate of substrate consumption is referred to $[GSH]_0$, results can be compared even to those obtained with other test systems. Accordingly, the unit of glutathione peroxidase activity defined by Eq. (1) can be adopted to the conditions of this procedure as follows:

$$U_k = (\Delta \log[GSH]/t) = 0.434(\Delta \ln[GSH]/t)$$
$$= (0.434/[GSH]_0)(d[GSH]/dt) \tag{3}$$

The term $d[GSH]/dt$ is equal to the apparent GSH turnover under regenerating conditions and, with regard to the stoichiometry of the reaction, can be replaced by twice the rate of decrease in NADPH concentration $\Delta[NADP]/t$. Considering $V_i$ and $V_s$, the volumes of the incubation mixture and the enzyme sample, respectively, the activity $(A)$ in the enzyme sample tested under the conditions described is calculated according to Eq. (4):

$$A = 0.868(\Delta[NADPH]/[GSH]_0 t)(V_i/V_s) \tag{4}$$

*Comments*

As the reaction rate depends on the steady-state level of GSH, any factor influencing GSH regeneration, e.g., by significantly decreasing glutathione reductase activity, will affect the determination. However, notwithstanding its susceptibility to disturbance, the coupled test system described is the method of choice for determination of glutathione peroxidase activity in biological material.

In order to illustrate how to cope with interfering factors present in crude biological material, two modifications of assay conditions are proposed.

*Modification for Liver*

As *t*-butyl hydroperoxide is a substrate for some glutathione *S*-transferases, $H_2O_2$ should be used in the test of tissue homogenate, e.g., from rat liver. In this case, however, catalase has to be blocked by the addition of sodium azide (1 m$M$ in the assay).

*Modification for Red Blood Cells*

As even traces of methemoglobin will falsify the determination of glutathione peroxidase in hemolysate, the conversion of hemoglobin into

COMPARISON OF ASSAY PROCEDURES FOR GLUTATHIONE PEROXIDASE

| Procedure | Buffer, pH | Temperature | $[GSH]_0$ (mM) | Definition of unit | Conversion factor[a] |
|---|---|---|---|---|---|
| This procedure | Phosphate, 7.0 | 37° | 1.0 | $\Delta\log[GSH]$/min | (1.00) |
| Temperature modified | | 25° | | | 2.25 |
| Tappel[8] | Tris–HCl, 7.6 | 37° | 0.23 | $\Delta\ \mu$mol NADPH/min | 1.70 |
| Awasthi et al.[9] | Phosphate, 7.0 | 37° | 4.0 | $\Delta\ \mu$mol GSH/min | 0.17 |
| Nakamura et al.[10] | Phosphate, 7.0 | 25° | 1.0 | $\Delta\ \mu$mol GSH/min | 0.53 |

[a] Factor for conversion of results obtained with the procedures cited into empirically equivalent activity data (in $U_k$) obtained with the procedure described here.

cyanomethemoglobin prior to assay is recommended.[7] The hemolysate is treated with a 1.2-fold excess of hexacyanoferrate(III) and a 12-fold excess of cyanide over heme concentration. t-Butyl hydroperoxide is favorably used in this assay.

## Comparison of Assay Procedures

The main differences between the assay procedure described here and those proposed by others are listed in the table. In order to compare results obtained by different procedures, appropriate empirical converting factors are given.

### General Notes of Precaution

As reversible inactivation can occur in glutathione peroxidase samples, the assay mixture should be preincubated in the presence of GSH prior to incubation with hydroperoxide.

The pH of incubation must be carefully controlled. Incubation at pH of maximal enzymic reaction, at about pH 8.7,[3] is not feasible due to an extremely high spontaneous reaction rate of GSH and hydroperoxides.

[7] W. A. Günzler, H. Kremers, and L. Flohé, Z. Klin. Chem. Klin. Biochem. 12, 444 (1974).
[8] A. L. Tappel, this series, Vol. 52, p. 506.
[9] Y. C. Awasthi, E. Beutler, and S. K. Srivastava, J. Biol. Chem. 250, 5144 (1975).
[10] W. Nakamura, S. Hosoda, and K. Hayashi, Biochim. Biophys. Acta 358, 251 (1974).

High concentrations of polyvalent anions, which may be contained in test samples or in the batch of glutathione reductase used, can inhibit glutathione peroxidase.

As differences in enzymic properties, e.g., in kinetics, of glutathione peroxidase can not be excluded in different tissues or species, optimal substrate concentrations, if in doubt, should be reassessed.

If the activity measured in crude samples is to be assigned to glutathione peroxidase unequivocally, a separation step and characterization, e.g., by correlating selenium content and enzymic activity during fractionation, should be considered.

# [13] Catalase *in Vitro*

## By Hugo Aebi

$$2H_2O_2 \xrightarrow{\text{catalase}} 2H_2O + O_2 \tag{1}$$

$$ROOH + AH_2 \xrightarrow{\text{catalase}} H_2O + ROH + A \tag{2}$$

Catalase exerts a dual function: (1) decomposition of $H_2O_2$ to give $H_2O$ and $O_2$ [catalytic activity, Eq. (1)] and (2) oxidation of H donors, e.g., methanol, ethanol, formic acid, phenols, with the consumption of 1 mol of peroxide [peroxidic activity, Eq. (2)].

## Kinetic Properties

The predominating reaction depends on the concentration of H donor and the steady-state concentration or rate of production of $H_2O_2$ in the system. In both cases the active catalase–$H_2O_2$ complex I is formed first. The decomposition of $H_2O_2$, in which a second molecule of $H_2O_2$ serves as H donor for complex I, proceeds exceedingly rapidly (rate constant $k \sim 10^7$ liters mol$^{-1}$ sec$^{-1}$) whereas peroxidative reactions proceed relatively slowly ($k \sim 10^2$–$10^3$).[1]

The kinetics of catalase do not obey the normal pattern. On the one hand it is not possible to saturate the enzyme with "substrate" within the feasible concentration range (up to 5 $M$ $H_2O_2$), and on the other there is a rapid inactivation of catalase at $H_2O_2$ concentrations above 0.1 $M$, when the active enzyme–$H_2O_2$ complex I is converted to the inactive complexes II or III. Measurements of enzyme activity at substrate saturation

[1] B. Chance, *Acta Chem. Scand.* **1**, 236 (1947).

Copyright © 1984 by Academic Press, Inc.
All rights of reproduction in any form reserved.
ISBN 0-12-182005-X

METHODS IN ENZYMOLOGY, VOL. 105

or determination of the $K_s$ is therefore impossible. In contrast to reactions proceeding at substrate saturation, the enzymic decomposition of $H_2O_2$ is a first-order reaction, the rate of which is always proportional to the peroxide concentration present. Consequently, to avoid a rapid decrease in the initial rate of the reaction, the assay must be carried out with relatively low concentrations of $H_2O_2$ (about 0.01 $M$).[2] As the activation energy for the decomposition of $H_2O_2$ catalyzed by catalase is very low (2500–7100 kJ/mol), there is only slight dependence on temperature ($Q_{10}$ = 1.05–1.12). The decomposition of $H_2O_2$ initially (approx. 0–30 sec) follows that of a first-order reaction with $H_2O_2$ concentration between 0.01 and 0.05 $M$. The rate constant ($k$) for the overall reaction is given by

$$k = (1/\Delta t)(\ln S_1/S_2) = (2 \cdot 3/\Delta t)(\log S_1/S_2) \tag{3}$$

where $\Delta t = t_2 - t_1$ = measured time interval and $S_1$ and $S_2$ = $H_2O_2$ concentrations at times $t_1$ and $t_2$. The constant $k$ can be used as a direct measure of the catalase concentration. In studies with purified enzyme preparations the specific activity ($k_1'$) is obtained by dividing $k$ by the molar concentration of catalase ($e$).

$$k_1' = k/e \qquad \text{(liters mol}^{-1} \text{ sec}^{-1}) \tag{4}$$

The value for $k_1'$ for pure catalase from human erythrocytes is $3.4 \times 10^7$ (liters mol$^{-1}$ sec$^{-1}$). This value is used to calculate the absolute content of enzyme in blood and tissues.[3]

## Assay Method

### Principle

In the ultraviolet range $H_2O_2$ shows a continual increase in absorption with decreasing wavelength. The decomposition of $H_2O_2$ can be followed directly by the decrease in absorbance at 240 nm ($\varepsilon_{240}$ = 0.00394 ± 0.0002 liters mmol$^{-1}$ mm$^{-1}$).[4] The difference in absorbance ($\Delta A_{240}$) per unit time is a measure of the catalase activity.

To avoid inactivation of the enzyme during the assay (usually 30 sec) or formation of bubbles in the cuvette due to the liberation of $O_2$, it is necessary to use a relatively low $H_2O_2$ concentration (10 m$M$). The $H_2O_2$ concentration is critical inasmuch as there is direct proportionality between the substrate concentration and the rate of decomposition. Due to the special situation in catalase the dependence of the $H_2O_2$ decomposi-

[2] R. K. Bonnichsen, B. Chance, and H. Theorell, *Acta Chem. Scand.* **1**, 685 (1947).
[3] R. K. Bonnichsen, this series, Vol. II, p. 781.
[4] D. P. Nelson and L. A. Kiesow, *Anal. Biochem.* **49**, 474 (1972).

tion on the temperature is small ($Q_{10} \sim 1.1$) so that measurements can be carried out between 0 and 37°; however, 20° is recommended. The pH activity curve relative to $V_0$ has a fairly broad pH optimum (pH 6.8–7.5): measurements are made at pH 7.0.[5]

### Reagents

Phosphate buffer 50 m$M$, pH 7.0: dissolve (a) 6.81 g $KH_2PO_4$, and (b) 8.90 g $Na_2HPO_4 \cdot 2H_2O$ in distilled water and make up to 1000 ml each. Mix solutions (a) and (b) in the proportion 1 : 1.5 (v/v)

Hydrogen peroxide 30 m$M$: dilute 0.34 ml 30% hydrogen peroxide with phosphate buffer to 100 ml

### Procedure

*Measurement in Blood.* Venous blood containing heparin or citrate is centrifuged and the plasma and leukocyte layers are removed. The erythrocyte sediment is washed three times with isotonic NaCl. A stock hemolysate is prepared containing ~5 g Hb/100 ml by the addition of four parts by volume of distilled water. A 1 : 500 dilution of this concentrated hemolysate is prepared with phosphate buffer immediately before the assay is performed and Hb (hemoglobin) content is determined in duplicate (e.g., by the method of Drabkin). For capillary blood, 0.1 or 0.02 ml is hemolyzed in 250 or 50 ml distilled water. If the hemoglobin content of the blood is required as reference point, it must be determined in a separate sample of blood.[6–8]

*Measurement in Tissues.* Catalase in tissues with relatively high activity, such as liver and kidney, can be determined spectrophotometrically if complete lysis of all organelles and clear (or only slightly colored) solutions or extracts can be obtained. A detergent (e.g., 1% Triton X-100) must be used in the preparation of the stock homogenate (1 + 9 or 1 + 19), otherwise too low values will result. Further dilutions can be made with phosphate buffer, pH 7.0 (1 : 100 to 1 : 500, depending on tissue and species). However, if the sample after lysis of the organelles cannot be diluted to this extent, the considerable UV absorption of Triton X-100 must be kept in mind. As an alternative digitonin (0.01%) or sodium cholate (0.25%) can be used. Normally, catalase activity of tissue samples is expressed on a milligram wet weight or milligram total N basis. A conven-

[5] B. Chance, H. Sies, and A. Boveris, *Physiol. Rev.* **59**, 527 (1979).

[6] H. Aebi, *in* "Exposés Annuels de Biochimie Médicale," 29ième série, p. 139. Masson, Paris, 1969.

[7] H. Aebi and H. Suter, *in* "Biochemical Methods in Red Cell Genetics" (J. J. Yunis, ed.), p. 255. Academic Press, New York, 1969.

[8] H. Aebi, S. R. Wyss, B. Scherz, and J. Gross, *Biochem. Genet.* **14**, 791 (1976).

ient method for the measurement of catalase activity in tissue extracts has been described by Cohen et al.[9]

## Assay Conditions

Wavelength, 240 nm; light path, 10 mm; final volume, 3.00 ml. Read the sample containing 2.00 ml enzyme solution or hemolysate and 1 ml $H_2O_2$ at 20° (~ room temperature) against a blank containing 1 ml phosphate buffer instead of substrate and 2 ml enzyme solution or hemolysate. The reaction is started by addition of $H_2O_2$. The initial absorbance should be approximately $A = 0.500$. Mix well with a plastic paddle and follow the decrease in absorbance with a recorder for about 30 sec.

## Stability of Enzyme

Catalase in intact erythrocytes and in concentrated hemolysates is stable up to 6 days when kept at 2°. However, there is a relatively rapid decline of activity in dilute hemolysates which is more likely due to decomposition of the enzyme into subunits than to proteolytic changes. At a concentration of 1.2 mg Hb/ml the activity decreases by 10–15% within 24 hr; at a concentration of 0.06 mg Hb/ml the loss of activity is 10% after 1 hr and 80–90% after 24 hr. Consequently, hemolysate samples should be analyzed within 5–10 min after dilution.

## Definition of Units and Specific Activity

It is not possible to define international catalase units (U) according to IUB recommendations due to the abnormal kinetics. Therefore, the use of a number of differently defined units and different methods of evaluation is acceptable for this enzyme (a selection of methods is listed in the table).[10] Use of the rate constant of a first-order reaction ($k$) is recommended. The rate constant related to the hemoglobin content ($k$/g Hb) can serve as a measure of the specific activity of erythrocyte catalase. Equation (3) applies in this case. If the decrease in absorbance is recorded, the value of $\log A_1/A_2$ for a measured time interval or the time required for a certain decrease in absorbance can be determined.

For a time interval of 15 sec the following relationship is obtained according to Eq. (3):

$$k = (2.3/15)(\log A_1/A_2) = 0.153(\log A_1/A_2) \quad (\text{sec}^{-1}) \quad (5)$$

To calculate $k$/ml or $k$/g Hb proceed as follows:

[9] G. Cohen, D. Dembiec, and J. Marcus, Anal. Biochem. **34**, 30 (1970).
[10] H. Aebi, in "Methods of Enzymatic Analysis" (H. U. Bergmeyer, ed.), 3rd Ed. Verlag Chemie, Weinheim, F. R. G., in preparation, 1983.

OTHER METHODS OF DETERMINATION

| Method/technique | References | Material |
|---|---|---|
| Determination of $H_2O_2$ removal | | |
| Titrimetric methods | | |
| Iodometric | Bonnichsen et al.[2] | Tissues, blood |
| Permanganometric | Bonnichsen[3] | Tissues, blood |
| Spectrophotometry | | |
| Substrate: $H_2O_2$ ($E_{240}$) | Bergmeyer[11] | Purified preparations |
| | Cohen et al.[9] | Tissues, organelles |
| Substrate: perborate ($E_{220}$) | Thomson et al.[12] | Tissue fractions |
| Photometry ($E_{405}$–$E_{415}$) | | |
| Vanadic acid | Warburg and Krippahl[a] | Cell cultures |
| Titanium tetrachloride | Pilz and Johann[b] | Blood |
| Fluorimetry | | |
| Scopoletin | Perschke and Broda[c] | Aqueous |
| Diacetyldichlorofluorescein | Keston and Brandt[d] | solutions |
| | | |
| Determination of $O_2$ production | | |
| Oxygen electrode | Ogata[13] | Blood |
| | Del Rio et al.[14] | Plant material |
| | Meerhof and Roos[15] | Blood |
| Polarography | | |
| Catalase and SOD | Rigo and Rotilio[16] | Tissue homogenate |
| | | hemolysate |
| Immunoprecipitation | Higashi et al.[e] | Blood, tissues |
| (with anti-catalase) | Ben-Yoseph and Shapira[17] | Blood |
| Screening techniques | | |
| Capillary tube | Fung and Petrishko[f] | Microbial cultures |
| "Siebtest" | Gross et al.[18] | Blood |
| Automated procedure | | |
| Technicon AutoAnalyzer | Lamy et al.[19] | Blood |
| | Leighton et al.[20] | Tissue fractions |

[a] O. Warburg and G. Krippahl, *Z. Naturforschung* **18b,** 340 (1963). [b] W. Pilz and J. Johann, *Z. Anal. Chem.* **210,** 358 (1965). [c] H. Perschke and E. Broda, *Nature* (*London*) **190,** 257 (1961). [d] A. S. Keston and R. Brandt, *Anal. Biochem.* **11,** 1 (1965). [e] T. Higashi, M. Yagi, and H. Hirai, *J. Biochem.* (*Tokyo*) **49,** 707 (1961). [f] D. Y. C. Fung and D. T. Petrishko, *Appl. Microbiol.* **26,** 631 (1973).

$$k/ml = ka \tag{6}$$
$$k/g\ Hb = k/ml(1000/b) = (2.3/15)(a/b)(\log A_1/A_2) \quad (\sec^{-1}) \tag{7}$$

where $A_1$ is $A_{240}$ at $t = 0$, $A_2$ is $A_{240}$ at $t = 15$ sec, $a$ is the dilution factor [Hb concentration in blood or erythrocyte sediment (mg Hb/ml)/Hb concentration in cuvette (mg Hb/ml)], and $b$ is the Hb content of blood or erythrocyte sediment (grams/liter).

For the difference in absorbance of 0.450–0.400 ($\log A_1/A_2 = 0.05115$) the following relation holds:

$$k = (2.3/\Delta t)(\log A_1/A_2) = 0.1175/\Delta t \quad (\sec^{-1}) \tag{8}$$

*Other Methods of Determination*

This contribution deals only with the catalytic (not the peroxidic) activity of catalase. It can be measured by following either the decomposition of $H_2O_2$ or the liberation of $O_2$. Accordingly, the remaining substrate concentration at a given moment of the reaction can be determined by UV spectrophotometry[9,11,12] or—at the end of the reaction period—by simple titration.[2,3] On the other hand, $O_2$ production can be followed with the oxygen electrode[13–15] or by polarography.[16] Alternatively, catalase can also be measured by immunoprecipitation.[17] The method of choice for biological material, however, is UV spectrophotometry. Titrimetric methods are suitable for comparative studies. For large series of measurements there are either simple screening tests which give a quick indication of the approximative catalase activity,[18] or automated methods (e.g., using the Technicon AutoAnalyzer) available[19,20] (see the table).

[11] H. U. Bergmeyer, *Biochem. Z.* **327**, 255 (1955).
[12] J. F. Thomson, S. L. Nance, and S. L. Tollaksen, *Proc. Soc. Exp. Biol. Med.* **157**, 33 (1978).
[13] M. Ogata, in "Symposium on Genetics and Biochemistry of Acatalasemia" (IX Annual Meeting of the Japan Society of Human Genetics), at Wakayama Medical College, 1964.
[14] L. A. Del Rio *et al., Anal. Biochem.* **80**, 409 (1977).
[15] L. J. Meerhof and D. Roos, *J. Reticuloendothel. Soc.* **28**, 419 (1980).
[16] A. Rigo and G. Rotilio, *Anal. Biochem.* **81**, 157 (1977).
[17] Y. Ben-Yoseph and E. Shapira, *J. Lab. Clin. Med.* **81**, 133 (1973).
[18] J. Gross, A. Hartwig, and A. Golding, *Z. Med. Labortech.* **16**, 336 (1975).
[19] J. N. Lamy *et al., Bull. Soc. Chim. Biol.* **49**, 1167 (1967).
[20] F. Leighton *et al., J. Cell Biol.* **37**, 482 (1968).

## [14] Assays of Lipoxygenase, 1,4-Pentadiene Fatty Acids, and $O_2$ Concentrations: Chemiluminescence Methods

*By* Simo Laakso, Esa-Matti Lilius, and Pekka Turunen

Lipoxygenases are iron-containing enzymes that catalyze the dioxygenation, by molecular oxygen, of *cis,cis*-1,4-pentadiene fatty acids. The reaction is a source of weak chemiluminescence due to dissociation of free radicals from the main path of hydroperoxidation.[1] Under alkaline conditions the intensity of light emission can be drastically amplified by the presence of luminol.[2] The kinetics of luminol chemiluminescence in

[1] A. Boveris, E. Cadenas, and B. Chance, *Photobiochem. Photobiophys.* **1**, 175 (1980).
[2] E.-M. Lilius and S. Laakso, *Anal. Biochem.* **119**, 135 (1982).

METHODS IN ENZYMOLOGY, VOL. 105

Copyright © 1984 by Academic Press, Inc.
All rights of reproduction in any form reserved.
ISBN 0-12-182005-X

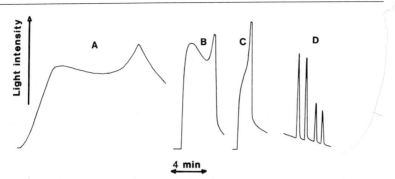

FIG. 1. Chemiluminescence tracings of luminol-supplemented lipoxygenase/linoleate reactions. The reactions contained air-saturated 0.2 $M$ sodium borate buffer, pH 9.0, 0.4 m$M$ linoleate, 0.04 m$M$ luminol and 0.02 (A), 0.08 (B), or 0.20 U/ml (C, D) of soybean lipoxygenase-1. The chemiluminescence responses of the anaerobic lipoxygenase reaction to added $O_2$ is shown by two consecutive additions of 2-$\mu$l and 1-$\mu$l samples of air-saturated water (D).

the reaction catalyzed by soybean lipoxygenase-1 (EC 1.13.11.12) is coupled to the changes in $O_2$ concentration. In air-saturated reaction mixtures the initial oxygenation phase emits light in proportion to the rate of the overall process measured by $O_2$ consumption or by accumulation of the conjugated diene products.[2] When dissolved $O_2$ is consumed to a certain critical level a second emission peak appears. This has been taken to indicate increased dissociation of enzyme-bound oxygen or fatty acid free radicals which react with the remaining oxygen prior to initiation of the chemiluminescence-silent anaerobic catalysis.[3] Consequently, the emission curve is two-phasic whenever the molar concentration of the substrate fatty acid exceeds that of dissolved $O_2$ (see Fig. 1). The former phase forms the basis for luminometric lipoxygenase assays whereas the latter serves to assay $O_2$ concentrations in samples added to the anaerobic mixtures of lipoxygenase-1 and linoleic acid.[4]

## Assay Methods

### Principle

The luminometric assay for lipoxygenase activity is scaled to 1.0 ml final volume and is performed with saturating concentrations of dispersed linoleic acid and luminol. The assay mixtures described below can be used

[3] G. A. Veldink, J. F. G. Vliegenthart, and J. Boldingh, *Prog. Chem. Fats Other Lipids* **15**, 131 (1977).
[4] S. Laakso and T. Huttunen, *J. Biochem. Biophys. Methods* **7**, 211 (1983).

as such for comparative assays with the $O_2$ electrode procedures or by spectrophotometry at 234 nm as previously described.[5] Consumption of $O_2$ via the chemiluminescent reaction of luminol is insignificant compared to the main process and cannot be observed with the polarographic $O_2$ assay or as an absorbance change at 234 nm. The assays of essential fatty acids and $O_2$ concentrations utilize the same reaction components. In the former the linoleic acid substrate is replaced by aqueous fatty acid samples. Soybean lipoxygenase-1 requires trace amounts of hydroperoxy fatty acid for maximum activity and this may cause a change in the reaction order and yield a two-phasic standard curve when very low concentrations of 1,4-pentadiene fatty acids are analyzed. Therefore, a range-finding procedure for the selection of the proper enzyme concentration is described. The determinations of $O_2$ concentrations include a preliminary step where the reaction mixture is made and maintained anaerobic by using a molar excess of linoleic acid over dissolved $O_2$. The resulting anaerobic reaction generates an oxygen reactive species which will spontaneously react to added oxygen with intense chemiluminescence. The procedures described below were developed using a LKB Wallac 1250 luminometer and the manufacturer's injection device. Also liquid scintillation counters, fluorometers, or spectrophotometers may be adapted to these assay purposes. However, it is essential that the procedures described below be stringently followed to obtain sufficient proportionality and reproducibility in the assays.

*Reagents*

Sodium borate buffer, 0.2 $M$, pH 9.0

Sodium linoleate, 10 m$M$. Weight out 140 mg of linoleic acid. Dissolve 0.180 ml of Tween 20 in 10 ml of $O_2$-free water. While bubbling nitrogen through the detergent solution the linoleic acid is added and the dispersion is homogenized by drawing back and forth in a Pasteur pipet. Add 1.0 ml 0.5 $M$ NaOH and mix under nitrogen stream to yield a clear solution. Make up to 50 ml total volume with $O_2$-free water. Divide into 2-ml portions in small test tubes, flush with $N_2$, and freeze immediately after capping

Sodium arachidonate, 10 m$M$. Weigh out 150 mg of arachidonic acid and continue in the manner described above

Luminol, 1 m$M$. Weigh out 9 mg of luminol (5-amino-2,3-dihydro-phthalazine-1,4-dione) and dissolve in 50 ml of the borate buffer, pH 9.0. Shield the vial from light and store at room temperature. The solution is prepared daily

[5] B. Axelrod, T. M. Cheesbrough, and S. Laakso, this series, Vol. 71 [53].

Lipoxygenase-1 from soybean, 80 U/ml. Dissolve 5.0 mg of the lyophilized powder containing about 160 U/mg (Sigma) in 10.0 ml of
0.2 $M$ sodium borate buffer, pH 9.0. Store in ice
Paraffin oil

*Procedure*

*Assay of Lipoxygenase.* The reaction is carried out at 25° in 3.5-ml
disposable Ellerman plastic tubes. The assay mixture contains $(0.960 - x)$
ml of the borate buffer, pH 9.0, 0.040 ml of the sodium linoleate substrate,
and $x$ ml of the enzyme. Prior to the addition of the enzyme the mixture is
stirred and placed in the luminometer to record the baseline chemiluminescence. The reaction is initiated by adding the enzyme and mixing
with a few strokes of a plastic puddler. The initial lag period appearing in
the emission curve permits the mixing of the reactants outside the measuring chamber of a luminometer without a loss of the straight-line portion
of the emission curve. The reaction is recorded until the anaerobic signal,
the second emission maximum (see Fig. 1), appears and the reaction rate
is determined as millivolts per minute from the positive slope of the initial
emission curve. Since the light output varies with the equipment used it is
convenient to define the lipoxygenase activity unit as the quantity of
enzyme that yields 1 $\mu$mol of conjugated diene per minute under standard
assay conditions. Therefore, spectrophotometric calibration of the
luminometric readings is required. This can be done by using luminol
supplemented reaction mixtures and a dual beam spectrophotometer
where the reference cuvette contains no enzyme. The extinction coeffi-
cient for the diene structure in the linoleate product is $2.5 \times 10^4 \ M^{-1} \ cm^{-1}$
at 234 nm. The quantity of enzyme must be in the range of 2–22 mU for
optimal proportionality.

*Assay of $O_2$ Concentrations.* The assay mixture is prepared in 3.5-ml
plastic Ellerman tube and consists of 0.860 ml of 0.2 $M$ sodium borate
buffer, pH 9.0, 0.080 ml of the sodium linoleate, 0.060 ml of soybean
lipoxygenase-1, and 0.040 ml of luminol. The mixture is covered with a
layer about 5 mm thick of oxygen-impermeable paraffin oil and placed in a
luminometer. For sample application an injection needle is pushed
through the cavities of the measuring chamber of the luminometer into the
aqueous phase of the reaction tube. Before $O_2$ assays the emission curve
is followed until all the oxygen is exhausted in the system and a chemiluminescence-silent state is established. The sample, 1–50 $\mu$l, is rapidly
pushed through the needle while light emission is continuously recorded.
$O_2$ content in the sample is proportional to the height of the emission peak
in millivolts and can be converted to micromoles of $O_2$ by comparing with

appropriate standards. The calibration curve is constructed from measurements with 1–50 $\mu l$ of air equilibrated water. The linear range of the assay is between 0.2 and 10 nmol of $O_2$ and both the standards and samples can be assayed with less than 10% individual variations until the volume of the original reaction mixture is increased by up to 50%.

*Assay of 1,4-Pentadiene Fatty Acids.* The same sample delivery systems as used in $O_2$ assays are applicable also to the analyses of essential fatty acids. The samples are injected into a reaction mixture containing 1.0 ml of 0.2 $M$ sodium borate buffer (pH 9.0), 0.040 ml of luminol, and 0.020–0.080 ml of the lipoxygenase-1 solution. Prior to assaying unknown samples the validity of the chosen enzyme concentration is checked by constructing chemiluminescence standards based on peak height for linoleic acid. For this, the linoleic acid stock solution is diluted $(1 : 10^3 – 1 : 10^5)$ in 0.2 $M$ sodium borate buffer (pH 9.0) and 0.040-ml aliquots from each dilution are immediately assayed for chemiluminescence. If a break in the curve appears below 100 pmol (usually between 20 and 60 pmol) of linoleic acid the enzyme content is gradually lowered and the procedure is repeated until a linear correlation is obtained. The resulting standard curve is equally applicable to the quantitation of arachidonic and linolenic acids as well as their mixtures. With the equipment used by the authors the detection limit was 2 pmol/ml at a signal-to-noise ratio of 2.

*General Comments.* The luminometric procedures described above offer high sensitivity combined with the ease of performance. Soya lipoxygenase-1 is well suited for these assays because of its high pH optimum against free fatty acids,[6] stability,[7] and availability as a commercial product. However, one must be aware of the fundamental limitations encountered in the lipoxygenase induced luminol chemiluminescence. The analyses are all restricted to alkaline pH region and, in practice, light amplification by the coupled luminol reaction becomes significant only above pH 8.0. Luminol oxidation occurs under a wide variety of reaction conditions[8] and therefore interferences may arise if the assays are employed in complex matrices. In the assays of essential fatty acids and $O_2$ concentrations internal standardization with pure substrates is entirely satisfactory and in lipoxygenase assays the $O_2$ electrode procedure serves to reveal possible absorption of free radicals by the sample constituents.

[6] G. S. Bild, C. S. Ramadoss, and B. Axelrod, *Lipids* **12,** 732 (1977).
[7] S. Laakso, *Lipids* **17,** 667 (1982).
[8] D. F. Roswell and E. H. White, this series, Vol. 67 [36].

## [15] High-Performance Liquid Chromatography Methods for Vitamin E in Tissues

*By* JUDITH L. BUTTRISS and ANTHONY T. DIPLOCK

Five decades of research have established that vitamin E is essential for the integrity and optimal function of all mammalian cells. Clinical signs of deficiency in experimental animals usually are of a degenerative nature and often involve the stability of biological membranes.

Free radical-catalyzed lipid peroxidation is a continual biological process which if unchecked may cause damage to cellular and intracellular membranes, resulting not only in changed membrane structure but also in destruction of the functional integrity of membrane-bound enzymes. Vitamin E, alone and in concert with the selenium-containing enzyme, glutathione peroxidase, can inhibit this process. This function of vitamin E may be aided by a specific physicochemical interaction between the phytyl side chain of $\alpha$-tocopherol and the fatty acyl chains of polyunsaturated phospholipids, particularly those derived from arachidonic acid, by providing a mechanism for anchoring vitamin E to membranes and facilitating the antioxidant behavior attributed to $\alpha$-tocopherol.[1] Other consequences could be a reduction in permeability of biological membranes containing relatively high levels of polyunsaturated fatty acids, particularly arachidonic acid, and prevention of the degradation of membrane phospholipids by membrane-bound phospholipases *in vivo*.[2]

Despite its ubiquity, vitamin E is often present in tissues at very low concentrations and, when specific subcellular organelles are of interest, this presents severe analytical problems. Being lipophilic, the vitamin is mainly located in subcellular membranes, such as those of nuclei, mitochondria, and microsomes, where it acts to maintain the integrity of the membrane against peroxidative damage.[3]

Vitamin E activity in food of plant origin derives from $\alpha$-, $\beta$-, $\gamma$-, and $\delta$-tocopherols and corresponding tocotrienols. In animal tissues, $\alpha$-tocopherol is predominant and has by far the highest biological activity. Some of the less active tocopherols, particularly $\gamma$-tocopherol, are

[1] A. T. Diplock and J. A. Lucy, *FEBS Lett.* **29,** 205 (1973).

[2] J. A. Lucy, *in* "Tocopherol, Oxygen and Biomembranes" (C. de Duve and O. Hayaishi, eds.). Elselvier, Amsterdam, 1978.

[3] K. Fukuzawa, H. Chida, A. Tokumura, and H. Tsukatani, *Arch. Biochem. Biophys.* **206,** 173 (1981).

Copyright © 1984 by Academic Press, Inc.
All rights of reproduction in any form reserved.
ISBN 0-12-182005-X

present in mixed diets and the $\gamma$ and $\delta$ forms occur mainly in vegetable oils.[4]

In 1971, Bunnell[5] summarized the methods existing at that time for estimation of vitamin E. They included colorimetry, fluorimetry, paper, column, and thin-layer chromatography and gas–liquid chromatography (GLC). Spectroscopic methods have the disadvantage that they do not differentiate the tocopherols and certain reducing substances can provide interference. Problems with GLC procedures include the requirement of relatively high temperatures and the difficulty in separating $\alpha$-tocopherol from other tocopherols and from cholesterol. Interference can also be attributed to other lipids, e.g., triglycerides, long-chain alcohols, and sterols. In contrast to these methods, high-performance liquid chromatography (HPLC) offers speed in analysis, sensitivity, selectivity, and simplicity.

In recent years, a variety of methods have appeared in the literature for the HPLC determination of tocopherols in biological samples. Many of these have concerned themselves primarily with $\alpha$-tocopherol, and serum or plasma has been the most frequently used form of sample.[6-10] For the determination of $\alpha$-tocopherol both normal and reverse phase methods have been successfully used.

As mentioned previously, $\alpha$-tocopherol has the highest biological activity, but the other tocopherols can also form a significant part of the dietary intake and may thus warrant measurement. For the separation of the four tocopherols, normal phase HPLC is the method of choice in order that $\beta$ and $\gamma$ isomers can be separated.[11-13] A variety of modifying organic solvents have been used in conjunction with the main solvent, hexane, and the modifiers include diisopropyl ether,[11,13] diethyl ether,[4,16] isopropanol,[12,15] and methanol.[6,14]

[4] J. S. Nelson and V. W. Fischer, in "Vitamins in Medicine" (B. B. Barker and D. A. Bender, eds.). Heinemann, London, 1980.

[5] R. H. Bunnell, Lipids 6, 245 (1971).

[6] G. T. Vatassery and W. J. Hagen, J. Chromatogr. 178, 525 (1979).

[7] A. P. De Leenheer, V. O. De Bevere, A. A. Cruyl, and A. E. Claeys, Clin. Chem. 24, 585 (1978).

[8] A. P. De Leenheer, V. O. De Bevere, M. G. M. De Ruyter, and A. E. Claeys, J. Chromatogr. 162, 408 (1979).

[9] J. G. Bieri, T. J. Tolliver, and G. L. Catignani, Am. J. Clin. Nutr. 32, 2143 (1979).

[10] C. H. McMurray and W. J. Blanchflower, J. Chromatogr. 178, 525 (1979).

[11] M. Matsuo and Y. Tahara, Chem. Pharm. Bull. 25, 3381 (1977).

[12] A. P. De Leenheer, V. O. De Bevere, and A. E. Claeys, Clin. Chem. 25, 425 (1979).

[13] L. Jansson, B. Nilsson, and R. Lindgren, J. Chromatogr. 181, 242 (1980).

[14] G. T. Vatassery, V. R. Maynard, and D. F. Hagen, J. Chromatogr. 161, 299 (1978).

[15] U. Manz and K. Philipp, Int. J. Vit. Nutr. Res. 51, 342 (1981).

Problems with these solvents include the risk of peroxide formation as is the case with diisopropyl ether and the high polarity of others such as methanol and isopropanol, requiring the addition of very small quantities of the modifier, of the order of 0.3%, an amount often below the accuracy of the mechanical mixing system of HPLC equipment.

With regard to the choice of column, most authors used a 5-$\mu$m[12,15,16] or 10-$\mu$m[11,13] silica packing, the exception being Vatassery and Hagen,[6] who used silica packing 37–50 $\mu$m in size. Resolution and capacity are improved by using the smallest particle possible, which means that the column does not need to be overloaded to produce a usable detection signal.

For specificity and increased sensitivity, fluorescence detection is recommended[6,13,17] to eliminate spurious, nonfluorescent components that would otherwise be measured by detectors that depend on ultraviolet absorption of the eluted compounds.

The estimation of tocopherols in serum is a relatively straightforward task but to attempt a similar process in tissues such as liver or heart is a more complex problem. Over the past few decades there have been a variety of approaches to the question, which tend to fall into one of two categories: solvent extraction of dried or frozen tissue, or direct alkaline digestion of the tissue.[18] These methods were often very drastic and involved boiling in potassium hydroxide, grinding, and other techniques that could cause destruction of tocopherol if they were not rigorously controlled.

Bayfield and Romalis[19] published a method for extraction of vitamin E from sheep liver, based on a saponification step followed by solvent extraction, with pyrogallol and hydroquinone playing essential parts in the procedure as antioxidants. Kormann[20] found it necessary to use 1.5% ascorbic acid in his liver extraction procedure and indeed most authors have found it essential to employ protective antioxidants. Hatam and Kayden[17] used 1% ethanoic ascorbic acid in the extraction of the vitamin from red and white blood cells, and in earlier studies[21] found the use of 2% ethanoic pyrogallol to be essential during both the saponification and the extraction stages. Other workers have used methanol instead of ethanol[20] but Diplock and co-workers[18] found that a decrease in yield was associ-

[16] J. N. Thompson and G. Hatina, *J. Liq. Chromatogr.* **2**, 327 (1979).

[17] L. J. Hatam and H. J. Kayden, *J. Lipid Res.* **20**, 639 (1979).

[18] A. T. Diplock, J. Green, J. Bunyan, and D. McHale, *Br. J. Nutr.* **20**, 95 (1966).

[19] R. F. Bayfield and L. F. Romalis, *Anal. Biochem.* **97**, 264 (1979).

[20] A. W. Korman, *J. Lipid Res.* **21**, 780 (1980).

[21] H. J. Kayden, C.-K. Chow, and L. K. Bjornson, *J. Lipid Res.* **14**, 533 (1973).

ated with the use of methanol because the lower temperature of the boiling solvent caused incomplete digestion.

Whereas most workers routinely employed a saponification step at either 70 or 100° for extraction of vitamin E from biological samples, Vatassery *et al.*[14] extracted rat brain tocopherols without the aid of a saponification step and a satisfactory result was obtained. Hatam and Kayden[17] also omitted saponification when measuring the vitamin in red and white blood cells. It should also be noted, since tocol is a frequently used internal standard for the determination of tocopherols by HPLC[8,12,13] that losses of up to 30% of tocol have been reported during a 70° saponification[17] while α-tocopherol itself remained unaffected.

HPLC Procedure

Using the work described previously, in particular the methods of De Leenheer *et al.*[12] and Jansson *et al.*[13] we have adapted a method for our own purposes, which uses a Varian 5000, connected in series with a Fluorichrom fluorescence detector and a UV50 (Varian Associates) ultraviolet absorption detector with a dual pen chart recorder. The UV detector was operated at 294 nm and the Fluorichrom was fitted with filters to provide excitation at 220 nm via an interference filter (Varian 220I) and emission at 360 nm by means of a band filter (Varian 7-60). The Varian 5000 was fitted with a 100-$\mu$l loop injector and a back-pressure restrictor, and was operated at ambient temperature at a maximum pressure of 350 atm with a minimum pressure programmed at 10 atm. The flow rate was 2 ml/min and the column was a Micropak Si-5 (4.5 mm i.d. × 15 cm length). Such a combination aids a fast separation.[22]

Methyl *t*-butyl ether was chosen as the modifying organic solvent and used isocratically at a concentration of 8–12% in hexane, the proportion varying with the number or isomers to be separated. All solvents used were of HPLC grade and supplied by Rathburn Chemicals Ltd., Walkerburn, Pebbleshire, Scotland. This particular solvent was used on the basis of its polarity (polarity index = 2.5) and the low risk of peroxide formation compared with other ethers of similar polarity.[23] Using this procedure, 12% modifier in hexane gave a separation in 4 min and 8% a separation in 6 min, the latter being of advantage when more than one tocopherol was being determined (Figs. 1 and 2).

The procedure was standardized by 100-$\mu$l injections of a range of solutions containing known amounts of α-tocopherol (Roche Products), γ-tocopherol (Kodak Ltd., Liverpool), and tocol (a generous gift from

---

[22] E. L. Johnson and R. Stevenson, "Basic Liquid Chromatography." Varian, London, 1978.

[23] C. J. Little, A. D. Dale, and J. A. Whatley, *J. Chromatrogr.* **169**, 381 (1979).

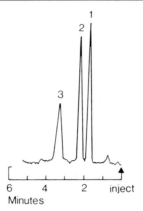

FIG. 1. HPLC separation of $\alpha$-tocopherol (peak 1), $\gamma$-tocopherol (peak 2), and tocol (peak 3) using hexane and methyl $t$-butyl ether (88 : 12), flow rate of 2.0 ml/min, ambient temperature, and fluorescence detection.

Eisai Company, Japan) in the range of 10–100 $\mu$g/ml. There were linear relationships between the peak heights for $\alpha$- or $\gamma$-tocopherol and tocol, and also for peak heights versus concentration for each tocopherol.

Using a known amount of tocol as an internal standard, the concentration of $\alpha$- and $\gamma$-tocopherols in a known sample was determined by comparing the tocopherol-to-tocol peak height ratio with a standard plot of peak height ratio against tocopherol concentration (Fig. 3). It was found to be possible to measure $\alpha$-tocopherol levels of 0.5 ng and $\gamma$-tocopherol levels of 1.0 ng in agreement with the work of De Leenheer et al.[12] Day-to-day and within-day variations in retention time and peak height were

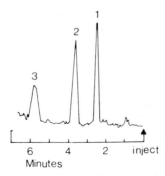

FIG. 2. HPLC separation of $\alpha$-tocopherol (peak 1), $\gamma$-tocopherol (peak 2), and tocol (peak 3), using hexane and methyl $t$-butyl ether (92 : 8), flow rate of 2.0 ml/min, ambient temperature, and fluorescence detection.

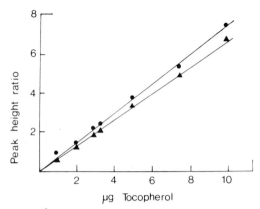

FIG. 3. Standard curve for α- and γ-tocopherol, where the peak height ratio is the ratio of tocopherol peak height to tocol peak height. ●, α-tocopherol; ▲, γ-tocopherol.

found to be very small, and could be regarded as negligible although the precaution of always including an internal standard was taken.

Extraction Procedure for Tissues

Three extraction procedures were examined, based on the methods discussed earlier.

1. To 1 ml of 25% rat liver homogenate was added 2 ml of 1% ascorbic acid (or 1% pyrogallol) in absolute alcohol, containing a known amount (10 μg) of α-tocopherol as internal standard. After equilibration for 2 min at 70°, 0.3 ml of saturated potassium hydroxide was added and the resultant mixture was incubated for a further 30 min. After cooling in ice, 1.0 ml distilled water was added, followed by 4 ml of hexane. After 2 min vigorous mixing and centrifugation at 1500 g for 5 min, the upper hexane layer was carefully removed and the residue reextracted with a further 2 ml of hexane. The two fractions were combined, filtered through a 0.45-μm filter and evaporated under nitrogen. The residue was redissolved in 300 μl spectroscopic grade hexane and injected into the HPLC apparatus in 100 μl portions.

2. The method was as in (1) with the exception that a 100° saponification step was used.

3. The method was in (1) but included no saponification step. After mixing with antioxidant in ethanol, the sample was extracted with hexane as described.

CAPACITY RATIOS ($k'$) OF CHROMATOGRAPHIC
PEAKS[a]

|  | $k'$ | |
|---|---|---|
|  | 8% | 12% |
| $\alpha$-Tocopherol | 1.55 | 0.80 |
| $\gamma$-Tocopherol | 2.77 | 1.33 |
| Tocol | 5.05 | 2.45 |

[a] $k'$ is a retention parameter and represents the ratio of the weight of the solute (sample) in the stationary phase to that in the mobile phase of the column.

The results showed that the highest recovery of $\alpha$-tocopherol (95%) is attained using freshly prepared 1% pyrogallol solution, rather than ascorbic acid solution together with a 70° saponification step. Saponification at 100° was found to be unnecessary and the yield without saponification was of the order of 75%. The use of ascorbic acid gave slightly lower yields with each method.

It has been reported that saponification can result in losses of up to 30% of added tocol.[17] In our studies we found that even greater losses occurred (up to 85%) when tocol was used as an internal standard, although losses of $\alpha$- and $\gamma$-tocopherol were of the order of 5 and 15%, respectively.

For this reason, two additional external standards were used. To one 10 $\mu$g tocol, 10 $\mu$g $\alpha$-tocopherol and 10 $\mu$g $\gamma$-tocopherol were added prior to the saponification and extraction stages, whereas the other received these after extraction. This procedure allowed calculation of the tocol and tocopherol losses due to saponification and extraction so that allowance could be made in the final calculations.

Capacity ratios ($k'$) can be calculated for chromatographic peaks at 8 and 12% modifier in hexane. These are shown in the table and are similar to those reported earlier.[12]

Based on these studies, we have adopted a method for extraction of tocopherols from liver homogenate and their subsequent HPLC determination. In summary, the tissue sample (1 ml), in 2 ml absolute alcohol containing 1% pyrogallol, was saponified at 70° for 30 min using 0.3 ml saturated potassium hydroxide and the tocopherols were then extracted using 4 ml hexane. For the HPLC determination of tocopherols, a normal phase procedure was adopted, involving a 5-$\mu$m silica column and methyl $t$-butyl ether in hexane (8–12%) as the eluting solvent.

Tocol was employed as an internal standard and $\alpha$- and $\gamma$-tocopherols were used as additional external standards since tocol is known to be lost during saponification.

This method was found also to be satisfactory for the extraction and measurement of tocopherols in subcellular liver fractions and in heart and kidney.

### General Comments

We are able to present a sensitive, rapid method suitable for the extraction and determination of tocopherols in a range of small tissue samples.

An aliquot of liver homogenate containing 0.25 g of tissue was found to be adequate for three determinations of the tocopherols and similar amounts of kidney and heart tissue were also required.

Quantities of $\alpha$- and $\gamma$-tocopherol of the order of 0.5 and 1.0 ng, respectively, were detectable by this method.

## [16] Vitamin E Analysis Methods for Animal Tissues

### By INDRAJIT D. DESAI

Vitamin E activity in animal tissues is essentially due to $\alpha$-tocopherol and more specifically $d$-$\alpha$-tocopherol ($R,R,R$-$\alpha$-tocopherol) in free form or as acetate ester. A variety of methods have been developed by various investigators for the analysis of vitamin E in animal tissues and can be referred to in a recent review,[1] but many of the earlier procedures involving column chromotography and gas–liquid chromatography are rather complicated and time consuming. The most commonly used methods are based on saponification and solvent extraction of lipids, removal or destruction of interfering substances, and determination of tocopherol spectrophotometrically or spectrofluorometrically as will be described in this report. The methods chosen are for common animal tissues such as blood and organ tissues. For more unusual materials such as platelet cells, cerebrospinal fluid, and adipose fat appropriate methods should be selected by referring to comprehensive reviews[1-3] on the subject which are

[1] I. D. Desai, in "Vitamin E: A Comprehensive Treatise" (L. J. Machlin, ed.), p. 67. Dekker, New York, 1980.

[2] J. G. Bieri, in "Lipid Chromatographic Analysis" (G. V. Marinetti, ed.), Vol. 2, p. 459. Dekker, New York, 1969.

Copyright © 1984 by Academic Press, Inc.
All rights of reproduction in any form reserved.
ISBN 0-12-182005-X

presently available. Most recently, a high-pressure liquid chromatography (HPLC) method for vitamin E (see [15] of this volume) has been introduced as a method of choice, but HPLC equipment is expensive and not readily available in every laboratory. The standard colorimetric and fluorometric methods are easily carried out using common laboratory equipment and are often adequate for routine analysis of vitamin E in animal tissues for biological research and for clinical testing. For more sophisticated and sensitive detection of commonly occurring $\alpha$-tocopherol along with trace amounts of other forms of tocopherol and tocotrienol in animal tissues, the HPLC method is highly recommended.

## Preparation of Tissues for Vitamin E Analysis

### Blood

Fresh venous blood is drawn with a suitable syringe and transferred in conical plastic centrifuge tubes containing 0.2 ml of 1% solution of disodium EDTA (ethylenediaminetetraacetic acid) per every 10 ml of whole blood. Mix the tubes immediately by gentle shaking. Blood can be refrigerated at 4° up to a maximum period of 24 hr, if necessary, or centrifuged immediately at 2500 rpm (500 $g$) for 10 min to separate the plasma and RBC (red blood cells). Plasma is transferred in a tube for saponification and extraction before vitamin E analysis or stored frozen at −20°. RBC are washed three times with 5 volumes of isotonic pH 7.4 buffer solution containing 1.42 g of anhydrous disodium phosphate, 7.27 g of sodium chloride, and 0.1 g of disodium EDTA per liter of distilled water as described by Kayden et al.[4] The washed RBC are resuspended in the above solution to a final hematocrit of 50%.

*Plasma: Saponification and Extraction.* Saponification of plasma or serum prior to hexane extraction is necessary to facilitate tocopherol extraction and to reduce the amount of saponifiable lipid compounds from interfering with the colorimetric or fluorometric assay of vitamin E. Kayden et al.[4] have described a simple and rapid procedure involving saponification in the presence of a large amount of added antioxidant pyrogallol to prevent destruction of tocopherols by alkali. Accordingly 1 ml of plasma is pipetted into a 15-ml glass-stoppered centrifuge tube to which is added 2 ml of 2% solution of pyrogallol in purified ethanol and mixed thoroughly. The mixture is heated at 70° for 2 min after which

[3] R. H. Bunnell, *in* "The Vitamins" (P. Gyorgy and W. N. Pearson, eds.), Vol. 6, p. 261. Academic Press, New York, 1967.
[4] H. J. Kayden, C. K. Chow, and L. K. Bjornson, *J. Lipid Res.* **14**, 533 (1973).

0.3 ml of saturated potassium hydroxide is added and mixed again. The mixture is further incubated at 70° for 30 min. The tubes are immediately cooled in an ice bath and 1 ml of distilled water and 4 ml of purified hexane are added. The tubes are shaken vigorously for 2 min and centrifuged at 1500 rpm for 10 min to separate the phases. The hexane extract can be subjected to TLC (thin-layer chromotography) treatment to separate tocopherols from small amounts of hexane soluble nontocopherol compounds or used directly for the colorimetric or fluorometric determination of vitamin E as will be described later.

*Red Blood Cells: Saponification and Extraction.* An improved method for the quantitative recovery of tocopherols for the determination of vitamin E in RBC recommends saponification and extraction steps in the presence of large amounts of added antioxidant such as pyrogallol or ascorbic acid as described by Kayden et al.[4] The procedure requires 2 ml of washed RBC resuspended in a pH 7.4 isotonic phosphate buffer with 1% EDTA and made up to a final hematocrit of 50%. The saponification is carried out to 50-ml glass-stoppered centrifuge tubes in the presence of 10 ml of a 2% solution of pyrogallol in purified absolute ethanol. The tubes are thoroughly mixed and placed in a 70° water bath for 2 min, after which 0.5 ml of saturated potassium hydroxide is added to each tube and incubated further for 30 min in the same water bath. The tubes are removed and immediately placed in an ice bath. After cooling to room temperature, 7.5 ml of distilled water and 22 ml of purified hexane are added to each tube and extraction is carried out by shaking the stoppered tubes vigorously for 2 min. The hexane extract separated by centrifugation at 1500 rpm for 10 min can be subjected to TLC on silica gel G for further purification, if required, or used directly for the colorimetric or fluorometric determination of vitamin E.

*Organ Tissues*

Analysis of vitamin E in organ tissues such as liver, kidney, heart, spleen, lungs is of interest in clinical and biological investigations and varieties of complex procedures are available, as recently reviewed,[1] depending on the nature of the tissue to be analyzed, the pattern of the tocopherols to be examined and the accuracy of quantitative estimate desired. A precise highly reproducible procedure for vitamin E analysis in organ tissue homogenates and subcellular fractions has been developed by Taylor et al.,[5] which is as follows.

*Preparation of Organ Tissues.* The organs are excised, weighed, and homogenized in 5 volumes of isotonic potassium chloride with a glass-

[5] S. L. Taylor, M. P. Lamden, and A. L. Tappel, *Lipids* **11**, 530 (1976).

Teflon homogenizer to obtain whole tissue homogenate. Subcellular fractions, if required for vitamin E analysis can be prepared according to the method of de Duve *et al.*[6] which involves centrifugal separation of nuclear-cell debris, mitochondrial, microsomal, and lysosomal fractions and soluble cell-free supernatant. The subcellular pellets are washed and suspended in isotonic potassium chloride.

*Saponification and Extraction.* A variety of saponification and extraction methods are available in the literature but the one which has been found most suitable by Taylor *et al.*[5] is a variation of the method of Kayden *et al.*[4] and is as follows. The saponification mixture containing 1.5 ml of tissue homogenate or suspensions of subcellular fractions, 1.0 ml of purified absolute ethanol and 0.5 ml of 25% ascorbic acid as antioxidant is preincubated at 70° for 5 min in 15-ml glass-stoppered centrifuge tubes. One milliliter of 10 $N$ potassium hydroxide is added to the above tubes and further incubation at 70° is carried out for 30 min. The tubes are then cooled by placing in an ice bath and 4 ml of purified hexane is added to extract nonsaponifiable material by Vortex mixing the tubes for 1 min. The hexane layer is separated by centrifuging the extracted mixture at 1500 rpm for 5 to 10 min and tocopherols are determined by colorimetric or fluorometric procedures. In the saponification procedure, it is important to include ethanol for optimal saponification and for separation of phases during extraction, and adequate amounts of ascorbic acid as an antioxidant to prevent tocopherol oxidation during saponification. Antioxidant pyrogallol is avoided in this procedure since it is converted to pyrogallin during the saponification treatment, and some pyrogallin is extracted in the hexane phase thus increasing the background fluorescence and thereby limiting the sensitivity of fluorometric assay of vitamin E.

Purification of Hexane Extract Prior to Vitamin E Determination

In some instances, depending on the nature of the tissue being examined and accuracy of tocopherol estimation desired, certain amounts of hexane-extracted interfering compounds may have to be eliminated by TLC on silica gel G for the spectrophotometric assay and sulfuric acid treatment for the fluorometric analysis of vitamin E.

*Thin-Layer Chromotography*

The TLC step may be necessary for the separation of nonspecific interfering compounds especially when total tocopherols are to be mea-

[6] C. de Duve, B. C. Pressman, R. Gianetto, R. Wattiaux, and F. Appelmans, *Biochem. J.* **60,** 604 (1955).

sured colorimetrically. However, separation and recovery of tocopherols by elution from the TLC plates are not satisfactory and better methods such as HPLC as described in the earlier section of this volume may have to be used for the estimation of individual tocopherols in animal tissues.

The TLC method recommended by Kayden et al.[4] requires 3 ml of hexane extract (20 ml if tissue being analyzed is RBC) to be pipetted in glass-stoppered test tubes and evaporated by flushing under nitrogen. The residue is dissolved in 20 to 50 $\mu$l of pure chloroform and spotted on a duplicate set of silica gel G plates. Pure $\alpha$-tocopherol standard is also spotted, keeping a 2 cm distance between spots. The plates are placed in a TLC chamber and developed using benzene–ethyl acetate 2 : 1 or benzene–ethanol 99 : 1 as a preferred solvent system. One developed plate is sprayed with 0.001% rhodamine 6G in purified methanol to identify tocopherol spots under UV light, and from the other plate an area of silica gel corresponding in $R_f$ to the $\alpha$-tocopherol standard is scraped with a razor blade and placed into a 7-ml centrifuge tube. The tocopherol from the scraped material is eluted by adding 1.5 ml of purified absolute ethanol and Vortex mixing it thoroughly before centrifuging it out at 2500 rpm for 5 min.

*Sulfuric Acid Treatment*

Taylor et al.[5] have successfully eliminated the presence of relatively large amounts of vitamin A in the hexane extract of animal tissues, especially liver, by the sulfuric acid treatment of Fox and Mueller[7] and thus solved the quenching problems in the fluorometric determination of tocopherols. A portion of hexane extract after saponification is treated in a glass-stoppered tube with 0.6 ml of 60% sulfuric acid by Vortex mixing for 30 sec. The mixture is centrifuged to separate the hexane phase which is used directly for fluorometric measurement of tocopherols. The relative fluorescence intensity is not affected by the sulfuric acid treatment and in fact the recovery of tocopherols is significantly improved by including this step in the fluorometric measurement of tocopherols in tissue homogenate and subcellular fractions of liver.

Improved Spectrophotometric Assay for Vitamin E

*Principle*

The method described here is a modified version of the classical Emmerie–Engle method[8] in which ferric ions are reduced to ferrous ions in

[7] S. H. Fox and A. Mueller, *J. Am. Pharm. Assoc. Sci. Ed.* **39,** 621 (1950).
[8] A. Emmerie and C. Engel, *Rec. Trav. Chim.* **57,** 1351 (1938).

the presence of tocopherols and the formation of pink-colored complex with a more sensitive reagent such as bathophenanthroline. Orthophosphoric acid is added as a chelating agent to reduce carotene interference by preventing its oxidation and stabilization of color by binding excess ferric ions and thus preventing their photochemical reduction. Absorbance of the stable chromophore is measured spectrophotometrically at 536 nm.

*Reagents*

Water, deionized water distilled in all glass distillation apparatus

Absolute ethanol, analytical grade or laboratory grade ethanol redistilled in all glass apparatus after adding pellets of potassium hydroxide and crystals of potassium permanganate

Hexane, analytical grade purified by distillation in all glass apparatus

Bathophenanthroline reagent, 0.2% solution of 4,7-diphenyl-1,10-phenanthroline in purified absolute ethanol

Ferric chloride reagent, 0.001 $M$ ferric chloride solution in purified absolute ethanol. Prepare fresh every time and keep in amber colored glass-stoppered bottle

Orthophosphoric acid reagent, 0.001 $M$ orthophosphoric acid solution in purified absolute ethanol

Vitamin E standard, external standards containing known amount (1 to 10 $\mu$g range) of $\alpha$-tocopherol per ml of purified absolute ethanol prepared and treated in the same manner as test samples for standardization, recovery tests, and calculation of vitamin E in unknown samples

*Procedure*

If the hexane extract of plasma, RBC or organ tissue preparations has to be directly used without TLC treatment, 3-ml aliquot of hexane extract is pipetted into suitable reaction tubes and evaporated to dryness under nitrogen. The residue is carefully dissolved in 1 ml of purified ethanol. If TLC was carried out as required for hexane extracts of RBC and organ tissue preparations, 1 ml of ethanolic eluate obtained from TLC on silica gel G is pipetted into 4-ml glass test tubes for the colorimetric assay of vitamin E. Tubes containing $\alpha$-tocopherol standard(s) treated exactly the same way as the test samples are also prepared. To each tube is added 0.2 ml of 0.2% bathophenanthroline reagent and the contents of the tubes thoroughly mixed. The assay should proceed very rapidly from this point on and care should be taken to reduce unnecessary exposure to direct light. Add 0.2 ml of ferric chloride reagent and Vortex mix. After 1 min

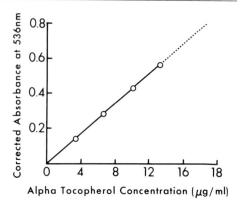

FIG. 1. A typical α-tocopherol absorbance at 536 nm using an improved spectrophotometric assay.

add 0.2 ml of orthophosphoric acid reagent and thoroughly mix the tubes once again. Read the absorbance of "blank," "test," and "standard" tubes at 536 nm using a suitable spectrophotometer set to zero with purified ethanol or deionized glass-distilled water.

*Calculations*

The following equation can be applied if the "standard" containing a known amount of α-tocopherol and "test" sample containing an unknown amount of α-tocopherol in tissue preparation are treated exactly in the same manner throughout the procedure, substituting water for tissue preparations in "standard" and reagent "blank."

$$C \text{ ``test''} = \frac{A \text{ ``test''} - A \text{ ``blank''}}{A \text{ ``standard''} - A \text{ ``blank''}} \times C \text{ ``standard''}$$

where $C$ is the concentration of α-tocopherol preferably expressed as micrograms per milliliter, and $A$ is the absorbance at 536 nm. A typical illustration of α-tocopherol absorbance at 536 nm using improved spectrophotometric assay is shown in Fig. 1.

Improved Spectrofluorometric Assay for Vitamin E

*Principle*

Spectrofluorometric assay for vitamin E as described by Taylor *et al.*[5] is based on the measurement of specific fluorescence of tocopherols at wavelengths of 286 and 330 nm ranges at which maximum excitation and emission signals are respectively obtained. An extract of the sample is prepared in an appropriate solvent such as hexane and the fluorescence is measured and compared against a standard of known concentration of tocopherols in the same solvent as that of the sample. Selection of a proper solvent is very important since different solvents can alter fluorescence measurements. Vitamin A interferes with $\alpha$-tocopherol fluorescence and should be destroyed by sulfuric acid treatment of the hexane extract.

### Reagents

Water, deionized water distilled in all glass distillation apparatus
Absolute ethanol, analytical grade or laboratory grade ethanol redistilled in all glass apparatus after adding pellets of potassium hydroxide and crystals of potassium permanganate
Hexane, analytical grade purified by distillation in all glass apparatus
Quinine sulfate standard, 1 $\mu$g/ml in 0.1 $N$ sulfuric acid
Vitamin E standard, external standards containing known amount (1 to 10 $\mu$g range) of $\alpha$-tocopherol per ml of purified absolute ethanol, prepared and treated in the same manner as test samples for standardization, recovery tests, and calculation of vitamin E in unknown samples

### Procedure

*Instrumentation.* Fluorescence is measured with a suitable fluorometer such as Aminco–Bowman spectrofluorometer linked to a ratio photometer. Normally, a slit combination of 3–1–3 mm (excitation–emission–photomultiplier turret slit) is used for $\alpha$-tocopherol determination in the 0.10 to 50 $\mu$g/ml range but use of a 5–5–5 slit combination may be necessary to increase the sensitivity to 0.05 $\mu$g/ml range. Fluorescence excitation and emission spectra are recorded on an X-Y recorder. A quinine sulfate standard (1 $\mu$g/ml in 0.1 $N$ sulfuric acid) is used to check wavelength standardization and to adjust the ratio photometer to make correction for instrumental sensitivity. The instrument is continually adjusted to give an arbitrary fluorescence intensity of 400 for the quinine sulfate standard.

*Measurement of $\alpha$-Tocopherol Fluorescence.* Hexane extracts of reagent "blank," "standard," and "test" samples are measured in Aminco–Bowman spectrofluorometer standardized as above for fluorescence at 286 nm excitation and 330 nm emission. Typical fluorescence spectra of $\alpha$-tocopherol in hexane and hexane extract of rat liver microsomes as reported by Taylor *et al.*[5] are illustrated in Fig. 2A and B,

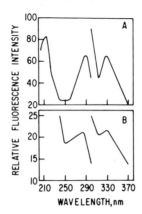

FIG. 2. Fluorescence spectra of (A) tocopherol (5 $\mu$g/ml) in hexane and (B) a hexane extract of saponifiable rat liver microsomal fraction. Fluorescence intensities are expressed relative to quinine sulfate standards. (From Taylor *et al.*, *Lipids* **11**, 530, 1976).

respectively. It should be noted that a slight shift in the excitation peak (Fig. 2B) is probably caused by quenching of the incident light by vitamin A and the usefulness of the 215 nm excitation peak is limited by severe quenching problems at this wavelength. The relative fluorescence intensity of hexane extract was found to be linear with rat liver microsome from 0 to 2 ml of microsomes and a correlation coefficient of 0.997 was found with duplicates run at each concentration.

*Calculations.* The following equation can be applied if the "blank," "standard," and "test" samples are equivalent with respect to their treatment all throughout the analysis and that "blank" and "standard" contained water instead of tissue preparation as present in "test" sample.

$$C \text{ "test"} = \frac{F \text{ "test"} - F \text{ "blank"}}{F \text{ "standard"} - F \text{ "blank"}} \times C \text{ "standard"}$$

where $C$ is the concentration of $\alpha$-tocopherol preferably expressed as micrograms per milliliter, and $F$ is fluorescence intensity.

General Remarks

The improved spectrophotometric and spectrofluorometric assays for vitamin E in animal tissues are simple, useful, and reproducible if properly conducted. The success of vitamin E assay by these procedures greatly depends on the cleanliness of the glassware and purity of solvents and chemicals to reduce interference from nonspecific reducing substances or fluorescing compounds, and preventing destruction of to-copherols by reducing unnecessary exposure to light, air, heat, alkaline conditions, and metal–ion contamination. All glassware should be soaked in a chromic acid solution and washed thoroughly with glass distilled water. Subsequent rinsing of the glassware should be carried out with redistilled ethanol followed by drying under heat or with redistilled ace-tone. All chemical reagents should be certified analytical grade and or-ganic solvents should be spectrograde. Use of silicone grease, wax pen-cils, and rubber or polyethylene should be completely avoided.

## [17] Simultaneous Determination of Reduced and Oxidized Ubiquinones

By Masahiro Takada, Satoru Ikenoya, Teruaki Yuzuriha, and Kouichi Katayama

Methods for the extraction and determination of ubiquinones (Q) have been described earlier in this series.[1–4] Recently, quantitative analyses of individual Q homologs in biologi-cal samples have been performed by high-performance liquid chromatog-raphy (HPLC) combined with an ultraviolet spectrometric detector (UVD) or by mass spectrometry (MS).[5,6] In addition, an electrochemical detector (ECD) for HPLC was confirmed to be simple and sensitive for the determination of Q.[7] However, only Q was determined by these meth-ods. For the determination of $QH_2$ (ubiquinol) and Q in mitochondria, submitochondrial particle and cell-free bacterial homogenates, the dual-wavelength spectrometric method which was developed by Chance,[8] Hatefi,[9] Pumphrey et al.,[10] Crane et al.,[2] and Kröger et al.[11] has been

[1] E. R. Redfearn, this series, Vol. 10, p. 381.
[2] F. L. Crane and R. Barr, this series, Vol. 18C, p. 137.
[3] P. J. Dunphy and A. F. Brodie, this series, Vol. 18C, p. 407.
[4] A. Kröger, this series, Vol. 53, p. 579.
[5] K. Abe, K. Ishibashi, M. Ohmae, K. Kawabe, and G. Katsui, *Vitamins* **51**, 111 (1971).
[6] S. Imabayashi, T. Nakamura, Y. Sawa, J. Hasegawa, K. Sakaguchi, T. Fujita, Y. Mori, and K. Kawabe, *Anal. Chem.* **51**, 534 (1979).
[7] S. Ikenoya, K. Abe, T. Tsuda, Y. Yamano, O. Hiroshima, M. Ohmae, and K. Kawabe, *Chem. Pharm. Bull.* **27**, 1237 (1979).
[8] B. Chance, *Rev. Sci. Instrum.* **22**, 619 (1951).
[9] Y. Hatefi, *Biochim. Biophys. Acta* **31**, 501 (1959).
[10] A. M. Pumphrey and E. R. Redfearn, *Biochem. J.* **76**, 61 (1960).
[11] A. Kroger and M. Klingenberg, *Biochem. Z.* **344**, 317 (1966).

METHODS IN ENZYMOLOGY, VOL. 105

Copyright © 1984 by Academic Press, Inc.
All rights of reproduction in any form reserved.
ISBN 0-12-182005-X

generally used. The method, however, cannot simultaneously measure the amounts of $QH_2$ and $Q$ in whole tissues owing to the presence of vitamin A and other interfering compounds which have an absorbance in the same spectral region as $Q$ and undergo an absorption change by chemical reduction. Moreover, the dual-wavelength spectrometric method cannot separately determine individual $Q$ homologs.

The analytical procedure described here was developed to provide a rapid, sensitive, and direct assay method for $QH_2$ and $Q$ in biological samples. This method is based on extraction from tissues, mitochondrial, microsomal fractions or plasma with organic solvents,[11] followed by quantitation by means of reversed-phase chromatography with UVD and ECD.[12,13]

## Apparatus

The HPLC system consisted of a YANACO L-2000 pump (Yanagimoto Manufactory Co., Ltd. Japan) with a Rheodyne loop injector. The UVD and ECD were a JASCO UVIDEC 100 (Japan Spectroscopic Co., Japan) and a YANACO VMD-101, respectively. The ECD was connected to the outlet of the UVD. Reversed-phase chromatography was carried out a Nucleosil C-18 column (15 cm × 4.0 mm i.d., Machery–Nagel Co., Germany, 5 $\mu$m). The mobile phase was prepared by dissolving 7.0 g of $NaClO_4 \cdot H_2O$ in 1000 ml of ethanol–methanol–70% $HClO_4$ (700 : 300 : 1). The flow rate was 1.2 ml/min. The HPLC measurements were performed at 30 ± 0.1°. In order to prevent oxidation by dissolved oxygen during separation, the mobile phase was deaerated by nitrogen gas bubbling.

## Preparation of Authentic $QH_2$ and $Q$

Coenzyme $Q_9(Q_9)$ and coenzyme $Q_{10}(Q_{10})$ were synthesized by Eisai Co. Ltd., Japan and Nisshin Chemicals, Japan, respectively. $Q_9$ or $Q_{10}$ solubilized with $n$-hexane was reduced by sodium borohydride aqueous solution. The $n$-hexane extract was washed three times with water, and then evaporated to dryness under a stream of nitrogen. Reduced $Q_9$ and $Q_{10}(Q_9H_2$ and $Q_{10}H_2)$ and $Q_9,Q_{10}$ were dissolved in ethanol and subjected to HPLC (Fig. 1a).

[12] K. Katayama, M. Takada, T. Yuzuriha, K. Abe, and S. Ikenoya, *Biochem. Biophys. Res. Comm.* **95**, 971 (1980).
[13] S. Ikenoya, M. Takada, T. Yuzuriha, K. Abe, and K. Katayama, *Chem. Pharm. Bull.* **29**, 158 (1981).

*Preparation of the Sample* [12,13]

Male guinea pigs were sacrificed by decapitation. The liver, heart, adrenal glands, kidneys, and brain were removed as quickly as possible, rinsed with ice-cold 0.15 $M$ NaCl, and homogenized at 4° with 4 vol (v/w) of aqueous water using a Polytron homogenizer (Hijiriseiko, Japan) at a setting 6 for 20 sec.

Mitochondria were isolated from liver of a male guinea pig and rat by differential centrifugation in 0.25 $M$ sucrose using the method of Johnson *et al.* [14] The isolated mitochondria were suspended in 20 m$M$ Tris–HCl (pH 7.4) containing 0.25 $M$ sucrose at a protein concentration of 0.54–4.2 mg/ml which was determined by the method of Lowry *et al.* [15]

The liver from a male guinea pig was homogenized in ice-cold 10 m$M$ Tris–HCl (pH 7.4)/0.25 $M$ sucrose/5 m$M$ MgCl$_2$/0.1 m$M$ EDTA. The microsomal fraction was obtained as a precipitate after the centrifugation (105,000 $g$ for 60 min) of the 15,000 $g$ (15 min) supernatant of the homogenate. The microsomes were suspended in the buffer described above at a protein concentration of 2.7 mg/ml.

*Extraction Procedure* [12,13]

One milliliter of the homogenate was poured into a test tube equipped with a glass stopper containing 7 ml of the mixture of ethanol and *n*-hexane (2 : 5 by volume), and then the tube was rapidly shaken for 10 min. [5,11] The mixture was centrifuged at 2000–3000 rpm for 3 min to separate it into two layers. The upper *n*-hexane layer was withdrawn using a Pasteur pipet. This extraction procedure was repeated three times. The combined *n*-hexane layer was evaporated to dryness under a stream of nitrogen. The resulting residue was dissolved in 0.5 ml of ethanol or isopropanol and subjected to HPLC.

One milliliter of the mitochondrial or microsomal suspensions was rapidly denatured by adding 7 ml of a mixture of ethanol/*n*-hexane (2 : 5), and then the quinone and quinol in the suspensions were extracted by the aforementioned method.

In experiments with plasma, 1-ml portions of plasma samples which had been freshly and quickly prepared were diluted with 1.0 ml H$_2$O, and then 14 ml of the ethanol/*n*-hexane (2 : 5) mixture was added to the diluted plasma. Subsequent procedures were as described above.

---

[14] D. Johnson and H. Lardy, this series, Vol. 10, p. 94.
[15] O. H. Lowry, N. J. Rosenbrough, A. L. Farr, and R. J. Randall, *J. Biol. Chem.* **193,** 265 (1951).

TABLE I
Chemical and Physical Data for Ubiquinones from
Guinea Pig Heart[a]

| Method | Expression | Peak B[b] | Peak D[b] |
|--------|-----------|-----------|-----------|
| HPLC[c] | ($t_R$, min) | 5.4 | 7.6 |
| TLC[d] | ($R_f$) | 0.29[e] | 0.40 |
| $E_p{}^f$ | (V vs Ag/AgCl) | 0.70 | −0.40 |
| UV | ($\lambda_{max}$, nm) | 290 | 275 |
| MS | ($m/e$) | M$^+$ 948,M$^+$-acetyl 905 M$^+$-diacetyl 862 | M$^+$ 862 |
| Identification | | $Q_{10}H_2$ | $Q_{10}$ |

[a] The data are from Ikenoya et al.[13]
[b] Peaks B and D are the same as Fig. 1b.
[c] HPLC conditions were the same as Fig. 1.
[d] HPTLC Silica Gel 60 F$_{254}$ (E. Merck, Germany), n-hexane/ isopropyl ether (1 : 1), UV$_{254\,nm}$.
[e] Diacetate of peak B.
[f] Peak potential obtained from the hydrodynamic voltammogram.

## Identification of Reduced and Oxidized Q[13]

Tentative identifications of $Q_{10}H_2$ and $Q_{10}$ present in the guinea pig heart were based on comparisons of the retention times, UV spectra, and hydrodynamic voltammograms of the chromatographic peaks with those of the corresponding authentic compounds. Typical chromatograms obtained from the guinea pig heart (used to identify the peaks) are shown in Fig. 1b. $Q_{10}H_2$ and $Q_{10}$ were clearly separated from each other, and had retention times consistent with those of the authentic compounds at 5.7 min (peak B) and 7.4 min (peak D), respectively. As shown in Table I, the UV spectra and hydrodynamic voltammograms of the corresponding peaks were in accord with those of the authentic compounds. Peak B, corresponding to $Q_{10}H_2$, was also characterized by the following experiments. Oxidation of the extract with PbO$_2$ or acetylation with acetic anhydride induced the disappearance of peak B, and the oxidation gave rise to an increase of the peak height of peak D corresponding to the decrease of peak B. Furthermore, we carried out a more rigorous identification of $Q_{10}H_2$ and $Q_{10}$ in the guinea pig heart by MS and thin layer chromatography (TLC). The mass fragmentation patterns of the compounds isolated by TLC were identical to those of authentic $Q_{10}$ and diacetate of $Q_{10}H_2$. In the rat heart, $Q_9H_2$ and $Q_9$ were identified as the main Q components. Therefore, based on a combination of data obtained by various tech-

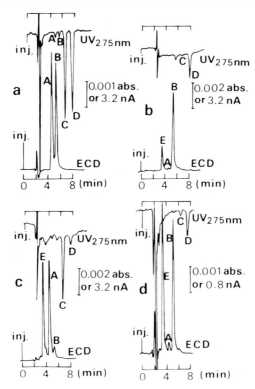

FIG. 1. Chromatograms of reduced and oxidized ubiquinones. Column, Nucleosil C-18 (5 μm), 15 cm × 4 mm i.d.; mobile phase, ethanol/methanol/70% HClO₄ (700 : 300 : 1, containing 0.05 M NaClO₄); flow rate, 1.2 ml/min; detection, UV, 275 nm, ECD, 0.7 V vs Ag/AgCl. (a) Standard (amount injected; 56 ng of each Q), (b) guinea pig heart, (c) rat heart, (d) mitochondrial fraction of guinea pig heart. Peak A, $Q_9H_2$; peak B, $Q_{10}H_2$; peak C, $Q_9$; peak D, $Q_{10}$; peak E, tocopherols and others. The data are from Ikenoya et al.[13]

niques, peaks A, B, C, and D in Fig. 1 were identified as $Q_9H_2$, $Q_{10}H_2$, $Q_9$, and $Q_{10}$, respectively.

## Chromatographic Detection[13]

Preliminary trials with measurement of the absorption at 290 nm, which is the maximum of $QH_2$, were totally unsuccessful for tissue samples because of interference by numerous UV-absorbing compounds having similar retention times, such as retinyl palmitate. When ECD in the anodic mode was, however, utilized for the determination of $Q_9H_2$ and $Q_{10}H_2$, the applied potential of 0.7 V vs Ag/AgCl was low enough to eliminate the retinyl palmitate contamination in the determination of

TABLE II

CONTENTS OF REDUCED AND OXIDIZED UBIQUINONES IN ANIMAL TISSUES[a]

| Animal | Tissue | Content ($\mu g/g$) | | | | | Total $QH_2$[c]/ total Q (%) |
|---|---|---|---|---|---|---|---|
| | | $Q_9H_2$ | $Q_9$ | $Q_{10}H_2$ | $Q_{10}$ | Total $Q$[b] | |
| Guinea pig | Heart (11) | $5.2 \pm 1.6$[d] | $7.7 \pm 1.5$ | $82.3 \pm 11.7$ | $114.7 \pm 10.7$ | $210.0 \pm 21.2$ | $41.6 \pm 3.1$ |
| | Liver (7) | $1.9 \pm 0.4$ | $1.5 \pm 0.5$ | $32.8 \pm 6.3$ | $24.2 \pm 2.4$ | $60.5 \pm 7.8$ | $57.2 \pm 3.8$ |
| | Kidney (5) | $6.6 \pm 0.7$ | $3.7 \pm 0.2$ | $109.6 \pm 20.4$ | $52.1 \pm 6.8$ | $172.1 \pm 27.0$ | $67.4 \pm 2.0$ |
| | Adrenal (4) | $3.6 \pm 0.3$ | $2.2 \pm 0.3$ | $48.8 \pm 6.9$ | $35.6 \pm 4.4$ | $89.9 \pm 11.0$ | $58.1 \pm 1.9$ |
| | Brain (4) | Trace | Trace | $10.4 \pm 2.7$ | $16.0 \pm 2.7$ | $23.9 \pm 0.9$ | $43.8 \pm 9.0$ |
| Rat | Heart (4) | $76.5 \pm 6.6$ | $156.9 \pm 7.7$ | $5.0 \pm 0.5$ | $13.3 \pm 1.1$ | $251.6 \pm 18.9$ | $32.4 \pm 2.5$ |
| | Kidney (4) | $71.5 \pm 2.0$ | $108.9 \pm 8.9$ | $10.1 \pm 0.4$ | $15.6 \pm 1.4$ | $206.1 \pm 19.5$ | $39.8 \pm 2.1$ |

[a] The data are from Ikenoya et al.[13]

[b] Total Q is the sum of $Q_9H_2$, $Q_9$, $Q_{10}H_2$, and $Q_{10}$.

[c] Total $QH_2$ is the sum of $Q_9H_2$ and $Q_{10}H_2$.

[d] Each value is represented as the mean ± SE of the numbers of animals in parentheses.

TABLE III

EFFECT OF RESPIRATORY INHIBITORS ON THE LEVELS OF
REDUCED UBIQUINONE IN GUINEA PIG LIVER MITOCHONDRIA[a]

| Condition | | $Q_{10}H_2$/total $Q_{10}$[b] |
|---|---|---|
| Inhibitor | Substrate | (%) |
| None | None | 18.9[c] |
| | 2 m$M$ succinate | 57.9 |
| | 2 m$M$ malate | 49.1 |
| | 4.5 m$M$ $\beta$-hydroxybutyrate | 60.3 |
| KCN | Succinate | 85.2 |
| | Malate | 82.2 |
| | $\beta$-Hydroxybutylate | 84.1 |
| NaN$_3$ | Succinate | 80.0 |
| Antimycin A | Succinate | 74.2 |
| Rotenone | Succinate | 54.2 |
| | Malate | 15.9 |
| Malonate | Succinate | 12.5 |
| | Malate | 47.3 |

[a] The data are from Takada et al.[16]
[b] Total $Q_{10}$ is the sum of $Q_{10}H_2$ and $Q_{10}$.
[c] Each value is represented as the average of two independent
determinations.

$Q_{10}H_2$. Ubichromenol, which was eluted at 5.9 min, was not present
(Fig. 1b).

## Determination of Reduced and Oxidized Q in Animal Tissues[12,13]

The chromatograms of reduced and oxidized Q homologs in guinea pig
and rat heart are shown in Figs. 1b and c, respectively. Peak E at the
retention time of 3.4 min was due to a mixture of tocopherols and other
electrochemically active compounds. Q components found in guinea pig
and rat were mainly $Q_{10}$ and $Q_9$ homologs. Table II shows the concentrations of reduced and oxidized Q homologs in some tissues. These is no
significant difference between the sum of QH$_2$ and Q obtained by our
method and those reported by Imabayashi et al.[6]

## Reduced Levels of Q in Mitochondrial and Microsomal Fractions[16]

Table III shows effects of inhibitors in electron transport on $Q_{10}H_2$
levels in the guinea pig liver mitochondria. In order to compare reduced
levels of Q homologs in mitochondrial respiratory chain, rat liver mitochondria which contain both $Q_9$ and $Q_{10}$, were examined to determine

TABLE IV
THE REDUCED LEVELS OF UBIQUINONES IN
RAT LIVER MITOCHONDRIA[a]

| Addition | $Q_9H_2$/total $Q_9$[b] (%) | $Q_{10}H_2$/total $Q_{10}$[c] (%) |
|---|---|---|
| None | 57.8[d] | 67.5 |
| Succinate | 74.9 | 81.3 |
| Succinate + ADP | 51.4 | 58.3 |
| Succinate + KCN | 83.3 | 89.7 |

[a] The data are from Takada et al.[16]
[b] Total $Q_9$ is the sum of $Q_9H_2$ and $Q_9$.
[c] Total $Q_{10}$ is the sum of $Q_{10}H_2$ and $Q_{10}$.
[d] Each value is represented as the average of two independent determinations.

TABLE V
REDUCTION OF MICROSOMAL UBIQUINONE[a]

| Reductant | Addition | $Q_{10}H_2$/total $Q_{10}$[b] (%) |
|---|---|---|
| None | None | 46.5[c] |
| NADH | None | 67.5 |
| | Rotenone | 65.0 |
| NADPH | None | 64.8 |
| | Rotenone | 66.2 |

[a] The data are from Takada et al.[16]
[b] Total $Q_{10}$ is the sum of $Q_{10}H_2$ and $Q_{10}$.
[c] Each value is represented as the average of two independent determinations.

levels of reduced quinones, $Q_9H_2$ and $Q_{10}H_2$ (Table IV). Table V shows the levels of $Q_{10}H_2$ in the guinea pig liver microsomes 15 min after adding NADH or NADPH.

*Plasma Levels of Reduced and Oxidized $Q$[16]*

The extractability of $Q_{10}H_2$ and $Q_{10}$ from human plasma was constant in the range of 92–98%. Similar results were obtained with plasma samples from guinea pig and rat. The $QH_2$ in the residue after evaporation of

[16] M. Takada, S. Ikenoya, T. Yuzuriha, and K. Katayama, *Biochim. Biophys. Acta.* **679**, 308 (1982).

TABLE VI
PLASMA LEVELS OF REDUCED UBIQUINONES[a]

| Animals | $Q_{10}H_2$ (%) | $Q_9H_2$ (%) | Content ($\mu$g/ml) | |
| --- | --- | --- | --- | --- |
| | | | Total $Q_{10}$[b] | Total $Q_9$[c] |
| Man (6) | $51.1 \pm 4.2$[d] | n.d.[e] | $0.92 \pm 0.12$ | n.d. |
| Guinea pig (4) | $48.9 \pm 3.1$ | n.d. | $0.35 \pm 0.02$ | n.d. |
| Rat (4) | Trace | $65.3 \pm 3.6$ | Trace | $0.27 \pm 0.03$ |

[a] The data are from Takada et al.[16] $QH_2$ and $Q$ in rat and guinea pig plasma are separated by the mobile phase prepared by dissolving 7.0 g $NaClO_4 \cdot H_2O$ in 1000 ml ethanol/methanol/$H_2O$/70% $HClO_4$ (500:500:10:1).
[b] Total $Q_{10}$ is the sum of $Q_{10}H_2$ and $Q_{10}$.
[c] Total $Q_9$ is the sum of $Q_9H_2$ and $Q_9$.
[d] Each value is represented as the mean $\pm$ SE of the numbers of animals in parentheses.
[e] Not detected.

the *n*-hexane extract was stable for approximately 5 hr when kept under nitrogen gas. Once the residue was dissolved in isopropanol, however, the solution had to be injected into the HPLC column as soon as possible (within 30 min), since the level of $QH_2$ in the solution gradually decreased and was several percent lower 12 h after dissolution.

Table VI shows the plasma levels of $QH_2$ in man, guinea pig and rat.

# [18] Assay of Carotenoids

By NORMAN I. KRINSKY and SUDHAKAR WELANKIWAR

There have been numerous articles and reviews written dealing with the analysis of carotenoid pigments,[1,2] including several excellent chapters which have appeared in earlier volumes of this series.[3–5] As evidence has accumulated that carotenoid pigments can function as effective

[1] B. H. Davies, *in* "Chemistry and Biochemistry of Plant Pigments" (T. W. Goodwin, ed.), p. 38. Academic Press, New York, 1976.
[2] E. DeRitter and A. E. Purcell, *in* "Carotenoids as Colorants and Vitamin A Precursors" (J. C. Bauernfeind, ed.), p. 815. Academic Press, New York, 1981.
[3] G. Britton and T. W. Goodwin, this series, Vol. 18C, p. 654.
[4] S. Liaaen-Jensen and A. Jensen, this series, Vol. 23, p. 586.
[5] R. F. Taylor and M. Ikawa, this series, Vol. 67, p. 233.

METHODS IN ENZYMOLOGY, VOL. 105

Copyright © 1984 by Academic Press, Inc.
All rights of reproduction in any form reserved.
ISBN 0-12-182005-X

quenchers of singlet oxygen[6,7] and oxygen-centered radicals,[8,9] interest has been rekindled in the utilization of these compounds as probes for mechanisms of oxygen radical generation, action, and damage. This contribution will deal with methods of extraction of carotenoid pigments, their separation by chromatographic procedures such as thin-layer chromatography (TLC) and by high-performance liquid chromatography (HPLC), the means for identifying these pigments, and, finally, the methods for quantitating the pigments present in biological samples. The entire field of carotenoids was reviewed in a monograph by Isler *et al.*[10] The biochemical aspects of carotenoids are reviewed in an excellent book by Goodwin.[11]

*Pigment Extraction*

There are a variety of techniques described in the literature for extracting carotenoid pigments from different tissue samples. Care must be exercised because of the facile oxidation which many carotenoid pigments can undergo as well as their lability to isomerization caused by light. Under these circumstances, it is usually prudent to work rapidly in dimly lit rooms using amber or red glassware, keep temperatures relatively low, and avoid excessive exposure to oxygen. The latter consideration involves storing and evaporating samples under nitrogen.

The extraction procedure involves macerating, grinding, or homogenizing the tissue with either alcohol–water or acetone–water mixtures. These solvents serve as effective means of denaturing proteins and permitting the release of carotenoid pigments from any binding or association that they may have with a protein component. The pigments themselves are then transferred to an organic solvent such as petroleum ether or diethyl ether. The transfer is facilitated and emulsions are prevented by the addition of relatively strong salt solutions (6–10% NaCl, w/v) to make the aqueous component approximately 50% water (v/v). The extract can then be reduced in volume for purposes of chromatographic separation, or investigated directly by spectrophotometric techniques. Solvent reduction can occur by evaporating the material at reduced pressure at temperatures not exceeding 40° or through reducing the sample volume by a

[6] C. S. Foote and R. W. Denny, *J. Am. Chem. Soc.* **90**, 6233 (1968).
[7] S. M. Anderson and N. I. Krinsky, *Photochem. Photobiol.* **18**, 403 (1973).
[8] J. E. Packer, J. S. Mahood, V. O. Mora-Arellano, T. F. Slater, R. L. Willson, and B. S. Wolfenden, *Biochem. Biophys. Res. Commun.* **98**, 901 (1981).
[9] N. I. Krinsky and S. M. Deneke, *J. Natl. Cancer Inst.* **69**, 205 (1982).
[10] O. Isler, H. Gutmann, and U. Solms, "Carotenoids." Birkhäuser, Basel, 1971.
[11] T. W. Goodwin, "The Biochemistry of the Carotenoids," 2nd ed. Chapman & Hall, London, 1980.

stream of nitrogen at a temperature not exceeding 40°. Both of these processes should be carried out in dim light.

In some cases, the extraction can be facilitated by saponifying the tissue. A common procedure is to use 6% KOH in MeOH (w/v) and carrying out the saponification under nitrogen at 40° for 10 min to 1 hr. The extraction procedure follows that described above with the exception that the organic phase is washed with water or dilute salt until free of alkali.

*Thin-Layer Chromatography (TLC)*

The ability to resolve carotenoid pigments by TLC using a wide variety of adsorbents and solvents has made this a favored technique for the resolution and characterization of carotenoid pigments. Despite its apparent simplicity, there are dangers that are associated with the process with respect to the accurate determination of the nature of the pigments involved. The dangers include the potential for photoisomerization, possible oxidation if the pigments are exposed to air when the TLC plate is dry, and chemical alterations to certain susceptible pigments if they are exposed to mild acid vapors, the kind normally present in most chemical and biochemical laboratories. The best separations utilize two-dimensional TLC and the literature abounds with examples of this process, using either silica gel, cellulose, $Al_2O_3$, $Ca(OH)_2$, or MgO. On glass plates, the carotenoid spots can be scraped into sintered glass funnels and the pigment extracted with either alcohol or acetone. On plastic sheets, the pigmented spot can be cut from the chromatogram, cut into small sections, immersed in either alcohol or acetone, and the pigment extracted via filtration. An excellent example of the speed and power of this type of analysis has appeared recently in an article describing the separation of carotenoid pigments from marine phytoplankton.[12] A more comprehensive review of carotenoid TLC is beyond the scope of this chapter, but the reader is referred to the excellent reviews of this technique by Davies[1] and by Taylor.[12a]

*High-Performance Liquid Chromatography (HPLC)*

Of even greater resolving power than that of TLC is the procedure of HPLC for the analysis of carotenoid pigments. This material was last reviewed in Volume 67 of this series[5] and additional information has appeared since that publication.[12a] In particular, the introduction of reverse-phase HPLC (rp-HPLC) has allowed a very rapid and selective separation of carotenoid pigments of varying polarities in a relatively

[12] S. W. Jeffrey, *Limnol. Oceanogr.* **26,** 191 (1981).
[12a] R. F. Taylor, *Adv. Chromat.* **22,** 157 (1983).

HPLC Systems Used for Carotenoid Analysis[a]

| Compounds and sources | Column packing(s) | Mobile phases | References |
|---|---|---|---|
| Isolated carotenoids | Spherisorb 5 $\mu$m | 0–40% acetone in hexane–MeOH (99:1) | Fiksdahl et al.[b] |
| Spinach chloroplasts | Nucleosil 50–5 5 $\mu$m | Isooctane–98% ethanol (90:10) | Stransky[c] |
| Spinach leaves | Silica gel SS-05 0.5 $\mu$m | 1–10% isopropanol in hexane | Iriyama et al.[d] |
| Green algae | Sorb-Sil 60-D 10$C_{18}^{k}$ | 50–100% MeOH in EtOH; 5–15% $H_2O$ in MeOH | Braumann and Grimme[e] |
| Tomatoes | Partisil-PXS-10/25 ODS-2[k]; Partisil-PXS-5 ODS[k] | 8–11.6% $CHCl_3$ in $CH_3CN$ | Zakaria et al.[f] |
| Isolated carotenoids | Nucleosil 10$C_{18}^{k}$ | 40–5% $H_2O$ in acetone | Matus et al.[g] |
| Green algae and spinach | LiChrosorb RP-8 10 $\mu$m[k]; Sil 60-RP 18 10 $\mu$m[k] | 75–100% $CH_3CN$–MeOH (75:25) in $H_2O$ | Braumann and Grimme[h] |
| Cheese | Lichrosorb Si 60 5 $\mu$m | Methyl ethyl ketone–hexane (10:90) | Stancher and Zonta[i] |
| Spinach carotenoids | LiChrosorb RP-18 10 $\mu$m[k] | $CH_3CN$–$CH_3OH$ (85:15) and hexane–MeOH (25:75) | Krinsky and Welankiwar[j] |

[a] For earlier citations, refer to Table VII of Taylor and Ikawa.[5]
[b] A. Fiksdahl, J. T. Mortensen, and S. Liaaen-Jensen, J. Chromatogr. **157**, 111 (1978).
[c] H. Stransky, Z. Naturforsch. **33C**, 836 (1978).
[d] K. Iriyama, M. Yoshiura, and M. Shiraki, J. Chromatogr. **154**, 302 (1978).
[e] T. Braumann and L. H. Grimme, J. Chromatogr. **170**, 264 (1979).
[f] M. Zakaria, K. Simpson, P. R. Brown, and A. Krstulovic, J. Chromatogr. **176**, 109 (1979).
[g] Z. Matus, M. Baranyai, G. Toth, and J. Szabolcs, Chromatographia **14**, 337 (1981).
[h] T. Braumann and L. H. Grimme, Biochim. Biophys. Acta **637**, 8 (1981).
[i] B. Stancher and F. Zonta, J. Chromatogr. **238**, 217 (1982).
[j] N. I. Krinsky and S. Welankiwar, this study.
[k] Reversed phase HPLC.

short period of time. The pigments are detected using a fixed or variable wave length detector. Choice of solvents is still very flexible and can be used as isocratic, step-gradient, or continuous-gradient solutions, depending on the separation desired and the equipment available. In addition, stopped-flow spectrophotometry can be useful in helping to characterize individual peaks as well as determining the purity of peak fractions. Finally, the HPLC can be coupled to a fraction collector permitting the isolation of the separated pigments for further characterization. The table summarizes some of the HPLC systems reported for carotenoid separation and identification which have appeared since the report of Taylor and Ikawa.[5]

We have use a relatively simple rp-HPLC system to rapidly resolve and characterize plant carotenoids. We feel this system should be applicable to many other types of carotenoid analyses and can be adapted or improved without an undue amount of work on the part of the experimenter. Our separation is depicted in Fig. 1, in which we readily separated the major carotenoid pigments of spinach in an 18 min run. The apparatus used was an Altex reciprocating pump, Model 110A, and an Hitachi–Perkin Elmer spectrophotometer, Model 155, set at 450 nm, was used to detect the pigments. The column was a 10 $\mu$m Lichrosorb RP-18 (4.6 mm × 25 cm). We used a step gradient of two degassed solvent mixtures, with the change indicated in Fig. 1 occurring at 11 min after

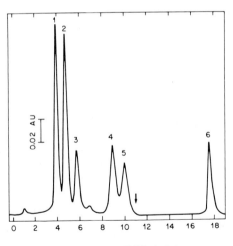

*RETENTION TIME (min)*

FIG. 1. HPLC of spinach carotenoids supplemented with zeaxanthin. A 10-$\mu$m LiChrosorb RP-18 column was eluted, at 2.0 ml/min, with acetonitrile–methanol (85 : 15) for 11 min, and then the solvent was changed to hexane–methanol (25 : 75). The detector was set at 450 nm. (1) Neoxanthin; (2) violaxanthin; (3) antheraxanthin; (4) lutein; (5) zeaxanthin; (6) $\beta$-carotene.

sample injection. Our first solvent consisted of acetonitrile–methanol (85 : 15) and this was changed to hexane–methanol (25 : 75) to remove the hydrocarbon fraction. The solvents were applied using a type 50 Rheodyne valve. The sample was injected into an Altex valve, Model 210, equipped with a 20-$\mu$l loop. For this particular analysis, the spectrophotometer range was set at 0.2 absorbancy units (AU). The samples were collected in a fraction collector in series with the detector and the individual pigments were characterized as described below. Using this procedure, we very readily separated neoxanthin, violaxanthin, antheraxanthin, lutein, zeaxathin, and $\beta$-carotene.

### Identification of Carotenoids

There are several techniques available for identifying carotenoid pigments. The primary identification is based on the spectral properties of the pigments, including not only the absorption spectrum, but also the quantitative analysis of absorption, usually depicted as either the $E_{1\,cm}^{1\%}$ or by the molar extinction coefficient, $\Sigma$. The absorption spectrum is a function of the number of conjugated double bonds present in the molecule, as well as other functional groups which may be allylic to the conjugated double-bond system. Extensive tables of $E_{1\,cm}^{1\%}$ values are available in the literature.[1,2]

In addition to the spectral properties, there are a number of relatively simple procedures which can be utilized to help characterize the nature of the carotenoid pigment. Differences in polarity can be readily determined by partitioning the compounds between an organic and aqueous phase. The most common system used for partitioning carotenoids is 95% aqueous methanol/petroleum ether,[13] although there are other solvent systems which yield much more information about the relative polarity of these compounds.[14,15] Systems have been described which permit the accurate determination of the relative polarity of carotenoid pigments working at submicrogram quantities of the pigments.[14]

Other techniques that have been used to facilitate characterization of carotenoid pigments involve the use of reducing agents such as NaBH$_4$ for the reduction of carbonyl groups.[16] If these carbonyl groups are in conjugation with the double bond system, the reduced product has a very marked change in spectrum and this is useful in identifying carbonyl

[13] F. J. Petracek and L. Zechmeister, *Anal. Chem.* **28**, 1484 (1956).
[14] N. I. Krinsky, *Anal. Biochem.* **6**, 293 (1963).
[15] B. P. Schimmer and N. I. Krinsky, *Biochemistry* **5**, 3649 (1966).
[16] N. I. Krinsky and T. H. Goldsmith, *Arch. Biochem. Biophys.* **91**, 271 (1960).

compounds. In addition, the reduction of a carbonyl group to an alcohol group changes the relative polarity of the molecule.

Another test for the nature of carotenoid pigments involves the marked hypsochromic shift when epoxide-containing carotenoids are treated with traces of acid.[17] The spectral shift can amount to 18–20 nm per conjugated epoxide and can be detected very readily, again at submicrogram quantities.[18] Acidic chloroform can be used to dehydrate carotenoids with tertiary hydroxyl groups, thereby altering their partition and chromatographic behavior.[18]

Another technique that has been used, particularly with characterizing naturally occurring cis-carotenoids, involves the appearance of a 340-nm band which is associated with cis-carotenoids. This band can be altered, as well as the major pigment peaks, by photoisomerization, usually carried out in the presence of iodine and visible light.

Finally, the position of standard carotenoids on TLC as well as the retention times of standard carotenoids on HPLC are immensely useful in identifying carotenoid pigments.

## Quantitation

The quantitative analysis of carotenoid pigments is based on their extinction coefficients. These are usually reported at the wave length of maximum absorption, but one can also readily determine the extinction coefficients at other wave lengths and use these to determine the quantity of carotenoid pigments present. This can be useful in the HPLC separation of carotenoids, where the detector is set at a given wavelength. In addition, the amount of carotenoids can be determined by calculating the area of eluted peaks.

Following saponification and extraction of frozen spinach, the major carotenoid pigments were isolated by column chromatography on deactivated $Al_2O_3$ (5% $H_2O$) and the individual fractions eluted with increasing concentrations of acetone in petroleum ether. Following the addition of zeaxathin, the pigments were separated by rp-HPLC, as described earlier. The resolution was quite adequate and permitted the isolation of individual fractions of each of the major peaks. Neoxanthin, violaxanthin, and antheraxanthin were all characterized by their hypsochromic displacements on the addition of traces of ethanolic HCl. In addition, their relative polarity was determined using either 60% aqueous methanol/petroleum ether–diethyl ether or 70% aqueous methanol/petroleum ether–

[17] P. Karrer, Helv. Chim. Acta **28**, 474 (1945).
[18] B. P. Schimmer and N. I. Krinsky, Biochemistry **5**, 1814 (1966).

diethyl ether. Lutein and zeaxanthin were characterized by their spectral properties, failure to display hypsochromic shifts upon addition of ethanolic HCl, and by their relative polarity values in 80% aqueous methanol/ petroleum ether. These five xanthophylls were separated in 11 min using an isocratic solvent of acetonitrile–methanol (85 : 15). To elute the hydrocarbon fraction, a less polar solvent is necessary in rp-HPLC, and, at 11 min, the solvent was changed to hexane–methanol (25 : 75) and this resulted in the elution of $\beta$-carotene at approximately 17 min. We have used this system for characterizing the effluents from column chromatography, resulting in a very rapid characterization and determination of purity, based on retention times. Under circumstances of detecting column effluents, the amounts required can be scaled down by a factor of 10, permitting the analysis of nanogram quantities of carotenoid pigments.

*Conclusions*

We have presented the general principles whereby carotenoid pigments can be extracted, separated, identified, and quantitated using both TLC and rp-HPLC. The latter system, in conjunction with a variety of tests carried out on the individual fractions obtained from chromatograph effluents, results in the rapid characterization and quantitation of a variety of carotenoids, particularly those from plant tissues. Similar systems are applicable to animal systems as well as *in vitro* systems to which carotenoids or tissue extracts have been added.

# [19] Uric Acid: Functions and Determination

By PAUL HOCHSTEIN, LINDA HATCH, and ALEX SEVANIAN

## Functions for Urate in Biological Systems

Uric acid has been known for almost 100 years as an end product of purine metabolism. However, it has only been in recent years that its capacity to act as a free-radical scavenger and a potentially important biological antioxidant has been recognized. It should be noted that many hydroxylated compounds, including purine bases and their derivatives, may act as scavengers of singlet oxygen, superoxide anions, and hydroxyl radicals. Uric acid may be unique in this regard because of both its high reactivity with such chemical species and its high concentration in biolog-

Copyright © 1984 by Academic Press, Inc.
All rights of reproduction in any form reserved.
ISBN 0-12-182005-X

ical fluids. It has been proposed that these attributes predict an important role for urate in oxidant and radical-induced aging and cancer.[1]

Urate may have yet additional functions, unrelated to its radical scavenging and consequent oxidation to allantoin. Recent experiments indicate that it forms complexes with iron.[2] This reaction results in the inhibition of chelated as well as free iron-dependent oxidation of ascorbic acid and the iron-dependent peroxidation of lipids in the presence of hydroperoxides. This latter inhibition of a Fenton-type reaction, without the apparent oxidation of urate, may serve to block the formation of free radicals and the initiation of damaging reactions in membranes as well as to other biological constituents. The finding that urate may complex with iron is supportive of earlier findings of Albert[3] and has implications for the suggestions by Mazur and his colleagues that xanthine oxidase activity may function in the mobilization of iron from ferritin.[4]

Finally, it has also been demonstrated that urate may regulate the synthesis of prostaglandins.[5,6] This effect is presumably mediated through its activity in scavenging hydroxyl radicals and may provide for additional roles of urate in a variety of normal and pathophysiological states.

## Formation of Urate

The high concentration, up to 450 $\mu M$, of urate in the plasma of man and certain primates is a consequence of the evolutionary loss of uricase (urate oxidase) and the development of efficient reabsorption mechanisms in the kidney. In mammalian tissues, free purine bases derived from nucleoside cleavage are ultimately oxidized by xanthine oxidase to yield uric acid. Purines are also derived from dietary sources and plasma levels of urate may be substantially increased in individuals on a high purine diet.

Xanthine oxidase activity is present in liver and small intestinal mucosa although traces of activity have been noted in heart and skeletal muscle as well as kidney and spleen.[7] In this connection, it is important to note that xanthine oxidase activity has been found to result from the

[1] B. N. Ames, R. Cathcart, E. Schwiers, and P. Hochstein, *Proc. Natl. Acad. Sci. U.S.A.* **78,** 6858 (1981).
[2] K. J. A. Davies, A. Sevanian, S. F. Muakkassah-Kelly, and P. Hochstein, to be published, 1984.
[3] A. Albert, *Biochem. J.* **54,** 646 (1953).
[4] A. Mazur and A. Carlton, *Blood* **26,** 317 (1965).
[5] N. Ogino, S. Yamamoto, O. Hayaishi, and T. Tokuyama, *Biochem. Biophys. Res. Commun.* **87,** 184 (1979).
[6] C. Deby, G. Deby-Dupont, F. X. Noel, and L. Lavergue, *Biochem. Pharmacol.* **30,** 2243 (1981).
[7] R. W. E. Watts, J. E. M. Watts, and J. E. Seegmiller, *J. Lab. Clin. Med.* **666,** 688 (1965).

modification of native xanthine dehydrogenase by limited proteolysis[8] or by the oxidation of sulfhydryl groups.[9] It seems likely that the conversion of xanthine dehydrogenase, which utilizes $NAD^+$, to xanthine oxidase, which utilizes $O_2$, may have important physiological consequences. At the present time, although uric acid formation seems to be predominantly a hepatic process, the importance of its formation in other tissues should not be minimized. It is of interest that uric acid ribonucleoside has been reported to be present in beef erythrocytes as well as in liver[10] and that an antioxidant function for ribosyluric acid has been described.[11]

Urate is present in the plasma, at pH 7.4, as a sodium salt (uric acid has a p$K$ of 5.8). A small portion of uric acid, less than 5%, may be bound to plasma proteins. Its concentration in males approaches its maximum solubility (about 7 mg/100 ml) and individuals with higher levels are susceptible to gout[12] if not increased intelligence.[13] In a normal male about 500 mg of urate per day is excreted from a pool of about 1200 mg. The table lists the plasma concentrations of urate and some other substances with antioxidant activities.

### Determination of Urate

In the past, the most commonly used methods for the determination of urate utilized its capacity to act as a reducing agent.[14] However, these colorimetric methods based, for example, on the reduction of sodium tungstate have inherent difficulties. These are related to the formation of turbidity, nonlinearity, and, most important, the interference by other reducing substances, e.g., glutathione.

Uric acid has a characteristic absorption spectrum with a maximum at 292 nm and a molar extinction coefficient of 12,500 $cm^2$/mol at pH 9.4. In the presence of uricase, urate is converted to allantoin which has no absorption at this wavelength. The decrease in optical density at 292 nm is a direct measure of the amount of uric acid consumed in samples.[15] This method avoids problems associated with the precipitation of protein and,

[8] M. G. Battelli, E. Della Corte, and F. Stripe, *Biochem. J.* **126,** 747 (1972).

[9] E. Della Corte and F. Stripe, *Biochem. J.* **126,** 739 (1972).

[10] R. C. Smith and C. M. Stricker, *J. Anim. Sci.* **41,** 1674 (1975).

[11] R. C. Smith, *Fed. Proc. Fed. Am. Soc. Exp. Biol.* **41,** 1287 (1982).

[12] J. B. Wyngaarden and W. N. Kelly, *in* "Metabolic Basis of Inherited Disease" (J. B. Stanbury, J. B. Wyngaarten, and D. S. Fredrickson, eds.). McGraw-Hill, New York, 1978.

[13] K. S. Park, E. Inouye, and A. Asaka, *Jpn. J. Hum. Genet.* **25,** 193 (1980).

[14] O. Folin, *J. Biol. Chem.* **101,** 111 (1933).

[15] E. Praetorius, *Scand. J. Clin. Lab. Invest.* **1,** 222 (1949).

PLASMA CONCENTRATIONS OF URATE AND
OTHER ANTIOXIDANTS[a]

| Substance | mg/100 ml | $\mu M$ |
|---|---|---|
| Urate | | |
| Males | 2.6–7.5 | 160–450 |
| Premenopausal females | 2.0–5.7 | 120–340 |
| Ascorbate | | |
| Normal adult | 0.7–2.5 | 40–140 |
| Normal adult + 2 g supplement | 1.7–2.8 | 100–160 |
| Vitamin E | 0.5–1.6 | 10–40 |
| Carotenoids | 0.09–0.12 | 2 |

[a] From Ref. 1.

providing that appropriate blanks are utilized to correct for reactions due to other substances, is sensitive and accurate.

Since the conversion of uric acid to allantoin by uricase involves the formation of $H_2O_2$, the reaction may be followed by coupling peroxide formation to the oxidation of o-dianisidine[16] or to the oxidation of scopoletin.[17] These methods combine the specificity of uricase activity with the ease of colorimetric or fluorometric detection.

In recent years rapid and sensitive methods for the determination of urate by high-pressure liquid chromatography (HPLC) have been used.[18] We have developed an HPLC method by which a variety of purines and their metabolites may be measured in biological samples or in a number of buffer systems.

For example, when urate measurements in serum or urine are required, a 500-$\mu l$ aliquot is applied to an aminopropyl-$NH_2$, 500 mg Bond Elute column (Analytichem International, Harbor City, California) previously conditioned with 1.0 ml acetonitrile. Vacuum-assisted elution of the sample fluid is followed with a 500 $\mu l$ rinse of the column with acetonitrile. These eluents are discarded. Collection of a final 1.0 ml wash with 0.1 $M$ $NaH_2PO_4$ permits quantitative recovery of uric acid and ascorbic acid as well as a number of related compounds.

Aqueous samples or those prepared as above are injected into a 4.6 mm × 30 cm, $\mu$Bondapak–$NH_2$ column (Waters Assoc., Milford, Massa-

[16] G. F. Domack and H. H. Schlicke, Ann. Biochem. 22, 219 (1968).
[17] P. L. Bloch and G. F. Lata, Ann. Biochem. 38, 1 (1970).
[18] E. J. Weinman, D. Steplock, S. C. Sansom, T. F. Knight, and H. O. Senekjian, Kidney Int. 19, 83 (1981).

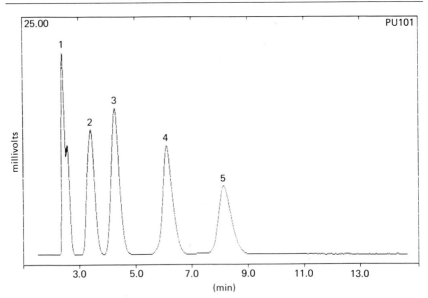

FIG. 1. Chromatographic tracing of uric acid, ascorbic acid, and other purine metabolites. Samples were prepared in 10 m$M$ Tris buffer, pH 7.4, at the following concentrations: (1) caffeine, 2.28 m$M$–2.38 min; (2) xanthine, 1.26 m$M$–3.40 min; (3) inosine, 0.63 m$M$–4.24 min; (4) urate, 1.39 m$M$–6.07 min; (5) ascorbate, 1.26 m$M$–8.09 min. High-pressure chromatography was performed using a Perkin–Elmer series 4 instrument and an LC-85 detector. The relative absorbances at 248 nm are recorded in millivolts where 0.04 absorbance units full-scale is equivalent to 10 mV. Elution conditions are described in the text. $T_0$ (using hexane) was found to be 2.07 min.

chusetts) which is eluted with 75 : 25 acetonitrile : NaH$_2$PO$_4$ (0.04 $M$) at a flow rate of 1.5 ml/min. Urate, ascorbate, and a number of other purines are detected in the eluent by spectrophotometric measurement at 248 nm.

A typical chromatogram for a sample containing a mixture of ascorbic acid, uric acid and related purine metabolites is shown in Fig. 1. This mixture was prepared in 0.01 $M$ Tris–HCl, pH 7.0, containing the following compounds: 1.26 m$M$ ascorbate, 1.26 m$M$ xanthine, 0.63 m$M$ inosine, 2.28 m$M$ caffeine, and 1.39 mM urate. A recovery of > 95% of all components was obtained when the mixture was applied to the Bond Elute preparative column described above. Interfering compounds which can be a potential problem in analyzing biological specimens include: theophylline, recovery time (rt) = 2.6 min; tyrosine, rt = 2.2 min; and dopamine, rt = 2.4 min or related metabolites.

# [20] Pulse Radiolysis Methodology

## By Klaus-Dieter Asmus

Oxygen-centered radicals are known to play significant roles in many chemical and biochemical reaction mechanisms. The most interesting among these radicals are probably the superoxide anion $O_2^-$ and its conjugate acid $HO_2$, the hydroxyl radical $\cdot OH$ and its conjugate base $O^-$, and organic peroxy and oxy radicals, $RO_2$ and $RO\cdot$. Reactions of these radicals with a substrate, or radical–radical reactions (combination, disproportionation) most often proceed with rates which are only controlled by the diffusion of the reactants, i.e., with absolute rate constants for a bimolecular process in the order of $10^9$–$10^{10}$ $M^{-1}$ $\sec^{-1}$ in solution. This in turn yields half-lives for these reactions in the nano- to millisecond time range whenever the concentrations of the reacting species exceed micromolar concentrations. These times are considerably shorter than those detectable by conventional or even rapid-mixing techniques. For direct observation of such reactions and of the radicals themselves a technique is therefore desirable which provides (1) the possibility of producing ("mixing") one (usually the radical) or sometimes even both of the reacting species in a time which is short compared with the time of their reaction; (2) a measurable physical property of either one of the reacting species or their immediate products such as optical absorption, conductivity, spin, etc.; (3) a high enough concentration of these species, which essentially depends on the magnitude of the measurable parameter, e.g., extinction coefficient, specific conductance, etc.; and (4) a time resolution high enough to allow recording of the above-mentioned physical properties as a function of time.

All these requirements are best met by any technique which provides an *in situ* generation of at least one of the reaction partners. Among these pulse radiolysis is very powerful and versatile.[1-5] It is based on a sudden creation of entirely new chemical species by exposure of a chemical system to a short pulse of high-energy radiation. Pulse radiolysis is therefore

[1] M. S. Matheson and L. M. Dorfman, *J. Phys. Chem.* **32,** 1870 (1960).
[2] R. L. McCarthy and A. MacLachlan, *Trans. Faraday Soc.* **56,** 1187 (1960).
[3] J. P. Keene, *Nature (London)*, **188,** 843 (1960).
[4] M. S. Matheson and L. M. Dorfman, "Pulse Radiolysis." M.I.T. Press, Cambridge, Massachusetts, 1969.
[5] J. H. Baxendale and F. Busi, "The Study of Fast Processes and Transient Species by Electron Pulse Radiolysis" (NATO Advanced Study Institute Series). Reidel, Dordrecht, 1982.

METHODS IN ENZYMOLOGY, VOL. 105
Copyright © 1984 by Academic Press, Inc.
All rights of reproduction in any form reserved.
ISBN 0-12-182005-X

in principle similar to corresponding photolysis techniques such as laser flash. The two methods differ only in the amount of external energy deposited into the irradiated system. In radiolysis the energy can be absorbed by any molecular or atomic constituent, which makes this technique more versatile but of course also less selective.

*Formation of Primary Radicals*

Pulse radiolysis is best suited for investigations in the liquid phase although it has also successfully been applied to gaseous and solid-state systems. The energy carriers are usually accelerated electrons with energies in the MeV range; i.e., any incident particle has an energy which is almost a million times higher than necessary for a bond rupture or ionization process. The net result of the energy absorption in an irradiated system can be summarized in the following points.[4-8]

1. Any high-energy electron loses its energy in a consecutive series of about $10^4$ individual steps with an average deposit of 60–100 eV for each interaction with an atomic constituent of the penetrated matter.
2. For high-energy electrons the individual energy deposits have an average effective radius of $\sim$30 Å and are $\sim$1000 Å apart from each other.
3. Each energy deposit results on average in one or two ionizations and excitations. Interaction depends on the electron density of the matter, and therefore in dilute solutions will essentially occur with the solvent.
4. The whole energy distribution occurs within less than $10^{-14}$ sec for an individual high energy electron.

In aqueous solutions, for example, the energy is used for ionization

$$H_2O \rightsquigarrow H_2O^+ + e^- \tag{1}$$

and excitation

$$H_2O \rightsquigarrow H_2O^* \tag{2}$$

The initial interaction products, $H_2O^+$, $e^-$, and $H_2O^*$, are then converted within $10^{-12}$ sec or less into highly reactive primary radical species,

[6] A. Henglein, W. Schnabel, and J. Wendenburg, "Einführung in die Strahlenchemie." Verlag Chemie, Weinheim, 1969.
[7] J. W. T. Spinks and R. J. Woods, "An Introduction to Radiation Chemistry." Wiley, New York, 1964 and 1976.
[8] A. J. Swallow, "Radiation Chemistry." Longman, London, 1973.

namely hydrated electrons, $e_{aq}^+$, hydrogen atoms, H·, and hydroxyl radicals, ·OH, via

$$H_2O^+ \rightarrow H_{aq}^+ + \cdot OH \tag{3}$$
$$e^- \rightarrow e_{aq}^- \tag{4}$$
$$(e^- + H_2O^+) \rightarrow H_2O^* \rightarrow H\cdot + \cdot OH \tag{5}$$

(Most of the excited water molecules probably suffer collisional deactivation.) Corresponding processes can be formulated for other solvents.

In conclusion, the absorption of a pulse of high-energy electrons results in the formation of reactive radical species which are practically homogeneously distributed throughout the irradiated volume.[9] Furthermore, their generation is completed within the duration of even the shortest, i.e., subnanosecond pulses available from an accelerator.

The radiation chemical yields of freely diffusing $e_{aq}^-$, ·OH, and H· in aqueous solutions are $G \approx 2.75$, 2.75, and 0.6 (species per 100 eV absorbed energy), respectively, which for an absorbed dose of 10 Gy (1000 rad) correspond to the same numbers in micromoles. The same order of magnitude applies also for the yield of primary species in other irradiated liquid system.[5–8]

*Preparation of Solutions and Formation of Oxygen-Centered Radicals*

Solutions are generally prepared from analytically pure compounds and specially purified solvents to avoid unwanted reactions of the primary radical species with possible impurities. In the case of aqueous systems the water should have gone through a multiple distillation process, preferably in quartz vessels with at least one oxidative step (e.g., distillation from $K_2Cr_2O_7$ or $KMnO_4$ solutions). Over the last years also other purifications such as Millipore filtration and ion exchange have successfully been applied. A critical test for the purity of water is generally the lifetime of the hydrated electron in a deoxygenated solution ($\geq 10$ $\mu$sec at a dose of 2–3 Gy), and an associated conductivity signal (if available) with the same decay kinetics as the optical signal in slightly acidic solutions (pH 4–5).[10–12]

Selective investigation of oxygen centered radicals can be achieved by

---

[9] The homogeneity no longer applies for accelerated particles of higher mass, e.g., protons, heavy ions, etc. Their energy is released in tracks (overlapping spurs).

[10] E. J. Hart and M. Anbar, "The Hydrated Electron." Wiley (Interscience), New York, 1970.

[11] K.-D. Asmus, G. Beck, A. Henglein, and A. Wigger, *Ber. Bunsenges. Phys. Chem.* **70,** 869 (1966).

[12] G. Beck, *Int. J. Radiat. Phys. Chem.* **1,** 361 (1969).

special composition of the irradiated systems.[4-8] Thus it is possible to convert hydrated electrons into hydroxyl radicals via

$$N_2O + e_{aq}^- \rightarrow N_2 + \cdot OH + OH^- \tag{6}$$

In $N_2O$ saturated solutions ($2.4 \times 10^{-2}$ $M$) this reaction is very fast ($t_{1/2} \approx$ 3.4 nsec based on $k_6 = 8.8 \times 10^9$ $M^{-1}$ sec$^{-1}$ [13]) and thus leads to a chemical system which provides almost only $\cdot OH$ radicals as reactive species (90%; the remaining 10% are H$\cdot$ atoms). High concentrations of other good electron scavengers will of course lower the yield of reaction (6) due to competition. This is of particular relevance in acid solutions where conversion of hydrated electrons into H$\cdot$ atoms via

$$H_{aq}^+ + e_{aq}^- \rightarrow H\cdot \tag{7}$$

comes into play.

Investigations with $O^-$ radical anions require the presence of high base concentrations since the p$K$ of the

$$\cdot OH \rightleftharpoons H^+ + O^- \tag{8}$$

equilibrium is 11.9.[14]

Superoxide radicals are best obtained by the reactions

$$e_{aq}^- + O_2 \rightarrow O_2^- \tag{9}$$
$$H\cdot + O_2 \rightarrow HO_2 \tag{10}$$

Their relative yields are determined by the acid–base equilibrium

$$HO_2 \rightleftharpoons H_{aq}^+ + O_2^- \tag{11}$$

with a p$K$ of 4.9.[15] Additional superoxide formation and simultaneous removal of $\cdot OH$ radicals is achieved by addition of formate (or formic acid) (e.g., $10^{-1}$ $M$) to the solution which results in the reaction sequence

$$HCOO^- + \cdot OH \rightarrow CO_2^- + H_2O \tag{12}$$
$$CO_2^- + O_2 \rightarrow O_2^- + CO_2 \tag{13}$$

Peroxy radicals $RO_2$ result from addition of $O_2$ to the organic radical R$\cdot$. If the latter is formed via an $\cdot OH$ radical-induced process, the solutions are usually treated with $N_2O/O_2$ mixtures [$N_2O$ in excess to avoid reaction (9)]. For most qualitative investigations solutions may be made up from mixtures of $N_2O$ and air or $O_2$-saturated solutions. Whenever exact $O_2$ concentrations have to be known and a gas head space is present

[13] S. Gordon, E. J. Hart, M. S. Matheson, J. Rabani, and J. K. Thomas, *Discuss. Faraday Soc.* **36**, 139 (1963).

[14] J. Rabani and M. S. Matheson, *J. Phys. Chem.* **70**, 761 (1966).

[15] D. Behar, G. Czapski, J. Rabani, L. M. Dorfman, and H. A. Schwarz, *J. Phys. Chem.* **74**, 3209 (1970).

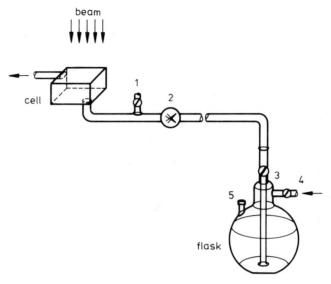

Fig. 1. Scheme of irradiation cell, flow system, and flask for storage of solution. 1–4, stopcocks (2 remote controlled); 5, rubber seal covered outlet.

above the solution, direct analytical $O_2$ measurements in the solution are necessary.

If R· is formed via a reductive process ($e_{aq}^-$, H·, etc.) the situation is more difficult. Scavenging of the ·OH radicals, e.g., by $t$-butanol

$$(CH_3)_3COH + ·OH \rightarrow ·CH_2(CH_3)_2COH + H_2O \tag{14}$$

or other suitable compounds, most often leads to the formation of radicals which themselves add or react with $O_2$ and thus result in interferences with the $RO_2^-$ reaction of interest. Without scavengers possible interferences by ·OH have to be taken into consideration.

Oxy radicals, RO·, finally are also mostly formed via ·OH-initiated processes as in the formation of phenoxy and semiquinone radicals from phenols and quinones, respectively. All investigations with these species are therefore also usually made with $N_2O$ saturated solutions.

Saturation with $N_2O$, $O_2$, or other gases is generally achieved by bubbling the solution for ~1 hr/1 liter. A commonly used flask is shown in Fig. 1. The gas is introduced into the solution through stopcock 3 and a porous bubbler. Volatile solutes have to be prepared separately via the same procedure prior to their introduction into the solution. The latter is generally done by a syringe equipped with a thin needle to penetrate through a gas-tight rubber seal attached to the flask (5 in Fig. 1).

Solute concentrations are chosen, whenever possible so that $c \cdot k$, i.e., the product of solute concentration times rate constant for the interesting reaction, exceeds the corresponding products $c \cdot k$ of any possible competing reactions by at least a factor of 10.

### Pulse Radiolysis Equipment

The equipment necessary for pulse radiolysis experiments consists of[4] a pulsed accelerator, physical dosimeter, irradiation cell, sample storage and flow system, and detection and signal recording system.

*Accelerators.* Irradiations are generally carried out with pulsed electron beams from a Van de Graaff or linear accelerator. Some laboratories also work with small cyclotrons or discharge machines (Febetron). Setting up and handling of this rather sophisticated machinery are generally done by qualified physicists and/or electronics engineers and thus do not need to be discussed further in this chapter.

*Physical Dosimeter.* To determine the intensity of the electron beam, i.e., to provide a measure for the dose per pulse, the accelerator is usually equipped with a thin foil (e.g., Al, Cu) from which secondary electrons are emitted upon penetration by the high energy beam electrons. The yield of these secondary electrons is practically linearly proportional to the total energy per pulse. It is displayed for recording purposes (e.g., as a digital signal, etc.).

*Irradiation Cell.* The irradiated volume is usually rather small. Any design of an irradiation cell has to take into consideration the depth of penetration of the accelerated electrons. In an aqueous solution, for example, 1 MeV electrons are already completely thermalized after ~5 mm, and 10 MeV electrons after ~50 mm. Accordingly the cell cover through which the beam penetrates into the sample has to be kept as thin as possible, particularly for low-energy electrons.

The shape and material of the cell depends also on the physical parameter to be measured. For *optical absorption* measurements the size of the cell is usually a few milliliters with a 0.2- to 0.3-mm-thick cover. The material should be quartz since normal glassware is not transparent for UV light and thus would not allow to detect radical species absorbing in the <400 nm wavelength range. Also, normal glass turns brown when exposed to already small doses of irradiation.

The same considerations apply to optical *emission* and *light scattering* techniques. The latter is particularly suited for time-resolved observations of molecular conformations (e.g., of biomacromolecules).[16]

---

[16] W. Schnabel, *in* "Developments in Polymer Degradation-2" (N. Grassie, ed.), p. 35. Applied Science Publ., London, 1979.

Cells for *conductivity*[11,12] and *polarography*[17,18] measurements may be of different material, but are usually also made of quartz to allow simultaneous optical measurements. Electrodes in a conductivity cell are usually made of platinum; recently glassy carbon has successfully been introduced.[19] The most common electrode material for polarography is mercury, and the whole arrangement is similar to the conventional steady-state polarography.

A special geometric design of irradiation cells is required for *electron spin resonance* (ESR)[20] and *microwave absorption*[21] measurements. It is essentially controlled by the wavelength of the electromagnetic waves.

*Sample Storage and Flow System.* The irradiation cell is usually connected via a flow system with a large sample of unirradiated solution which is stored, for example, in a flask (typically of 0.5–2 liter volume). This is schematically shown in Fig. 1. The solution is pressed through a glassy or resistant plastic capillary flow system into the cell. The pressurizing gas is to be chosen according to the chemical system investigated, i.e., usually $N_2$, Ar, $N_2O$, or mixtures thereof with $O_2$. Stopcocks 3 and 4 are necessary to prevent access of air to the solutions during and after preparation of the solutions. The remote controlled stopcock 2 allows stopping of the flow if desired. Gas bubbles which may occasionally emerge from the solution are collected in a trap associated with stopcock 1. The cell design with an inlet at the bottom and outlet at the diagonal top also prevents trapping of bubbles and ensures rapid and complete exchange of an irradiated sample by fresh solution.

Whenever the total sample volume must be kept low, e.g., for very expensive solutes at high concentrations, other designs are of course possible. Thus small syringes of a volume of 10 ml or less may directly be connected with the irradiation cell or solutions may be prepared in a cell itself.

*Detection and Signal Recording Systems.* Only the two most common detection systems, namely, optical and conductivity, shall be discussed here in some greater detail. For *optical* measurements[1-6] the irradiation cell is penetrated by a beam of intense light rectangularly to the electron beam. The latter is most often a Xe high-pressure lamp (e.g., Osram XBO 450) for investigations in the visible and near UV. Other light sources with

[17] J. Lilie, G. Beck, and A. Henglein, *Ber. Bunsenges. Phys. Chem.* **75**, 458 (1971).

[18] A. Henglein, *in* "Electroanalytical Chemistry—A Series of Advances" (A. J. Bard, ed.), Vol. 9. Dekker, New York, 1976.

[19] E. Janata, *Radiat. Phys. Chem.* **19**, 17 (1982).

[20] R. W. Fessenden and R. H. Schuler, *in* "Advances in Radiation Chemistry" (M. Burton and J. L. Magee, eds.), Vol. 2. Wiley (Interscience), New York, 1970.

[21] J. M. Warman, *in* "The Study of Fast Processes and Transient Species by Electron Pulse Radiolysis" (J. H. Baxendale and F. Busi, eds.). Reidel, Dordrecht, 1982.

different characteristics (deuterium, tungsten lamps, lasers etc.) may, however, also be chosen depending on the wavelength or wavelength ranges of interest. A particularly high light intensity can be obtained by using a pulsed lamp (e.g., Oriel Model 6140). It increases the sensitivity of the optical detection by about two orders of magnitude. To minimize possible photolysis the light pulse duration is usually limited to a few milliseconds triggered in association with the electron pulse. In addition, a mechanical shutter can interrupt the light beam.

Light focusing through the irradiation cell and into a monochromator is achieved by an appropriate lense system. Optical filters are necessary to cut out optical overtones if grating monochromators are used. If the light is dispersed by a prism slit width considerations apply in particular for higher wavelengths in the red and infrared section.

After passing through the monochromator, the light beam is converted into an electrical signal by a photodetector, and finally after amplification the time resolved signal is displayed on an X-Y recording system (e.g., oscilloscope).

Modern pulse radiolysis is generally equipped with an on-line computer which assists not only in the quantitative analysis of the data but also allows improvement in the sensitivity by pulse sampling and signal averaging. The overall detection limit is thus kept at a very low level, and optically absorbing transients with molar extinction coefficients in the order of 100 $M^{-1}$ cm$^{-1}$ at concentrations of some $10^{-6}$ $M$ can still be observed.

A similarly high sensitivity can be achieved with the *conductivity* detection system.[12,19,22,23] In this case the electron beam penetrates the area between two electrodes to which a voltage of typically 20–100 V is applied. Any generation or destruction of charged species in the irradiated solution leads to a change in conductance and is recorded as an electrical signal.

Two different conductivity methods are in common use, namely, a dc[11,12,19,22,23] and an ac[22–25] method. Both are complementary. The dc method is particularly suited for investigations on a time scale $\leq 10$ $\mu$sec (down into the nanosecond and upper picosecond region). The ac method is applied for longer times (up to seconds) since it minimizes polarization effects and furthermore has only a time resolution of $\sim 2$ $\mu$sec.

[22] K.-D. Asmus, *Int. J. Radiat. Phys. Chem.* **4**, 417 (1972).
[23] K.-D. Asmus and E. Janata, *in* "The Study of Fast Processes and Transient Species by Electron Pulse Radiolysis" (J. H. Baxendale and F. Busi, eds.), p. 91. Reidel, Dordrecht, 1982.
[24] J. Lilie and R. W. Fessenden, *J. Phys. Chem.* **77**, 674 (1973).
[25] M. Kelm, J. Lilie, A. Henglein, and E. Janata, *J. Phys. Chem.* **78**, 882 (1974).

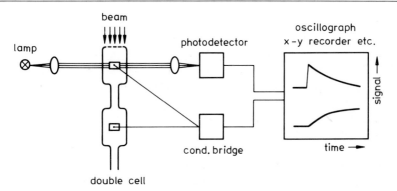

FIG. 2. Scheme of double-cell arrangement, and detection and recording system for combined optical and conductivity measurements.

A special ac double-cell arrangement suitable for low energy electrons is shown in Fig. 2. Only the upper part of the cell is irradiated while the lower, unirradiated part can be used as a reference. Before the pulse both cells are balanced with a Wheatstone bridge. The measured entity is then the radiation-induced disbalance rather than an absolute change as in dc measurements.

The signals displayed in Fig. 2 represent two typical time-resolved curves from a simultaneous optical/conductivity experiment. The upper, optical signal indicates an initial rapid formation and consecutive slow decay of an absorbing transient. The lower, conductivity signal indicates that the formation of the optically absorbing species is not accompanied by a net change in charge. However, the decay of the absorbing species evidently leads to an increase in conductance, which in solutions with a pH below neutral could, for example, be indicative for an anion/proton pair formation.

## Chemical Dosimetry

For the determination of the yield of transient species at any given time it is necessary to convert the measured electrical signals into concentrations and to determine the absorbed dose per pulse. This is achieved by calibration of the signal from the secondary electron emission, i.e., the physical dosimeter, with a chemical system. The latter should be of defined stoichiometry and known yields and properties of the involved species.

A very suitable dosimetry system is a deoxygenated aqueous solution containing 2-propanol (0.2 $M$) and tetranitromethane ($10^{-3}$ $M$).[22,23,26] Irradiation leads, via the following processes

$$H_2O \rightsquigarrow e_{aq}^-, H_{aq}^+, H\cdot, \cdot OH, \ldots \tag{15}$$

$$e_{aq}^- + C(NO_2)_4 \rightarrow C(NO_2)_3^- + NO_2 \tag{16}$$

$$\cdot OH/H\cdot + (CH_3)_2CHOH \xrightarrow{85.5\%} (CH_3)_2\dot{C}OH + H_2O/H_2 \tag{17}$$

$$(CH_3)_2\dot{C}OH + C(NO_2)_4 \rightarrow C(NO_2)_3^- + NO_2 + H_{aq}^+ + (CH_3)_2CO \tag{18}$$

to the formation of $C(NO_2)_3^-$ (nitroform) ions which are stable and exhibit a strong optical absorption at 350 nm with $\varepsilon = 1.5 \times 10^4$ $M^{-1}$ cm$^{-1}$. The radiation chemical yield of the nitroform formation amounts to

$$G[C(NO_2)_3^-] = G(e_{aq}^-) + 0.85G(\cdot OH + H\cdot) = 5.6$$

The same solution is also suited for calibration of the conductivity signals. The $H_{aq}^+/C(NO_2)_3^-$ ion pair which is formed as only ionic species in the above reaction sequence has a known equivalent conductance at 18° of $\Delta\Lambda = 360$ $\Omega^{-1}$ cm$^2$ (315 and 45 for $H_{aq}^+$ and $C(NO_2)_3^-$, respectively).

Other also very common dosimetry systems are based on the formation of $(SCN)_2^-$ radical anions in $N_2O$ saturated solutions of KSCN, and the oxidation of $Fe^{2+}$ and $Fe(CN)_6^{4-}$ to the corresponding Fe(III) species, also in $N_2O$ saturated solutions of the respective ions. The hydrated electron itself may also be used for dosimetry since its yield ($G \approx 2.75$ at 1 $\mu$sec after generation) and both its extinction coefficient ($\varepsilon = 1.86 \times 10^4$ $M^{-1}$ cm$^{-1}$ at 720 nm)[10] and equivalent conductivity (180 $\Omega^{-1}$ cm$^2$)[12] are known. The best dosimetry solution in this case is deoxygenated ($N_2$ saturated), slightly acidic water containing $10^{-1}$ $M$ alcohol as $\cdot OH$ scavenger.

*Analysis of Optical and Conductivity Signals*

The change in optical density at any given wavelength and time results in a measurable voltage drop $\Delta U$ across a working resistance $R_a$ from which the optical density (OD) is calculable via

$$OD = \frac{\Delta U}{R_a I_0} \times 100 \tag{19}$$

with $I_0$ the initial prepulse light-induced current. The correlation of OD with the molar concentration $c$, the $G$ value and the absorbed dose $D$ (in Gy) is given by

$$E = 2 - \log(100 - OD) = \varepsilon dc \tag{20}$$

($E$ is extinction, $\varepsilon$ is the extinction coefficient in $M^{-1}$ cm$^{-1}$, and $d$ is the optical pathlength in cm) and

$$G = (cN_L \times 100)/(D\rho \times 6.24 \times 10^{18}) = (c/D\rho) \times 9.66 \times 10^6 \tag{21}$$

[26] K.-D. Asmus, H. Möckel, and A. Henglein, *J. Phys. Chem.* **77**, 1218 (1973).

($\rho$ is the specific density of the irradiated system in g ml$^{-1}$ and $N_L$ is Avogadro's number).[6]

The correlation between the conductivity signal and the overall change in concentration of the conducting species, $\Delta c_i$, is given by[12,22,23]

$$\Delta U_s = [UR/(10^3 \times C_k)] \sum_i \Delta(c_i|z_i|\Lambda_i) \tag{22}$$

In this equation $\Delta U_s$ represents the measured voltage drop across the working resistance $R$ (typically 50–1000 $\Omega$), $U$ is the voltage applied to the electrodes (typically 20–100 V), $C_k$ is the cell constant, $z_i$ is the charge number, and $\Lambda_i$ is the equivalent (specific) conductivity (in $\Omega^{-1}$ cm$^2$) of the $i$th species. The most critical parameter is $R$. Increase in $R$ leads to a higher sensitivity, i.e., larger $\Delta U_s$ signal. On the other hand, $R$ has to be much smaller than the inner resistance of the solution. This limits the total ion concentration of a solution (which controls the inner resistance) to an upper limit of about $10^{-2}$ $M$ monovalent ions or to 2.5 $\leqslant$ pH $\leqslant$ 11.5. The working resistance $R$ also controls the time resolution, $\tau$, of the dc technique via $\tau = RC$ where $C$ is the cell capacitance. Another important aspect which should be considered in the conductivity data analysis is the generally strong temperature dependence of the equivalent conductivities of most ions. Exact temperature measurements or a temperature-controlled setup is therefore advised for the experiments.

For most practical purposes one usually relies on relative measurements which reduces analysis to the ratio

$$\Delta U_s/(\Delta U_s)_{ref} = \Delta(c_i\Lambda_i)/\Delta(c_j\Lambda_j)_{ref} = G_i \, \Delta\Lambda_i/(G_j \, \Delta\Lambda_j)_{ref} \tag{23}$$

A defined reference system where $\Delta c_j$, $G_j$, and $\Delta\Lambda_j$ are known is provided by any chemical dosimetry system presented in the previous section.

*Absorption Spectra and Kinetics*

The superoxide radicals $O_2^-$ as well as most of the oxy radicals derived from aromatic compounds exhibit characteristic and easily detectable optical absorptions. The spectra of $O_2^-$ and $HO_2$ are reported to show maxima at 245 and 230 nm with $\varepsilon = 2000$ and 1230 $M^{-1}$ cm$^{-1}$, respectively.[15] The absorptions of the aromatic oxy radicals (phenoxy, semiquinone) are generally found at higher wavelengths in the near-UV and visible range.

Assignment of reactions of nonabsorbing radicals, i.e., $\cdot$OH, $O^-$ and most of the peroxy radicals $RO_2$ is often possible through direct observation of an immediate reaction product if the latter has an absorption or its formation is associated with a change in conductivity. Rate constants are generally derived from kinetic analysis of absorption vs time and conductivity vs time curves, preferably at various substrate concentrations.

The situation is slightly more complicated if neither the oxygen-centered radical nor any of its immediate products exhibit any measurable property. In this case the reactivity toward a substrate S, e.g.,

$$\cdot OH + S \rightarrow products \tag{24}$$

can be deduced and quantified via the competing effect of S on the yield of a known and directly observable process, e.g., on the formation of the $(SCN)_2^-$ radical anion from the reaction sequence

$$\cdot OH + SCN^- \rightarrow SCN\cdot + OH^- \tag{25}$$
$$SCN\cdot + SCN^- \rightleftharpoons (SCN)_2^- \tag{26}$$

for which at high $SCN^-$ concentrations ($>10^{-3}$ $M$) reaction (25) is the rate-determining step ($k_{25} = 2.8 \times 10^{10}$ $M^{-1}$ $cm^{-1}$, $\lambda_{max}$ of $(SCN)_2^-$ at 475 nm).[27] The unknown rate constant $k_{24}$ is then calculable via

$$\frac{[(SCN)_2^-]_0}{[(SCN)_2^-]} = 1 + \frac{k_{24}[S]}{k_{25}[SCN^-]} \tag{27}$$

(The index "o" refers to the $(SCN)_2^-$ yield in the absence of S.) The choice of a particular reference reaction for any substrate–oxy radical reaction usually depends on the actual chemical system to be investigated.

## Availability of Pulse Radiolysis Equipment

Pulse radiolysis is a rather sophisticated and expensive experimental technique and therefore available only at a few, selected places. Most often it is found at national institutions and special research centers, but also at a number of hospitals and university institutes throughout the world. All these places usually not only provide the equipment, technical staff, scientific know-how, and experience, but are generally open for consideration of any reasonable research project on either a collaborative or a service basis. The assistance of experienced pulse radiolysists is particularly advised if special problems are to be investigated with one of the less common detection techniques available at only one or two places. Optical measurements are, however, possible at practically every pulse radiolysis center and an increasing number of places also provide conductivity facilities. Both the latter techniques are rather straightforward and most researchers will probably find their efforts rewarded after the first few experiments.

[27] J. H. Baxendale, P. L. T. Bevan, and D. A. Stott, *Trans. Faraday Soc.* **64**, 2389 (1968).

## [21] Electron and Hydrogen Atom Transfer Reactions: Determination of Free Radical Redox Potentials by Pulse Radiolysis

*By* L. G. FORNI and R. L. WILLSON

The pulse radiolysis technique is proving to be particularly useful for the kinetic and thermodynamic study of free radical reactions in solution.[1-4] A large variety of free radical reactions relevant to biochemistry, biology, and medicine can be observed directly. Electron transfer cascades, the catalytic action of glutathione and related thiols in linking hydrogen transfer and electron transfer reactions, and the protection of vitamin E by vitamin C are just three examples:

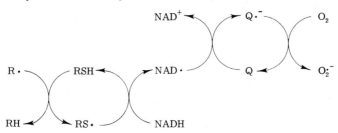

In the study of hydrogen atom and electron transfer reactions, most of the experiments to date, particularly those linked to biology, have been undertaken in systems in which water is the principal component. Under such conditions the principal reactive species formed immediately after a radiation pulse of 10 J/kg (10 Gr) are the hydroxyl radical(2.7 $\mu M$), the solvated electron(2.7 $\mu M$), and the hydrogen atom(0.6 $\mu M$). Although in principle such a pulse of ionizing radiation could lead to a vast assortment of transient species, by using selective free radical scavengers, individual free radical reactions can be studied in isolation. Changes in light transmission through a solution exposed to a short (say 200 nsec) pulse of ionizing radiation are monitored by means of a photomultiplier linked to an oscilloscope and digital recorders. By taking a series of measurements

[1] See "Pulse Radiolysis" (M. Ebert, J. P. Keene, A. J. Swallow, and J. H. Baxendale, eds.). Academic Press, New York, 1965.
[2] R. L. Willson, *in* "Biochemical Mechanisms of Liver Injury" (T. F. Slater, ed.), p. 123. Academic Press, New York, 1978.
[3] L. M. Dorfman, "Techniques of Chemistry," Vol. VI, Part II, p. 436. Wiley Interscience, New York, 1974.
[4] K.-D. Asmus, see this volume.

Copyright © 1984 by Academic Press, Inc.
All rights of reproduction in any form reserved.
ISBN 0-12-182005-X

at different wavelengths, transient absorption spectra at times of the order of a microsecond can be readily obtained. Before such pulse radiolysis experiments can be undertaken the following information is therefore required: (1) the solubility of the compounds to be studied and their stability in solution in the presence of other components (some multisolute systems are extremely photosensitive and must be exposed to the minimum of light), (2) the ground-state absorption spectra of the components: kinetic spectrophotometric analysis relies on the ability of the solution to transmit light before the radiation pulse and studies at wavelengths where the ground state absorption is high are limited due to a correspondingly high signal-to-noise ratio, and (3) the rate constants of $e_{aq}^-$, OH·, and to a lesser extent H·, with the components of the solution under study, and with any selective free radical scavengers to be used. Many of these rate constants have now been documented.[5-9] Where such information does not exist values can be obtained by using well-established methods: (a) e(aq): by following the decay of its absorption in the region of 600 nm, in neutral or alkaline solutions containing 1 m$M$ $t$-butanol; (b) OH·: by following either the appearance of a product free radical absorption or the bleaching of the parent compound in solutions saturated with nitrous oxide (where no strong absorptions can be monitored competition methods can be employed using thiocyanate ion, phenylalanine, ferrocyanide ion, or 2,2'-azino-di[3-ethylbenzethiazoline-6-sulfonate] (ABTS), as reference solutes[10,11]; and (c) H·: by following either the appearance of a product free radical absorption or the bleaching of the compound of interest in the presence of $t$-butanol, which reacts rapidly with hydroxyl radicals but comparatively slowly with H atoms, and at low pH, where solvated electrons are converted into H atoms on reaction with protons (where no convenient absorptions exist competition methods can again be employed using benzoquinone or ferricyanide as reference solutes).

The choice and concentrations of the selective free radical scavengers to be used depend on the type of reaction to be studied and on the relative rates of reaction of the compounds of interest compared to those for the other solutes under study. In principle in any multisolute system the proportion of a particular radical reacting with a particular component will depend on the rate constants of the individual reactions and the

[5] E. J. Hart and M. Anbar, "The Hydrated Electron." Wiley, New York, 1970.
[6] M. Anbar, M. Bambenek, and A. B. Ross, *Natl. Bur. Standards* NSRDS-NBS 43 (1970).
[7] L. M. Dorfman and G. E. Adams, *Natl. Bur. Standards* NSRDS-NBS 46 (1973).
[8] M. Anbar, Farhataziz, and A. B. Ross, *Natl. Bur. Standards* NSRDS-NBS 51 (1975).
[9] Farhataziz and A. B. Ross, *Natl. Bur. Standards* NSRDS-NBS 59 (1977).
[10] R. L. Willson, C. L. Greenstock, G. E. Adams, R. Wageman, and L. M. Dorfman, *Int. J. Radiat. Phys. Chem.* **3**, 211 (1971).
[11] B. S. Wolfenden and R. L. Willson, *J. Chem. Soc. Perkin Trans.* **II**, 805 (1982).

respective solute concentrations. Typical scavengers used in studies of electron and hydrogen transfer reactions together with the absolute rate constants of the initial interactions, that must be used when designing the composition of solutions to be irradiated are available.[2,8-11]

## Experimental Design

### Electron Transfer Reactions: Radical Oxidized

Studies of electron transfer reactions in which the radical is the electron donor are often undertaken in the presence of $t$-butanol, which reacts rapidly with hydroxyl radicals but not with solvated electrons or hydrogen atoms. The $t$-butanol radical absorbs below 300 nm, and is relatively resistant to further oxidation.[12] On pulse radiolysis of an aqueous $t$-butanol solution containing two other components, A and B, any changes in absorption observed above 300 nm, initially, can be attributed to one or both of the competing reactions:

$$e_{(aq)} + A \rightarrow A^{\cdot} \tag{1}$$
$$e_{(aq)} + B \rightarrow B^{\cdot} \tag{2}$$

The relative yields of $A^{\cdot}$ and $B^{\cdot}$ formed initially are given by

$$[A^{\cdot}]/[B^{\cdot}] = k_1[A]/k_2[B] \tag{3}$$

Subsequently, however, the radical $B^{\cdot}$ may react with A to give $A^{\cdot}$ and B or alternatively the reverse may occur. The extent of either reaction will depend upon the equilibrium constant of the reaction and the relative concentrations of A and B.

$$A^{\cdot} + B \rightarrow A + B^{\cdot} \tag{4}$$

Since the equilibrium constant is related to $G$, the relative one-electron redox potential $\Delta E^1$ of the redox couples, $E^1(A/A^{\cdot})$ and $E^1(B/B^{\cdot})$, can be determined using the equation

$$\Delta G = -RT \ln K = -n\Delta EF. \tag{5}$$

where the gas constant $R = 8.31$ J $K^{-1}$ $mol^{-1}$, the Faraday constant $F = 9.65 \times 10^4$ C $mol^{-1}$, and $T$ is the absolute temperature. Thus, provided the $E^1$ for one particular couple is known, $E^1_7$ of other couples can be obtained by measuring the equilibrium constants of electron transfer reactions in which they are involved. For example, on pulse radiolysis of solutions containing $t$-butanol and duroquinone and oxygen in relatively low concentrations, the durosemiquinone absorption can be observed immedi-

[12] G. E. Adams and R. L. Willson, *J. Chem. Soc. Faraday Trans.* **65**, 2981 (1969).

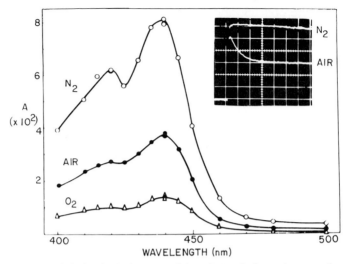

FIG. 1. Pulse radiolysis of solutions containing *t*-butanol, duroquinone, and oxygen.

ately after the radiation pulse.[13] The absorption decays rapidly over a period of a few microseconds and then plateaus at a level dependent on the relative duroquinone-to-oxygen concentration: the higher the oxygen concentration, $[O_2]$, the lower the transient absorption after 100 $\mu$sec; the higher the duroquinone concentration, $[DQ]$, the higher the absorption (Fig. 1). The magnitude of the absorption immediately after the pulse $[DQ^{\cdot-}]_0$ depends on the relative extent of the reactions analogous to (1) and (2) and after 100 $\mu$sec the absorption $(DQ^{\cdot-})_{100}$ depends on the position of the equilibrium

$$DQ + O_2^{\cdot-} \leftrightarrow DQ^{\cdot-} + O_2 \tag{6}$$

Since both radicals are relatively long-lived in the absence of other solutes the total radical concentration $[DQ^{\cdot-}] + [O_2^{\cdot-}]$ remains approximately constant and

$$[O_2^{\cdot-}]_{100} = [DQ^{\cdot-}]_0 - [DQ^{\cdot-}]_{100} \tag{7}$$

At equilibrium

$$[DQ^{\cdot-}]_{100}/[O_2^{\cdot-}]_{100} = K_c([DQ]/[O_2]) \tag{8}$$

where $K_c$ is the equilibrium constant. A plot of $[DQ^{\cdot-}]_{100}/[O_2^{\cdot-}]_{100}$ against $[DQ]/[O_2]$ is linear (Fig. 2), with the slope equal to the equilibrium constant, $K_c = 2.3 \times 10^{-2}$, and corresponding to $\Delta G = 9.2$ kJ mol$^{-1}$ and $E_7^1(O_2/O_2^{\cdot-}) - E_7^1(DQ/DQ^{\cdot-}) = 0.095$ V.

[13] K. B. Patel and R. L. Willson, *J. Chem. Soc. Faraday Trans.* **69,** 814 (1973).

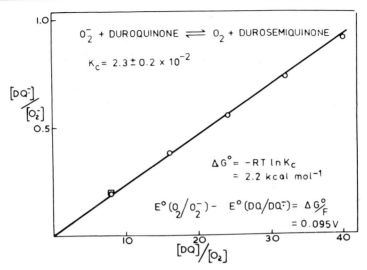

FIG. 2. Plot of $[DQ^-]_{100}/[O_2^-]_{100}$ vs $[DQ]/[Q]$.

Since this initial study, many one-electron radical equilibria involving semiquinones have been observed by other workers.[14-17] In 1974 the versatility of the technique was enhanced when $E_7^1(DQ/DQ^-) = -0.25$ V was calculated from spectroscopic data available in the literature.[18] Using this value, a value $E_7^1(O_2/O_2^-) = -0.16$ V was derived. If the oxygen concentration is taken as 1 atm rather than 1 $M$, this corresponds to a value $E_7^1(O_2/O_2^-) = -0.33$ V. A similar value has since been derived from similar studies using dimethylbenzoquinone[19] and the validity of this method has been further supported by measurements of equilibria involving benzyl viologen $(BV^{2+})$.[20,21] The one-electron oxidation potential $E_7^1(BV^{2+}/BV^{+}) = -0.35$ V has been determined by electrochemical methods. Pulse radiolysis studies of the equilibria

$$BV^+ + DQ \leftrightarrow BV^{2+} + DQ^- \qquad (9)$$
$$BV^+ + AQ \leftrightarrow BV^{2+} + AQ^- \qquad (10)$$

have provided values for $E_7^1(DQ/DQ^-)$ and $E_7^1(AQ/AQ^-)$ which are consistent with direct measurements of the equilibrium

[14] D. Meisel and P. Neta, *J. Am. Chem. Soc.* **97**, 5198 (1975).
[15] D. Meisel and P. Neta, *J. Phys. Chem.* **79**, 2459 (1975).
[16] D. Meisel and G. Czapski, *J. Phys. Chem.* **79**, 1503 (1975).
[17] Y. A. Ilan, G. Czapski, and D. Meisel, *Biochim. Biophys. Acta* **430**, 209 (1976).
[18] P. M. Wood, *FEBS Lett.* **44**, 22 (1974).
[19] Y. A. Ilan, D. Meisel, and G. Czapski, *Isr. J. Chem.* **12**, 891 (1974).
[20] P. Wardman and E. D. Clarke, *J. Chem. Soc. Faraday Trans.* **72**, 1377 (1966).
[21] P. Wardman and E. D. Clarke, *Biochem. Biophys. Res. Commun.* **69**, 942 (1976).

$$AQ^{\cdot-} + DQ \leftrightarrow AQ + DQ^{\cdot-} \qquad (11)$$

An alternative method of study is to measure the rate of approach to equilibrium rather than the equilibrium concentrations of the radicals involved. In these types of studies the concentrations of the compounds of interest are much greater than the associated radical concentrations and in the case of an equilibrium in which two species A and B are in equilibrium $A^{\cdot-}$ and $B^{\cdot-}$ the first-order rate constant characterizing the rate of approach to equilibrium is given by

$$k_{obs} = k_{1f}[A] + k_{1b}[B] \qquad (12)$$

where f and b refer to the forward and backward processes. A plot of $k_{obs}$/[B] against [A]/[B], is linear with slope $k_{1b}$ and intercept $k_{1f}$: hence the equilibrium constant can be calculated. Many other one-electron oxidation potentials have since been determined using these pulse radiolysis equilibrium methods (see the table).[22–26] In some cases the formate or isopropanol radicals, rather than $e_{aq}^-$, can be used to generate the initial radical species involved in the equilibrium. Experiments can be undertaken in the presence of either excess acetone and isopropanol or nitrous oxide and formate rather than $t$-butanol.

*Electron Transfer Reactions: Radical Reduced*

In the above experiments we have been concerned with electron transfer reactions in which the radical is the reducing species. Analogous equilibria can be established in which the radical is the oxidizing species. Halide radical anions and in alkaline solution the acetaldehyde radical derived from ethylene glycol can be used to generate the radicals involved. The halide radical anions $(Cl_2)^{\cdot-}$, $(Br_2)^{\cdot-}$, $(I_2)^{\cdot-}$, and $(SCN_2)^{\cdot-}$, absorb strongly in the visible and their reactions with other compounds can be followed directly.[27–32] In the case of the reaction between tryptophan and $(SCN_2)^{\cdot-}$ equilibrium conditions can be observed attributable to the overall reaction

[22] P. Wardman, *Curr. Top. Radiat. Res. Q.* **11**, 347 (1977).

[23] R. F. Anderson, *Ber. Bunsenges. Phys. Chem.* **80**, 969 (1976).

[24] J. A. Farrington, E. J. Land, and A. J. Swallow, *Biochim. Biophys. Acta* **590**, 273 (1980).

[25] R. F. Anderson, *Biochim. Biophys. Acta* **590**, 277 (1980).

[26] B. A. Swingan and G. Powis, *Arch. Biochem. Biophys.* **209**, 119 (1981).

[27] G. E. Adams, J. E. Aldrich, R. H. Bisby, R. B. Cundall, J. L. Redpath, and R. L. Willson, *Radiat. Res.* **49**, 278 (1972).

[28] J. F. Ward and I. Kuo, *Adv. Chem. Ser.* **81**, 368 (1968).

[29] E. J. Land and A. J. Swallow, *Biochim. Biophys. Acta* **234**, 34 (1971).

[30] R. L. Willson, *Biochem. Soc. Trans.* 1082 (1974).

[31] J. L. Redpath and R. L. Willson, *Int. J. Radiat. Biol.* **23**, 51 (1973).

[32] J. L. Redpath and R. L. Willson, *Int. J. Radiat. Biol.* **27**, 389 (1975).

ONE-ELECTRON OXIDATION POTENTIALS
DETERMINED BY PULSE RADIOLYSIS[a]

| Compound | $E^0$ (V) |
|---|---|
| Cytochrome $c$ | 0.26[a] |
| Benzoquinone | 0.1 |
| 2-Methylbenzoquinone | 0.02 |
| 2,3-Dimethylbenzoquinone | −0.07 |
| Oxygen (1 $M$) | −0.162 |
| 2,3,5-Trimethylbenzoquinone | −0.17 |
| 4-Nitropyridine | −0.19 |
| Duroquinone | −0.24[a] |
| FAD | −0.24 |
| Adriamycin | −0.29 |
| Riboflavin | −0.29 |
| Benzyl viologen | −0.35[a] |
| $p$-Nitroacetophenone | −0.36 |
| Misonidazole | −0.38 |
| Anthraquinone 2-sulfonate | −0.38 |
| Methyl viologen (Paraquat) | −0.45[a] |
| Nitrobenzene | −0.49 |
| Metronidazole | −0.49 |
| NAD$^+$ | −0.93 |

[a] Also determined electrochemically.

$$(SCN_2)^{\bar{\cdot}} + Trp \leftrightarrow 2SCN^- + Trp^{\dot{+}} \tag{13}$$

The overall equilibrium constant was again derived from measurements of the equilibrium concentrations of the radicals at different solute concentrations.[33] For the intermediate reaction

$$SCN\cdot + Trp \leftrightarrow SCN^- + Trp^{\dot{+}} \tag{14}$$

$K_c = 1.5 \times 10^7$ has been determined corresponding to $E^0(CNS\cdot/CNS^-) - E^0(Trp^{\dot{+}}/Trp) = 0.41$ V.

Recent studies have shown the occurrence of similar equilibria involving iodine radical anions and the phenothiazine promethazine, PZH$^+$.[34]

$$(I_2)^{\bar{\cdot}} + PZH^+ \leftrightarrow 2I^- + PZH^{2\dot{+}} \tag{15}$$

From measurements of the equilibrium radical concentrations as well as the rate of approach to equilibrium at different iodide and promethazine concentrations, a mean value for the equilibrium constants $K_c = 90$ mol/liter can be obtained corresponding to $E^0(PZH^{2\dot{+}}/PZH^+) - E^0(I\cdot/I^-) =$

[33] M. L. Posener, G. E. Adams, P. Wardman, and R. B. Cundall, *J. Chem. Soc. Faraday Trans.* **72**, 2231 (1973).
[34] D. Bahnemann, K.-D. Asmus, and R. L. Willson, *J. Chem. Soc. Perkin* **II**, in press, 1983.

$-0.415$ V. Using a value $E^0(PZH^{2+}/PZH^+) = 0.865$ V determined electrochemically a value $E^0(I\cdot/I^-) = 1.28$ V is obtained.

In studies with ethylene glycol at pH $\sim$ 13, tetramethylphenylenediamine, hydroquinone, and various phenol derivatives can be used as reference couples.[35,36]

$$HOCH_2CH_2OH + OH\cdot \rightarrow HOCH_2\dot{C}HOH + H_2O \qquad (16)$$
$$HOCH_2\dot{C}HOH + OH^- \rightarrow HOCH_2\dot{C}HO^- + H_2O \qquad (17)$$
$$HOCH_2\dot{C}HO^- \rightarrow \dot{C}H_2CHO \qquad (18)$$

Relative redox potentials have been placed on an absolute basis using a value for $E(Q^-/Q^{2-}) = 23$ mV derived from spectroscopic measurements.[37,38] A number of other systems can also be used to generate oxidizing radicals. The peroxy radical derived from $CCl_4$ and thiyl radicals derived from organic disulfides or thiols can be particularly useful for generating radicals from reducing compounds which are relatively insoluble in water. Irradiated solutions containing $CCl_4$ are generally air saturated and contain between 10 and 50% $t$-butanol. Alternatively studies can be carried out in the presence of excess isopropanol and acetone.[39-42]

$$e_{(aq)}^- + CCl_4 \rightarrow CCl_3^- + Cl^- \qquad (19)$$
$$(CH_3)_2\dot{C}OH + CCl_4 \rightarrow CCl_3^- + Cl^- + (CH_3)_2CO \qquad (20)$$
$$CCl_3^- + O_2 \rightarrow CCl_3OO\cdot \qquad (21)$$

In the case of thiols, solutions containing excess nitrous oxide or excess acetone and isopropanol are used. With disulfides excess $t$-butanol is also normally present.[43-45] Using these systems the absolute rate constants for a wide variety of electron transfer reactions can be determined,[46-48] e.g.,

[35] K. M. Bansal, M. Gratzel, A. Henglein, and E. Janata, *J. Phys. Chem.* **77**, 16 (1973).
[36] S. Steenken, *J. Phys. Chem.* **83**, 595 (1979).
[37] S. Steenken and P. Neta, *J. Phys. Chem.* **83**, 1134 (1979).
[38] S. Steenken and P. Neta, *J. Phys. Chem.* **86**, 3661 (1982).
[39] J. E. Packer, T. F. Slater, and R. L. Willson, *Life Sci.* **23**, 2617 (1978).
[40] J. E. Packer, T. F. Slater, and R. L. Willson, *Nature (London)* **278**, 737 (1979).
[41] J. E. Packer, J. S. Mahood, R. L. Willson, and B. S. Wolfenden, *Int. J. Radiat. Biol.* **39**, 135 (1980).
[42] J. E. Packer, R. L. Willson, D. Bahnemann, and K.-D. Asmus, *J. Chem. Soc. Perkin Trans. II* 296 (1980).
[43] G. E. Adams, G. S. McNaughton, and B. D. Michael, *in* "Chemistry of Ionization and Excitation" (G. R. A. Johnson and G. Scholes, eds.), p. 281. Taylor & Francis, London, 1967.
[44] G. E. Adams, G. S. McNaughton, and B. D. Michael, *Trans. Faraday Soc.* **64**, 902 (1968).
[45] M. Z. Hoffman and E. Hayon, *J. Am. Chem. Soc.* **94**, 7950 (1972).
[46] J. E. Packer, J. S. Mahood, V. O. Mora-Arellano, T. F. Slater, R. L. Willson, and B. S. Wolfenden, *Biochem. Biophys. Res. Commun.* **98**, 901 (1981).
[47] L. G. Forni, J. Monig, V. O. Mora-Arellano, and R. L. Willson, *J. Chem. Soc., Perkin II;* 961 (1983).
[48] L. G. Forni and R. L. Willson, to be submitted.

$$CCl_3OO\cdot + \beta\text{-carotene} \rightarrow CCl_3OO^- + \beta\text{-carotene}^{\dot{+}} \qquad (22)$$
$$CCl_3OO\cdot + PZH^+ \rightarrow CCl_3OO^- + PZH^{\dot{2}+} \qquad (23)$$
$$e_{aq}^- + GSSG \rightarrow GS\cdot + GS^- \qquad (24)$$
$$GS\cdot + PZH^+ \rightarrow GS^- + PZH^{\dot{2}+} \qquad (25)$$
$$GS\cdot + \text{red cyt } c \rightarrow GS^- + \text{ox cyt } c \qquad (26)$$

In these, as in many other instances, the appearance of a product corresponding to a net loss of one electron strongly supports the occurrence of an electron transfer reaction. However, in other instances a product corresponding to the net loss of a hydrogen atom is formed and the occurrence of an electron transfer reaction may be less obvious.[40,48–50]

$$CCl_3OO\cdot + NADH \rightarrow CCl_3OOH + NAD\cdot \qquad (27)$$
$$GS\cdot + NADH \rightarrow GSH + NAD\cdot \qquad (28)$$
$$GS\cdot + AH^- \rightarrow GSH + A^{\dot{-}} \qquad (29)$$
$$RO\cdot + AH^- \rightarrow ROH + A^{\dot{-}} \qquad (30)$$

Indeed in the reaction of peroxy radicals with vitamin E ($\alpha$-tocopherol) hydrogen transfer may well occur.[40]

$$CCl_3OO\cdot + \text{Vit E–OH} \rightarrow CCl_3OOH + \text{Vit E–O}\cdot \qquad (31)$$

In the reaction of carbon-centered radicals with ascorbate or thiols hydrogen atom transfer is strongly indicated. Although the rate constants of many hydrogen transfer reactions involving OH radicals and organic compounds have been known for many years, little information has been available concerning the rates of hydrogen atom transfer to organic radicals. However in the case of ascorbate ($AH^-$), the rate of hydrogen transfer to the isopropanol radical can be observed by monitoring the buildup of the ascorbyl radical at 360 nm[31,50]:

$$CH_3\dot{C}OHCH_3 + AH^- \rightarrow CH_3CHOHCH_3 + A^{\dot{-}} \qquad (32)$$

The analogous reactions of thiols can be followed indirectly by observing the formation of disulfide radical anions formed when the product thiyl radical reacts with excess thiolate ion also present.

$$CH_3\dot{C}OHCH_3 + RSH \rightarrow CH_3CHOHCH_3 + RS\cdot \qquad (33)$$
$$RS\cdot + RS^- \leftrightarrow RSSR^{\dot{-}} \qquad (34)$$

Experiments must be undertaken at or near the $pK_a$ of the thiol in order to prevent reaction (34) being non-rate limiting. An alternative approach is to include ABTS in the irradiated system. This reacts rapidly with many thiyl radicals and the product radical cation has a very strong absorption maximum at 415 nm:

[49] B. H. J. Bielski, D. A. Comstock, and R. A. Bowen, *J. Am. Chem. Soc.* **93**, 5624 (1971).
[50] R. H. Schuler, *Radiat. Res.* **69**, 417 (1977).

$$RS\cdot + ABTS \rightarrow RS^- + ABTS^{\cdot +} \qquad (35)$$

Since ABTS does not react with the isopropanol radical and by making the concentration of ABTS sufficiently high to prevent reaction (35) from becoming rate limiting, the appearance of $ABTS^{\cdot +}$ corresponds to the occurrence of the hydrogen atom transfer reaction (33).

## [22] Spin Trapping

### By EDWARD G. JANZEN

The method of detecting reactive short-lived free radicals by the spin-trapping technique was first demonstrated in the late 1960s. A number of reviews are available: for early work see Perkins,[1] Lagercrantz,[2] and Janzen[3]; for reviews of later work see Perkins,[4] Janzen,[5–7] Finkelstein et al.,[8] and Kalyanaraman.[9]

At this time the spin-trapping method involves the addition of an organic compound (spin trap) to the solution under investigation. The spin trap is capable of rapidly trapping free radicals to form more persistent radicals (spin adducts) detectable by electron spin resonance (ESR). The most commonly used spin traps are nitrones and nitroso compounds which usually give nitroxide spin adducts:

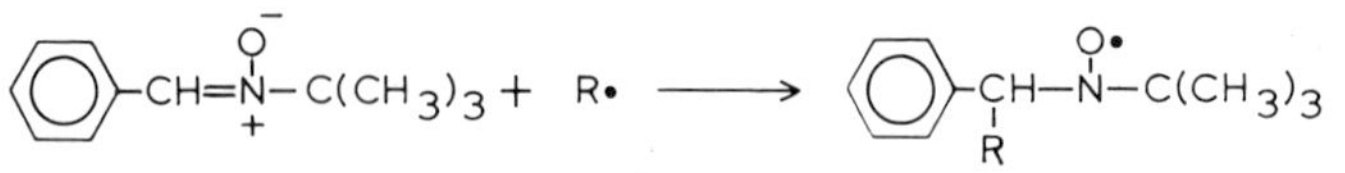

[1] M. J. Perkins, in "Essays in Free Radical Chemistry," p. 97. Special Publication No. 24, Chemical Society, London, 1970.
[2] C. Lagercrantz, J. Phys. Chem. **75**, 3466 (1971).
[3] E. G. Janzen, Acc. Chem. Res. **4**, 31 (1971).
[4] M. J. Perkins, Adv. Phys. Org. Chem. **17**, 1 (1980).
[5] E. G. Janzen, C. A. Evans, and E. R. Davis, in "Organic Free Radicals," p. 433. (ACS Symposium Series No. 69). American Chemical Society, 1978.
[6] E. G. Janzen, Free Radicals Biol. **4**, 115 (1980).
[7] E. G. Janzen and E. R. Davis, in "Free Radicals and Cancer" (R. A. Floyd, ed.), p. 397. Dekker, New York, 1982.
[8] E. Finkelstein, G. M. Rosen, and E. J. Rauckman, Arch. Biochem. Biophys. **200**, 1 (1980).
[9] B. Kalyanaraman, Rev. Biochem. Toxicol. **4**, 73 (1982).

Copyright © 1984 by Academic Press, Inc.
All rights of reproduction in any form reserved.
ISBN 0-12-182005-X

An immediate requirement of the technique is that the spin trap and spin adducts are soluble in the medium of interest and that free diffusion of the spin trap to the location of the free radical event is allowed. Also the environment should permit high mobility of the spin adduct so that the ESR spectrum consists of a pattern of sharp lines.

Ideally the detection of the ESR spectrum of a nitroxide spin adduct is proof that a radical has been trapped in the solution under investigation. However, some chemical artifacts of the system can sometimes produce the same result. A good example is the case of the hydroxyl radical adduct which could be produced from water. Thus if the spin trap is exposed to a strong oxidizing agent so that one-electron oxidation occurs the unstable radical cation so formed can react with a nucleophile and produce the spin adduct indicative of hydroxyl radical trapping:

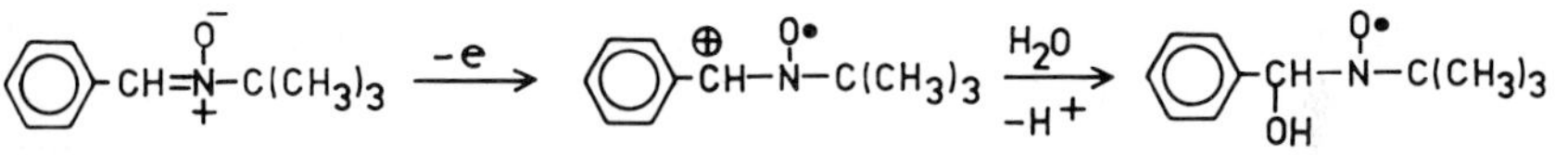

Since nitronyl spin traps are not too easily oxidized[10] this situation is not very likely in a biochemical system. However coordination of the nitronyl oxygen to an electrophile such as a metal center[11] can simulate this effect if one-electron oxidation happens later:

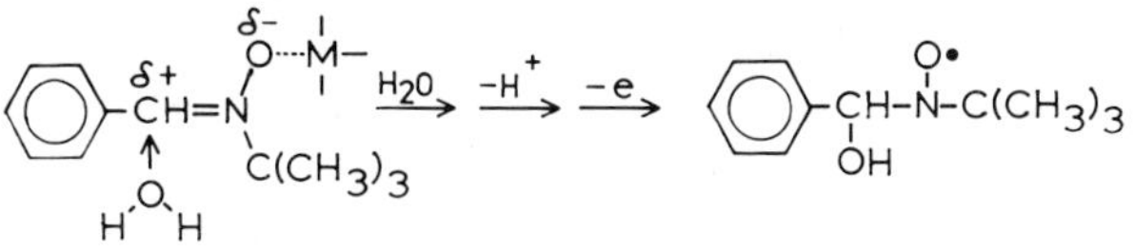

A more obvious artifact is the possibility that the radical trapped is not the primary radical formed in the radical event. A good example is the case of hydroxyl and hydroperoxyl free radicals. Hydroxyl radicals are produced from the photolysis of hydrogen peroxide but in the presence of a large amount of hydrogen peroxide the hydroperoxyl radical is formed from a secondary reaction:

$$HOOH + h\nu \rightarrow 2HO\cdot$$
$$HO\cdot + HOOH \rightarrow HOO\cdot + H_2O$$

In the presence of relatively high concentrations of a spin trap the hydroxyl radical spin adduct would be detected but at lower concentrations of the spin trap in the presence of relatively high concentrations of hydro-

[10] G. L. McIntire, H. N. Blount, H. J. Stronks, R. V. Shetty, and E. G. Janzen, *J. Phys. Chem.* **84,** 916 (1980).
[11] E. G. Janzen and B. J. Blackburn, *J. Am. Chem. Soc.* **91,** 4481 (1969).

gen peroxide both radicals or even only the hydroperoxyl radical would be detected.[12]

As the discussion above suggests spin trapping is a kinetic method, i.e., the success of the spin trapping experiment depends critically on the rate conditions which exist in the system. For a favorable rate of spin adduct formation the rate of spin trapping must be much faster than the rates of other reactions of the radical. Moreover the ideal spin trap would give spin adducts which are perfectly stable and unreactive to other reagents in the environment, even free radicals. Although these conditions are rarely met the technique is still quite successful. An examination of the reasons why may be helpful.

If we assume *only* the following reactions occur

$$\text{Radical Source} \xrightarrow{k_1} \text{R·} \qquad\qquad (1)$$
$$\text{R·} + \text{ST} \xrightarrow{k_2} \text{RSA·} \qquad\qquad (2)$$

(where ST = spin trap and RSA· = spin adduct).

The increase in concentration of spin adduct = $d[\text{RSA·}]/dt$ = $k_2[\text{R·}][\text{ST}]$. Under steady state-conditions for the radicals, R·,

$$\frac{d[\text{R·}]}{dt} = 0 = k_1\,[\text{Radical Source}] - k_2[\text{R·}][\text{ST}]$$

Then,

$$\frac{d[\text{RSA·}]}{dt} = k_1\,[\text{Radical Source}]$$

Thus the ESR signal strength is directly proportional to the number of radicals produced, i.e., spin trapping can be a quantitative tool for counting radical events in the system of interest.

One example comes very close to demonstrating this limiting case: when di-*t*-butylperoxyoxalate thermally decomposes at room temperature

$$C_4H_9OOCC OOC_4H_9 \xrightarrow{\Delta} 2\,C_4H_9O\cdot\ +\ 2\,CO_2$$

in the presence of 0.1 *M* phenyl *t*-butylnitrone (PBN) in benzene the rate of formation of spin adduct is exactly equal to the known rate of decomposition of the initiator,[13] i.e., every radical produced is trapped and the magnitude of the ESR signal is directly proportional to the number of

---

[12] J. R. Harbour, V. Chow, and J. R. Bolton, *Can. J. Chem.* **52**, 3549 (1974); E. G. Janzen, D. E. Nutter, Jr., E. R. Davis, B. J. Blackburn, J. L. Poyer, and P. B. McCay, *Can. J. Chem.* **56**, 2237 (1978).

[13] E. G. Janzen and C. A. Evans, *J. Am. Chem. Soc.* **95**, 8205 (1973); see also E. G. Janzen, C. A. Evans, and Y. Nishi, *J. Am. Chem. Soc.* **94**, 8236 (1972).

$$C_4H_9O\cdot + PBN \longrightarrow \begin{matrix} C_6H_5 \\ \diagdown \\ C_4H_9O \diagup \end{matrix} \begin{matrix} O\cdot \\ | \\ CH-N-C(CH_3)_3 \end{matrix}$$

radicals created in the thermal decomposition of the peroxide. This case is made favorable by the fact that the oxy radicals have little else to do (except dimerize) since the rate of cleavage or reaction with solvent is

$$(CH_3)_3CO\cdot \longrightarrow CH_3\cdot + CH_3\overset{\overset{\textstyle O}{\|}}{C}CH_3$$

relatively slow. The rate of spin trapping is enhanced by having a relatively large concentration of PBN present. Also the spin adduct is quite stable and does not appear to react significantly with the original radical produced. It is quite likely that most alkoxyl radicals behave similarly, i.e., the ESR signal of an alkoxyl spin adduct can probably be taken as a measure of the number of alkoxyl radicals produced as long as reduction of the nitroxide to the hydroxylamine thus decreasing the signal strength has not occurred.

$$C_6H_5-\underset{\underset{\textstyle OR}{|}}{\overset{\overset{\textstyle O\cdot}{|}}{C}}H-N-C(CH_3)_3 \xrightarrow{+e} \xrightarrow{+H^+} C_6H_5-\underset{\underset{\textstyle OR}{|}}{C}H-\overset{\overset{\textstyle OH}{|}}{N}-C(CH_3)_3$$

In addition to reactions (1) and (2) other reactions of R· may be possible:

$$R\cdot \xrightarrow{k_3} \text{nonradical products or} \\ \text{nontrapped radical products} \tag{3}$$

$$R\cdot + YH \xrightarrow{k_{4b}} RH + Y\cdot \quad \text{(abstraction)} \tag{4a}$$

$$R\cdot + X{=}Y \xrightarrow{k_{4b}} R\text{-}X\text{-}Y\cdot \quad \text{(addition)} \tag{4b}$$

[let both reactions (4a and 4b) give Y· with rate constant $k_4$]. Clearly if reaction (3) is the only other reaction and $k_3$ is relatively large the spin trapping result will not be quantitative. The only remedy is to increase the concentration of the spin trap if possible. If reactions (4a) or (4b) are important new radicals are produced which may in turn be trapped leading to mixtures of spin adducts (RSA· and YSA·)

$$Y\cdot + ST \xrightarrow{k_5} YSA\cdot \tag{5}$$

The relative concentrations of these spin adducts will depend on the relative concentrations of the spin trap and YH or X=Y and the rate constants $k_2$ and $k_{4a}$ or $k_{4b}$. Thus from steady-state considerations for R· and Y·

$$\frac{d[\text{RSA}\cdot]/dt}{d[\text{YSA}\cdot]/dt} = \frac{k_2[\text{ST}]}{k_4[\text{YH or X}{=}\text{Y}]}$$

Although $k_4$ is relatively small when YH is a hydrocarbon (e.g., for a long-chain alkyl group $k_4 \gtrsim 10^4$), $k_4$ can be quite large and compete favorably with $k_2$ if YH is a compound with an activated carbon–hydrogen bond (e.g., alcohols[14]) or if X=Y is an olefin[5] ($k_4 \simeq 10^6$). In the case of hydroxyl radicals probably all reactions (2–5) occur because of the high reactivity of this species. Dependable kinetic data from spin trapping experiments in such systems will thus be difficult to obtain particularly in biochemical systems.

A further complication in spin trapping experiments comes from the fact that the spin adducts themselves can be fairly reactive in some cases and in the absence of other reactions will disporportionate to a new nitrone and the hydroxylamine:

$$2RSA\cdot \xrightarrow{k_{6a}} R \text{ nitrone} + ST(R)NOH \tag{6a}$$

$$2YSA\cdot \xrightarrow{k_{6b}} Y \text{ nitrone} + ST(Y)NOH \tag{6b}$$

The rates of disproportionation are particularly fast for spin adducts of cyclic nitrones such as 5,5-dimethylpyrroline $N$-oxide (DMPO):[15]

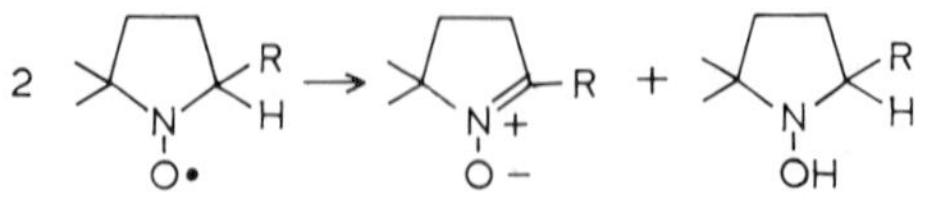

The new nitrone may trap radicals to give more stable spin adducts (since there are no $\beta$-hydrogens to disproportionate) and the hydroxylamine functionality may induce radical decomposition of peroxides. Thus in the case of peroxy spin adducts (e.g., when X=Y is oxygen) the disproportionation products are certainly unstable and will produce new radicals:

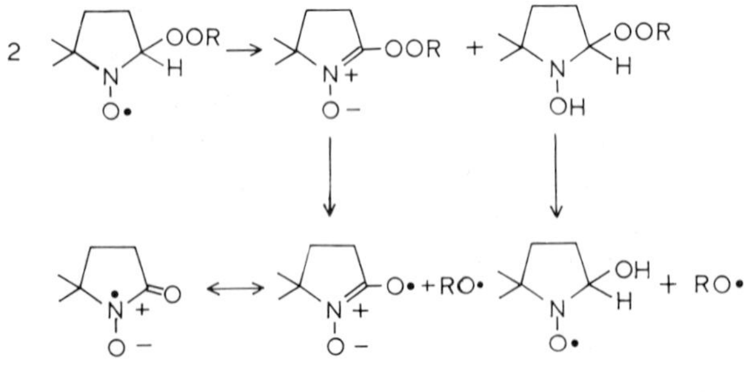

An example of this behavior has been found in spin trapping $O_2^{\cdot-}/HOO\cdot$ with DMPO.[16] A kinetic expression suitable for the case where the dispro-

[14] E. G. Janzen, D. E. Nutter, Jr., and C. A. Evans, *J. Phys. Chem.* **79**, 1983 (1975).

[15] K. U. Ingold, in "Free Radicals," Vol. 1, p. 37. Wiley, New York, 1973; D. L. Haire and E. G. Janzen, *Can. J. Chem.* **60**, 1514 (1982).

[16] E. Finkelstein, G. M. Rosen, and E. J. Rauchman, *Mol. Pharmacol.* **16**, 676 (1979).

portionation of the spin adduct is significant has been developed[17] but the case where the disproportionation product gives more radicals has not been treated.

Assignment of observed spectra to the correct structure of a new spin adduct is not a trivial matter. The argument that *because* a given radical is expected to be present in the system under investigation, the observed spectrum must be due to this spin adduct is *not* sufficient proof of the structure of the spin adduct. The best methods for assigning the structures of new spin adducts involve designing alternate routes for the production of the same spin adduct:

for PBN spin adducts.

For oxy radicals the most characteristic features of the ESR spectra are that both the nitroxide nitrogen and the $\beta$-hydrogen hyperfine splittings are smaller than for comparable adducts of carbon-centered radicals. However, this rule is not completely dependable since for electron-withdrawing carbon-centered radicals the ESR parameters of the spin adducts can be very similar to those of oxy adducts; for example, the hyperfine splitting constants for the trichloromethyl and alkoxyl adducts of PBN are very similar:

$a_N = 14.1; a_\beta^H = 1.8$ G
in $CCl_4$ (Ref. 18)

$a_N = 14.22; a_\beta^H = 1.95$ G
in benzene (Ref. 13)

However a significant solvent effect exists for both spin adducts.[19] The best approach of course is to use isotopic labeling to provide extra hyperfine splitting from the atom in the $\beta$-position (C-13 in the case of the trichloromethyl adduct: $a_{\beta\text{-}C\text{-}13} = 9.68$ G[20] and O-17 in the case of the t-butoxy adduct: $a_{\beta\text{-}O\text{-}17} = 5.05$ G[21]).

Alkylperoxy adducts always give smaller nitrogen and $\beta$-hydrogen

[17] P. Schmid and K. U. Ingold, *J. Am. Chem. Soc.* **99**, 6434 (1977); **100**, 2493 (1978); Y. Maeda and K. U. Ingold, *J. Am. Chem. Soc.* **101**, 4975 (1979).

[18] J. L. Poyer, R. A. Floyd, P. B. McCay, E. G. Janzen, and E. R. Davis, *Biochim. Biophys. Acta* **539**, 402 (1978).

[19] E. G. Janzen, G. A. Coulter, U. M. Oehler, and J. P. Bergsma, *Can. J. Chem.* **60**, 2725 (1982).

[20] J. L. Poyer, P. B. McCay, E. K. Lai, E. G. Janzen, and E. R. Davis, *Biochem. Biophys. Res. Commun.* **94**, 1154 (1980).

[21] J. A. Howard and J. C. Tait, *Can. J. Chem.* **56**, 176 (1978).

HYPERFINE SPLITTING CONSTANTS FOR ROO·/PBN AND RO·/PBN ADDUCTS

| | $a_N$ | $a_\beta^H$ | $T$ (°C) | | $a_N$ | $a_\beta^H$ | $T$ (°C) | Reference |
|---|---|---|---|---|---|---|---|---|
| $n$-C$_5$H$_{11}$OO· | 13.44 | 1.39 | −80 | $n$-C$_5$H$_{11}$O· | 13.89 | 2.21 | RT | 22 |
| $n$-C$_{18}$H$_{37}$OO· | 13.50 | 1.17 | −80 | $n$-C$_{18}$H$_{37}$O· | 13.89 | 2.18 | RT | 22 |
| | 13.50 | 1.61 | RT | | | | | |
| $s$-BuOO· | 13.50 | 1.40 | −80 | $s$-BuO· | 13.94 | 1.91 | RT | 22 |
| $t$-BuOO· | 13.42 | 0.95 | −20 | $t$-BuO· | 13.62 | 1.72 | RT | 21 |
| | ($a_\beta^{0-17} = 2.9$ G) | | | | ($a_\beta^{0-17} = 5.05$ G) | | | |
| CCl$_3$OO· | 13 | 1.63 | −50 to 0 | CCl$_3$O· | — | — | — | 23 |
| HOO· | 14.8 | 2.75 | RT | HO· | 15.30 | 2.75 | RT | 12 |

hyperfine splitting constants than alkoxy adducts (although the latter cannot always be strictly compared since the temperatures at which the spectra were obtained are not the same—in general alkylperoxy adducts are not stable at room temperature; see the table and references cited therein (Refs. 12,21–23).

## Method of Janzen[24] and Co-workers

The mixing and sample cells shown in Fig. 1 are used for all room temperature experiments. A sample of the solid spin trap is introduced into one arm of the sample cell and weighed. The system to be studied is introduced into the second arm either as a solid or in solution. Enough solvent is added to make up the desired concentration. The mixing cell is capped with septa and the flat cell attached. Long needles are passed through the two septa and N$_2$ outgassing commences immediately. After 15 min of gentle bubbling the needles are carefully removed as the flat cell is closed with a Teflon stopper. The reaction begins when the contents of the two arms are mixed. When the solution is considered to be homogeneous the mixing cell is turned over to allow the solution to run down into the flat cell. A certain amount of shaking may be necessary to fill the flat cells completely and eliminate trapped bubbles. The flat cell is mounted in the ESR cavity and the search for a signal is begun.

This arrangement has been found to be satisfactory for all aqueous solution samples. The rubber septa and the ground-glass joints are adequate seals toward leakage of solution out of or air seepage into the cell for hours or even days. However, for nonaqueous solvents the septa sometimes swell and pop out (e.g., benzene) or the silicon grease in the

[22] M. V. Merritt and R. A. Johnson, *J. Am. Chem. Soc.* **99**, 3713 (1977).

[23] A. Tomasi, E. Albano, K. A. K. Lott, and T. F. Slater, *FEBS Lett.* **122**, 303 (1980).

[24] G. A. Russell, E. G. Janzen, and E. T. Strom, *J. Am. Chem. Soc.* **86**, 1807 (1964); G. A. Russell, E. G. Janzen, A. G. Bemis, E. J. Geels, A. J. Maye, S. Mak, and E. T. Strom, *Adv. Chem. Ser.* **51**, 112 (1965).

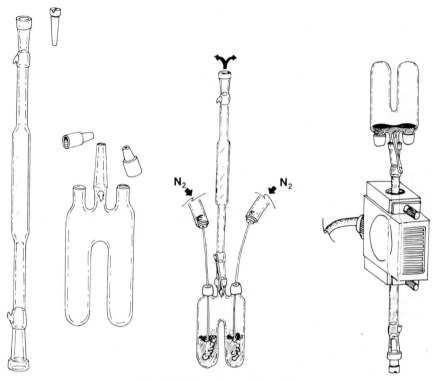

FIG. 1. Mixing and sampling cells.

ground glass joint dissolves (e.g., ethers) and the solution sometimes runs out.

The scan rate of the spectrometer is calibrated in gauss by the use of Fremy's salt is commercially available as $NaO_3SN(OH)SO_3Na$. This salt is dissolved in saturated $NaHCO_3$ and exposed to air. A violet solution is produced which gives a strong 3-line spectrum due to "Fremy's salt":

$$\text{Na}^+ \ ^-\text{O-}\underset{\underset{\text{O}}{\|}}{\overset{\overset{\text{O}}{\|}}{\text{S}}}\text{-}\underset{}{\overset{\overset{\text{H}}{\overset{\text{O}}{|}}}{\text{N}}}\text{-}\underset{\underset{\text{O}}{\|}}{\overset{\overset{\text{O}}{\|}}{\text{S}}}\text{-O}^- \ ^+\text{Na} \quad \xrightarrow{[\text{O}]} \quad \text{Na}^+ \ ^-\text{O-}\underset{\underset{\text{O}}{\|}}{\overset{\overset{\text{O}}{\|}}{\text{S}}}\text{-}\overset{\bullet}{\text{N}}\text{-}\underset{\underset{\text{O}}{\|}}{\overset{\overset{\text{O}}{\|}}{\text{S}}}\text{-O}^- \ ^+\text{Na}$$

The spacings are 13.03 G. The solution is stable for many days (but not indefinitely) if sealed from further exposure to air.

The spacings of a spin adduct spectrum are measured as indicated in Fig. 2. The nitrogen hyperfine splitting is measured eight times and averaged (four times for each of the maxima and minima) and the $\beta$-hydrogen hyperfine splitting is measured six times and averaged. By this method the best reproducibility over 15 years of work is about ±0.05 G for the same spin adduct obtained in the same solvent. As noted earlier large changes in splitting constants can be observed if the ESR spectra are obtained in different solvents.[19]

Computer simulations are done on a microprocessor with a program written in BASIC.[25]

It should be cautioned that some spin traps may be mutagenic.[26]

*Method of McCay, Poyer, and Lai*[27,28]

*In vitro*.[27] Microsomes are prepared[29] from livers of adult male albino rats derived from the Holtzman Sprague–Dawley strain. Food is withheld 24 hr before killing and the livers are homogenized with an all-glass grinding tube in cold 0.15 M potassium phosphate buffer, pH 7.5 (5.0 ml/g of liver, wet weight). The homogenate is centrifuged at 8000 g for 15 min. The supernatant fraction is collected, transferred to ultracentrifuge tubes, and centrifuged at 105,000 g for 90 min. The pellet obtained is resuspended in the same volume of fresh phosphate buffer and centrifuged again at 105,000 g for 60 min. After a small amount of buffer is added the microsomal portion of the pellet is floated off the glycogen layer. The microsomes are resuspended in the original volume of phosphate buffer and centrifuged again at 105,000 g for 60 min. The supernatant fraction is decanted and the tubes containing the microsomal pellets are covered and stored at −20° until ready for use.

The systems incubated with $CCl_4$ contained approximately 2 mg microsomal protein per milliliter of incubation system 0.14 M PBN and 20 μl $CCl_4$ in 0.05 M phosphate buffer, pH 7.4, final volume 1 ml. Systems were incubated at 24° for 15 min. The NADPH-generating system where utilized was composed of 5 mM glucose 6-phosphate, 0.3 mM NADP, and 0.5 Kornberg units glucose-6-phosphate dehydrogenase per milliliter of reaction system. After addition of all of the components, the mixture is vortexed, placed in a Pasteur pipet with a sealed tip, and centrifuged. This

[25] U. M. Oehler and E. G. Janzen, *Can. J. Chem.* **60**, 1542 (1982).

[26] M. J. Hampton, R. A. Floyd, E. G. Janzen, and R. V. Shetty, *Mutat. Res.* **91**, 279 (1981).

[27] J. L. Poyer, R. A. Floyd, P. B. McCay, E. G. Janzen, and E. R. Davis, *Biochim. Biophys. Acta* **539**, 402 (1978); J. L. Poyer, P. B. McCay, E. K. Lai, E. G. Janzen, and E. R. Davis, *Biochem. Biophys. Res. Commun.* **94**, 1154 (1980).

[28] E. K. Lai, P. B. McCay, T. Noguchi, and K-L Fong, *Biochem. Pharmacol.* **28**, 2231 (1979); J. L. Poyer, P. B. McCay, C. C. Weddle, and P. E. Downs, *Biochem. Pharmacol.* **30**, 1517 (1981).

[29] C. C. Weddle, K. R. Hornbrook, and P. B. McCay, *J. Biol. Chem.* **251**, 4973 (1976).

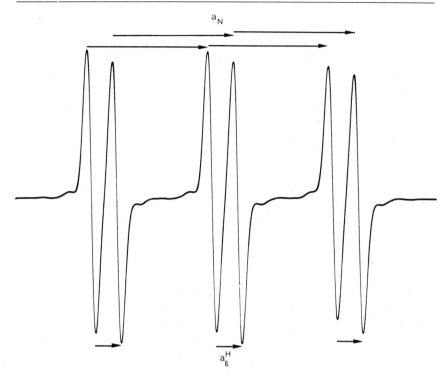

$a_N$

$a_\beta^H$

FIG. 2. Spin adduct spectrum.

technique is used to concentrate the microsomes within the magnetic field of the spectrometer.

*In vivo*.[28] In *in vivo* experiments with $CCl_4$, 70 $\mu l$ $CCl_4$ is administered orally by stomach tube in 2 ml of a solution containing 1 ml corn oil homogenized with 1 ml of a 0.14 $M$ solution of PBN in 0.02 $M$ phosphate buffer, pH 7.4. The rats have fasted for 20 hr. After 2 hr the animals are killed, the livers are immediately removed and homogenized directly in a choloroform–methanol (2 : 1) mixture, and the lipids extracted.[30] The chloroform layer containing the lipids (including the lipid-soluble PBN-radical adduct) is concentrated by evaporation under vacuum to 1 ml or less in volume and degassed with $N_2$ for 15 min prior to examination by ESR.

The halocarbon can also be administered by inhalation. Immediately following the administration of the PBN-buffer–corn oil emulsion a dose of 0.5% (v/v) administered in the breathing air for 2 hr using a Foregger

[30] J. Floch, M. Lees, and G. H. Sloan Stanley, *J. Biol. Chem.* **226**, 497 (1957).

inhalator equipped with a Fluotec attachment. The dose is checked using a Hewlett–Packard gas–liquid chromatograph with integrator.

Acknowledgments

The work at the University of Guelph was supported by the Natural Sciences and Engineering Research Council Canada. Grateful acknowledgment is hereby made. The author is thankful for an invitation to spend November 1982 at the Oklahoma Medical Research Foundation Research Laboratory where this chapter was written. Help from Gregory A. Coulter, who checked the manuscript, and Uwe M. Oehler, who made the drawings, is appreciated.

## [23] Spin Trapping of Superoxide and Hydroxyl Radicals

*By* GERALD M. ROSEN and ELMER J. RAUCKMAN

A free radical is by definition a species containing an unpaired electron, and is therefore paramagnetic. Paramagnetism forms the basis for the detection of free radicals by electron paramagnetic resonance (EPR) spectrometry, whereby the magnetic moment exerted by the unpaired electron is directly detected. This high degree of selectivity for only paramagnetic species renders EPR useful in complex biological systems.

The theoretical lower limit of free radical detection by EPR in aqueous solutions using existing instruments is approximately 10 n$M$.[1] In most circumstances, however, if hyperfine splittings are to be resolved, the practical limit of detection is about 1 $\mu M$.[1] Thus, it is only possible to detect radicals which accumulate to these measurable concentrations. There are many examples of such radicals being directly detected in biological systems. Such studies have recently been reviewed by Mason.[2]

Many free radicals of biological interest are highly reactive and never reach a concentration high enough to be detected by EPR. An example of this is the hydroxyl radical, which reacts with itself or with most organic molecules at diffusion controlled rates.[3] Its rate of reaction is limited mainly by the frequency with which it collides with other species. Thus, the direct detection of hydroxyl radicals by EPR in a biologic system is impossible.

For short-lived radicals of lesser reactivity compared to the hydroxyl radical, there are various means of detection using EPR. A simple method

[1] D. C. Borg, *Free Radicals Biol.* **1**, 69 (1976).
[2] R. P. Mason, *Rev. Biochem. Toxicol.* **1**, 151 (1979).
[3] L. M. Dorfman and G. E. Adams, *Nat. Bur. Standards* NSRDS No. 46 (1973).

Copyright © 1984 by Academic Press, Inc.
All rights of reproduction in any form reserved.
ISBN 0-12-182005-X

is to slow the rate of disappearance of the radical by rapidly freezing the sample. This has the disadvantage that the radical is no longer in a fluid environment, and the resultant anisotropic effects can obscure the identification of the radical. This technique is further limited by the concentration of the radical present before freeezing and by the length of time required to freeze the sample, which is about 5 to 10 msec.[4] One can improve the sensitivity of free radical detection in biological samples by lyophilization; this decreases microwave absorption by water and increases signal intensity. Artifactual radicals, however, such as that due to ascorbate, are often seen in lyophilized samples exposed to air.[1] Continuous flow EPR, in conjunction with signal averaging techniques, improves the detectability of short lived radicals, and enabled Yamazaki and Piette to detect the ascorbate semiquinone free radical in the EPR studies of ascorbate oxidase.[5] However, such studies are time consuming and require large quantities of enzyme.

In theory, spin trapping can overcome many of these difficulties. This technique consists of using a spin trap, i.e., a compound that forms a stable free radical by reacting covalently with an unstable free radical. Thus, the radical species is "trapped" in a long-lived form which can be observed at room temperature using conventional EPR equipment. The hyperfine splitting of the adduct provides information which can aid in the identification of the original radical. Since the stable free radical accumulates, spin trapping is an integrative method of measuring free radicals and is inherently more sensitive than procedures which measure only instantaneous or steady-state levels of free radicals.

Nitrones and nitroso compounds are the spin traps most commonly used, however, only nitrones detect oxygen centered radicals like superoxide and hydroxyl radicals at room temperature.[6] Both nitrones and nitroso compounds react covalently with numerous free radicals to produce a nitroxide ($N\overset{\cdot}{-}O$) spin adduct. The spectrum of a nitroxide gives a characteristic triplet which can exhibit further hyperfine splittings if nuclei having a magnetic moment are bonded nearby. The magnitude and nature of this interaction is dependent upon the nuclear quantum spin number as well as resonance, inductive, and steric effects.[7] Unless a conjugated system is present, magnetic nuclei farther than three bond lengths away from the nitroxide, where the unpaired electron is localized,

[4] J. R. Bolton, D. C. Borg, and H. M. Swartz, in "Biological Applications of Electron Spin Resonance" (H. M. Swartz, J. R. Bolton, and D. C. Borg, eds.), p. 63. Wiley (Interscience), New York, 1972.
[5] I. Yamazaki and L. H. Piette, *Biochim. Biophys. Act* **50,** 62 (1961).
[6] J. A. Wargon and F. Williams, *J. Am. Chem. Soc.* **94,** 7917 (1973).
[7] E. G. Janzen, *Acc. Chem. Res.* **4,** 31 (1971).

will not cause further resolvable splittings. With nitrone spin traps, the trapped radical is bonded to the $\alpha$-carbon, and magnetic nuclei present in the trapped radical are far away from the nitroxide nitrogen. Thus, hyperfine splitting due to the original radical is less readily resolved. Spin traps possessing a $\alpha$-hydrogen, such as 5,5-dimethyl-1-pyrroline $N$-oxide (DMPO), will yield adducts with hyperfine splitting due to both the $\alpha$-hydrogen and the nitroxide nitrogen. In these spin-trapped adducts, the magnitudes of $A_N$ and $A_H$ are very sensitive to the nature of the trapped radical, and this can serve as a means to help identify the trapped species.[7-9] In this chapter we discuss the spin trapping of the biologically important free radicals: superoxide and hydroxyl.

### Chemistry of Nitrone Spin Traps

Nitrones are highly reactive compounds, which can participate in a wide variety of reactions other than radical trapping. Thus, it is not surprising to note that nitroxides can be generated from nitrones by methods other than radical trapping. For this reason, a knowledge of nitrone chemistry is essential in understanding how artifactual radicals are generated from these compounds.

Nitrones can be reduced or oxidized into a variety of products.[10] Interconversions between nitrones, hydroxylamines, oximes, imines, hydroxamic acids, nitroxides, and nitroso compounds are possible, depending upon the conditions and reagents used. Metal ions commonly encountered in biological systems, such as iron and copper, can often carry out or catalyze such reactions. For example, aqueous ferric chloride is known to oxidize DMPO and related nitrones into the corresponding hydroxamic acid.[11] Oxidation of the hydroxamic acid would produce the corresponding nitroxide 5,5-dimethylpyrrolidone-(2)-oxyl-(1) (DMPOX). This compound has also been observed in a biochemical system containing hematin and cumene hydroperoxide.[12] However, in this case DMPOX arises by spin trapping cumene hydroperoxyl radicals followed by base-catalyzed rearrangement.[13] We have also found that DMPOX can be produced from DMPO by the action of oxidizing agents such as lead dioxide.[13] Thus, the production of DMPOX from DMPO is undoubtedly a common artifact in many oxidizing systems.

[8] E. G. Janzen and J. I. P. Liu, *J. Magn. Reson.* **9,** 510 (1973).
[9] E. G. Janzen, C. A. Evans, and J. I. P. Liu, *J. Magn. Reson.* **9,** 513 (1973).
[10] J. Hamer and A. Macaluso, *Chem. Rev.* **64,** 473 (1964).
[11] J. F. Elsworth and M. Lamchen, *J. S. Afr. Chem. Inst.* **24,** 196 (1971).
[12] R. A. Floyd and L. A. Soong, *Biochem. Biophys. Res. Commun.* **74,** 79 (1977).
[13] G. M. Rosen and E. J. Rauckman, *Mol. Pharmacol.* **17,** 233 (1980).

Chelated iron can also produce radicals from nitrones. In phosphate buffer, iron–EDTA oxidizes DMPO into a nitroxide, $A_N = 15.3$, $A_H = 22.0$.[14] The spectrum is due to an oxidative product of DMPO itself, as the same signal is observed in Tris buffer containing iron, without EDTA. Only trace amounts of iron are required to generate enough of this species to be observed by EPR. The iron present in phosphate buffer as an impurity is usually sufficient to produce a detectable signal. We have suggested that the spectrum is due to formation of a DMPO dimer.[15]

Copper ions can also give rise to artifacts in spin trapping experiments. For example, the air oxidation of hydroxylamines is greatly accelerated by cupric salts.[10] Therefore, we recommend that buffers used in spin trapping be passed through a Chelex 100 (Bio-Rad, Richmond, CA) column to remove polyvalent metal ion impurities. The use of diethylenetriaminepentaacetic acid (DETAPAC), a chelating agent which renders iron and copper incapable of oxidizing DMPO, is highly recommended.[14,16]

Nitrones are prone to hydrolysis in aqueous solution, forming an aldehyde and a hydroxylamine.[10] The hydrolysis is pH dependent, being more rapid at low pH,[10] and is structure specific. Acyclic nitrones are very susceptible to hydrolysis, while aryl nitrones are less so, and cyclic nitrones are reportedly very resistant to hydrolysis.[10] For example, one report claimed that there was little decomposition of an aqueous DMPO solution stored in the dark for 5 months,[11] as measured by its UV absorbance. In contrast, the half-life of the aryl nitrone, $\alpha$-(4-pyridyl 1-oxide)-$N$-$t$-butylnitrone (4-POBN) is 13.8 min at pH 2, although it is stable for 32 hr at neutral pH.[17]

Hydrolysis of nitrones can also give rise to nitroxides. Janzen et al.[17] have shown the addition of water across the double bond of 4-POBN, followed by oxidation, produces 4-POBN-OH which is the same species generated by the spin trapping of the hydroxyl radical by 4-POBN. Thus, under certain conditions, hydrolysis and air oxidation can lead to the erroneous assumption that the hydroxyl radical has been spin trapped.

## Synthesis of Spin Traps

Nitrones used in spin trapping experiments should be of the highest purity, and should be free from nitroxide or hydroxylamine impurities.

[14] E. Finkelstein, G. M. Rosen, E. J. Rauckman, and J. Paxton, *Mol. Pharmacol.* **16**, 676 (1979).

[15] R. F. C. Brown, V. M. Clark, M. Lamchen, and A. Todd, *J. Chem. Soc.* 2116 (1959).

[16] G. R. Buettner, L. W. Oberley, and S. W. H. G. Leuthauser, *Photochem. Photobiol.* **28**, 693 (1978).

[17] E. G. Janzen, Y. Y. Wang, and R. V. Shetty, *J. Am. Chem. Soc.* **100**, 2923 (1978).

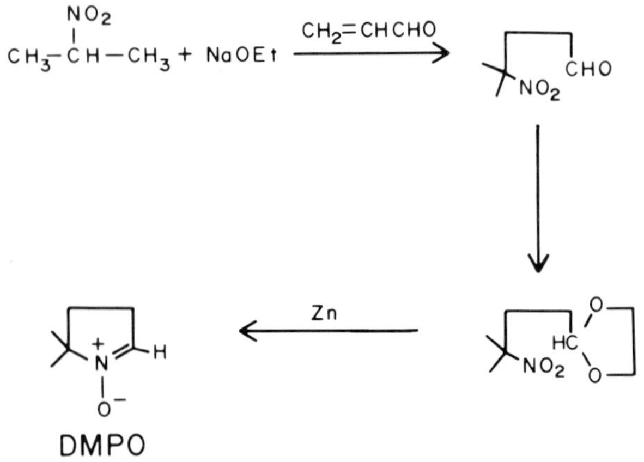

FIG. 1. The synthesis of 5,5-dimethyl-1-pyrroline $N$-oxide (DMPO).

Commercially available aryl nitrones such as 4-POBN are usually of sufficient purity and appear to be stable for a long period of time; however, DMPO and related spin traps have shorter shelf lives as they are more susceptible to decomposition by light and oxygen. Storage of cyclic nitrones like DMPO should always be at $-20°$, under nitrogen and away from light.

Commercially available DMPO usually requires further purification. The method of choice, especially for large quantities, is fractional vacuum distillation. DMPO purified by this method is a colorless solid, mp 25°. An alternative method of purification is column chromatography using charcoal-Celite, or filtration of an aqueous DMPO solution through charcoal, as described by Buettner and Oberley.[18] Charcoal-Celite behaves as a true reverse phase chromatographic medium, in that polar solvents elute only the nitrone, while nonpolar solvents will elute both nitrone and impurities.[19] Elution of DMPO can be conveniently monitored by its UV absorbance (DMPO, $\varepsilon_{max(234)} = 7700$ $M^{-1}$ cm$^{-1}$.[10]

The synthesis of DMPO is a three-step procedure which depends upon the reduction of 4-methyl-4-nitro-1-pentanal with zinc (Fig. 1).

*Preparation of 4-Methyl-4-nitro-1-pentanal*

A solution of sodium methoxide was prepared by dissolving 3.7 g (0.16 mol) sodium metal in 250 ml anhydrous methanol. To this mixture was

[18] G. R. Buettner and L. W. Oberley, *Biochem. Biophys. Res. Commun.* **83**, 69 (1978).
[19] G. M. Rosen, E. Finkelstein, and E. J. Rauckman, *Arch. Biochem. Biophys.* **215**, 367 (1982).

added 106.8 g (1.2 mol) 2-nitropropane (Aldrich Chemical Company, Milwaukee, WI). The addition was at such a rate that the temperature of the reaction mixture did not exeed 10°. After the additon of 2-nitropropane was completed, the temperature of the reaction was reduced to −20°, and a solution of 9 g (0.16 mol) of freshly distilled acrolein (Aldrich Chemical Company, Milwaukee, WI) and 26.7 g (0.3 mol) 2-nitropropane was added dropwise over a 3-hr period such that the temperature was kept at or below −20°. After the addition was completed, the mixture was stirred for 30 min, acidified with gaseous HCl (to pH 1.0), maintaining a temperature in the reaction flask no greater than 10°, and then dried with anhydrous sodium sulfate. Evaporation of the solution to dryness gave a liquid which was vacuum distilled to yield 9.5 g (41%) 4-methyl-4-nitro-1-pentanal, bp 85–88° at 3 mm Hg, lit. ref. 88.3–89.5° at 33 mm Hg.[20]

*Preparation of 2-(3-Methyl-3-nitrobutyl)-1,3-dioxolane*

To 9.5 g (0.066 mol) 4-methyl-4-nitro-1-pentanal in 50 ml dry benzene was added 4.26 g (0.069 mol) dry ethylene glycol and a catalytic amount (0.2 g) *p*-toluenesulfonic acid. This reaction was refluxed under a Dean–Stark trap until 1.1 ml water was collected. The benzene solution was cooled, treated with aqueous sodium hydrogen carbonate, dried over anhydrous sodium sulfate, and evaporated to dryness to give, after distillation, 9.2 g (75%) 2-(3-methyl-3-nitrobutyl)-1,3-dioxolane, bp 102–105° at 0.5 mm Hg, lit. ref. bp 105° at 0.5 mm Hg.[21]

*Preparation of 5,5-Dimethyl-1-pyrroline-N-oxide*

To a rapidly stirred solution of 9.2 g (0.057 mol) 2-(3-methyl-3-nitrobutyl)-1,3-dioxolane and 2.76 g (0.052 mol) ammonium chloride in 56 ml water at 10° was added 12.9 g (0.197 mol) zinc dust over a 20-min period of time. During the addition, the temperature was maintained below 15°. The reaction was stirred for 15 min, filtered, and the filter cake washed with hot water (at approximately 70°). The filtrate and washings were acidified with concentration hydrochloric acid, let stand overnight, and then heated to 75° for 1 hr before evaporation to one-third the original volume. The solution was then made alkaline, evaporated to near dryness, and extracted with chloroform. The chloroform solution was dried over anhydrous sodium sulfate and evaporated to dryness. Vacuum distillation gave

[20] H. Schechter, D. E. Ley, and L. Zeldin, *J. Am. Chem. Soc.* **74**, 3664 (1951).
[21] R. Bonnett, V. M. Clark, A. Giddy, and A. Todd, *J. Chem. Soc.* 2087 (1959).

5,5-dimethyl-1-pyrroline $N$-oxide, bp 64–66° at 0.6 mm Hg, lit. ref. bp 66–67° at 0.6 mm Hg.[22]

### Experimental Considerations

Spin trapping is easily conducted directly in an EPR cell so that the progress of the reaction may be continuously monitored by observing the EPR spectrum. A flat quartz cell is most frequently used because of its large surface area, and small volumes (less than 0.4 ml is sufficient). Generally, the reactants are mixed in a test tube, poured into the cell and immediately placed in the EPR cavity. Since spin trapping is an integrative technique, it invariably takes several minutes to reach sufficiently high levels that the nitroxide can be observed spectrometrically.

Optimal conditions for spin trapping superoxide requires high concentrations of DMPO since the rate constants for the reaction of DMPO with superoxide are small.[23] Concentrations of DMPO between 10 and 100 m$M$ are recommended. Under these conditions, we have not found that DMPO inhibits the activity of such enzymes as cytochrome $P$-450. Nevertheless, because of the high concentrations required, it is important to determine whether or not DMPO is a promoter or an inhibitor of the enzymic process under study.

As expected, the spin trapping of hydroxyl radicals by DMPO is quite rapid with a rate constant of $1.8 \times 10^9 \ M^{-1} \ sec^{-1}$, and thus, lower concentrations of this spin trap may be employed. Because of the disparity between the rates of spin trapping superoxide and hydroxyl radical, care must be taken to prevent the formation of hydroxyl radical by artifactual means.

The pH of the buffer will greatly affect the ability of DMPO to spin trap superoxide. This is due to two separate phenomena. First, the rate of spontaneous dismutation of superoxide is pH dependent, with superoxide having a longer half-life at higher pH. Second, the second order rate constant for the reaction of HOO· with DMPO is $6.6 \times 10^3 \ M^{-1} \ sec^{-1}$, while that for OO$^-$ is only $10 \ M^{-1} \ sec^{-1}$.[23] Thus, reaction of superoxide with DMPO favors acidic conditions. Unfortunately, most biological systems in which the detection of superoxide is desirable have pH optima in the range of 7.0–8.0.

Detection of DMPO–OH does not necessarily mean that the hydroxyl radical has been spin trapped. One method of verification is to utilize the

---

[22] R. Bonnett, R. F. C. Brown, V. M. Clark, I. O. Sutherland, and A. Todd, *J. Chem. Soc.* 2094 (1959).

[23] E. Finkelstein, G. M. Rosen, and E. J. Rauckman, *J. Am. Chem. Soc.* **102**, 4994 (1980).

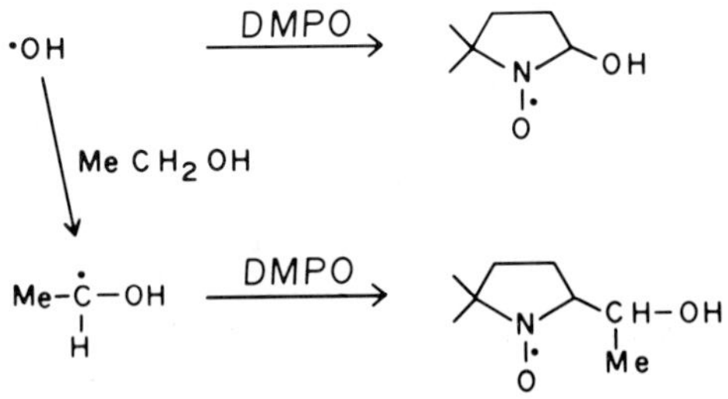

FIG. 2. Reaction of hydroxyl radical with either ethanol or DMPO to give $\alpha$-hydroxyethyl radical or DMPO–OH. DMPO then spin traps $\alpha$-hydroxyethyl radical, which has an EPR spectrum distinctively different from that of DMPO–OH.

ability of spin trapping techniques to distinguish between different radical species. For example, hydroxyl radicals react with ethanol to produce $\alpha$-hydroxyethyl radicals.[24] These secondary radicals then react with the spin trap to produce an adduct with an EPR spectrum distinguishable from that of the hydroxyl radical. In some biological systems, ethanol cannot be used, so dimethyl sulfoxide or sodium formate are excellent agents to employ as sources for secondary radical trapping experiments. Thus, if the production of DMPO–OH is due to the spin trapping of hydroxyl radicals, the addition of ethanol (dimethyl sulfoxide or sodium formate) should both inhibit the production of DMPO–OH and result in the appearance of a new signal due to the spin trapping of $\alpha$-hydroxyethyl radical. This is demonstrated in Fig. 2.

The quartet spectrum of DMPO–OH is not unique to DMPO–OH, since any nitroxide with hyperfine splitting constants with $A_N = A_H = 14.4$ G has the same or a similar spectrum.[25,26] Therefore, verification of hydroxyl radical trapping by independent means is a necessity.

Spin trapping of oxygen-centered radicals other than superoxide can likewise result in spectra similar to the spectrum of DMPO–OOH. For example, the spectrum of benzyloxyl radical adduct of DMPO is similar to that of DMPO–OOH (Fig. 3). Verification of superoxide trapping can be obtained by proper design of experiments. For example, in homogenous preparations, the addition of superoxide dismutase will inhibit the formation of DMPO–OOH. Also certain sulfhydryl agents, such as diethyldithiocarbamate rapidly reduce DMPO–OOH to DMPO–OH.[19] Finally,

[24] E. G. Adams and P. Wardman, *Free Radicals Biol.* **3**, 53 (1977).
[25] F. P. Sargent and E. M. Gardy, *Can. J. Chem.* **54**, 275 (1976).
[26] B. Kalyanaraman, E. Perez-Reyes, and R. P. Mason, *Biochim. Biophys. Acta* **630**, 119 (1980).

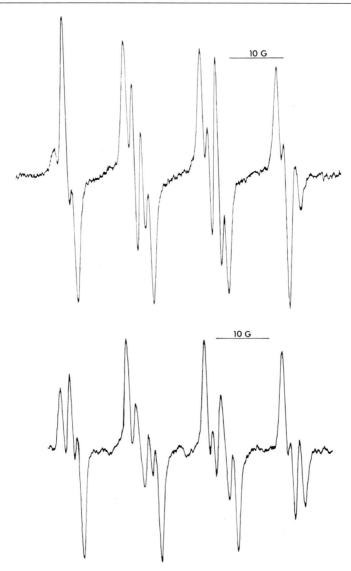

FIG. 3. Similarity between superoxide and benzyloxy radical adduct of DMPO, in aqueous solution. The upper scan is DMPO–OOH produced by the reaction of DMPO with superoxide generated by a xanthine/xanthine oxidase-generating system. The lower spectrum is that of benzyloxy radical adduct of DMPO, generated by thermal decomposition of benzoyl peroxide in a 1 : 1 mixture of ethanol and water, which was dissolved in water.

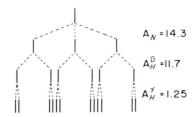

$A_N = 14.3$

$A_H^\beta = 11.7$

$A_H^\gamma = 1.25$

FIG. 4. Computed (stick) spectrum for DMPO–OOH which shows the formation of the 12-line spectrum.

as discussed earlier, the stability of DMPO–OOH is pH dependent and is greater at acidic pH. Thus, the chemical properties of DMPO–OOH can serve to distinguish it from other species.

During our studies into the mechanism of oxygen radical production by human neutrophils, we observed that, no matter how diligent we were in removing metal ion impurities from our reaction mixtures, we invariably spin trapped a small but quantifiable amount of hydroxyl radical. Upon extensive investigation, we discovered that during the spin trapping of superoxide by DMPO, *de novo* production of hydroxyl radical takes place.[27] It appears that once DMPO–OOH is formed, one of its decomposition products is hydroxyl radical which is then spin trapped by DMPO to give DMPO–OH. Thus, there will be a background level of approximately 3% DMPO–OH relative to DMPO–OOH. Therefore, the detection of hydroxyl radical by means of the spin trap DMPO must be interpreted with caution if the level of hydroxyl radical is less than 3% of the rate of superoxide generation.

Analysis of Spectrum

The EPR spectrum of DMPO–OOH displays three hyperfine splittings, one from the nitroxide nitrogen (with a multiplicity of 3) and two from nonequivalent protons (with a multiplicity of 2 each). The result of these hyperfine splittings is to give a 12-line spectrum as depicted in Fig. 4. Because an EPR spectrum is displayed as the first derivative (having both positive and negative components), and because the lines have finite widths and are extensively overlapped, the resulting spectrum appears to have lines of variable intensity. This observation is misleading in that all the lines are of equal size.

The EPR spectrum of DMPO–OH displays only two hyperfine splittings as shown in Fig. 5. The splittings for the nitrogen and $\beta$-hydrogen

[27] E. Finkelstein, G. M. Rosen, and E. J. Rauckman, *Mol. Pharmacol.* **21**, 262 (1982).

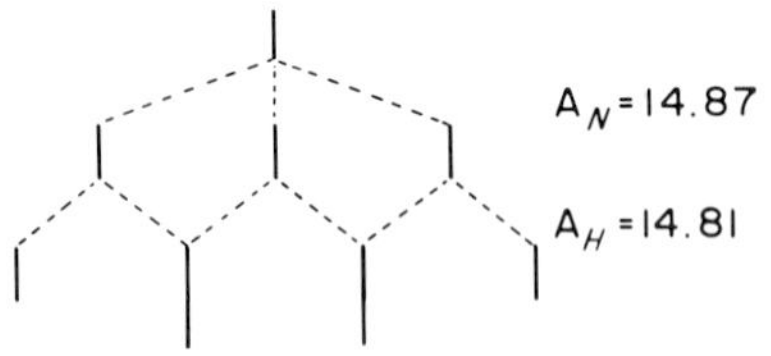

$A_N = 14.87$

$A_H = 14.81$

FIG. 5. Computer (stick) spectrum for DMPO–OH which illustrates the formation of the 6-line (with the overlap of two lines spectrum.

are equal causing overlap of the central lines. This results in a four line spectrum with a $1:2:2:1$ intensity ratio. Because DMPO–OOH rapidly decomposes into DMPO–OH and because the time required to obtain an EPR spectrum is measured in minutes, the observed spectrum is invariably a combination of DMPO–OOH and DMPO–OH. Determining relative ratios of each species can only be conducted with the assistance of a computer.

A potential problem with spin trapping in biological systems is the reduction, either enzymatically or chemically, of the nitroxide into its hydroxylamine, which cannot be detected by EPR. Nitroxides can be reduced by various biological systems such as ascorbic acid,[28] sulfhydryl agents,[29] the mitochondrial electron transport chain,[30] cytochrome $P$-450,[31,32] and bacterial electron transport systems.[33] Although superoxide and other radical species have been detected by spin trapping techniques in hepatic microsomes,[34–37] none of these studies has considered whether or not reduction actually has taken place.

In general, superoxide and hydroxyl radical adducts of aryl nitrones are less stable than their cyclic nitrone counterparts. For example, the half-lives of DMPO–OH and 4-POBN–OH are 2.6 hr and 23 sec, respec-

[28] I. C. P. Smith, in "Biological Applications of Electron Spin Resonance" (H. M. Swartz, J. R. Bolton, and D. C. Borg, eds.), p. 483. Wiley (Interscience), New York, 1972.

[29] J. M. Jallow, A. Difranco, F. Leterrier, and L. Piette, Biochem. Biophys. Res. Commun. 74, 1186 (1977).

[30] V. K. Koltover, L. M. Reichman, A. A. Jasaitis, and L. A. Blumenfeld, Biochim. Biophys. Acta 234, 306 (1971).

[31] A. Stier and E. Sackman, Biochim. Biophys. Acta 311, 400 (1973).

[32] G. M. Rosen and E. J. Rauckman, Biochem. Pharmacol. 26, 675 (1977).

[33] J. S. Goldberg, E. J. Rauckman, and G. M. Rosen, Biochem. Biophys. Res. Commun. 79, 198 (1977).

[34] R. C. Sealy, H. M. Swartz, and P. L. Olive, Biochem. Biophys. Res. Commun. 82, 680 (1978).

[35] G. M. Rosen and E. J. Rauckman, Proc. Natl. Acad. Sci. U.S.A. 78, 7346 (1981).

[36] C. S. Lai and L. H. Piette, Biochem. Biophys. Res. Commun. 78, 51 (1977).

[37] B. Kalyanaraman, R. P. Mason, E. Perez-Reyes, and C. F. Chignell, Biochem. Biophys. Res. Commun. 89, 1065 (1979).

tively.[14] The half-lives of DMPO–OH and 4-POBN–OOH are approximately 8 min and 30 sec, respectively.[14]

An alternative integrative EPR technique for the detection of superoxide in biological systems is the "spin-exchange" of superoxide with hydroxylamines.

$$\text{>N—OH} + \text{HOO·} \rightarrow \text{>N}\overset{+}{-}\text{O} + H_2O_2$$

The second-order rate constant for reaction of certain hydroxylamines with superoxide is much greater than for the covalent reaction of superoxide with DMPO.[19] Thus, this technique offers an alternative to the spin trapping method for the detection of superoxide.

# [24] Reaction of ·OH

By GIDON CZAPSKI

## History

The first evidence for a free OH radical was observed in the absorption spectra of $H_2$–$O_2$ flames by W. W. Watson in 1924.[1] In later studies other spectral lines of OH were observed in the gas phase, in pyrolyzed vapor and in electric discharges of water vapor or $H_2/O_2$ mixtures. As a result of these observations, Bonhoeffer and Haber[2] proposed that OH· and H atoms are the chain carriers in the chain reaction of $H_2$–$O_2$ combustion.

It was first suggested by Haber and Willstätter[3] that in aqueous media the OH· radical plays a role in the mechanism of decomposition of $H_2O_2$ by catalase. Later Haber and Weiss[4] proposed that OH· is an intermediate in the Fenton reaction. Since then, OH· is known to participate in many reactions in aqueous solutions and is known to be formed in chemical, photochemical, electrochemical, and radiolytic reactions. More recently, in the last decade it was shown that OH· is formed also in biological systems and is involved as an active intermediate in physiological processes.

[1] W. W. Watson, *Astrophys. J.* **60**, 145 (1924).
[2] K. F. Bonhoeffer and F. Haber, *Z. Phys. Chem. (Leipzig)* **A137**, 263 (1928).
[3] F. Haber and R. Willstätter, *Ber. Dsch. Chem. Ges.* **64**, 2844 (1931).
[4] F. Haber and J. Weiss, *Naturwissenschaften* **20**, 948 (1932).

Copyright © 1984 by Academic Press, Inc.
All rights of reproduction in any form reserved.
ISBN 0-12-182005-X

## The OH· in Aqueous Media

The early work of Haber and Willstatter[3] and that of Haber and Weiss[4] assumed that OH· is an intermediate in the decomposition of $H_2O_2$ by catalase and by $Fe^{2+}$ salts. It is very difficult to detect OH· directly due to its high reactivity and short lifetime.

However, it was found that OH· recombines with $k_{OH+OH} = 6 \times 10^9$ $M^{-1}$ $sec^{-1}$ and that OH· reacts with most organic compounds with rates not far from the diffusion-controlled rate constants. Therefore, even in pure systems, OH· has a very short lifetime and will have very low concentrations; consequently it is hard to detect directly. When short-pulse methods, with high intensity (like pulse radiolysis or flash photolysis techniques), are used to produce the OH· it is possible to obtain initial concentrations of OH· as high as $10^{-5}$–$10^{-3}$ $M$ which will decay to less than $10^{-6}$ $M$ in $10^{-4}$ sec.

## Methods of Generation of OH·

We will mention only the most common methods for the generation of OH· in aqueous media.

*Radiolysis.* In radiolysis of aqueous solutions, two radicals are mainly formed, OH· and $e_{aq}^-$. In the presence of either $N_2O$ or $H_2O_2$, radiolysis produces OH· almost solely.

*Photolysis.* Several systems yield OH· by photolysis, the most practical being the photolysis of $H_2O_2$ which yields two OH· radicals. This photolysis occurs at wavelengths below 350 nm.

*Chemical Reactions.* The most common source for OH· is the Fenton reaction

$$Fe^{2+} + H_2O_2 \rightarrow Fe^{3+} + OH^- + OH·$$

This same reaction takes place also with other transition metal ions at their lower valences, like $Ti^{3+}$, $Cr^{2+}$, and $Cu^+$, and with various complexes of these ions.

## Physical Properties of OH· in Aqueous Media

*The pK of OH·.* The OH· radical is a very weak acid and dissociates, yielding the alkali form $O^-$.

$$OH· + OH^- \rightarrow O^- + H_2O$$

The pK of the OH· was determined to be 11.85,[5] which is practically identical with the pK of $H_2O_2$, the dimer of OH·.

[5] J. Rabani and M. S. Matheson, *J. Phys. Chem.* **70**, 761 (1966); J. L. Week and J. Rabani, *J. Phys. Chem.* **70**, 2100 (1966).

RATE CONSTANTS FOR OH· REACTIONS WITH VARIOUS FREE
RADICALS IN AQUEOUS SOLUTIONS[a]

| Reaction | pH | $k_{OH·}$ + radical $(M^{-1} sec^{-1})$ |
|---|---|---|
| OH· + OH· → $H_2O_2$ | 0.4–3 | $6 \times 10^9$ |
| OH· + $O^-$ → $HO_2^-$ | >12 | $\leq 2.6 \times 10^{10}$ |
| OH· + $e_{aq}^-$ → $OH^-$ | Basic | $3 \times 10^{10}$ |
| OH· + H· → $H_2O$ | Acid | $\sim 2 \times 10^{10}$ |
| OH· + $HO_2^-$ → $H_2O_3$ | Acid | $\sim 4$–$10 \times 10^9$ |
| OH· + $HO_2^-$ → $H_2O$ + $O_2$ | Acid | $\sim 2$–$5 \times 10^9$ |
| OH· + $O_2^-$ → $OH^-$ + $O_2$ | 2.7–6.7 | $1 \times 10^{10}$ |
| $O^-$ + $e_{aq}^-$ → $2OH^-$ | Basic | $\sim 2 \times 10^{10}$ |
| $O^-$ + $O^-$ → $O_2^{2-}$ | 12–13 | $\sim 10^9$ |
| $O^-$ + $O_3^-$ → $O_4^{2-}$ or $O_2$ + $O_2^{2-}$ | 13–13.7 | $8 \pm 2 \times 10^8$ |

[a] From Farhataziz and Ross.[8]

*The spectra of OH· and $O^-$.* Both forms of the hydroxyl radical, OH·
and $O^-$, have absorptions in the UV. The absorption peaks of OH· and $O^-$
are at about 240 nm.[6] The OH· absorbance is about twice that of $O^-$, and
the absorption peak for OH· is $\varepsilon_{240} = 600\ M^{-1}\ cm^{-1}$ and for $O^-$, $\varepsilon_{240} = 240$
$M^{-1}\ cm^{-1}$.[6]

*The Standard Electrode Potential.* The OH· is one of the strongest
oxidizing agents. At 25° its standard electrode potential is $E° = 1.83$ V.[7]

*Chemical Properties*

The OH radical is very reactive and undergoes various reactions.
*Radical–Radical Reactions.* OH· recombines very rapidly with itself
to yield $H_2O_2$ and it also reacts very fast with many other radicals.
The table gives some of these reactions and their rate constants.[8]
From these values, one can easily calculate that OH· or $O^-$ will have very
short lifetimes and if formed at any concentration will decay in less than
$10^{-4}$ sec to submicromolar concentrations even in the absence of solutes
which react with OH·. In the presence of such solutes, the OH· will decay
even faster.
If the OH· radicals are continuously generated they will achieve
steady-state concentrations very quickly, which will depend both on the
rate of their formation and on the rate of their decay with itself or with
other solutes.

[6] G. Hug, *Natl. Bur. Standards* NSRDS No. 69 (1981).
[7] H. A. Schwarz, *J. Chem. Educ.* **58**, 101 (1981).
[8] Farhataziz and A. B. Ross, *Natl. Bur. Standards* NSRDS No. 59 (1977).

Under usual steady-state methods, even in pure systems, OH· will be at concentrations below $10^{-8}$ $M$. In biological systems its steady-state concentration is far below $10^{-10}$ $M$.

*Reactions of OH· and O⁻ Radicals*

The reactions of hydroxyl radicals can be classified into various classes.

*Radical–Radical Reactions.* Radical–radical reactions are either recombination reaction such as

$$2OH· \rightarrow H_2O_2$$
$$H· + OH· \rightarrow H_2O$$
$$2O^- \rightarrow O_2^{2-}$$
$$e_{aq}^- + OH· \rightarrow OH^-$$
$$OH· + HO_2^- \rightarrow H_2O_3$$

or dismutation reactions, such as electron or H atom transfer reactions

$$OH· + HO_2^- \rightarrow H_2O + O_2$$
$$OH· + O_2^{\cdot-} \rightarrow OH^- + O_2$$

or even possibly an atom and charge transfer such as in

$$O^- + O_3^{\cdot-} \rightarrow O_2 + O_2^{2-}$$

although this reaction most probably proceeds via an $O_4^{2-}$ intermediate in which $O_2$ then splits off.

*H· Abstraction Reaction.* The OH· radical reacts with most C–H bonds with rather high rate constants and H atom abstraction is one of the most common reaction pathways, forming an organic radical

$$R—H + OH· \rightarrow R· + H_2O$$

The abstraction pathway is not restricted only to organic solutes. OH· can abstract an H atom from $HCO_3^-$, $H_2$, $NH_3$, $H_2S$, etc. These abstraction reactions take place also with the dissociated form—the O⁻.

*OH· Addition Reactions.* In several reactions, mainly with organic compounds having double bonds or with aromatic compounds, OH· prefers to add to the aromatic ring or to the double bond and yields an addition product.

$$RCH{=}CH_2 + OH· \rightarrow R\dot{C}H—CH_2OH \quad \text{or} \quad \varphi H· + OH· \rightarrow \varphi HOH·$$

These reactions can also take place with O⁻.

*Electron Transfer Reactions.* The hydroxyl radical is a powerful oxidant, one of the most powerful oxidants in aqueous solution ($E^0 = 1.83$ V).

Therefore, one expects the electron transfer reaction of the following types to be quite common:

$$OH\cdot + S \rightarrow OH^- + S^+$$

or

$$O^- + S \rightarrow O^{2-} + S^+$$

Such reactions represent the majority of reactions taking place between OH· or O$^-$ with inorganic solutes, and even with some organic solutes.

*Mixed Pathways.* In several instances OH· may react with a molecule in more than one reaction mode.

It is very common with organic molecules that more than one reaction product is formed. Even if the only reactions are H atoms abstraction reactions, several radicals may be formed, depending on which H atom is abstracted.

In some cases one may get in parallel more than one kind of reaction product (i.e., transfer, addition, abstraction, or recombination) as, for example,

$$OH\cdot + HO_2^- \begin{cases} \rightarrow H_2O_3 \\ \rightarrow H_2O + O_2 \end{cases}$$

$$OH\cdot + CH_3CHOHCH_3 \begin{cases} \rightarrow \dot{C}H-CHOH-CH_3 \\ \rightarrow CH_3-COH-CH_3 \end{cases}$$

$$\underset{(p\text{-toluidine})}{O^- + CH_3-C_6H_4-NH_2} \begin{cases} \rightarrow \dot{C}H_2-C_6H_4-NH_2 + OH^- \\ \rightarrow CH_3-C_6H_4-\dot{N}H + OH^- \end{cases}$$

$$OH\cdot + CH_2=CH_2 \begin{cases} \rightarrow \dot{C}H_2-CH_2OH \\ \rightarrow CH_2=\dot{C}H \end{cases}$$

## Rate Constants of OH and O$^-$ Radicals

There are several compilations which give the rate constants of OH· and O$^-$ in aqueous solutions.[8,9] We find that many of these rates are very fast, having rates as high as diffusion-controlled reactions. For example, with most organic and biological molecules the rate constant is above $10^7$ $M^{-1}$ sec$^{-1}$ and generally even higher ($10^8$–$10^{10}$ $M^{-1}$ sec$^{-1}$).[8,9]

## Rate Measurements

There are basically two methods of measuring rate constants of OH· with solutes, the direct method and the indirect one.

[9] L. M. Dorfman and G. E. Adams, *Natl. Bur. Standards* NSRDS No. 46 (1973).

In the direct method, OH· is formed by pulse radiolysis or flash photolysis in the presence of a solute. One can directly follow the decay of the OH· radical with the solute, or the formation of the reaction product by following changes in physical properties such as their optical or ESR spectra, or their conductivity. From these measurements the rate constant can be determined.

In the indirect method, one measures the relative rates of reaction of OH· with two solutes from analysis of the products, with knowledge of the rate constant for the reaction of one solute with OH·, the other rate constant may be determined.

## Detection of OH·

The identification of the formation of the OH· radical in a system is not an easy task, since, as mentioned previously, the radical is very reactive and thus possesses a very short lifetime and is usually found at very low concentrations. Therefore, techniques such as recording the ESR or optical spectra of the radical will generally fail.

There are two main methods of identifying OH· in such systems, both indirect ones.

*Spin Trapping.* The addition of a solute which reacts with OH· and yields a long-lived, OH· adduct free radical, characterized by a known specific ESR spectrum.

*Competition Kinetics.* The addition of OH· scavengers to such systems and the analysis of the reaction products yield relative rate constants of the reaction of OH· with these scavengers.

## OH· in Biology

The formation of OH· in biological systems through ionizing radiation has been known for a long time. However, the possible participation and formation of OH· in physiological processes has only recently come to light following the discovery of superoxide dismutase (SOD) and investigations of the role of $O_2^-$ in physiological processes. Today, many normal physiological processes, such as phagocytosis, are known to involve the formation of OH· radicals. In addition, many toxic agents seem to generate OH· and their toxicity is apparently due to the subsequent reactions of the OH·.

It appears that the main pathway for the formation of OH· in biological systems is through the following sequence of reactions.

Fe(III) or Cu(II) + reducing agent (like $O_2^-$, vitamin C, etc.) → Fe(III) or Cu(I).

Fe(II) or Cu(I) + $H_2O_2$ → Fe(III) or Cu(II) + OH· + $OH^-$ when $O_2^-$ is the reducing agent this sequence is a metal-catalzyed reaction of $O_2^-$ with $H_2O_2$

$$O_2^- + H_2O_2 \rightarrow O_2 + OH^- + OH\cdot$$

which is sometimes referred to as Haber–Weiss reaction.

## [25] Electron Spin Resonance Spin Destruction Methods for Radical Detection

### By ROLF J. MEHLHORN and LESTER PACKER

Nitroxides ($R_2NO$:) react rapidly with many free radicals to form diamagnetic products and thus have potential utility for studies of free radical reactions. A typical rate constant is $7.8 \times 10^8 \ M^{-1} \ sec^{-1}$ for the reaction of 2,2,5,5-tetramethyl-3-pyrrolin-1-yloxy-3-carboxamide with methyl radicals.[1] Nitroxides and a nitrone "spin trap"[2] as tools for studies of free radical reactions are compared schematically (Fig. 1).

Generally, much lower concentrations of nitroxides than of the nitrone are sufficient for electron spin resonance (ESR)-detectable radical trapping. The basic strategy for using nitroxides as radical traps consists of accumulating diamagnetic products from the reaction

$$R_2NO\cdot + R'\cdot \rightarrow R_2NOR'$$

and characterizing $R_2NOR'$ by chemical, chromatographic, and/or biophysical methods. This reaction can be monitored continuously by observing the loss of the nitroxide's paramagnetic signal in an ESR spectrometer.

A complication in detecting free radicals with nitroxides is that other diamagnetic products can arise from reactions with biomolecules. One of these is the hydroxylamine which can arise during reactions of nitroxides with metabolic free radical intermediates in electron-transport processes (e.g., mitochondrial respiration[3]),

$$QH\cdot + R_2NO\cdot \rightarrow R_2NOH + Q$$

and with certain nonradical reductants like ascorbic acid,

$$AH_2 + R_2NO\cdot \rightarrow AH\cdot + R_2NOH$$

[1] S. Nigam, K.-D. Asmus, and R. L. Willson, *J. Chem. Soc. Farady Trans. I* **72**, 2324 (1976).
[2] A. T. Quintanilha and L. Packer, *Proc. Natl. Acad. Sci. U.S.A.* **74**, 570 (1977).
[3] E. G. Janzen, *Free Radicals Biol.* **4**, (1980).

Copyright © 1984 by Academic Press, Inc.
All rights of reproduction in any form reserved.
ISBN 0-12-182005-X

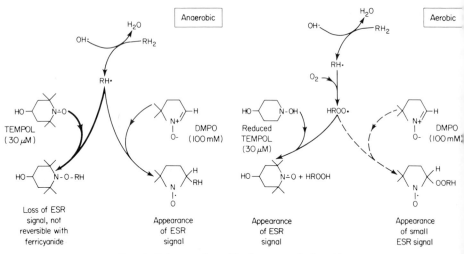

FIG. 1. Spin trapping with nitrones and nitroxides.

The third common diamagnetic species that can be formed from nitroxides in biological systems is the amine which can result from reactions with thiyl radicals.[4]

$$R'S\cdot + R_2NO\cdot \rightarrow R_2NH + products$$

The applicability of nitroxides to spin-trapping research requires that the three probable diamagnetic reaction products of nitroxides be separable and that the desired product, $R_2NOR'$, be sufficiently stable for further analyses. The stability requirement is met for many alkyl and aryl radical products.[5] Other products, like those with hydroxyl radicals, peroxy radicals, and presumably other oxygen-centered radicals are not stable[6] and this selectivity confers further diagnostic value upon nitroxides as radical traps. Separation of the three possible diamagnetic products can often be achieved by simple solvent partitioning. The p$K$ value of stable $R_2NOR'$ species is generally below 4,[1] that of $R_2NOH$ is about 7 (unpublished data), and that of $R_2NH$ above 9. Hence, the desired product can often be concentrated by means of a separatory funnel, organic solvents, and aqueous solutions of varying pH. The concentration of $R_2NOH$ in the system can also be estimated by treating the diamagnetic products with 1 m$M$ ferricyanide in aqueous solution. Ferricyanide oxidizes $R_2NOH$ to

[4] K. Murayama and T. Yoshioka, *Bull. Chem. Soc. Jpn.* **42**, 1942 (1969).
[5] D. W. Gratan, D. J. Carlsson, J. A. Howard, and D. M. Wiles, *Can. J. Chem.* **57**, 2834 (1979).
[6] R. J. Mehlhorn and L. Packer, *Can. J. Chem.* **60**, 1452 (1982).

$R_2NO\cdot$ without affecting the other products, and the paramagnetic $R_2NO\cdot$ species can be quantitated by ESR.[6]

## Characterization of Free Radical Reaction

Often, it is important to establish only whether or not free radicals are involved in some reaction. Nitroxides can be useful for this purpose if it can be shown that loss of an ESR signal occurs which is not reversible with ferricyanide oxidation, and that the nonreoxidizable reaction product is not the secondary amine, $R_2NH$. To ensure efficient trapping, i.e., maximal accumulation of free radicals by nitroxides it is advisable to avoid competing formation of peroxy radicals by conducting experiments anaerobically. A convenient procedure for performing such experiments in an ESR instrument is to place the reaction mixture into gas permeable tubing[7] under a stream of nitrogen gas. This procedure proved useful for light-generated radicals when the reaction mixture was alternately made aerobic and anaerobic to consume all of the hydroxylamine reaction products that can accumulate via electron- or hydrogen-transfer reactions.[6]

Nitroxides can also provide evidence for free radical chain reactions involving peroxy intermediates if it can be shown that oxygen consumption in an aerobic process is decreased in the presence of nitroxides. The nitroxide concentration should exceed that of oxygen by a factor of about 10 to compete effectively for reactive radicals.

## Specific Nitroxides for Radical Detections

A major advantage of nitroxides for radical detection schemes is that many different compounds have previously been developed for spin-labeling purposes[8] and these offer the opportunity for defining specific locations in cells and other biological environments where free radicals are generated. Among the most interesting questions about radical-mediated damage in cells are where are radicals generated, and how far from the site of generation do the radicals propagate? In particular, to what extent do radicals cross membranes? To answer these questions nitroxide traps localized inside cells, inside membranes, and inside or outside of membrane vesicles can be employed.

Figure 2 shows some examples of nitroxides and analogous compounds which can be applied to study sites where radicals are generated in cells and the extent to which radicals diffuse away from these initiation sites. Highly permeable nitroxides like TEMPOL will equilibrate among

[7] W. Z. Plachy and D. A. Windrem, *J. Magn. Reson.* **27**, 237 (1977).
[8] L. P. Berliner, ed., "Spin-Labeling." Academic Press, New York, (1976).

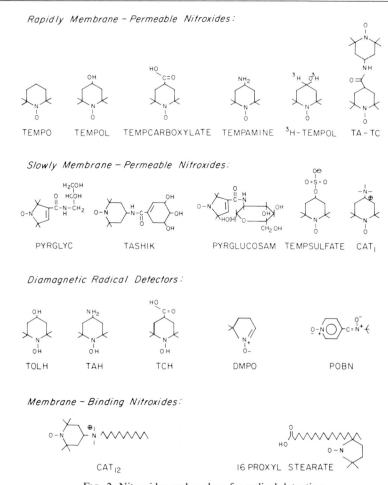

FIG. 2. Nitroxides and analogs for radical detection.

different subcellular compartments rapidly. The half-time for equilibration of TEMPOL across red cell membranes is less than 100 msec; hence aqueous diffusion will impose the rate-limiting step on equilibration in most cells. TEMPAMINE and TEMPCARBOXYLATE will accumulate in acidic and alkaline subcellular compartments, respectively,[9] and may therefore provide insights about free radical processes in these compartments. The biradical, TA-TC, is an example of a water-soluble nitroxide with adequate membrane penetration to react with intramembrane radicals. The amide linkage is rigid so that attachment of this biradical to an

[9] R. J. Mehlhorn, P. Candau, and L. Packer, this series, Vol. 88, p. 751.

intramembrane site should cause a significant immobilization of the nitroxide which should be evident in its ESR spectrum (see next section).

Slowly membrane-permeable probes shown in Fig. 2 provide the opportunity for selectively probing internal aqueous environments of membrane-enclosed compartments.[10] Half-times for the movement of these probes into envelope residues of halobacteria range from 40 sec for PYRGLYC to 280 days for $CAT_1$.[10] Hence, time-dependent loading of subcellular compartments can be attempted with these probes without resorting to membrane lysis. Subsequent removal of extracellular probes by centrifugation or filtration assures that only intracellular aqueous radicals are being trapped. The diamagnetic hydroxylamines like TOLH react with peroxy radicals whereas nitroxides do not; hence they can provide complementary information about aerobic reactions. TAH and TCH have the potential to probe compartments of different pH or, perhaps, of inferring radical charge, since collisions between nitroxides and radicals at low ionic strengths should be sensitive to Coulomb effects. The nitrone traps DMPO and DOBN form nitroxides upon undergoing addition reactions with reactive radicals. However, once formed, the products are subject to the reactions already described for nitroxides. Indeed, nitroxide spin-loss data should be directly relevant to a kinetic analysis of nitrone-trapping data.

Two examples of many available membrane-binding nitroxides are shown in Fig. 2. $CAT_{12}$ is essentially impermeable[11] and intercalates among membrane lipids such that the nitroxide projects into the polar–headgroup interface. Hence $CAT_{12}$ can be used to study inner or outer membrane surfaces as potential sources of radicals. This information is particularly interesting for metabolic radicals generated during electron-transport reactions. The lipid spin label, 16-proxyl stearate,[12] will intercalate among lipids as well but its nitroxide group is expected to be deeply buried within the hydrophobic membrane core.

*Characterization of Trapped Radicals*

Radioactive nitroxides provide a sensitive tool for analyzing reaction products with free radicals. It is likely that free radical chain reactions in biological systems will produce a variety of molecular species, so the potential of radioactive tracers for subsequent analysis is considerable. Concentration of reaction products can be pursued by conventional cell fractionation, protein purification, or chromatographic means. For small

[10] R. J. Mehlhorn and L. Packer, *Proc. N.Y. Acad. Sci.,* in press (1983).
[11] R. J. Mehlhorn and L. Packer, this series, Vol. 56, p. 515.
[12] J. F. Keana, T. D. Lee, and E. M. Bernard, *J. Am. Chem. Soc.* **98,** 3052 (1976).

molecules, pH-dependent solvent partitioning may be sufficient to effect significant concentration of reactions products. Subsequent analysis of these products could be conducted with autoradiograms.

Another procedure for tagging a nitroxide is with a paramagnetic label, i.e., to use a biradical nitroxide as the radical trap. This affords the opportunity of limited characterization of reaction pathways in terms of ESR spectral changes. For example, reaction of a water-soluble biradical with a membrane-bound radical species would convert a freely tumbling biradical to a relatively immobilized monoradical. The successful utilization of the biradical technique hinges upon the availability of membrane-penetrating nitroxide reagents having adequate water solubility so that substantial spectral changes can occur.

*Oxygen Involvement*

As shown in Fig. 1, under aerobic conditions peroxy radicals are formed which react with $R_2NOH$, thus regenerating $R_2NO\cdot$. Superoxide radicals also oxidize $R_2NOH$,[13] hence the hydroxylamine is a transient species when free radicals are generated aerobically. The final diamagnetic reaction products when all nitroxides have been consumed aerobically are either $R_2NOR'$ or $R_2NH$.[6]

Nitroxides have particular interest for studies of oxygen addition reactions because of the chemical resemblance between ground state oxygen $\cdot O{-}O\cdot$ and $R_2N{-}O\cdot$.

$$R'\cdot + R_2NO\cdot \rightarrow R_2NOR'$$
$$R'\cdot + \cdot O{-}O\cdot \rightarrow R'OO\cdot \rightarrow \text{chain reaction}$$

Hence, analysis of nitroxide reaction products can, in principle, yield information about molecules which will react with $O_2$ to form peroxy radicals. Generally $O_2$ reacts more rapidly with carbon-centered radicals than $R_2NO\cdot$, consistent with steric restraints on the latter reaction.[14]

Acknowledgments

Research supported by National Institute for Aging Grant AG04818 and the National Foundation for Cancer Research.

[13] G. M. Rosen, E. Finkelstein, and E. J. Rauckman, *Arch. Biochem. Biophys.* **215,** 367 (1982).
[14] R. L. Willson, *Trans. Faraday Soc.* **67,** 3008 (1971).

## [26] Low-Level Chemiluminescence as an Indicator of Singlet Molecular Oxygen in Biological Systems

*By* ENRIQUE CADENAS and HELMUT SIES

Excited oxygen species can emit low-level (ultraweak) chemiluminescence. Singlet ($^1\Delta_g$) molecular oxygen, abbreviated $^1O_2$, is an excited state of molecular oxygen about 93 kJ/mol above the triplet ground state, $^3O_2$. Dimol emission of $^1O_2$ has bands at 634 and 703 nm {$[^1\Delta_g][^1\Delta_g]$, (0,0) and (0,1) transitions, respectively} [reaction (1)]

$$^1O_2 + {}^1O_2 \rightarrow 2\,{}^3O_2 + h\nu \tag{1}$$

whereas excited carbonyls, denoted as RO*, emit between 380 and 460 nm. Singlet molecular oxygen, therefore, can be followed in the near-infrared using appropriate spectral windows.

Because of its high reactivity, $^1O_2$ may be classed as one of the aggressive oxygen species known to arise in biological systems. Chemically, the reactions include addition to olefins (ene reactions) leading to allylic hydroperoxides, additions to diene systems leading to endoperoxides, and further types of reactions leading to dioxetanes or oxidation of certain heteroatoms (see Refs. 1 and 2 and references therein).

Photon counting has been applied to biological systems, and recently evidence has accumulated that low-level chemiluminescence in the near-infrared provides useful information on oxidative processes in cells and tissues. Advantages of the technique are that it is noninvasive and provides continual monitoring. Major biological sources of $^1O_2$ can be grouped into (1) those systems involving interaction of organic oxygen radicals such as lipid peroxy radicals, ROO·, that arise from lipid peroxidation; and (2) those which involve reactions of reduced oxygen intermediates like $O_2^{-}$, $H_2O_2$, HO·. A third group might be represented by those systems which probably involve an enzymatic activation of oxygen, i.e., cyclooxygenase activity during prostaglandin biosynthesis. Photosensitization reactions represent a fourth category, which is not considered here.

It is obvious that $^1O_2$ in general will arise in very low yield and must be considered as a side-product of major radical processes. This implies

[1] H. H. Wasserman and R. W. Murray, eds., "Singlet Oxygen." Academic Press, New York, 1979.
[2] W. Adam and G. Cilento, eds., "Chemical and Biological Generation of Excited States." Academic Press, New York, 1982.

METHODS IN ENZYMOLOGY, VOL. 105
Copyright © 1984 by Academic Press, Inc.
All rights of reproduction in any form reserved.
ISBN 0-12-182005-X

possible pitfalls in the interpretation of low-level chemiluminescence measurements, and extrapolation from such measurements to other reactions must be made with due caution in each case. Supporting evidence from chemical systems capable of identifying singlet oxygen should be collected wherever possible.[3]

Detection of Chemiluminescence

*Single-Photon-Counting Apparatus*

The apparatus used by Tarusov *et al.*[4] in 1961 was improved to be applicable in a wider range of objects and toward higher efficiency (Fig. 1).[5,6] The system uses a red-sensitive photomultiplier cooled to $-25°$ by a thermoelectric cooler (EMI Gencom, Plainview, NY) in order to reduce the dark current. Suitable photomultipliers with S-20 response are EMI 9658AM,[5,6] RCA4832,[7] and Hamamatsu HTV R374.[8] The photomultiplier is connected to an amplifier-discriminator (EG&G Princeton Applied Res., Princeton, NJ) adjusted for single-photon counting. This, in turn, is connected to a frequency counter and/or a chart recorder or oscilloscope. Efficient light-gathering from the sample is established by using a Lucite rod as optical coupler[5,6] or an ellipsoidal light reflector.[8] A shutter allows a continuous operation of the photomultiplier in order to remain dark adapted. Either the animal for the exposed organ measurements *in situ* (Fig. 1A) or the isolated perfused organ (Fig. 1B) or a thermostated cuvette for cellular or subcellular and enzyme suspensions (Fig. 1C) is placed inside a light-tight box. Additions are made from the outside using precision micropumps and light-shielded tubing.

For spectral analysis of emitted light, optical filters may be placed into the light path. These can be interference filters[7] (e.g., Jenaer Glaswerke Schott, Mainz, Germany; transmission ranging from 45 to 55%, half-bandwidth 12–22 nm) or cutoff filters consisting of colored glass[8] (Toshiba Electric Co., Japan; transmission from 45 to 65%) or Wratten gelatin[9] (Eastman Kodak, Rochester, NY) filters. Filters may be in a circular

[3] C. S. Foote, *in* "Biological and Clinical Aspects of Oxygen" (W. A. Caughey, ed.), p. 603. Academic Press, New York, 1978.

[4] B. N. Tarusov, A. I. Polidova, and A. I. Zhuravlev, *Radiobiologiza* **1**, 150 (1961).

[5] A. Boveris, E. Cadenas, R. Reiter, M. Filipkowski, Y. Nakase, and B. Chance, *Proc. Natl. Acad. Sci. U.S.A.* **77**, 347 (1980).

[6] A. Boveris, E. Cadenas, and B. Chance, *Fed. Proc. Fed. Am. Soc. Exp. Biol.* **40**, 195 (1981).

[7] C. F. Deneke and N. I. Krinsky, *Photochem. Photobiol.* **25**, 299 (1977).

[8] H. Inaba, Y. Shimizu, Y. Tsuji, and A. Zamagishi, *Photochem. Photobiol.* **30**, 169 (1979).

[9] B. R. Andersen, T. F. Lint, and A. M. Brendzel, *Biochim. Biophys. Acta* **542**, 527 (1978).

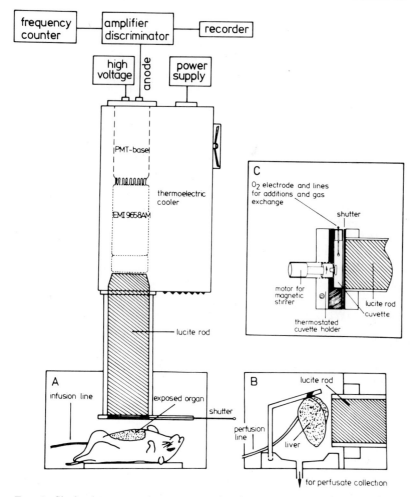

FIG. 1. Single-photon-counting apparatus for the measurement of low-level chemi-luminescence. The Lucite rod used as optical coupler is placed in front of the exposed liver *in situ* (A) or the perfused liver (B) or a cuvette (C). PMT, photomultiplier tube. Modified from Refs. 5 and 6.

array[8] (with automatic insertion through a filter drive controller) or exchangeable by slits.

Data are expressed in counts per second, sometimes also referred to the area exposed (in square centimeters). Although geometry is an important factor, reproducibility is surprisingly good, even with exposed organ surfaces in different animals. Conversion of counts per second to photons per second requires careful calibration of the photon counter.[10]

*Exposed Organ Surfaces in Situ and in Isolated Perfusion.*[5,11,12] The organ surface is exposed to the Lucite rod as close as possible. Spontaneous chemiluminescence of the liver in the red region (>610 nm) is about 3–10 cps and it may increase, for example, to 1000 cps upon addition of menadione (2-methyl-1,4-naphthoquinone)[13,14] or even higher to 5000 cps when oxidative stress is imposed by excessive doses of organic hydroperoxides.[5] A similar increase of hydroperoxide-stimulated chemiluminescence is observed with the perfused lung.[11,12] Steady-state conditions of chemiluminescence recording can be obtained for extended periods of time, so that small changes and their reversibility can be checked.

*Isolated Cells, Subcellular Organelles, and Enzyme Reactions.* The thermostatted cuvette (Fig. 1C) is equipped with a magnetic stirrer to maintain homogeneous suspension, and the lid contains ports for an oxygen electrode and for required tubings, e.g., for injection of compounds or for gas exchange. The latter may be of particular interest when studies on $O_2$ dependence are performed. Usually, hypoxic conditions can be obtained from a gas-mixing device capable of partitioning the relative inflow of $O_2/CO_2$ (19/1, v/v) and $N_2/CO_2$ (19/1, v/v) or $CO/CO_2$ (19/1, v/v). Maintenance of steady-state conditions of $O_2$ concentration in the cuvette can be afforded by an $O_2$-stat system.

Responses to additions are often in the form of a burst of photoemission; therefore additions are made through tubing and magnetic stirring, in order to avoid intervals without recording.

The table offers a list of different biological systems in which the observed chemiluminescence would be consistent with $^1O_2$ involvement.

### Liquid Scintillation Counter in the "Out-of-Coincidence" Mode

Though widely used, for example, with phagocytic cells[15] or in studies with microsomes,[16,17] the precise nature of the photoemissive species

[10] H. H. Seliger, this series, Vol. 57, p. 560.
[11] E. Cadenas, I. D. Arad, A. Boveris, A. B. Fisher, and B. Chance, *FEBS Lett.* **111**, 413 (1980).
[12] E. Cadenas, I. D. Arad, A. B. Fisher, A. Boveris, and B. Chance, *Biochem. J.* **192**, 303 (1980).
[13] H. Wefers and H. Sies, *Arch. Biochem. Biophys.* **224**, 568 (1983).
[14] E. Cadenas, H. Wefers, A. Müller, R. Brigelius, and H. Sies, *Agents Actions Suppl.* **11**, 203 (1982).
[15] M. A. Trush, M. E. Wilson, and K. Van Dyke, this series, Vol. 57, p. 462.
[16] J. R. Wright, R. C. Rumbaugh, H. D. Colby, and P. R. Miles, *Arch. Biochem. Biophys.* **192**, 344 (1979).
[17] M. T. Smith, H. Thor, P. Hartzell, and S. Orrenius, *Biochem. Pharmacol.* **31**, 19 (1982).

LOW-LEVEL CHEMILUMINESCENCE CONSISTENT WITH INVOLVEMENT OF
SINGLET MOLECULAR OXYGEN IN BIOLOGICAL SYSTEMS[a]

| Biological system | Conditions examined |
|---|---|
| Exposed organ surface | |
| Liver | |
| *In situ* (anesthetized) | Hyperbaric oxygen[b] |
| *In vitro* (perfused) | *t*-Butyl hydroperoxide infusion[c] |
| | Glutathione depletion[d,e] |
| | Menadione redox cycling[d,e] |
| Brain (*in situ,* anesthetized) | Hyperbaric oxygen[b] |
| Lung (perfused) | *t*-Butyl hydroperoxide/$H_2O_2$ infusion[f,g] |
| Isolated cells | |
| Hepatocytes | Hyperoxia/glutathione depletion[h] |
| Macrophages | Phagocytic activity[i] |
| Microsomal fraction | |
| Liver (rat) | NADPH/Fe-induced lipid peroxidation[j] |
| | *t*-Butyl hydroperoxide metabolism[k] |
| | Oxene donor metabolism[l] |
| | Redox cycling (paraquat[m], menadione[d]) |
| Seminal vesicle (ram) | Arachidonate metabolism[n] |
| Homogenates | |
| Liver | Hyperoxia, *t*-butyl hydroperoxide[o] |
| Brain | Hyperoxia[o] |
| Isolated enzymes | |
| Xanthine oxidase | Acetaldehyde metabolism[f] |
| Lipoxygenase | Linoleate oxidation[q] |
| Cyclooxygenase | Arachidonic acid-dependent formation of $PGG_2$ and $PGH_2$[n] |
| Isolated cytochrome *P*-450 | Oxene donor metabolism[l] |

[a] Systems listed were spectrally characterized by either distribution of light emission at 634 and 703 nm versus 668 nm or by cut-off filters to monitor light emission beyond 610 nm. Work using the scintillation counting method and luminol-mediated chemiluminescence is not included.

[b] A. Boveris, E. Cadenas, and B. Chance, *Fed. Proc. Fed. Am. Soc. Exp. Biol.* **40,** 195 (1981).

[c] A. Boveris, E. Cadenas, R. Reiter, M. Filipkowski, Y. Nakase, and B. Chance, *Proc. Natl. Acad. Sci. U.S.A.* **77,** 347 (1980).

[d] H. Wefers and H. Sies, *Arch. Biochem. Biophys.* **224,** 568 (1983).

[e] E. Cadenas, H. Wefers, A. Müller, R. Brigelius, and H. Sies, *Agents Actions Suppl.* **11,** 203 (1982).

[f] E. Cadenas, I. D. Arad, A. Boveris, A. B. Fisher, and B. Chance, *FEBS Lett.* **111,** 413 (1980).

[g] E. Cadenas, I. D. Arad, A. B. Fisher, A. Boveris, and B. Chance, *Biochem. J.* **192,** 303 (1980).

[h] E. Cadenas, H. Wefers, and H. Sies, *Eur. J. Biochem.* **119,** 531 (1981).

[i] E. Cadenas, R. P. Daniele, and B. Chance, *FEBS Lett.* **123,** 225 (1981).

[j] K. Sugioka and M. Nakano, *Biochim. Biophys. Acta* **423,** 203 (1976).

[k] E. Cadenas and H. Sies, *Eur. J. Biochem.* **124,** 349 (1982).

[l] E. Cadenas, H. Sies, H. Graf, and V. Ullrich, *Eur. J. Biochem.* **130,** 117 (1983).

[m] E. Cadenas, R. Brigelius, and H. Sies, *Biochem. Pharmacol.* **32,** 147 (1983).

[n] E. Cadenas, H. Sies, W. Nastainczyk, and V. Ullrich, *Hoppe-Seyler's Z. Physiol. Chem.* **364,** 519 (1983).

[o] E. Cadenas, A. I. Varsavsky, A. Boveris, and B. Chance, *Biochem. J.* **198,** 645 (1981).

[p] H. Inaba, Y. Shimizu, Y. Tsuji, and A. Yamagishi, *Photochem. Photobiol.* **30,** 169 (1979).

[q] A. Boveris, E. Cadenas, and B. Chance, *Photobiochem. Photobiophys.* **1,** 175 (1980).

detected with this convenient technique remains to be clarified. Most scintillation counters are equipped with photomultipliers sensitive only up to approximately 600 nm, so that $^1O_2$ dimol emission (occurring at 634 and 703 nm) is likely to escape detection with ordinary instruments. Thus, it is probable that excited carbonyls, RO*, contribute largely to the signal, rather than $^1O_2$.

### Luminol-Mediated Chemiluminescence

Luminol (5-amino-2,3-dihydrophthalazine-1,4-dione) is employed to amplify chemiluminescence signals. Luminol is oxidized by several oxygen intermediates, e.g., $O_2^-$ $H_2O_2$, HO·, $^1O_2$, to an electronically excited aminophthalate anion that, upon relaxation to the singlet ground state, emits photons. Therefore, luminol-mediated chemiluminescence lacks specificity for the initial oxygen species responsible, a fact sometimes apparently overlooked in the literature. However, the method is useful for detecting cellular production and efflux of the above-mentioned oxygen intermediates capable of reacting with luminol, so that the method is excellently suited for screening. Emittance is at around 450 nm, so that the liquid-scintillation counter method is applicable in addition to single-photon counting.

### Identification of Singlet Oxygen

As chemiluminescence is not unique to $^1O_2$, physical and chemical identification of the photoemissive sources is required. Deneke and Krinsky[7] suggested similar relative intensities at 634 and 703 nm ($^1O_2$ dimol emission) and low intensity at 668–670 nm as a suitable criterion. The lower emission at 668–670 nm would allow differentiation between a broad band of red chemiluminescence and specific emission centered at 634 and 703 nm. However, it should be noted that this is given for a chemically defined reaction, such as that of $HOCl/H_2O_2$ mixtures, which are known to produce $^1O_2$.[18] When applied to biological systems, several variables can affect the intensity and position of these $^1O_2$ dimol emission peaks, thereby weakening the certainty of $^1O_2$ involvement in the system. For example, it was reported that the presence of certain amino acids, such as tryptophan, can shift $^1O_2$ emission arising from the $HOCl/H_2O_2$ system to shorter wavelengths.[9] However, during the redox cycling of paraquat in isolated rat liver microsomes the positions were found as

[18] A. U. Khan and M. Kasha, *J. Am. Chem. Soc.* **92**, 3293 (1970).

predicted, with a 634/703 nm ratio of 1, and the intensity of the 668 nm band ~0.3.[19]

Further support is given by the enhancement or decrease of light emission at these specific wavelengths by DABCO (1,4-diazabicyclo[2.2.2]octane) or azide, respectively. The former is regarded as an enhancer of $^1O_2$ dimol emission in aqueous solutions,[7] though the mechanism of action remains obscure, especially since this enhancing effect is unusual as compared with other tertiary amines which are known to quench $^1O_2$.[20] The latter, azide, is known as a quencher of $^1O_2$. Neither compound changes the spectral characteristics of light; they just modify its intensity.[7] Additional evidence might be obtained by enhancement of light intensity by a solvent deuterium effect. The lifetime of $^1O_2$ is 10-fold longer in $D_2O$ relative to $H_2O$.[21]

While intensities of chemiluminescence at 634 and 703 nm are used as an indicator of $^1O_2$ dimol emission,[7] the observation of $^1O_2$ monomol emission at 1268 nm[22,23] would provide further proof recently, germanium diodes have been applied to biological samples.[23a,b]

The detection of $^1O_2$ production by electron spin resonance (ESR) relies on the generation of stable nitroxide radicals upon reactions of $^1O_2$ with sterically hindered amines, such as 2,2,6,6-tetramethylpiperidine.[24] The specificity of this method along with its sensitivity were tested with systems generating $^1O_2$ by means of a chemical reaction ($H_2O_2$/HOCl) or by energy transfer from triplet state sensitizers. Although this technique allows a specific study of free radical generation without interference from other by-products, its applicability to biological systems remains to be explored.

Identification of $^1O_2$ by detection of the oxidation product after reaction with quenchers is problematic because quenchers may also act as radical scavengers. The available methods and a careful critique were discussed by Foote.[3] Oxidation of 1,3-diphenylisobenzofuran is not specific enough, but the oxidation product of cholesterol, 5$\alpha$-hydroperoxide,[25] seems to be specific for $^1O_2$ and is distinct from the products formed in radical autoxidation, a mixture including 7$\alpha$- and 7$\beta$-hydroperoxides,

[19] E. Cadenas, R. Brigelius, and H. Sies, *Biochem. Pharmacol.* **32,** 147 (1983).
[20] C. Ouannès and T. Wilson, *J. Am. Chem. Soc.* **90,** 6527 (1968).
[21] D. R. Kearns, *Chem. Rev.* **71,** 395 (1971).
[22] A. U. Khan, *J. Am. Chem. Soc.* **103,** 6517 (1981).
[23] A. U. Khan and M. Kasha, *Proc. Natl. Acad. Sci. U.S.A.* **76,** 6047 (1979).
[23a] J. R. Kanofsky, *J. Biol. Chem.* **258,** 5991 (1983).
[23b] E. Lengfelder, E. Cadenas, and H. Sies, *FEBS Lett.,* in press (1983).
[24] Y. Lion, M. Delmelle, and A. van de Vorst, *Nature (London)* **263,** 442 (1976).
[25] Schenck, G. O., *Angew. Chem.* **69,** 579 (1957).

epoxides, etc. However, the low efficiency of the method and the laborious analytical procedure have restricted its use.

Possible Sources of Chemiluminescence

*Chemiluminescence Associated with Lipid Peroxidation*

Based on different evidence, the generation of $^1O_2$ during lipid peroxidation is attributed to the breakdown of lipid peroxides.[26–28] It seems that $^1O_2$ is regarded as a consequence of lipid peroxidation rather than as an initiator of the process.

This hypothesis[26]—based on Russell's mechanism[29,30]—assumes the self-reaction of lipid peroxy radicals (ROO·) with formation of a cyclic intermediate which decomposes in a triplet state carbonyl compound [reaction (2)] or oxygen in a singlet state [reaction (3)].

$$ROO· + ROO· \rightarrow ROH + O_2 + RO^* \tag{2}$$
$$ROO· + ROO· \rightarrow ROH + RO + {}^1O_2 \tag{3}$$

An excited carbonyl group is known to be formed during the autoxidation of fatty acids, and its triplet or singlet state seems to emit around 420–450 nm[31,32] [reaction (4)].

$$RO^* \rightarrow RO + h\nu \tag{4}$$

Malondialdehyde, a molecule also containing carbonyl groups, is ruled out as a possible source of photoemissive species during lipid peroxidation, since it absorbs at 265 nm and does not emit light in the region 300–700 nm.[26]

If $^1O_2$ is formed during lipid peroxidation, its reaction with unsaturated lipids to form a dioxetane intermediate is feasible [reaction (5)].

$$>C{=}C< + {}^1O_2 \rightarrow \underset{O-O}{\overline{\underline{\phantom{--}}}} \rightarrow >C{=}O + >C{=}O^* \tag{5}$$

Decomposition of dioxetanes might yield excited carbonyl compounds which emit in the blue region upon relaxation to the ground state [reaction (4)].

Russell's mechanism invoked to interpret the formation of $^1O_2$ was, however, subject to criticism.[33] A decomposition of hydroperoxides by

[26] K. Sugioka and M. Nakano, *Biochim. Biophys. Acta* **423**, 203 (1976).
[27] Yu. A. Vladimirov, M. V. Korchagina, and V. I. Olenev, *Biofizika* **16**, 994 (1971).
[28] M. M. King, E. F. Lai, and P. B. McCay, *J. Biol. Chem.* **250**, 6496 (1975).
[29] G. A. Russell, *J. Am. Chem. Soc.* **79**, 3871 (1957).
[30] J. A. Howard and K. U. Ingold, *J. Am. Chem. Soc.* **90**, 1058 (1968).
[31] R. F. Vassil'ev, *Opt. Spectrosc.* **18**, 131 (1965).
[32] R. A. Lloyd, *Faraday Soc. Trans.* **61**, 2182 (1965).
[33] N. I. Krinsky, *in* "Singlet Oxygen" (H. H. Wasserman and R. W. Murray, eds.), p. 597. Academic Press, New York, 1979.

metals could yield alcohols, ketones, and alkyl radicals which are capable of generating chemiluminescence. Chemiluminescence arising from transition metal ion- or heme compound-catalyzed decomposition of hydroperoxides has been observed.[34–36]

The relationship between low-level chemiluminescence and lipid peroxidation can be summarized as follows:

Chemiluminescence of systems undergoing lipid peroxidation is accompanied by an accumulation of malondialdehyde, as was observed in liver and brain homogenates,[37,38] isolated hepatocytes,[39] microsomal fractions,[16,26,40,41] and during oxidation of unsaturated fatty acids in model systems.[42] Although the amount of malondialdehyde accumulated usually correlates well with the chemiluminescence intensity observed,[16,17] it should be noted that malondialdehyde and light emission species are formed by different pathways and at different times during the process of lipid peroxidation.[26]

Light emission is proportional to the square of the concentration of lipid hydroperoxide accumulated in Fe-induced lipid peroxidation of mitochondrial and microsomal membranes,[26,43] consistent with the bimolecular reaction of lipid hydroperoxides as a source of light emission.

Autoxidation of fatty acids is an effective source of light emission.[44] Chemiluminescence is linearly related to the degree of unsaturation of fatty acids.[45]

Chemiluminescence of liver homogenates is enhanced in animals exposed to hepatotoxic agents that increase lipid peroxidation (carbon tetrachloride and hydrazine) and is decreased on *in vitro* addition of vitamin E and glutathione.[37] Glutathione-depleted hepatocytes show a higher chemiluminescence intensity and malondialdehyde accumulation than controls.[39]

Chemiluminescence and the rate of glutathione disulfide release (see Ref. 46) are correlated.[5]

[34] E. Cadenas, A. Boveris, and B. Chance, *Biochem. J.* **187,** 131 (1980).
[35] E. Cadenas, A. I. Varsavsky, A. Boveris, and B. Chance, *FEBS Lett.* **113,** 141 (1980).
[36] R. J. Hawco, C. R. O'Brien, and P. J. O'Brien, *Biochem. Biophys. Res. Commun.* **76,** 354 (1977).
[37] N. R. DiLuzio and T. E. Stege, *Life Sci.* **21,** 1457 (1977).
[38] E. Cadenas, A. I. Varsavsky, A. Boveris, and B. Chance, *Biochem. J.* **198,** 645 (1981).
[39] E. Cadenas, H. Wefers, and H. Sies, *Eur. J. Biochem.* **119,** 531 (1981).
[40] E. Cadenas and H. Sies, *Eur. J. Biochem.* **124,** 349 (1982).
[41] R. M. Howes and R. H. Steele, *Res. Commun. Pathol. Pharmacol.* **2,** 619 (1971).
[42] J. A. Neifakh, *Biofizika* **16,** 584 (1971).
[43] Y. M. Petrenko, D. L. Roshchupkin, and Yu. A. Vladimirov, *Biofizika* **20,** 617 (1975).
[44] S. N. Orlov, V. S. Danilov, A. A. Khrushchev, and Yu. N. Shvetsov, *Biofizika* **20,** 620 (1975).
[45] E. Cadenas, A. Boveris, and B. Chance, *Biochem. J.* **188,** 577 (1980).
[46] H. Sies and T. P. M. Akerboom, this volume [59], p. 445.

References listed in the table provide further information on the relationship between chemiluminescence and lipid peroxidation.

## Chemiluminescence Not Associated with Lipid Peroxidation

Chemiluminescence of biological reactions not associated with peroxidation of membrane lipids can involve, for example, the photoemission observed during phagocytosis of leukocytes and macrophages involving NADPH oxidases, enzymatic reactions such as the oxidation of substrates by xanthine oxidase, and redox cycling of compounds such as menadione or paraquat. In most cases, and based on convincing or unconvincing evidence, this photoemission is ascribed to the generation of $^1O_2$. A common factor in these systems is the occurrence of free radicals of oxygen, such as $O_2^-$, $H_2O_2$, $HO\cdot$, etc. $^1O_2$ would arise in these reactions in a secondary fashion, as a product of the interaction of reduction intermediates of oxygen.

We list below several reactions which are hypothesized to evolve $^1O_2$. Only in a few cases, such as the detection of monomol emission of $^1O_2$ in a system generating $O_2^-$, there is strong supporting evidence.

The spontaneous disproportionation of $O_2^-$ (or electron transfer reaction of the $O_2^-$) was regarded as a source of $^1O_2$[22,47] [reaction (6)], unlike the superoxide dismutase-catalyzed disproportionation.

$$O_2^- + O_2^- \rightarrow H_2O_2 + {}^1O_2 \tag{6}$$

However, it was recently suggested that the uncatalyzed dismutation of $O_2^-$ might be a poor source of $^1O_2$.[48]

The interaction of $O_2^-$ and $H_2O_2$ through a Haber–Weiss reaction [reaction (7)] was also regarded as a source of $^1O_2$[49] as was the reaction of $HO\cdot$ and $O_2^-$ [reaction (8)].[50,51]

$$O_2^- + H_2O_2 \rightarrow HO\cdot + HO^- + {}^1O_2 \tag{7}$$
$$O_2^- + HO\cdot + H^+ \rightarrow H_2O + {}^1O_2 \tag{8}$$

Recently, spontaneous disproportionation of $H_2O_2$ was indicated as generating $^1O_2$ by monitoring the formation of 5α-hydroperoxide of choles-

[47] A. U. Khan, *J. Phys. Chem.* **80**, 2213 (1976).

[48] C. S. Foote, F. C. Shook, and R. B. Akaberli, *J. Am. Chem. Soc.* **102**, 1503 (1980).

[49] E. W. Kellogg and I. Fridovich, *J. Biol. Chem.* **250**, 8812 (1975).

[50] J. P. Henry and A. M. Michelson, *in* "Superoxide Anion and Superoxide Dismutase" (A. M. Michelson, J. M. McCord, and I. Fridovich, eds.), p. 283. Academic Press, New York, 1977.

[51] R. M. Arneson, *Arch. Biochem. Biophys.* **136**, 352 (1970).

terol, a specific trap of $^1O_2$[52] [reaction (9)]. Conversely, catalase leads to the formation of $^3O_2$.[53]

$$H_2O_2 + H_2O_2 \rightarrow 2H_2O + {}^1O_2 \tag{9}$$

### Acknowledgments

Supported by Deutsche Forschungsgemeinschaft, Schwerpunktsprogramm "Mechanismen toxischer Wirkungen von Fremdstoffen." E. C. held an Alexander-von-Humboldt Stiftung fellowship.

[52] L. L. Smith and M. J. Kulig, J. Am. Chem. Soc. 98, 1027 (1976).
[53] D. J. T. Porter and L. L. Ingraham, Biochim. Biophys. Acta 334, 97 (1974).

# [27] High-Pressure Liquid Chromatography–Electrochemical Detection of Oxygen Free Radicals

By ROBERT A. FLOYD, C. ANN LEWIS, and PETER K. WONG

The presence of superoxide, hydrogen peroxide, and a catalytic amount of iron complexed to an appropriate liquid such as di- or triphosphate nucleotides will lead to the production of hydroxyl free radicals (OH),

$$O_2^- + H_2O_2 \xrightarrow{[Fe]} \dot{O}H + OH^- + O_2$$

Hydroxyl free radicals are extremely reactive and thus would be present at any specific time in at most very low amounts in biological systems even under conditions of severe oxidative stress. Detection of hydroxyl free radicals per se by electron paramagnetic resonance (EPR) in biological systems is virtually impossible, however the use of spin-trapping techniques offers possibilities. For example, the reaction of DMPO (5,5-dimethylpyrroline N-oxide) with OH proceeds rapidly,[1]

$$\dot{O}H + DMPO \xrightarrow{k = 3.2 \times 10^9} DMPO—\dot{O}H$$

thus yielding a spin adduct which decays in aqueous solution slowly and possesses a unique EPR spectrum.[2,3] Spin-trapping OH in biological

[1] E. Finkelstein, G. M. Rosen, and E. J. Rauckman, J. Am. Chem. Soc. 102, 4994 (1980).
[2] J. R. Harbour, V. Chow, and J. R. Bolton, Can. J. Chem. 52, 8549 (1974).
[3] R. A. Floyd, L. M. Soong, M. A. Stuart, and D. L. Reigh, Photochem. Photobiol. 28, 857 (1978).

Copyright © 1984 by Academic Press, Inc.
All rights of reproduction in any form reserved.
ISBN 0-12-182005-X

systems presents one complicating factor, however, in that the nitroxyl moiety of the DMPO–ȮH is readily reduced apparently by endogenous processes, thus rendering the spin-adduct diamagnetic and thus nondetectable by EPR,

$$\text{DMPO—ȮH} \xrightarrow{\text{Ḣ}} \text{DMPO—OH}_2$$

Since either the DMPO-hydroxyl free radical adduct in either the nitroxyl form (DMPO–ȮH) or in the reduced, hydroxyl amine form (DMPO–OH$_2$), can be oxidized electrochemically it appeared reasonable to attempt to detect and quantify the ȮH adduct by HPLC–electrochemical methods. The sensitivity afforded by electrochemical detection (picomole range) combined with the separation methods of HPLC seemed ideal for detection and quantification of ȮH production in biological systems. The method would be effective even if the ȮH adducts of DMPO were in the nitroxide or hydroxylamine form.

## Methods

### DMPO Preparation, Storage, and Use

The spin trap DMPO can be synthesized[4] or purchased from Aldrich or Sigma Chemical Company. The DMPO should be a creamy white amorphous solid at 4° but melt into an oil at slightly above room temperature. The oil (0.5 ml) is diluted with water up to a total volume of 5 ml. All work with DMPO should be conducted in subdued light conditions. The diluted DMPO is then passed through 0.5 g of charcoal (Fisher Chemical Company Carbon Decolorizing Neutral, Norit) which has been washed with deionized water several times. The charcoal as a thin layer rests on a 5-cm-diameter Whatman filter paper in a vacuum filter funnel. The diluted DMPO is passed by vacuum through the charcoal, and the volume restored to 5 ml if loss has occurred during filtration. The charcoal treatment removes traces of compounds that give an EPR signal. They apparently are due to a polymer of DMPO which has formed and is incompletely removed by vacuum distillation.

After charcoal filtration, the total 5 ml of diluted DMPO is then divided into five 1-ml aliquots each stored in a small dark brown bottle. Each lot of DMPO stock is bubbled with nitrogen gas for 15 min and then capped and stored at −20° until used. The concentration of DMPO can be determined by optical spectroscopy. The absorptivity of DMPO in aqueous

[4] R. Bonnett, R. F. C. Brown, V. M. Clark, I. O. Sutherland, and A. Todd, *J. Chem. Soc.* 2094 (1959).

solution is $\varepsilon_{226\,nm} = 7.22$ m$M^{-1}$ cm$^{-1}$. The concentration of the DMPO stock solutions is about 800 m$M$.

## Generation of DMPO-Hydroxyl Free Radical Adduct

Hydroxyl free radicals can be generated by any one of several different methods all of which yield the same results, but for the present purposes we routinely used the simplest system possible, namely, the UV photolysis of $H_2O_2$. UV photolysis of $H_2O_2$ has been utilized before to generate DMPO-hydroxyl free radical adduct.[2] The conditions we used were as follows: to a 10 ml beaker (2 cm diameter, 3 cm height) the following were added, 20 $\mu$l of DMPO stock, 35 $\mu$l of 0.2 $M$ (pH 7.6) Tris–HCl buffer, 10 $\mu$l of 30% $H_2O_2$, and 35 $\mu$l of $H_2O$. Irradiation was conducted with a mineral lamp (UVS-11, UltraViolet Products Inc., San Gabriel, California) by placing the lamp directly atop the beaker. Irradiation was usually continued for 6 min, which was the time period that gave the maximum amount of DMPO-hydroxyl free radical adduct.

Immediately after cessation of irradiation, 50 $\mu$l of the sample was placed in a heat-sealed transfer pipet for determination of the amount of O·H adduct of DMPO present utilizing EPR as has been described before.[3] Almost simultaneously a portion of the sample, usually 35 $\mu$l, was injected into a HPLC unit having an electrochemical detection attachment. The HPLC unit was custom constructed and consisted of a Gilson Model 302 pump and a Gilson 802 manometric damping module. The column eluate was monitored with an ampometric detector constructed by Dr. C. Leroy Blank (Dept. of Chemistry, University of Oklahoma), and equipped with a Bioanalytical System detector cell (Bioanalytical Systems Inc., West Lafayette, Indiana). The detector potential was set at +0.6 V versus a Ag/AgCl reference electrode. The output of the detector was registered on a Linear model 585 recorder (Linear Instrument, Irvine, California).

Two columns have been used successfully and they are (1) Hibar RT LiChrosorb-RP-18, 10 $\mu$m, 250 × 4 mm i.d. and (2) Altex Ultrasphere 3 $\mu$m-ODS 75 × 4.6 mm i.d. The longer column was used in the beginning but later the shorter column was found to be more effective since DMPO–O·H is decaying with a finite lifetime and thus the more rapid elution of this compound was desirable. Elution was accomplished with a flow rate of 0.5 ml/min. The mobile phase consisted of citric acid (monohydrate) 0.03 $M$, sodium acetate (anhydrous), 0.05 $M$, sodium hydroxide, 0.05 $M$, and glacial acetic acid, 0.02 $M$, resulting in a final pH of 5.1. The solvent was filtered through a Millipore filter of 0.45 $\mu$m pore size. The half-life of DMPO–O·H in the pH 7.6 Tris buffer was 16–18 min and was 18–20 min in the HPLC elution solvent.

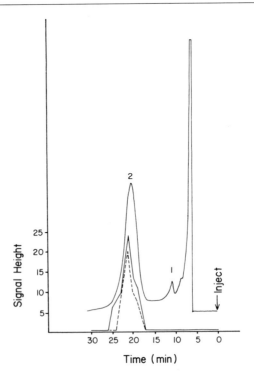

Time (min)

FIG. 1. High-pressure liquid chromatography–electrochemical detection (HPLC–ED) trace of the hydroxyl free radical adduct of DMPO and the electron paramagnetic resonance (EPR) signal height (bottom two traces) of the elution fractions from the column. The Hibar column was used. Elution fractions were subjected to EPR analysis as described in the text and in Fig. 4. The EPR signal height is presented on the ordinate and the time after injection is presented on the abscissa.

In order to determine the elution position of DMPO–OH from the HPLC column, 250-$\mu$l samples of the eluate coming directly from the column was taken and rapidly frozen in a dry ice-acetone bath. This prevents the decay of DMPO–OH such that the radical content can be determined at any time after collection. The content of DMPO–OH in each fraction was determined using EPR by inserting the eluate fractions in quartz tubes of 2.16 mm i.d. (Wilmad Glass Co., Buena, NJ) and running the spectra at $-30°$. The characteristic $1:2:2:1$ spectrum of DMPO–OH was displayed at this temperature, except that there was a slight broadening of the transitions. The peak-to-peak height of the second, from low to high field, transition was assayed as a measure of the amount of radical present.

Results and Discussion

Figure 1, top trace, demonstrates the results we obtain from the HPLC-electrochemical detection system when an aliquot of a sample prepared by UV photolysis of $H_2O_2$ in the presence of DMPO is injected. Two peaks are readily apparent labeled peak 1 and peak 2. Peak 1 grows at the expense of peak 2 as the time after irradiation before injection is increased. Also peak 2 disappears and peak 1 is dramatically enhanced if the DMPO–OH solution is treated with ascorbate (Fig. 2). Ascorbate reduces DMPO–OH into DMPO–OH$_2$. Peak 1 is not due to ascorbate, however, since this compound elutes with the solvent front. The EPR signals of a sample prepared by UV photolysis of $H_2O_2$ in the presence of DMPO and a sample which has been similarly prepared but subsequently reduced by ascorbate is shown in Fig. 3. Ascorbate completely eliminates

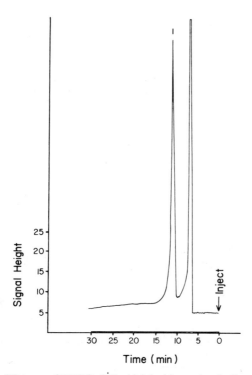

Time ( min )

FIG. 2. HPLC–ED trace of DMPO–OH which had been chemically reduced with ascorbate (top trace) and the EPR signal height (bottom trace, which is zero height) of the elution fractions from the column. The conditions were as given in Fig. 1 and as described in the text.

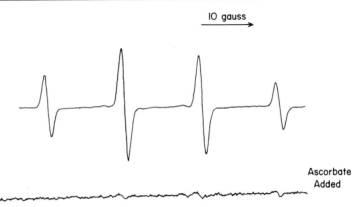

FIG. 3. Electron paramagnetic resonance traces of the UV-$H_2O_2$-photolyzed production of DMPO–$\dot{O}$H (top trace) and the same which had been reduced with ascorbate (bottom trace). The conditions are described in the text. The EPR parameters are 2 G modulation at 100 kHz frequency, gain $6.3 \times 10^2$, incident microwave power 15 mW, microwave frequency approximately 9.14 GHz, scan time 100 G/4 min, response time 3 sec, sample temperature about 25°.

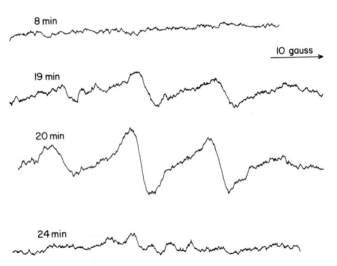

FIG. 4. Electron paramagnetic resonance traces of elution fractions obtained from the HPLC column of DMPO–$\dot{O}$H produced by UV-$H_2O_2$ photolysis. The time after injection is noted. The EPR parameters are 4 G modulation at 100 kHz frequency, gain $5 \times 10^4$, incident microwave power 15 mW, microwave frequency approximately 9.14 GHz, scan time 100 G/ 30 min, response time 10 sec, sample temperature −30°.

the presence of the EPR signal. The amount of DMPO–OH present in the $H_2O_2$–UV photolysis sample is about 20 $\mu M$ based on calculation of stable free radicals used as standards.[5] Figure 1 also demonstrates in the lower two traces the amount of EPR-detectable DMPO–OH present as a function of eluate fraction. The traces in Fig. 4 illustrate the EPR spectra of eluate fractions taken at 8, 19, 20, and 24 min. It is quite apparent that the 19 min spectrum and even more so the 20 min spectrum had DMPO–OH present, but this was not true for the 8 and 24 min spectra. We found a linear relationship between the amount of DMPO–OH present as measured by EPR versus the height of peak 2.

The results presented here as well as others not shown illustrate that the hydroxyl free radical adduct of DMPO (DMPO–OH) and its one-electron reduction product (DMPO–OH$_2$) can be separated out and detected by HPLC-electrochemical detection. Peak 2 in the traces is due to DMPO–OH and peak 1 is due to DMPO–OH$_2$. Regarding the sensitivity of the HPLC-electrochemical detection method, very low sensitivity settings were used in the present experiments. That is, the settings were 10 nA/V for the detector and 1 V for the recorder. It is possible to increase the sensitivity by greater than a factor of $10^4$ using the electrochemical methods, hence detecting conservatively 10 p$M$ of DMPO–OH or DMPO–OH$_2$. It is clear from the 19 and 20 min traces of Fig. 4 that we are near the limits of detection by EPR. The estimated DMPO–OH present in the 20 min trace of Fig. 4 is 0.1 $\mu M$.

It is clear since biological systems tend to reduce DMPO–OH to the diamagnetic DMPO–OH$_2$ and since HPLC-electrochemical detection offers extreme sensitivity as well as selective separation that this method has excellent potential of being very useful as a tool for studying oxygen free radicals in biological systems.

### Acknowledgments

Thanks are due to Mrs. Anita Hill for typing this manuscript and to Mrs. Sandra K. Nank who helped in certain experiments. National Institutes of Health Grant AG02599 provided partial support of this project.

---

[5] R. A. Floyd, *Can. J. Chem.* **60**, 1577 (1982).

## [28] Survey of the Methodology for Evaluating Negative Air Ions: Relevance to Biological Studies

### By Ronald Pethig

Since the discovery of electrically charged, molecular sized, "ions" in the atmosphere there has been considerable interest in their possible biological effects. Evidence from several laboratories indicates that air ions can produce a variety of biochemical and physiological responses in man, animals, plants, and microorganisms, and the data up to 1978 have been comprehensively reviewed by Kotaka.[1] Recent studies include the effects of air ions in altering the concentrations of serotonin and cyclic nucleotides in the cerebral cortex of rats,[2] and physiological and psychological effects on humans.[3] In general, positive air ions produce deleterious effects and negative ions beneficial ones, but despite the extensive data available there is no real understanding of the biological mechanisms involved. The purpose of this chapter is to aid such studies and the advancement of an understanding by providing details of how air ions can be produced and monitored under laboratory conditions.

### Natural Generation of Air Ions

This results directly from the initial ionization of atmospheric molecules such as oxygen and nitrogen. Since the energy required to ionize oxygen and nitrogen molecules is of the order 13.5 eV (1300 kJ/mol) and 14.5 eV (1400 kJ/mol), respectively, then the primary sources of natural air ion generation are cosmic and ultraviolet radiation, as well as natural radioactivity. In thunderstorms, corona discharges and electrical breakdowns of the air produce large numbers of ions locally. The average rate of natural generation of air ion pairs is around 10 $cm^{-3}$ $sec^{-1}$ and near the earth's surface in clean air the steady-state concentration is around 1000 ± 500 $cm^{-3}$. Due to the negative polarity of the atmospheric electric field (~1.3 V $cm^{-1}$) the number of positive ions exceeds the number of negatively charged ones by a factor of around 20%.[4]

If, after the ionization event, the free electrons initially produced have

[1] S. Kotaka, *CRC Crit. Rev. Microbiol.* 109 (1978).
[2] M. C. Diamond, J. R. Connor, E. K. Orenberg, M. Bissell, M. Yost, and A. Krueger, *Science* **210,** 652 (1980).
[3] J. M. Charry and F. B. W. Hawkinshire, *J. Personality Soc. Psychol.* **41,** 185 (1981).
[4] J. A. Chalmers, "Atmospheric Electricity," 2nd Ed. Pergamon, Oxford, 1967.

METHODS IN ENZYMOLOGY, VOL. 105
Copyright © 1984 by Academic Press, Inc.
All rights of reproduction in any form reserved.
ISBN 0-12-182005-X

sufficient energy (>360 kJ/mol) then oxygen radicals can be produced by the dissociative attachment process:

$$O_2 + electron \rightarrow O + O^-$$

This in turn can lead to the formation of $CO_3^-$ and $O_3^-$ ions by collisionally aided electron attachment processes such as

$$CO_2 + O^- \rightarrow CO_3^-$$

and

$$O_2 + O^- \rightarrow O_3^-.$$

Ions such as $O_2^-$ can also be produced in relatively small concentrations by the charge transfer process

$$O_2 + O^- \rightarrow O_2^- + O$$

and such products can in turn lead to the formation of harmful neutral molecules such as ozone, nitric oxide and nitrous oxide by such processes as

$$O_2 + O \rightarrow O_3$$
$$2O + N_2 \rightarrow 2NO$$
$$NO + O \rightarrow NO_2$$

The $NO_3^-$ ion, which is stable in terms of it surviving further reactions, can be formed by the collision process

$$NO_2 + O^- \rightarrow NO_3^-$$

For less energetic free electrons the most dominant ions produced are $O_2^-$ and $CO_4^-$ and they result from collisionally-aided electron capture as in the following process in which the electron loses about 42 kJ/mol of energy (corresponding to the electron affinity of the oxygen molecule):

$$O_2 + electron \rightarrow O_2^-$$

By combining with carbon dioxide, $CO_4^-$ is formed as follows:

$$O_2^- + CO_2 \rightarrow CO_4^-$$

and very small quantities of $NO_2^-$ can be produced by the charge transfer process:

$$O_2^- + NO_2 \rightarrow O_2 + NO_2^-$$

In general, collisionally aided electron attachments that produce such ions as $O_2^-$ and $CO_4^-$ are greatly facilitated by the presence of water vapor in the atmosphere.

Positive ions (e.g., $N_2^+$) produced after the initial ionization event can also become involved in various ion–molecule interactions to produce $N_2H^+$, $H_2O^+$, $NO^+$, $(H_2O)H^+$, and $(H_2O)O_2^+$, for example.

These various ions, which have been produced within a time period of the order $10^{-8}$ sec or less, are of molecular size and carry a net positive or negative electronic charge of $1.6 \times 10^{-19}$ C. Such small ions create large electric fields ($\gtrsim 6 \times 10^9$ V m$^{-1}$) which polarize nearby polar molecules such as $H_2O$ or $CO_2$ and attract them to form larger ionic molecular aggregates of the form $(H_2O)_nH^+$, $(H_2O)_nO_2^+$, $(H_2O)_nOH^-$, or $(H_2O)_nCO_4^-$, for example, where $n$ can have a value up to 10 or so. The time taken for the formation of such molecular aggregates is of the order of half a microsecond. These ions are termed *small ions*, they have a diameter of around $1.5 \pm 0.5 \times 10^{-3}$ $\mu$m and an electrical mobility of the order 1 cm$^2$ V$^{-1}$ sec$^{-1}$ and larger.

Naturally occurring aerosol particles in the atmosphere have diameters mostly ranging from $10^{-2}$ to 40 $\mu$m (approximate size of cloud droplets) with the log-normal distribution centered around 0.1 $\mu$m.[5] In countryside air the total aerosol count ranges from 400 to 18,000 cm$^{-3}$ with the average 2600 cm$^{-3}$, and in urban areas particle counts exceeding $10^5$ or $10^6$ cm$^{-3}$ are frequently encountered. These air-borne particles enable air ions to increase in size by collision-aided charge transfer or straightforward ion–particle attachment processes. As the air ions increase in size, their mobility decreases due to frictional drag, so that for particle diameters of 0.5 $\mu$m the mobility is of the order $2 \times 10^{-5}$ cm$^2$ V$^{-1}$ sec$^{-1}$. Biological activity is usually attributed to small- and medium-sized ions, and in this author's opinion the ions of greatest interest are likely to be those associated with oxygen radicals.

### Air Ion Generators for Laboratory Use

Methods of generating air ions include using thermionic electron emission from hot wires or photoelectrodes, $\beta$ and $\gamma$ radiation from radioisotopes, and corona discharges. Thermionic and photo-induced electron emission tends to be inefficient in producing large air ion concentrations and so the use of radioisotopes and corona discharges is generally preferred.

The radioactive materials commonly used are tritium and polonium which are impregnated into or contained in titanium, zirconium or stainless steel foils. If around 500 V is applied to the foil, then depending on the polarity of this voltage either the positive or the negative ions produced by the emitted high energy particles are repelled into the surrounding

[5] C. Junge, *J. Meteorol.* **12**, 13 (1955).

atmosphere. Generation rates of $10^9$ ions per second and greater are possible, and tritium with a half-life of 12.3 years and 0.02 MeV $\beta$ energy tends to be used more than polonium which has a half life of only 0.4 years and a large $\beta$-energy of 5.3 MeV. In their most basic form such ion sources can cause problems when used for biological studies. Apart from direct exposure to radioactive emission the test objects may also be subjected to a significant electric field associated with the voltage used to separate the ion polarities. The control of ion concentrations and polarities is also not reliable, especially if the ion source is located near the test object in an attempt to achieve a maximum dosage of small ions. Bachman et al.[6] have described methods, using nonionized gases as ion concentration dilutants and ducts, to control ionized air streams. Ion concentrations of the order $10^6$ cm$^{-3}$ were achieved using 50 mCi krypton and tritium sources contained in titanium foil.

The most common method of air ion generation employs corona discharges produced when high voltages are applied to wires or pin electrodes. The electric field strength at the tip of a pin electrode can be approximated[7] by the relationship

$$E \cong 2V/a \, \log_e(4x/a) \tag{1}$$

where $V$ is the electrode voltage, $a$ is the radius of curvature of the electrode tip, and $x$ is the distance to the nearest earth plane. Fields in excess of $10^8$ V/m are readily attainable at electrode tips and as such greatly exceed the normal breakdown voltage ($\sim 3 \times 10^6$ V/m) for air. Complete electrical breakdown of air requires the creation of a large number of very energetic electrons, and since the field at an electrode tip rapidly diverges then only those air molecules nearest the tip become ionized. If the electrode is at a negative voltage, then the negative ions produced are accelerated away from the electrode, while the positive ones are accelerated toward the electrode where they become neutralized by electrons "leaking" from ground potential.

Generation rates greater than $10^{11}$ ions per second are readily achievable with corona discharges, and pin electrodes made of tungsten or stainless steel, for example, are more commonly used than wire electrodes. One type of electrode, preferred by the author, is composed of some 40,000 carbon fibers of which a dozen or so can act as corona sources at any one time. This type of electrode is less likely to become inefficient as a result of dirt accumulating at the electrode tip or the tip becoming blunted, which are effects commonly experienced with pin electrodes. If sufficient ionization energy is available, then physiologically harmful

[6] C. H. Bachman, M. E. Crandell, and F. X. Hart, *Rev. Sci. Instrum.* **39**, 1168 (1968).
[7] C. F. Eyring, S. S. Mackeown, and R. A. Millikan, *Phys. Rev.* **31**, 900 (1928).

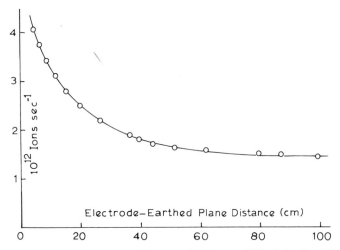

FIG. 1. The variation of the generation rate of air ions at 60% relative humidity and 20° emitted by a "Sphereon" air ionizer as a function of distance from an earthed plane. The fiber electrode voltage was $-6.8$ kV.

gases such as ozone and nitrogen oxides can be formed. In this author's experience, based on gas analysis measurements, then provided a pin or fiber electrode is maintained at a voltage lower than around 7.5 kV concentrations of less than 1 part per $10^9$ in air of ozone, nitric oxide or nitrous oxide are normally produced.

Once air ionization is initiated the generation rate of the desired air ions depends strongly on the electric field, especially at the higher field levels. From Eq. (1) it is to be expected that the generation rate of air ions should depend on the distance to the nearest earth plane, and an example of this is shown in Fig. 1 for a "Sphereon" air ionizer (Hygeia Ltd.) which uses a carbon fiber electrode at a dc voltage of $-6.8$ kV as the negative ion source. In this ionizer model the electrode can be removed and so the ion generation rate $g$ can be measured by the current $I$ being delivered to the electrode ($g = I/q$) as a function of its distance from a large conducting earth plane. Results similar to those shown in Fig. 1 were obtained by placing the ionizer and other commercially available models (which use one-, two-, or four-pin electrodes at negative voltages ranging from 4.8 to 6.0 kV) in various-sized Faraday cages and measuring the total current collected by the cages. Taking into account the electrode voltages, the carbon fiber electrode was found to be more efficient than the pin electrodes in generating air ions.

Unlike the situation for natural air ions, when air ion generators are used the concentration of the ions can be so large that there are significant space charge, charge density gradient and local electric field effects which

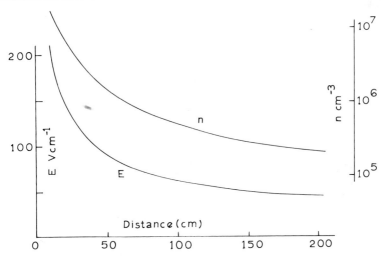

FIG. 2. Variation of the air ion density and electrostatic field generated by an isolated "Sphereon" electrode as a function of radial distance. The plots are theory [Eqs. (2) and (3) with $g = 2 \times 10^{12}$ ions $sec^{-1}$ and $\mu = 0.5$ $cm^2$ $V^{-1}$ $sec^{-1}$].

result in strong ion–ion repulsions and an increased precipitation of charged aerosol particles. An approximation of the relationship between the local air ion density $n$ and electric field $E$ can be obtained by assuming that the generator acts as a point ion source, and that the nearest earth plane is at least 2.5 m away. The problem is then reduced to solving Poisson's equation, and assuming spherical symmetry it can be shown[8] that

$$n = (3\varepsilon_0 g/8\pi\mu q s^3)^{1/2} \tag{2}$$

and

$$E = (gq/6\pi\varepsilon_0\mu s)^{1/2} \tag{3}$$

where $\varepsilon_0$ is the permittivity of free space, $g$ is the generation rate of ions of charge $q$ and mobility $\mu$, and $s$ is the distance from the electrode tip. These results shown plotted in Fig. 2 for the case $g = 1.5 \times 10^{12}$ ions/sec typical of the "Sphereon" carbon fiber electrode, indicate that the air ion concentration decreases with distance $s$ from the generator as a 3/2 power law and that the local electric field decreases with distance as a half-power law. From Eqs. (2) and (3) it follows that there is a relationship between the air ion concentration and the electric field strength of the form

$$n = 3\varepsilon_0 E/2sq \tag{4}$$

[8] R. Pethig, *J. Bioelect.* **2**, 15 (1983).

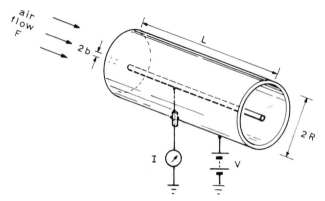

FIG. 3. Basic outline of the Gerdien tube for the measurement of air ion density.

## Measurement of Air Ion Concentration

A common method for determining air ion concentrations is from a measurement of the electrical conductivity of the atmosphere. Such measurements must be made with care to ensure that the technique either does not influence the apparent atmospheric conductivity or that any such effect is well quantified. In particular this means that the ion population (typically less than 1 part in $10^{14}$ of the total atmospheric molecular content) must not be substantially altered during the measurement process, or if it is that this is accomplished in a controlled manner. Such measurements are usually made with a Gerdien tube,[9] the basic outline of which is shown in Fig. 3. The tube has a length typically of the order 50 cm and an internal diameter of around 12 cm. A central axial rod of external diameter around 2 cm acts as one electrode and the outer tube as the other. When a dc voltage of the order 50 to 200 V is applied to these two electrodes as atmospheric air is pumped through the tube, the air acts as an electrical conductor and the resulting current $I$ is determined using an electrometer. To measure the effect of the total population of air ions, the air flow through the tube must be large enough to ensure that considerably more air ions pass right through the tube than are collected by the electrodes. If this condition is not satisfied then the depletion of ions as the air passes through the tube will alter the overall conductivity of the air. In other words, the radius of the tube must be large compared to the average distance that an air ion moves under the influence of the applied electric field $E$. For the tube of Fig. 3 and with a volumetric air flow of $F$ cm³

[9] H. Gerdien, *Phys. Z.* **6**, 800 (1905).

$sec^{-1}$, then the time $t$ taken by an air ion to pass through the central portion of the tube is given by

$$t = \pi L(R - b)^2/F \qquad (5)$$

For an air ion of mobility $\mu$, the distance it will move under the influence of an electric field $E$ during time $t$ is $\mu Et$. From Eq. (5) the condition to be met is

$$F \gg \pi \mu LE(R - b) \qquad (6)$$

The result is that the current flowing between the electrodes is proportional to the applied voltage $V$ and to the conductivity $\sigma$ of the air. In the cylindrical electrode arrangement of Fig. 3, the electrical field at any radius $r$ is given by

$$E = V/r \log_e(R/b)$$

and the resulting current $I$ for the tube of length $L$ is given by

$$I = 2\pi\sigma VL/\log_e(R/b)$$

For an air ion concentration of $10^3$ cm$^{-3}$ and an average air ion mobility of 0.5 cm$^2$ V$^{-1}$ sec$^{-1}$, the conductivity of the air will be $8 \times 10^{-17}$ mho/cm. For an applied voltage of 200 V and with $R = 6$ cm, $b = 1$ cm, and $L = 50$ cm, then the measured current will be $2.8 \times 10^{-12}$ A and capable of being measured by most standard electrometers. The atmosphere to be sampled should be brought immediately into the tube without the use of long air intakes, so as to avoid ion losses and the effect of such structures in altering the local electric fields. By altering the various parameters of Eq. (6) the ion monitoring tube can be arranged so as to determine the populations of various ion mobility ranges, and the polarity of the air ions being monitored is determined by the polarity of the applied voltage $V$. The rate of air flow through the tube should be limited to that which ensures laminar flow, but should be large enough to overcome effects due to any convection currents. Mounting the tube vertically can provide the greatest mechanical stability combined with the least area of insulating surface. The latter factor is important in minimizing effects due to contamination and the accumulation of surface charges. Surrounding the tube and electrometer by a grounded electrostatic screen, with just one ground connection, aids in maintaining sensitivity of electrometer readings.

Equation (4) indicates that under the conditions assumed (i.e., a single ion emitter generating a spherically symmetrical population of air ions) the air ion concentration can be determined from a measurement of the local electrostatic field. Since knowledge of such fields may also be of general importance in studies of the biological effects of air ions, it will be

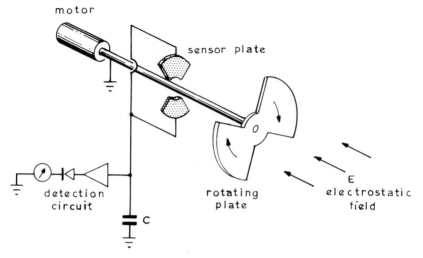

FIG. 4. Basic construction of a field mill for the measurement of electrostatic fields.

of value to give a basic description of a suitable method of measurement. One such method employs a field mill, whose basic design is shown in Fig. 4. A metal sensor plate is connected to ground potential through a capacitor C, and a detection circuit monitors the voltage developed across this capacitor. If the sensor field is exposed to an electrostatic field an induced charge $Q$ appears on the plate and produces a voltage $Q/C$ across the capacitor. Biasing currents in the detection circuit and the interception of air ions by the sensor plate can lead to a steadily drifting reference level, and this effect can be minimized by introducing an earthed rotating plate in front of the sensor. As the rotor turns the sensor is alternatively shielded from and exposed to the electric field; the peak voltage across the capacitor is given by Secker[10] as

$$\hat{V} = A\pi f \varepsilon_0 RE \qquad \text{for } (2\pi f RC)^2 \ll 1 \qquad (7)$$

or

$$\hat{V} = A\varepsilon_0 E/2C \qquad \text{if } (2\pi f RC)^2 \gg 1 \qquad (8)$$

where $A$ is the sensor plate area, $R$ is the input resistance to the amplifier, and $f$ is the rate at which the rotor modulates the electric field exposed to the sensor. Equation (8) offers the most useful application since the voltage is independent of the rotor speed and with a set value for $C$ (usually not less than around 150 pF) the voltage is directly proportional to the

[10] P. E. Secker, *J. Electrostat.* **1,** 27 (1975).

local electrostatic field. For a two-blade rotor operating at 100 rps Eq. (8) is valid for amplifier input resistances of around 50 MΩ and more. The lack of importance of a precise or constant rotor speed allows the rotor motor to be clockwork driven, which can simplify construction and enable the field sensor to be remotely located from the main detector electronics. By incorporating phase-sensitive detection and the use of a rotor-controlled chopped reference signal, the polarity of the electrostatic field can be determined. Although home-made models can be readily constructed and used effectively, a range of commercially produced field mills is available. The results obtained using an Electrostatic Fieldmeter (Industrial Development Bangor Ltd.) Model 107 to determine the field strength at various distances in front of a "Sphereon" ionizer located on a 1-m high wooden plinth are shown in Fig. 5. By comparison with the theoretical plot of Eq. (3) with $g = 2 \times 10^{12}$ ions sec$^{-1}$ and $\mu = 0.02$ cm$^2$ V$^{-1}$ sec$^{-1}$, it can be seen that the assumptions used in deriving Eqs. (2) and (3) provide only an order of magnitude estimate of practical situations. In particular, the assumption of spherical symmetry of air ion population is not correct for most commercial air ionizers. The "Sphereon," which produces a more spatially uniform air ion distribution than other ionizers tested by the author, typically produces an electrostatic field (and hence an air ion concentration) in front of the ionizer that is double that pro-

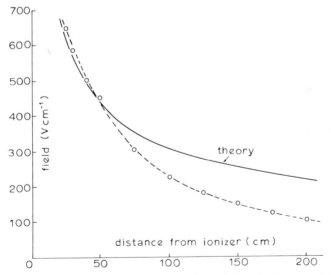

FIG. 5. Variation of the electrostatic field with distance from the front of a "Sphereon" ionizer. The theoretical line is derived from Eq. (3) with $g = 2 \times 10^{12}$ ions sec$^{-1}$ and $\mu = 0.02$ cm$^2$ V$^{-1}$ sec$^{-1}$.

duced behind it. The more rapid decline in the field strength with distance shown by the experimental results of Fig. 5, as compared with theory, results from distortion of the field by the floor, the wooden plinth, and the ionizer shape itself. The experimental results also suggest that ion–aerosol attachments occur increasingly with distances exceeding 50 cm from the ionizer, and that the air ions produced by the ionizer have an average mobility somewhat lower than 0.02 cm$^2$ V$^{-1}$ sec$^{-1}$.

## Concluding Remarks

In the design and performance of experiments to investigate the biological effects of air ions, considerable care is required to ensure that the generation and population of the ions are accomplished in a controlled and reproducible manner. Excessive voltages at ion-emitting electrodes can result in the production of deleterious gases such as ozone and oxides of nitrogen, and changes in relative humidity and aerosol content can have large effects on the physical characteristics of the air ion population. Such influences, as well as the proximity of nearby laboratory apparatus, structures, and personnel in altering the local electrostatic fields, can complicate interpretations of the observations or otherwise of reproducible biological effects of air ions. As an aid to such studies it is possible to construct or purchase ion density and electrostatic field-measuring equipment.

### Acknowledgments

Thanks are due to Mr. E. L. Head, Principal, Institute of Pneumotherapy, Cranbrook, for valuable discussions; Mr. D. W. Poirot, who assisted with some of the experimental studies; Industrial Development Bangor (UCNW) Ltd., for the loan of several field mill models; and Hygeia Ltd., Christchurch, which generously supplied gas analysis equipment and various air ion generators.

## [29] Detection of Oxidative Mutagens with a New Salmonella Tester Strain (TA102)

By DAVID E. LEVIN, MONICA HOLLSTEIN,
MICHAEL F. CHRISTMAN, and BRUCE N. AMES

Damage to DNA is likely to be a major cause of cancer and other diseases.[1,2] The *Salmonella* mutagenicity test,[3,4] along with other short-term assays,[5] is being used extensively to survey a variety of substances in our environment for mutagenic activity. The test measures back-mutation in several specially constructed mutants of *Salmonella*. A homogenate of rat liver (or other mammalian tissue) is added to the bacterial suspension as an approximation of mammalian metabolism.[3,4] A detailed methods paper has recently appeared.[4] Using this system, over 80% of the organic carcinogens tested have been detected as mutagens.[6–8]

Among the classes of carcinogens that have gone undetected as mutagens in the *Salmonella* assay are chemical oxidants. Oxygen radicals may be the most important class of mutagens contributing to aging and cancer,[9] yet a number of oxidants known to generate reactive oxygen species are not detected as mutagens in the standard *Salmonella* mutagenicity assay. A new *Salmonella* tester strain, TA102, has been described recently[10] that detects a variety of oxidants and other agents as mutagens which are not detected by the standard tester strains. Among the oxidants detected are hydrogen peroxide and other peroxides, X rays, bleomycin, neocarzinostatin, streptonigrin, and other quinones, and phenylhydrazine. TA102 differs from the other standard tester strains used in muta-

[1] B. N. Ames, *Science* **204**, 587 (1979).
[2] H. H. Hiatt, J. D. Watson, and J. A. Winsten, eds., "Origins of Human Cancer." Cold Spring Harbor Laboratory, Cold Spring Harbor, New York, 1977.
[3] B. N. Ames, J. McCann, and E. Yamasaki, *Mutat. Res.* **31**, 347 (1975).
[4] D. M. Maron and B. N. Ames, *Mutat. Res.* **113**, 173 (1983).
[5] M. Hollstein, J. McCann, F. A. Angelosanto, and W. W. Nichols, *Mutat. Res.* **65**, 133 (1979).
[6] J. McCann, E. Choi, E. Yamasaki, and B. N. Ames, *Proc. Natl. Acad. Sci. U.S.A.* **72**, 5135 (1975).
[7] J. McCann and B. N. Ames, *Proc. Natl. Acad. Sci. U.S.A.* **73**, 950 (1976).
[8] B. N. Ames and J. McCann, *Cancer Res.* **41**, 4192 (1981).
[9] B. N. Ames, *in* "Mutagens in Our Environment" (M. Sorsa and H. Vainio, eds.), p. 3. Liss, New York, 1982.
[10] D. E. Levin, M. Hollstein, M. F. Christman, E. A. Schwiers, and B. N. Ames, *Proc. Natl. Acad. Sci. U.S.A.* **79**, 7445 (1982).

METHODS IN ENZYMOLOGY, VOL. 105

Copyright © 1984 by Academic Press, Inc.
All rights of reproduction in any form reserved.
ISBN 0-12-182005-X

genicity screening in that it has A : T base pairs at the site of reversion, whereas all of the other tester strains have G : C base pairs at their reversion sites. It is likely that this difference is responsible for the unique sensitivity of TA102 to reversion by chemical oxidants.

## Materials and Methods

### TA102

*hisG428,* an ochre mutant of *Salmonella typhimurium,*[11] was selected from 70 histidine-requiring mutants screened as the most sensitive to reversion by chemical oxidants. The tester strain constructed from this mutant, TA102,[10] carries the *hisG428* mutation on the multicopy plasmid, pAQ1, to increase its sensitivity to reversion. TA102 also carries the *rfa* (deep-rough) mutation to increase permeability to large molecules,[12] and the R-factor resistance plasmid pKM101, which confers error-prone repair to the cell.[13]

Tester strain cultures are inoculated from frozen permanents[4] into 18 × 150 mm culture tubes containing 5 ml of nutrient broth (Oxoid No. 2), which are then shaken at 37° for 12–16 hr in a gyrorotary incubator (New Brunswick Scientific Co.). This fully grown culture has 1–2 × 10⁹ bacteria per milliliter. Fresh overnight cultures should be prepared daily and can be used for mutagenicity testing up to 8 hr after overnight growth.

### Mutagens

Chemical sources were as follows: hydrogen peroxide from Mallinckrodt; cumene hydroperoxide from Pfaltz and Bauer; *t*-butyl hydroperoxide, bleomycin sulfate, paraquat, diquat, danthron, and phenylhydrazine from Sigma; *t*-butylperoxymaleic acid from ICN (K & K Laboratories); emodin and mitomycin C from Aldrich; streptonigrin from Flow Laboratories); methyl ethyl ketone peroxide was the gift of A. L. Tappel; neocarzinostatin was the gift of M. Hofnung. Mutagens were dissolved in distilled water except for streptonigrin, danthron, emodin, *t*-butylperoxymaleic acid, cumene hydroperoxide, methyl ethyl ketone peroxide (in dimethyl sulfoxide), and phenylhydrazine (in 95% ethanol).

[11] P. E. Hartman, Z. Hartman, R. C. Stahl, and B. N. Ames, *Adv. Genet.* **16**, 1 (1971).
[12] B. N. Ames, F. D. Lee, and W. E. Durston, *Proc. Natl. Acad. Sci. U.S.A.* **70**, 782 (1973).
[13] J. McCann, N. E. Spingarn, J. Kobori, and B. N. Ames, *Proc. Natl. Acad. Sci. U.S.A.* **72**, 979 (1975).

OXIDANTS THAT REVERT TA102[a]

| Mutagen | His+ Revertants per plate on TA102 | |
| --- | --- | --- |
| | Plate incorporation | Preincubation |
| (Spontaneous) | (283) | (292) |
| Streptonigrin (100 ng) | 1114 | 1163 |
| Mitomycin C (250 ng) | 2682 | 7617 |
| Bleomycin (1 μg) | 1183 | 1128 |
| Neocarzinostatin (50 ng) | 119 | 387 |
| Neocarzinostatin (500 ng) | 1148 | Toxic |
| Phenylhydrazine (250 μg) | 1096 | 1153 |
| Hydrogen peroxide (100 μg) | 643 | 592 |
| t-Butyl hydroperoxide (100 μg) | 3001 | 3116 |
| t-Butylperoxymaleic acid (100 μg) | 287 | 681 |
| Cumene hydroperoxide (100 μg) | 1890 | 2044 |
| Methyl ethyl ketone peroxide (200 μg) | 261 | 237 |
| X Rays (2.5 krad) | 1132 | N.T. |
| Paraquat (10 ng) | 40 | 48 |
| Diquat (10 ng) | 36 | 31 |

[a] The number of spontaneous revertants per plate was subtracted. Values were taken from the linear region of dose–response curves. N.T., not tested.

## The Mutagenicity Test

*The Standard Assay with Preincubation.* The assay[3,4,14] involves pre-incubating the test compound and the bacterial tester strain in buffer or rat liver homogenate (S-9 Mix) for 20 min at 37°. Soft agar is then added to the preincubation mix which is poured onto the surface of a minimal agar plate. The plates are then incubated at 37° for 48 hr, and then revertant colonies (His+) are counted. The mutagenic oxidants described in the table do not require metabolic activation by exogenously added liver homogenate. The complete mutagenicity assay does involve the use of liver homogenate and as some quinones require the addition of liver for mutagenicity, we also describe its use.

*Procedure.* Deliver 0.1 ml or less of the test solution (in DMSO, water, or other appropriate solvent[15]) into sterile 13 × 100-mm culture tubes placed in an ice bath. Add 0.5 ml of either S-9 Mix (see "S-9 section") or

[14] T. Matsushima, T. Sugimura, M. Nagao, T. Yahagi, A. Shirai, and M. Sawamura, in "Short-Term Test Systems for Detecting Carcinogens" (K. H. Norpoth and R. C. Garner, eds.), p. 273. Springer-Verlag, Berlin and New York, 1980.

[15] D. Maron, J. Katzenellenbogen, and B. N. Ames, *Mutat. Res.* **88**, 343 (1981).

0.02 *M* sodium phosphate buffer (pH 7.4) and then 0.1 ml of an overnight culture of the tester strain. Vortex the tube and incubate at 37° for 20 min. The tubes should be shaken at moderate speed during the incubation. A Thermolyne Dri-Bath attached to a Model G2 New Brunswick Scientific Co. laboratory rotator makes a convenient incubation apparatus which can be placed in the fume hood. The Dri-Bath is fitted with aluminum blocks, with 20 wells each, for 13-mm tubes.

Soft agar is boiled and maintained at 45° in a heating block to keep it molten. The soft agar (0.6% Difco agar and 0.5% NaCl) contains a trace of histidine (0.05 m*M*) to allow the bacteria to undergo several rounds of cell division. Add 2 ml of molten soft agar to the preincubation mixture. Mix the test components by vortexing the soft agar for 3 sec at low speed before pouring onto minimal glucose agar plates (1.5% Difco agar and 2% glucose in Vogel-Bonner minimal medium E[4,16]). Sterile, disposable plastic petri plates (100 × 15 mm) are used. To achieve a uniform distribution of the top agar on the surface of the plate, quickly tilt and rotate the plate; then place it on a level surface to harden. The mixing, pouring, and distribution should take less than 20 sec and the plates should be left to harden for several minutes. Within an hour the plates should be inverted and placed in a dark, vented, 37° incubator. After 48 hr the revertant colonies on the test plates and the control plates are counted, and the presence of a background lawn of bacterial growth on all plates is confirmed. A lawn that is thin compared to the lawn on the negative control plate is evidence for bacterial toxicity. A negative control containing the bacteria and the solvent is required to establish the number of colonies that arise spontaneously. The number of spontaneous revertants per plate for TA102 falls within the range of 240–320. A reproducible dose–response curve with an increase in the number of revertants over the spontaneous value is evidence for mutagenicity. Most mutagens demonstrate a linear dose–response at levels that are not excessively toxic to the bacteria. All of the mutagenic oxidants tested on TA102 demonstrate a linear dose–response. Mutagenicity data for several diagnostic oxidants appear in the table. Two oxidants, paraquat and diquat, are not mutagenic in this assay, though they are thought to be active through formation of a semi-quinone which generates superoxide radicals. They are quite toxic to the bacteria and it must be determined whether there is some other toxicity factor that masks mutagenicity.

*The Standard Assay without Preincubation.* The standard assay[3,4] without preincubation involves adding, in order, the test compound, the bacterial tester strain, and the S-9 Mix, if required, to soft agar, which is

---

[16] H. J. Vogel and D. M. Bonner, *J. Biol. Chem.* **218**, 97 (1956).

then briefly mixed and poured directly onto the minimal agar plate. This assay requires somewhat less time and fewer manipulations than the modification with preincubation, but is somewhat less sensitive for some mutagens.

The mutagenic activity of the diagnostic oxidants described above was determined both with and without preincubation. In all cases the addition of preincubation resulted in equal or greater sensitivity than without it (see the table).

*X Rays.* X-Ray mutagenesis was done by irradiating standard pour plates already seeded with bacteria. The X-ray source was a Machlett OEG 60 tube with a beryllium window, operated at 50-kV peak and 25 mA, to give a dose rate of 250 rad/sec (1 rad = 0.01 Gr). Plates were irradiated in 5-sec increments to a maximum of 30 sec.

*The Spot Test.* This is a variation of the assay in which the mutagen is left out of the soft agar and is applied directly to the surface of the top agar. A few crystals of solid mutagens or up to 10 $\mu$l of liquid mutagens can be added to the agar surface. The spot test is a considerably less sensitive, semiquantitative assay that is appropriate for initial screening purposes. The standard assay is preferable for generating a dose–response curve and for general testing.

### Rat Liver S-9

For general mutagenesis screening we recommend liver homogenates from rats induced with a polychlorinated biphenyl (PCB) mixture, Aroclor 1254.[3,4] Details of S-9 preparation have been reported previously.[3,4] S-9 preparations can be purchased from the AMC Cancer Research Center and Hospital, c/o Dr. Elias Balbinder, 6401 W. Colfax Ave., Lakewood, Colorado 80214; from Litron Laboratories, 1351 Mt. Hope Ave., Suite 207, Rochester, New York 14620; from Microbiological Associates, c/o Dr. Steven Haworth, 5221 River Rd., Bethesda, Maryland 20816; or from Litton Bionetics, c/o Dr. David Brusick, 5516 Nicholson Lane, Kensington, Maryland 20795. S-9 is also available in Europe from Dr. E. Fresenius KG, Chem-Pharm. Industry, Postfach 1809, D-6370 Oberursel, Federal Republic of Germany. Commercial S-9 preparations should be tested for sterility and enzyme activity.

For general screening, we recommend a concentration of 20 $\mu$l of S-9 per plate. At this concentration, the S-9 mix will contain 0.04 ml of S-9 fraction per milliliter of Mix. If a compound is negative using this concentration of S-9, it should be retested with 50 $\mu$l of S-9 per plate.

Two anthracene quinones tested for mutagenicity required S-9 activation to demonstrate their mutagenic effects. Danthron (1,8-dihydroxyan-

thraquinone) and emodin (1,3,8-trihydroxy-6-methylanthraquinone) induced 1140 and 420 revertants for 30 and 50 $\mu$g, respectively, on TA102
using 50 $\mu$l of S-9 per plate in a preincubation assay. Without S-9, neither
of these agents was mutagenic.

*The S-9 Mix.* The components of the standard S-9 mix are 8 m$M$
MgCl$_2$, 33 m$M$ KCl, 5 m$M$ glucose 6-phosphate, 4 m$M$ NADP, 100 m$M$
sodium phosphate, pH 7.4, and S-9 at a concentration of 0.04 ml/ml of
Mix. S-9 Mix is prepared fresh for each mutagenicity assay and may be
kept on ice for several hours without significant loss of activity.

Acknowledgments

This work was supported by D.O.E. Contract DE-AT03-76EV70156 to B.N.A. and by
N.I.E.H.S. Center Grant ES01896. D.E.L. was supported by N.I.E.H.S. Training Grant
ES07075, M.H. was supported by a postdoctoral fellowship from the Monsanto Fund, and
M.F.C. was supported by a predoctoral fellowship from the University of California,
Berkeley.

# [30] Determination of the Mutagenicity of Oxygen Free Radicals Using Microbial Systems*

*By* Hosni M. Hassan and Carmella S. Moody

Increased oxygen tension has been known, for more than two decades,
to cause chromosomal breaks and mutations both in eukaryotes[1-3] and in
prokaryotes.[4,5] Recently, it has been shown that physiological concentrations of oxygen ($\sim$5% O$_2$) are mutagenic in certain oxygen-sensitive histidine auxotrophs of *Salmonella typhimurium* strain TA100.[6] A basic understanding of these deleterious effects of oxygen came with the advent of

* Paper Number 8836 of the Journal Series of the North Carolina Agricultural Research
Service, Raleigh, NC 27650. The use of trade names in this publication does not imply
endorsement by the North Carolina Agricultural Research Service of the products named,
nor criticism of similar ones not mentioned.
[1] A. D. Conger and L. M. Fairchild, *Proc. Natl. Acad. Sci. U.S.A.* **38**, 289 (1952).
[2] M. Moutschen-Dahmen, J. Moutschen, and L. Ehrenberg, *Hereditas* **45**, 230 (1959).
[3] W. E. Kronstad, R. A. Nilan, and C. F. Konzak, *Science* **129**, 1618 (1959).
[4] W. O. Fenn, R. Gerschman, D. L. Gilbert, D. E. Terwilliger, and F. W. Cothran, *Proc. Natl. Acad. Sci. U.S.A.* **43**, 1027 (1957).
[5] G. D. Gifford, *Biochem. Biophys. Res. Commun.* **33**, 294 (1968).
[6] W. J. Bruyninckx, H. S. Mason, and S. A. Morse, *Nature (London)* **274**, 606 (1978).

Copyright © 1984 by Academic Press, Inc.
All rights of reproduction in any form reserved.
ISBN 0-12-182005-X

the theory of oxygen toxicity.[7] This theory states that the partially re-duced intermediates of oxygen—the superoxide anion ($O_2^-$), hydrogen peroxide ($H_2O_2$), and the hydroxyl radical (OH·)—are the damaging agents. In accordance with this theory, it has been shown that ionizing radiation generates oxygen free radicals,[8–10] that superoxide radicals can indirectly cause DNA strand scission, *in vitro,*[11] and that oxygen free radicals generated by paraquat (PQ) are mutagenic.[12] Hydrogen peroxide has been shown to liberate DNA bases, cause DNA strand breakage, and alter the chemical composition of the bases.[13,14] Hydroxyl radicals have also been shown to cause DNA strand breaks, *in vitro.*[11,15] In this chapter, methods for assessing the mutagenicity of oxygen free radicals in micro-bial systems will be presented. These methods are usually rapid, simple, and inexpensive.

### Methods for Generating Oxygen Free Radicals

The different procedures for generating oxygen free radicals in biologi-cal systems are discussed in detail in [53] of this volume. It should be reemphasized here that choosing the right system for generating the de-sired partially reduced oxygen species (i.e., $O_2^-$, $H_2O_2$, or OH·) at the desired target is very important. We should bear in mind that the cell envelope of *Escherichia coli* is impermeable to the superoxide anion,[16] and that the OH· will react indiscriminantly with organic molecules. Fur-thermore, most, if not all, aerotolerant organisms possess a powerful system for the detoxication of superoxide radicals and hydrogen peroxide (i.e., superoxide dismutases and hydroperoxidases, respectively). These detoxifying enzymes prevent the interaction between $O_2^-$ and $H_2O_2$ and therefore eliminate OH· formation by the iron-catalyzed Haber–Weiss reaction.[17–19] These cellular defenses must be overcome in order to study

[7] J. M. McCord, B. B. Keele, Jr., I. Fridovich, *Proc. Natl. Acad. Sci. U.S.A.* **68**, 1024 (1971).
[8] J. F. Ward, *Adv. Radiat. Biol.* **5**, 182 (1975).
[9] K. C. Smith and J. E. Heys, *Radiat. Res.* **33**, 129 (1968).
[10] J. J. Van Hemmen, *Nature (London)* **231**, 79 (1971).
[11] K. Brawn and I. Fridovich, *Arch. Biochem. Biophys.* **206**, 414 (1981).
[12] C. S. Moody and H. M. Hassan, *Proc. Natl. Acad. Sci. U.S.A.* **79**, 2855 (1982).
[13] H. J. Rhaese and E. Freese, *Biochim. Biophys. Acta* **155**, 476 (1968).
[14] H. Massie, H. Samis, and M. Baird, *Biochim. Biophys. Acta* **272**, 539 (1972).
[15] R. A. Floyd, *Biochem. Biophys. Res. Commun.* **99**, 1209 (1981).
[16] H. M. Hassan and I. Fridovich, *J. Biol. Chem.* **254**, 10846 (1979).
[17] F. Haber and J. Weiss, *Proc. R. Soc. London Ser. A* **147**, 332 (1934).
[18] J. M. McCord and E. D. Day, Jr., *FEBS Lett.* **86**, 139 (1978).
[19] H. M. Hassan and I. Fridovich, *in* "Enzymatic Basis of Detoxication" (W. B. Jakoby, ed.), Vol. 1, p. 311. Academic Press, New York, 1980.

the mutagenicity of oxy-radicals. Conditions known to overcome these defenses include, but are not limited to (1) the use of mutants deficient in superoxide dismutases, hydroperoxidases, or both, (2) the use of hyperbaric oxygen, (3) the use of growth conditions or chemical inhibitors that prevent the induction and the synthesis of these detoxifying enzymes, and (4) the use of redox-active compounds that are known to exacerbate the flux of oxy-radicals inside the cell and overwhelm their natural defenses. The above listed approaches are discussed in [53] of this volume, which should be consulted before designing experiments to study the mutagenicity of oxy-radicals.

The ultimate proof for the mutagenicity of oxy-radicals should come from experiments clearly showing that molecular oxygen is essential[12] (i.e., no effect under anaerobic conditions) and that higher levels of the detoxifying enzymes (i.e., superoxide dismutases and hydroperoxidases) offer protection.[12]

### Methods for Assessing the Mutagenicity of Oxygen Free Radicals in Microbial Systems

During the past 13 years several short-term tests have been developed for the evaluation of the toxicity and the mutagenicity of environmental chemicals.[20–26] The majority of these assays utilize microbial systems which allows rapid screening of many compounds. Furthermore, there is a very high correlation between mutagenicity and carcinogenicity of chemicals.[21,23,27,28] In these tests both forward- and back-mutations have been used. Strains of *S. typhimurium*,[22,24] *E. coli*,[25] *Bacillus subtilis*,[26] and *Saccharomyces cerevisiae*[25] are commonly used. Any of these systems can equally be used for testing the mutagenicity of oxy-radicals; however, only methods using the *Salmonella typhimurium* strains commonly known as the Ames' tester strains will be mentioned here. These strains are histidine auxotrophs which are capable of reverting to histidine proto-

[20] F. J. DeSerres, *Mutat. Res.* **33**, 11 (1975).

[21] B. N. Ames, F. D. Lee, and W. E. Durston, *Proc. Natl. Acad. Sci. U.S.A.* **70**, 782 (1973).

[22] B. N. Ames, J. McCann, and E. Yamasaki, *Mutat. Res.* **31**, 347 (1975).

[23] J. McCann, N. E. Spingarn, J. Kobori, and B. N. Ames, *Proc. Natl. Acad. Sci. U.S.A.* **72**, 979 (1975).

[24] D. E. Levin, M. Hollstein, M. F. Christman, E. A. Schwiers, and B. N. Ames, *Proc. Natl. Acad. Sci. U.S.A.* **79**, 7445 (1982).

[25] I. de G. Mitchell, *Agents Action* **10**, 287 (1980).

[26] I. Morimoto, F. Watanabe, T. Osawa, T. Okitsu, and T. Kada, *Mutat. Res.* **97**, 81 (1982).

[27] E. E. Slater, M. D. Anderson, and H. S. Rosenkranz, *Cancer Res.* **31**, 970 (1971).

[28] B. N. Ames, W. E. Durston, E. Yamasaki, and F. D. Lee, *Proc. Natl. Acad. Sci. U.S.A.* **70**, 2281 (1973).

trophy upon exposure to chemical mutagens. Five strains are generally recommended for screening of mutagens. Strains TA1535 and TA100 are used to detect base-pair substitution mutagens while strains TA1537, TA1538, and TA98 are used to detect frameshift mutagens.[23] TA1537 has a cytidine-rich region at the site of the mutation[21] while TA1538 has an alternating G·C-rich region.[29] The strains also have a deletion in the gene for excision repair and a mutation that prevents formation of the lipopolysaccharide coat on the bacterial surface.[21] Strains TA98 and TA100 carry an ampicillin resistance transfer factor that reportedly increases the sensitivity of these strains for detecting certain mutagens.[23] Recently, a new tester strain (TA102) has been developed and found to be more sensitive toward oxidative mutagens. This strain harbors a multicopy plasmid which carries an A·T-rich region at the mutation site.[24] These tester strains can be obtained from B. N. Ames, Department of Biochemistry, University of California, Berkeley.[21]

The use of mammalian liver S-9 fraction for the activation of some chemical mutagens has been recommended[22,30]; however, it is not required when testing for the mutagenicity of oxy-radicals.[12] The different procedures for performing the mutagenicity tests are listed below.

## The Plate Incorporation Assay

This procedure was originally devised by Ames et al.[21,22,28] for testing chemical mutagens, and has recently been revised.[30] In this procedure the tester strain, the test compound, and the liver S-9 fraction (if required) are mixed in a soft top agar before pouring onto minimal agar plates. Revertant colonies (His$^+$) are normally counted after 2–3 days of incubation at 37°.

### Materials

Sterile top agar (0.6% agar, 0.5% NaCl)

Sterile stock solution of L-histidine·HCl/D-biotin, 0.5 mM each

Petri plates containing approximately 20 ml of sterile Vogel–Bonner (VB) glucose minimal agar medium[31] containing, per liter, $MgSO_4 \cdot 7H_2O$, 0.2 g; citric acid · $H_2O$, 2.0 g; $K_2HPO_4$, 10 g; $NaNH_4PO_4 \cdot 4H_2O$, 3.5 g; glucose, 20 g; and agar, 15 g

Oxygen radical generator—this could be a redox-active compound[32]

[29] K. Isono and J. Yourno, Proc. Natl. Acad. Sci. U.S.A. **71**, 1612 (1974).
[30] D. M. Maron and B. N. Ames, Mutat. Res. **113**, 173 (1983).
[31] H. J. Vogel and D. M. Bonner, J. Biol. Chem. **218**, 97 (1956).
[32] H. M. Hassan and I. Fridovich, Arch. Biochem. Biophys. **196**, 385 (1979).

(i.e., paraquat, pyocyanine, streptonigrin, etc.) or phagocytic leu-kocytes[33]

The tester strains (TA100, TA98 and TA102) grown overnight at 37° and 200 rpm in 50–100 ml of VB glucose minimal medium with an excess amount of histidine (0.1 mM), or grown in Oxoid nutrient broth No. 2

Anaerobic chamber where $O_2$ is kept at <5 ppm (Coy Laboratory Products Inc., Ann Arbor, Michigan or a similar product)

*Procedure*. Melt the top agar in a steam bath and add the L-histidine/D-biotin (10% v/v) to yield 0.05 mM of each, final concentration. Distribute 2-ml portions of the molten top agar into sterile-capped test tubes and keep at 45° in a water bath. Asceptically centrifuge the cells from the overnight culture and resuspend in an equal volume of sterile cold glucose minimal medium lacking histidine (this step minimizes the amount of histidine carried over). Add the desired concentrations of the compound or the system chosen for generating oxy-radicals, (add 0.5 ml of S-9 fraction, if required) followed by 0.1 ml of the test organism. Mix by gentle swirling, and pour evenly onto the surface of the glucose minimal agar plates. Run triplicate plates for each concentration tested, and include both positive and negative controls to ascertain the fidelity of the tester strains.[22] Incubate the plates at 37° for 48–72 hr before counting the colonies. Repeat the same protocol using the anaerobic chamber to demonstrate that oxygen is essential for mutagenicity. A range of concentrations of the oxy-radical generator should be tested for a dose–response effect, both in the presence and absence of the S-9 fraction. A background lawn of growth should be seen on the control and treated plates. A lack of the background lawn indicates that the compound tested is toxic, and colonies appearing on these plates are probably not true revertants. These colonies are His⁻ survivors that have grown in colony size because of the excess histidine available per surviving cell. In this test, a twofold increase in the number of revertants over the control numbers is considered to be significant.[22]

This method has been used successfully to demonstrate the mutagenicity of oxy-radicals generated by phagocytic leukocytes[33] and by several redox-active compounds.[34] Paraquat (an oxy-radical generator) was found to be slightly mutagenic in the standard Ames test, however, because of its extreme toxicity the background lawn was eliminated and a dose–response effect was not seen. Therefore, a modified assay (see Liquid incubation assays) where the number of revertants is related to the viable cells, was necessary to assess its extreme mutagenicity.[12]

[33] S. A. Weitzman and T. P. Stossel, *Science* **212**, 546 (1981).
[34] T. Yamaguchi, *Agric. Biol. Chem.* **45**, 327 (1981).

## The Spot Test

The spot test is valuable for the rapid screening of different oxy-radical generators or other chemical mutagens.[22] It allows preliminary determination of the most suitable tester strain, the toxicity of the compound as determined by the size of the zone of inhibition and whether the S-9 fraction is needed for mutagen activation. The test is performed as in the plate incorporation assay except that the mutagen is added as solid, crystals or a small drop onto the center of the top agar. Mutagenicity or the lack of it should be confirmed by using the plate incorporation assay.

## Preincubation Assay

This procedure was originally developed for testing the mutagenicity of azo dyes.[35] It follows the same basic procedure originally recommended by Ames et al.[22,28] except that the test compound is preincubated with the S-9 fraction (when required) and $1-2 \times 10^8$ bacterial tester strain cells in a total volume of 0.7 ml at 37° for 20 min. After the preincubation period, 2 ml of the molten top agar is added, mixed, and poured onto the surface of the glucose minimal agar plates. Plates are incubated at 37° and handled as indicated above.

## Liquid Incubation Assays

The literature is replete with reports that attest to the inability of the Ames Salmonella test to detect the mutagenicity of many well known carcinogens. For example, ionizing radiation was shown to be nonmutagenic in the Ames test.[36,37] However, a recent report[38] has shown that γ irradiation causes a dose-related increase in the number of induced mutants per $10^8$ survivors while the number of induced mutants per plate is small. It was also noted by Ames and co-workers[24] that the standard tester strains were unable to detect the mutagenicity of "a variety of oxidants and other agents," and therefore prompted the development of the new tester strain TA102.

There seems to be a positive correlation between the toxicity of the test compound and the tendency to obtain false-negative results in the standard Ames test. Recently, Mitchell[39] outlined the problems encountered with the plate assay methods and developed a liquid culture method

[35] T. Yahagi, M. Degawa, Y. Seino, T. Matsushima, M. Nagao, T. Sugimura, and Y. Hashimoto, Cancer Lett. 1, 91 (1975).
[36] M. Fox, Nature (London) 268, 488 (1977).
[37] S. J. Rinkus and M. S. Legator, Cancer Res. 39, 3289 (1979).
[38] F. P. Imray and D. G. MacPhee, Int. J. Radiat. Biol. 40, 111 (1981).
[39] I. de G. Mitchell, Mutat. Res. 54, 1 (1978).

MUTAGENICITY TESTING OF PARAQUAT BY PLATE
INCORPORATION ASSAY

| Paraquat (mM) | His+ revertants per plate[a] | | | | |
|---|---|---|---|---|---|
| | TA98 | TA100 | Strain TA1535 | TA1537 | TA1538 |
| 0 | 31 | 154 | 26 | 15 | 13 |
| 0.05 | 40 | 172 | 30 | 1 | 15 |
| 0.1 | 34 | 83 | 28 | 0 | 0 |
| 0.5 | 4 | 17 | 1 | 2 | 1 |
| 1.0 | 2 | 12 | 1 | 1 | 0 |
| 5.0 | 1 | 2 | 3 | 3 | 0 |

[a] Three plates per dose level were tested. Colonies below
1 mm diameter were His− and were not counted.

that was more sensitive and accurate in detecting the mutagenicity of highly toxic and mutagenic compounds. In this method the frequency of mutation ($Q$) is expressed per $10^9$ colony-forming units instead of per plate, thus eliminating the possibility of underestimating the mutagenicity of toxic compounds.

In an attempt to determine the mutagenicity of oxygen free radicals we used paraquat (as a generator of $O_2^-$) in the standard Ames test (see the table). The data showed that paraquat is either nonmutagenic or weakly mutagenic. We questioned these negative results because (1) oxygen free radicals have been shown to cause DNA strand scissions *in vitro*[11,40]; (2) increased oxygen tension causes chromosomal aberrations *in vivo*[41]; and (3) paraquat causes DNA damage in mammalian cells in culture,[42] induced antifertility effects in male mice,[43] induces gene conversion in yeast,[44] and is mutagenic in *S. typhimurium* using the 8-azaguanine forward mutation test.[45] Furthermore, it is known that paraquat is very toxic in prokaryotes[46–48] and in eukaryotes[49–51] (see also [53] and [69] of this volume).

[40] S. A. Lesko, R. J. Lorentzen, and P. O. P. Ts'o, *Biochemistry* **19**, 3023 (1980).
[41] H. Joenje, F. Arwert, A. W. Eriksson, H. de Koning, and A. B. Oostra, *Nature (London)* **290**, 142 (1981).
[42] W. E. Ross, E. R. Block, and R-Y. Chang, *Biochem. Biophys. Res. Commun.* **91**, 1302 (1979).
[43] A. Pasi, J. W. Embree, Jr., G. H. Eisenlord, and C. H. Hine, *Mutat. Res.* **26**, 171 (1974).
[44] J. M. Parry, *Mutat. Res.* **21**, 83 (1973).
[45] R. Benigni, M. Bignami, A. Carere, G. Conti, L. Conti, R. Crebelli, E. Dogliotti, G. Gualandi, A. Novelletto, and V. A. Ortali, *Mutat. Res.,* **68**, 183 (1979).
[46] H. M. Hassan and I. Fridovich, *J. Biol. Chem.* **253**, 8143 (1978).
[47] I. Fridovich and H. M. Hassan, *Trends Biochem. Sci.* **4**, 113 (1979).

This toxic effect was also evident by the large zones of inhibition seen in the spot assay (data not shown) and by the absence of a heavy background lawn seen in the plate incorporation assay (data not shown). In view of these facts and our knowledge of the mode of action of paraquat and of its ability to induce superoxide dismutase in *E. coli*[32,52] and in *S. typhimurium*,[12] a reevaluation of the mutagenicity of this compound using a modified liquid incubation assay[12] was deemed necessary. This assay allowed cultivation of the tester strain under conditions where superoxide dismutase levels were kept at a minimum (glucose minimal medium) and allowed the data to be expressed as revertants per $10^8$ viable cells. The assay was used successfully in testing the mutagenicity of paraquat[12] and should be useful in determining the mutagenicity of other oxy-radical generators.

A general protocol for the liquid incubation assay is as follows. The *Salmonella* tester strain is incubated in glucose minimal medium containing an excess amount of histidine in the presence of different concentrations of the oxy-radical generator. After incubation under predefined conditions the cells are removed, washed free of the oxy-radical generator, and then plated on three different media: one for the determination of total viable count, one for determination of cells requiring division before their mutagenicity is expressed, and one for determination of the cells that already express the His⁺ phenotype at the end of the liquid incubation period.

*Materials*

VB minimal salts containing, per liter, $MgSO_4 \cdot 7H_2O$, 0.2 g; citric acid $\cdot$ $H_2O$, 2.0 g; $K_2HPO_4$, 10 g; and $NaNH_4PO_4 \cdot 4H_2O$, 3.5 g, sterilized at 120° for 20 min

Complete VB minimal medium containing excess histidine. Prepared by adding sterile 0.1 m$M$ L-histidine–HCl, 5 $\mu M$ D-biotin, and 2% glucose to VB-minimal salts

Complete VB-minimal agar plates. Prepared by adding 2% agar to VB-minimal salts and sterilized at 120° for 20 min before adding sterile 0.1 m$M$ L-histidine–HCl, 5 $\mu M$ histidine, 5 $\mu M$ D-biotin, and 2% glucose

[48] H. M. Hassan and C. S. Moody, *Can. J. Physiol. Pharmacol.* **60**, 1367 (1982).
[49] A. D. Dodge, *Endeavour* **30**, 130 (1971).
[50] J. A. Farrington, M. Ebert, E. J. Land, and K. Fletcher, *Biochim. Biophys. Acta* **314**, 372 (1973).
[51] J. S. Bus, S. Z. Cagen, M. Olgaard, and J. E. Gibson, *Toxicol. Appl. Pharmacol.* **35**, 501 (1976).
[52] H. M. Hassan and I. Fridovich, *J. Biol. Chem.* **252**, 7667 (1977).

VB minimal agar plates lacking histidine. Prepared by adding sterile 5 $\mu M$ D-biotin and 2% glucose to sterile VB minimal salts containing 2% agar

The tester strains (TA98, TA100, and TA102)

Oxygen radical generator (i.e., paraquat, pyocyanine, streptonigrin, etc.); see [53] of this volume

Trypticase soy/yeast extract (TSY) medium. Prepared by mixing 30 g of trypticase soy broth and 5 g yeast extract per liter of $H_2O$

Anaerobic chamber where $O_2$ is kept at <5 ppm

*Procedure.* Grow the desired tester strain overnight (15–17 hr) at 37° and 200 rpm in 100 ml of complete VB minimal medium (i.e., excess histidine). Use this to inoculate approximately 500 ml of fresh medium of the same composition (the total culture volume should be enough to allow distribution into 50-ml portions for each concentration of the test compound) to an initial optical density at 600 nm ($OD_{600}$) of 0.1. Reincubate the culture at 37° and 200 rpm until it reaches an $OD_{600}$ of approximately 0.2 to assure that the culture is actively growing before being exposed to the test compound. Subdivide the culture (50-ml portions) into sterile 250-ml flasks (flask to volume ratio is 5 : 1) containing different concentrations of the test compound. The rat liver homogenates (S-9 fraction) should be added at this time, if required for activation of the compound (we found that this activation step is not required when paraquat is used as the oxy-radical generator).[12] Incubate the cultures at 37° and 200 rpm for the length of time required for optimum mutagenesis, as determined from a time course study. Different compounds and different tester strains will probably require different times of exposure for the expression of maximum mutagenesis. The optimum incubation time for paraquat in TA98 or TA100 was found to be 4–5 hr.[12]

Determine the $OD_{600}$ at the end of the incubation period. Collect the cells by centrifugation at 12,000 $g$ and 4° for 20 min. Wash the cells once by resuspending in 50 ml of sterile VB minimal salts to remove the toxic test compound and minimize the amount of histidine carried over to the selection plates, before recentrifugation. From the final $OD_{600}$ data, calculate the volume of sterile VB minimal salts needed to adjust the $OD_{600}$ to 2.0. This step equalizes the densities of the cultures that did not grow to the same extent as the control culture. Plate aliquots (0.1 and 0.2 ml) of the above cell suspensions onto each of four VB selection plates containing limited histidine and onto four plates lacking histidine. Determine the total viable counts of the same cell suspensions by plating in duplicate 0.1-ml aliquots of appropriate dilutions onto complete VB minimal agar plates containing excess histidine. Incubate at 37° for 72–96 hr before counting the colonies. Use the viable cells per milliliter and the number of

His$^+$ revertants per milliliter to calculate the number of revertants per $10^8$ cells. Positive and negative mutagens should be included as controls.

The effect of oxygen removal (anaerobiosis) should be tested to ensure that the mutagenicity of the test compound is due to its ability to generate oxygen radicals. This is performed by repeating the above experiment completely under strict anaerobic conditions (i.e., growth, exposure to the test compound, washing the cells, plating, and incubation). The effects of different levels of superoxide dismutase and catalase on the mutagenicity of the test compounds should be examined. This is best performed by incubating the cells and the test compound in a rich (TSY) medium where the induction of the detoxifying enzymes by the oxy-radicals is optimal.[32,46,52] Assay the cells for the level of superoxide dismutase using the cytochrome $c$ method.[53] Correlate the level of the enzyme with the degree of mutagenicity of the test compound. The effects of different hydroxyl radical scavengers on the mutagenicity of the test compound may also be explored.

### Acknowledgments

This work was supported, in part, by the NIH Biomedical Research Support Grant 88204 and NSF Grant PCM-8213853. We wish to thank Irwin Fridovich for the critical reading of this manuscript and Ann Farmer for her expert typing.

[53] J. D. Crapo, J. M. McCord, and I. Fridovich, this series, Vol. 53, p. 382.

## [31] Assay of Rate of Aging of Conidia of *Neurospora crassa*

By KENNETH D. MUNKRES and CHERYL A. FURTEK

This chapter describes a method developed in our laboratory during the past 4 years for the assay of rate of aging and viability of conidia of *Neurospora crassa*. Once all of the materials have been assembled, the procedure is simple and can be quickly mastered by undergraduate students having previously only been exposed to elementary microbiological techniques.

Certain aspects of the method have already been in practice many years in various *Neurospora* laboratories; however, we believe that the present narrative is essential because (1) some of our techniques are unique; (2) there is no single reference which, in our opinion, describes

Copyright © 1984 by Academic Press, Inc.
All rights of reproduction in any form reserved.
ISBN 0-12-182005-X

such a method in sufficient detail; and (3) the method has been the foundation for our subsequent studies of the biochemical genetics of conidial longevity. The results of those subsequent studies, briefly presented elsewhere,[1-8] are particularly relevant to this volume on free radicals in biological systems because it now appears that the genetic control of several antioxygenic (free radical protective) enzymes, such as superoxide dismutase and catalase, is important in the genetical control of the cell's longevity, at least under the particular environment examined.

Now it is fairly certain that the genetic and biochemical basis of the complex phenomenon of aging can ultimately be understood in terms of the fundamental genetical theorem stating that the phenotype is an expression of the complex interaction of genotype and environment. At the outset of a biochemical genetic investigation, it is initially desirable to define and hold constant at least the major environmental variables—that is particularly essential in studies of the population genetics of quantitative characteristics such as viability.

Aging is herein defined as the time-dependent loss of conidial viability, or, in other words, survival or plating efficiency under a carefully controlled constant set of environmental conditions. We feel justified in this definition, at least as a first approximation, because gerontologists generally think that such so-called "mortality kinetics" may reflect the underlying kinetics of the microscopic aging process or processes.

It is already well known, at least by mycologists, that the viability of fungal spores is best preserved by cool and dry conditions; however such knowledge tells little about the aging process, other than the rather trivial inferences that low entropy and slow metabolism favor longevity.

Longevity of conidia in the wild is probably impossible to determine—they are readily dispersed by wind and water. Since *Neurospora* clones and species are most abundant in the tropical and semitropical zones, it appears that the organism prefers hot, sunny, and humid conditions; therefore, we have chosen controlled laboratory conditions resembling the "native habitat" for measurement of aging rates; namely at 30°, 85–100% relative humidity in continuous cool white fluorescent light of incident fluence of 24 J m$^{-2}$. The light intensity is that expected year round in

[1] K. D. Munkres, *in* "Age Pigments" (R. S. Sohal, ed.), p. 83. Elsevier, Amsterdam, 1981.
[2] K. D. Munkres, C. A. Furtek, and E. Goldstein, *Age* **3**, 108 (1980).
[3] K. D. Munkres and R. S. Rana, *Age* **3**, 108 (1980).
[4] K. D. Munkres, C. A. Furtek, and R. S. Rana, *Age* **4**, 135 (1981).
[5] K. D. Munkres, R. S. Rana, E. Goldstein, and C. A. Furtek, *Age* **4**, 135 (1981).
[6] C. A. Furtek, K. D. Munkres, and R. S. Rana, *Age* **5**, 133 (1982).
[7] R. S. Rana, K. D. Munkres, and C. A. Furtek, *Age* **5**, 133 (1982).
[8] K. D. Munkres, C. A. Furtek, and R. S. Rana, *Age* **5**, 134 (1982).

the tropics and is equivalent to that of a bright sunny day at noon in June in the northern hemisphere.

Most wild-type strains of *Neurospora* bear both macro- and microconidia, the former being multinucleate and about 12 $\mu$m in diameter, and the latter mononucleate and 2 $\mu$m in diameter. We were initially concerned that these two types of cells might have intrinsically different rates of aging; however, because of observations too detailed for the scope of this chapter, we believe that is not so.

Other laboratories have described age-dependent changes of biochemical and biological parameters in conidia (reviewed by Munkres[1]). Ontogenetics, of course, fails to reveal cause and effect relations. Furthermore, quantitative comparisons of the interlaboratory observations are difficult because of differences in strains, environments, and, in some cases, because little or no survival data are reported. Moreover, not all relevant environmental variables were reported or controlled. In addition, our studies indicate that some strains may be heterogeneous for longevity determinant genes.[2,4,6]

We hope that the present experimental procedures, although somewhat more involved and detailed than those of others, will be widely adopted so that the foregoing problems may be avoided.

## Materials and Methods

### Culture Media and Growth Conditions

Stocks are grown on 1-ml slants of VM (Vogel's minimal medium N[9] containing 1.5% sucrose and 1.5% agar) in 10 × 75-mm glass test tubes. After inoculation with conidia, the slants are incubated 4 days at 35° in the dark followed by 3 days at room temperature (20–25°) in continuous fluorescent light of about 20–24 J m$^{-2}$.

Conidial viability is determined by incubation in 9-cm petri dishes containing at least 20 ml of VSS (Vogel's medium containing 0.1% sucrose, 1.0% sorbose, and 1.5% agar).[9] Smaller volumes of media or plates with more than 100 colonies are undesirable because growth rate and development of conidiophores are inhibited.

A reserve of slants and plates is stored at 14° in closed containers. A few drops of Kelthane (Science Products Co., Chicago, Ill.) may be added to the containers to prevent foreign fungal growth brought in by mites.

All media are sterilized by autoclaving 15 min at 121°.

[9] R. H. Davis and F. S. deSerres, this series, Vol. 17A, p. 79.

Working stocks are grown 7 days on 25 ml of VM agar media in 125-ml Erlemneyer flasks and stored at −20°.

Surgical grade, rather than unpurified cotton, is used for plugging culture and media vessels because the latter contains oils which contaminate the media.

### Dilution Tubes

Distilled water is dispensed in 18 × 150-ml glass test tubes with an automatic syringe, the tubes are covered with metal caps (Bellco), sterilized by autoclaving, and stored at 14°. Racks are prepared with three ranks of 10 tubes each; the first and second ranks contain 9.9 ml and the third rank contains 9.0 ml.

### Cell Transfer and Plating

Conidia are transferred and plated in a LabConCo hood (K.C., Mo.) sterilized with a germicidal lamp. The hood contains a Vortex mixer (Scientific Products, McGraw Park, Ill.), petri plate turntables (Fisher Scientific Co., Pittsburgh, PA), glass cell spreaders in a beaker of 70% ethanol, platinum transfer needles, and stainless-steel spatulas (Fisher).

A rather uniform distribution of a conidial suspension on plates is achieved with the aid of a turntable and glass spreader. The latter is bent from soft glass rod (3 mm diameter) to form a triangular tip and is sterilized by a flame after being dipped in 70% ethanol. Stainless-steel spatulas for harvesting conidia are sterilized in the same manner.

### Incubators

Growth on slants and plates is at 35 ± 1° in the dark at ambient relative humidity in a LabConCo incubator. To minimize desiccation of the medium, plates with conidial suspensions are incubated upright in closed containers 1 day and then inverted.

A constant environment for conidial aging is obtained with a Biotronette Mark III incubator (Lab-Line Inst., Inc., Melrose Park, Ill.). (Fig. 1) Glass aquaria within the incubator chamber, containing distilled water of depth about 4 cm, are covered with glass plates sealed with stopcock grease and fitted with a shelf above the water 63 cm from the chamber lamps, and contain an Airguide (Airguide Inst. Co., Chicago, Ill.) temperature and relative humidity indicator.

The light source in the Biotronette is four 40-W cool white fluorescent and three 75-W tungsten lamps. The chamber is operated in the automatic

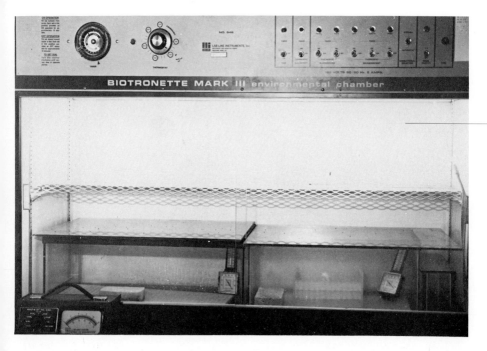

FIG. 1. Environmental chamber for aging.

mode with heat from the lamps and a heating cable and the fan in continuous operation.

Light intensity is measured with a YSI-Kettering model 65 radiometer (Yellow Springs Inst. Co., Yellow Spring, OH).

## Harvest, Dilution, and Plating of Conidia

One milliliter of water is put in the culture tube. A stainless-steel spatula with semicircular end and width slightly less than the tube's internal diameter is rotated in the tube to disperse conidia. The tube is set 15 sec on a Vortex mixer and 0.1 ml of the suspension is immediately withdrawn with a pipet and transferred to 9.9 ml of water in the dilution tube. The dilute suspension is stirred with a vortex mixer, and 0.1 ml is immediately withdrawn with a fresh pipet and put in 9.9 ml of water. After mixing this suspension, 1 ml is transferred to 9.0 ml of water. Finally 0.1 ml of the last dilution is spread on a plate with the aid of a turntable and glass rod.

All of the foregoing steps are usually performed within 1–2 hr. If it is not possible to plate immediately, the dilute suspension may be stored for at least 1 day at 5° without loss of viability.

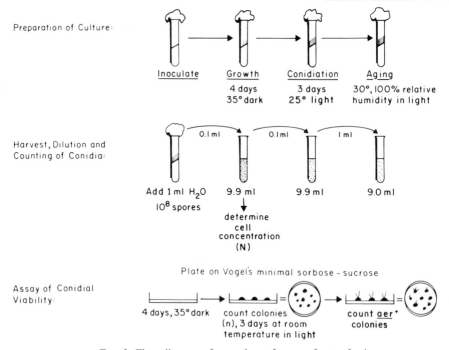

FIG. 2. Flow diagram of procedure of assay of rate of aging.

If the foregoing details are observed, a typical culture yields $10^8$ conidia and the plate contains about 100 conidia. The suspension is essentially devoid of hypha and hyphal fragments.

### Determination of Conidial Concentration

After 100-fold dilution, conidial concentration is determined in a Spencer–Neubauer hemocytometer in a phase microscope at 200-fold magnification. Since a culture yields about $10^8$ conidia, about 100 are within the 400 minor squares of the hemocytometer: at least 100 are counted. The coefficient of variation of three measurements is 15%. *Old conidia must be observed by phase microscopy;* they are translucent in directly transmitted light.

### Experimental Procedure

Figure 2 illustrates the major steps of the assay procedure. After growth for 4 days at 35° in the dark at ambient relative humidity followed by conidiation for 3 days at 25° in continuous light, most strains are

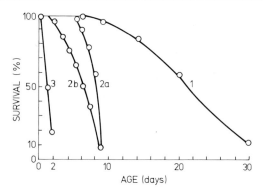

FIG. 3. Survivorship of selected strains. (1) Newly inbred Oak Ridge wild type (F₁); (2a, 2b) selected spontaneous short-lived mutants (*age*⁻); (3) recombinant from cross of mutants in 2a and 2b.

mature, i.e., all conidia are viable and no additional growth or conidiation occurs.

For the construction of survival curves, replicate cultures are incubated in the biotron at 30°, 100% relative humidity in continuous light. Periodically, a tube is withdrawn from the biotron and conidia are harvested, diluted, counted, and plated. The plates are incubated at 35° for 4 days, a time which is sufficient for full expression of survival. Survival is defined as the ratio of the number of colonies to the number of conidia. Mean life span is defined as the age of 50% survival.

Upon additional incubation for 3 days at room temperature in light, most wild-type colonies differentiate with the formation of aerial hyphae (conidiophores) bearing conidia. This phenotype, *aer*⁺, is of interest because a pleiotropic character of short-lived mutants (2, Fig. 3), designated *age*⁻, is *aer*⁻.

*Application of Procedure and Genetic Definitions*

Various laboratory wild types were newly acquired from the Fungal Genetics Stock Center, Humboldt University, Arcata, CA, i.e., Oak Ridge, Stanford, Yale, and Cornell. Their mean lifespan (mls) was observed to be 5–10 days. Fifty F₁ progeny, selected at random from each of the inbred crosses, Cornell × Cornell and Oak Ridge × Oak Ridge, exhibited an average mls of 22 days (Fig. 3). Subsequent inbreeding for six generations led to no further change of mls.

Certain auxotrophs from the Fungal Genetics Stock Center also exhibited short lifespans of 5–10 days, but their progeny were normal after one backcross to wild type.

The reason for such "degeneracy" of these strains is not known. We imagine that since they had been maintained many years in the vegetative state they may have accumulated nonspecific nonheritable mutations unfavorable for survival.

Spontaneous *heritable* mutations occur in ascospore or conidial populations of wild type at a frequency of 5–10%, with mean lifespans of 5–10 days (Fig. 3). Genetic and biochemical studies of such mutants have been reported.[1-7]

For some purposes, such as testing of antiaging drugs, mutant selection or genetic analysis, the construction of survivorship curves is too tedious; however, a spot test based on survival at 9 or 10 days is adequate. The choice of this time is based on the observation that it yields the maximum differential between wild-type and mutant survival (Fig. 3).

### Acknowledgments

This work was supported by the College of Agriculture and Life Sciences, the Graduate School, and the National Institute of Aging (AG-00930). Contribution No. 2622 from the Department of Genetics.

# Section III

# Assay of Modes of Biological Damage Imposed by $O_2$ and Reduced Species

### A. Lipid Peroxidation
*Articles 32 through 44*

### B. Plasma Membrane
*Articles 45 through 53*

### C. Microsomal Membranes
*Articles 54 and 55*

### D. Organelles
*Articles 56 through 58*

# [32] Chemistry of Lipid Peroxidation

*By* NED A. PORTER

## Diene Fatty Acids

Systematic study of the chemistry of fatty acid autoxidation was begun in the late 1940s.[1] The degradation of fats and oils in the air had long been known but it remained for Criegee[2] to provide evidence that hydroperoxides are the primary products of hydrocarbon oxidation before significant advances could be made in the study of autoxidation of natural substances. Subsequent investigation by Bolland and his collaborators[3] established the primary autoxidation products of methyl linoleate as hydroperoxides containing conjugated diene. A mechanism for autoxidation of diene fatty acids as written by these workers is shown in Scheme I. Only propagation steps of autoxidation are considered in this chapter.

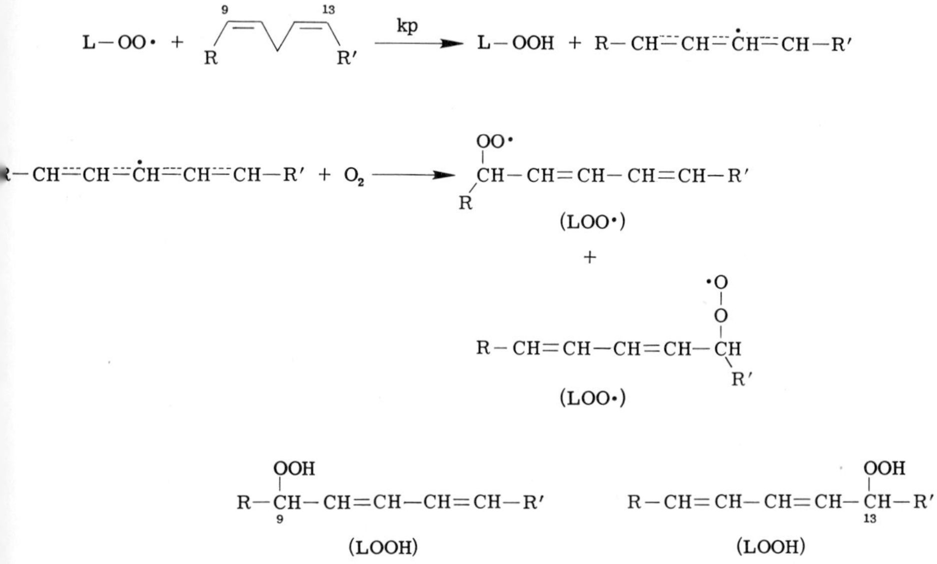

SCHEME I.

---

[1] For a review, see W. G. Lloyd, "Methods in Free Radical Chemistry" (E. S. Huyser, ed.), Vol. 4. Dekker, New York, 1973.
[2] R. Criegee, H. Pilz, and H. Flugare, *Ber.* **72,** 1799 (1939).
[3] L. Bateman, *Q. Rev.* **8,** 147 (1954); J. L. Bolland, *Q. Rev.* **3,** 1 (1949).

METHODS IN ENZYMOLOGY, VOL. 105

Copyright © 1984 by Academic Press, Inc.
All rights of reproduction in any form reserved.
ISBN 0-12-182005-X

Initiation of biological oxidation is considered extensively elsewhere in this volume. The two products of linoleate autoxidation described are the 9 and 13 substituted diene hydroperoxides, LOOH.

For a period of some 20 years little additional evidence concerning the chemical mechanism of fatty acid autoxidation was published. In 1967, Howard and Ingold[4] reported the absolute rate constant for hydrogen atom transfer from linoleate to peroxy radicals ($k_p = 62 \ M^{-1} \ \mathrm{sec}^{-1}$) and in the mid-1970s several research groups[5–8] began to study diene fatty acid autoxidation. Reexamination of the products of autoxidation of linoleic acid or linoleate esters revealed a more complex mixture than was originally proposed. Four major conjugated diene hydroperoxides are formed and may be purified by HPLC. These major products have been shown to be **1–4** (R = H, Me, or Et).

Two of the products have trans,cis-diene stereochemistry (**1** and **3**) while two of the products (**2** and **4**) have trans,trans-diene stereochemistry. In addition to these four major products (~97% of the product mixture) trace amounts of nonconjugated hydroperoxides have also been isolated and identified. It should be emphasized that these diene hydroperoxides are themselves unstable with respect to decomposition. Thermal- or metal-catalyzed decomposition of **1**, for example, presumably leads to the alkoxy radical **5** which can be a source of pentyl radicals **6** (and ultimately pentane) or epoxides derived from radical **7**. Product mixtures derived from linoleates after extensive oxidation are thus complex.

[4] J. A. Howard and K: U. Ingold, *Can. J. Chem.* **45**, 793 (1967).
[5] E. N. Frankel, *in* "Fatty Acids" (E. H. Pryde, ed.), p. 353. AOCS, Champaign, Illinois, 1970.
[6] N. A. Porter, B. A. Weber, H. Weenen, and J. A. Khan, *J. Am. Chem. Soc.* **102**, 5597 (1980).
[7] H. W.-S. Chan, G. Levett, and J. A. Matthew, *Chem. Phys. Lipids* **24**, 245 (1979).
[8] L. R. C. Barclay and K. U. Ingold, *J. Am. Chem. Soc.* **102**, 7794 (1980).

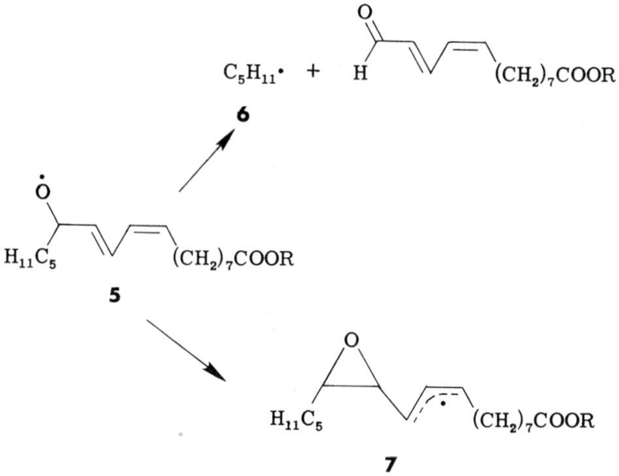

The primary products that lead to this complex mixture are, however, the four simple hydroperoxides **1–4** and the chain propagation sequence shown in Scheme I accounts for products formed at C-13 (**1** and **2**) and C-9 (**3** and **4**) of linoleic acid. Any mechanism that accounts for stereochemistry of the products must, however, be more complex than the original Bolland mechanism, and radical isomerization (Scheme II) has been proposed to account for formation of trans,cis- and trans,trans-dienes. The radical **8** serves as a precursor to the *trans,cis* products while **9** could lead to trans,trans products by oxygen addition and hydrogen abstraction propagation.

A mechanism more consistent with experimental observations is presented in Scheme III. In this scheme, the radical **8** is isomerized to **9** by reversible oxygen additions. Loss of oxygen from conformer **10a** leads to **8** while conformer **10b** leads to radical **9**, a source of trans,trans-diene products. The necessity for proposing Scheme III comes from the observation that the ratio of trans,cis/trans,trans products depends directly on the concentration of L-H or other hydrogen atom donors (R-H) present in the oxidation mixture. With high concentrations of L-H or R-H, conversion of **10** (**a** and **b**) to trans,cis products (**1** and **3**) competes effectively with loss of oxygen from **10b** ($k_\beta$) to give **9** (and consequently trans,trans

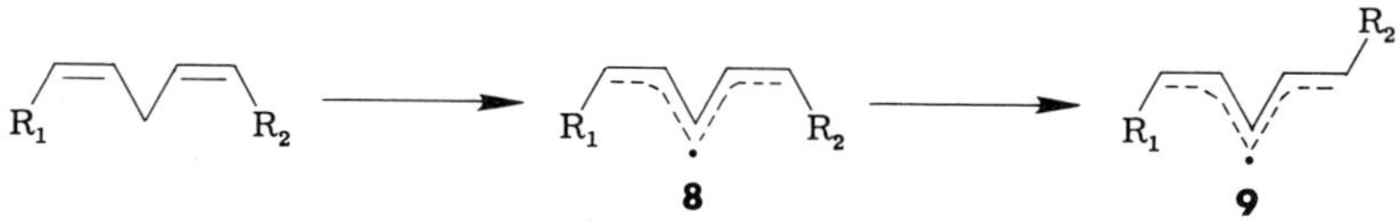

SCHEME II.

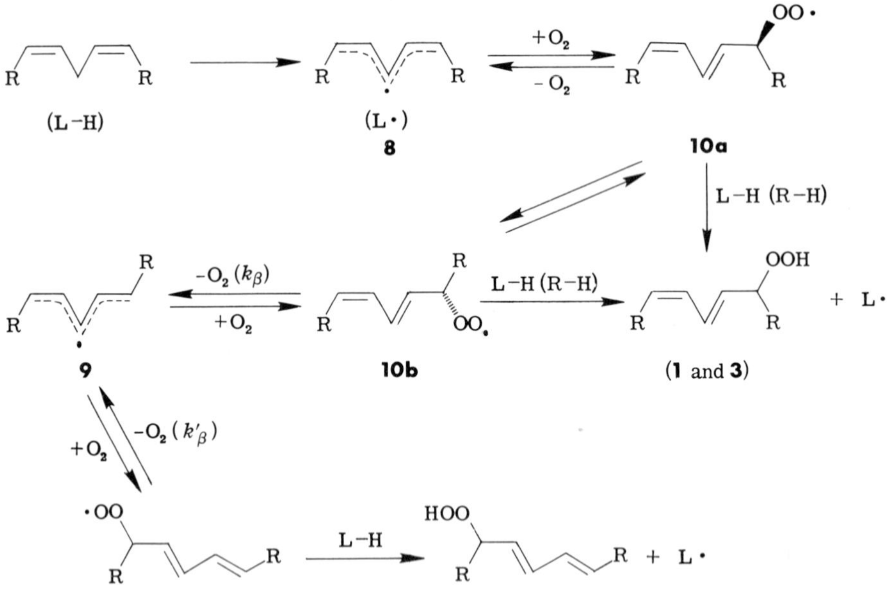

SCHEME III.

products). Thus, the distribution of autoxidation products depends directly on the concentration and effectiveness (relative $k_p$ values) of any H atom donors present in the reaction mixture. A more detailed discussion of this mechanism of diene fatty acid oxidation has been published elsewhere.[6]

It should be noted that singlet oxygen oxidation of diene fatty acids leads to a different set of products than does free radical autoxidation. Singlet oxygen oxidation of linoleate leads to the trans,cis-dienes **1** and **3**, none of the trans,trans-dienes **2** and **4**, and significant amounts of nonconjugated dienes **11** and **12**. These nonconjugated products are formed in only trace amounts in free radical chain autoxidation while they may be major products in singlet oxygen oxidation.

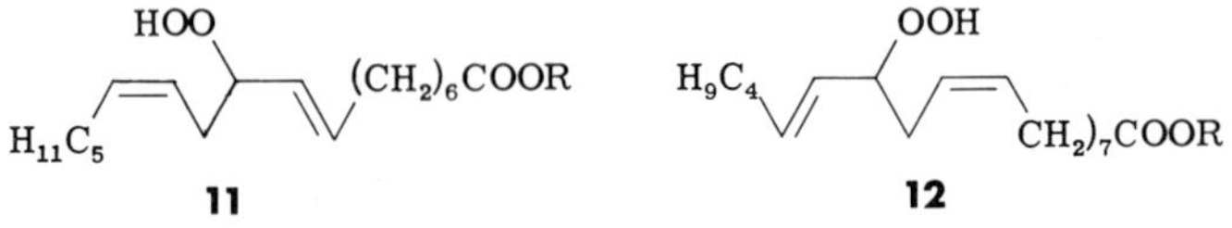

*Triene and Tetraene Fatty Acids*

Autoxidation of triene or tetraene fatty acids leads to a much more complex mixture of products than that formed in diene fatty acid autoxidation. Arachidonic acid, **13**, as an example, has three carbon centers flanked by two double bonds ($C_7$, $C_{10}$, $C_{13}$) while linoleate has only one such carbon ($C_{11}$). Six major hydroperoxide products are formed in arachidonic acid autoxidation and these products may be separated and analyzed by HPLC.[9] All six major products have trans,cis-diene stereochemistry and hydroperoxide substitution in these products is at C-5, C-8,

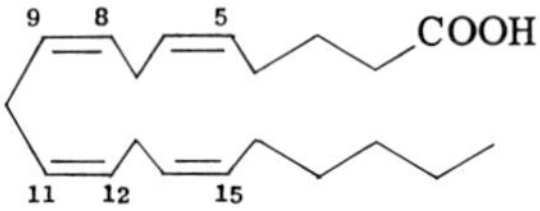

C-9, C-11, C-12, and C-15. Two of the products, **14** and **15,** formed from hydrogen atom abstraction at C-11 are shown here and products formed by abstraction at C-7 (the 5 and 9 substituted hydroperoxides) and C-10 (8 and 12 substituted hydroperoxides) have analogous structures. Autoxidation of arachidonic acid in benzene or chlorobenzene (0.24 $M$) solution[10] gives rise to 15-hydroperoxyeicosatetraenoic acid (15-HPETE) as the major product (40%) with 5-HPETE being the next most prevalent product hydroperoxide (27%). Other products are formed as follows (8-HPETE, 7%; 9-HPETE, 9%; 11-HPETE, 11%; 12-HPETE, 6%). Two factors, selective hydrogen abstraction and peroxy radical cyclization, are responsible for this unequal distribution of products. In solution, C-13 hydrogen atom is abstracted more readily than is a hydrogen atom at C-10 or C-7. The relative rates of H atom abstraction are C-7, 0.86; C-10, 0.70; C-13, 1.33. Products that derive from abstraction at C-13 (15-HPETE and 11-HPETE) are thus formed preferentially to products that derive from abstraction at C-7 and C-10. It is unclear as to why the C-13 hydrogens are more prone to abstraction than those at C-7 or C-10 but it appears that hydrogens near the tail of the molecule are generally more available for reaction, perhaps because of overall molecular conformation. A similar preference for 15-HPETE formation is observed in singlet oxygen oxida-

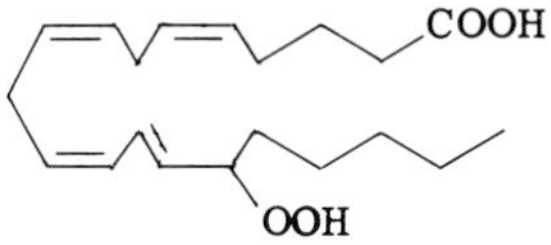

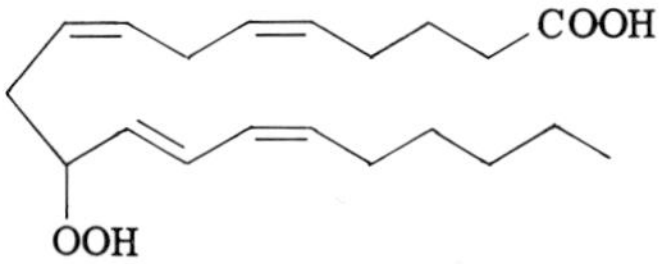

[9] N. A. Porter, R. A. Wolf, E. M. Yarbro, and H. Weenen, *Biochem. Biophys. Res. Commun.* **89,** 1058 (1979).

[10] N. A. Porter, L. S. Lehman, B. A. Weber, and K. J. Smith, *J. Am. Chem. Soc.* **103,** 6447 (1981).

tion of arachidonic acid[11] and conformational preferences may be important here also in determining product distribution. From these data it may also be noted that arachidonic acid is oxidized about 2.9 times as fast as is linoleate.

Peroxy radical cyclization is a second factor that determines product distribution in triene and tetraene oxidation. The peroxy radical leading to hydroperoxide 15-HPETE has only two mechanistic pathways available, hydrogen atom abstraction to propagate the chain and loss of oxygen to form a carbon radical ($\beta$ scission). In addition to these two pathways, a peroxy radical leading to hydroperoxide 11-HPETE has a third alternative. In particular, peroxy radicals have been shown to add to double bonds at adjacent centers to form five and six membered ring products.[12,13] In the case of peroxy radicals derived from arachidonic acid a host of products may ultimately form from cyclization. Monocyclic peroxides like 16, serial cyclic products such as 17, and bicyclic endoperoxides (18) (Scheme IV) have all been isolated from triene or tetraene fatty acid systems after autoxidation. The products 16, 17, and 18 exist in many diastereomeric forms with sixteen pairs of enantiomers possible for compound 17 alone. It should also be noted that products analogous to but different from 16, 17, and 18 may be formed from four different peroxy radicals that have adjacent double bonds available for cyclization. Thus, peroxy radicals leading to 8-HPETE, 9-HPETE, 11-HPETE, and 12-HPETE have competitive cyclization pathways available and cyclization products may form from these species. It is for this reason that the relative amounts of the 8-,9-,11-, and 12-hydroperoxides are smaller than the 5- and 15-HPETEs formed in arachidonic acid oxidation. The 5- and 15-peroxy radicals do not have a competitive cyclization pathway and are formed in undiminished yield. Comment should also be made about the fact that both trans,cis and trans,trans diene hydroperoxides are formed in linoleate oxidation while trans,cis-diene hydroperoxides always dominate in triene or tetraene oxidation. Hydroperoxide products with trans, trans-diene stereochemistry are formed by unimolecular peroxy radical $\beta$ scission ($k_\beta$) that competes with bimolecular hydrogen atom abstraction. The rate of $\beta$ fragmentation has been found[6] to be on the order of 150 sec$^{-1}$ while unimolecular cyclization occurs with a significantly greater rate constants,[10] $k_c \sim 800$ sec$^{-1}$. Thus, $\beta$ scission leading to trans,trans product never competes effectively with cyclization and/or hydrogen atom abstraction.

[11] N. A. Porter, J. Logan, and V. Kontoyiannidou, *J. Org. Chem.* **44**, 3177 (1979).

[12] N. A. Porter, M. O. Funk, D. W. Gilmore, R. Isaac, and J. R. Nixon, *J. Am. Chem. Soc.* **98**, 6000 (1976).

[13] E. D. Mihelich, *J. Am. Chem. Soc.* **102**, 7141 (1980).

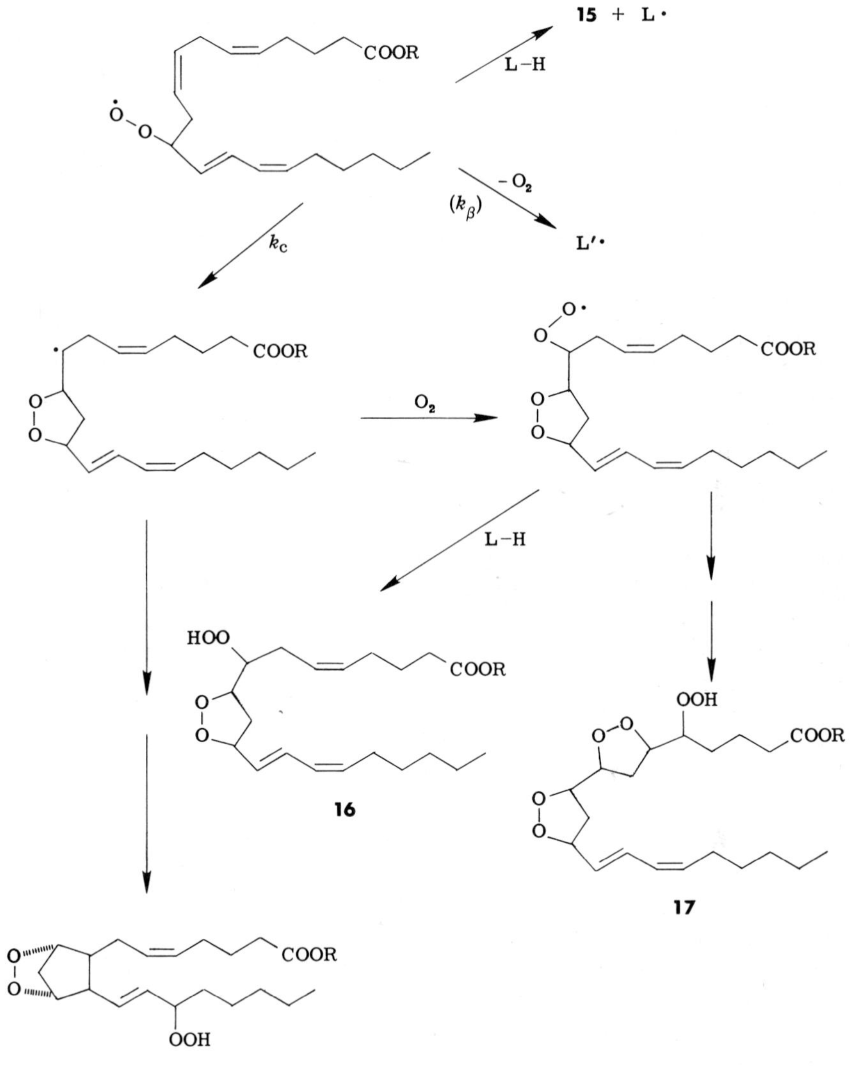

SCHEME IV.

Products such as **17** and **18** probably account for a significant fraction of the mixture formed in triene and tetraene oxidation and it seems likely that these compounds form malondialdehyde upon decomposition. Malondialdehyde is, of course, detected in the thiobarbituric acid assay (TBA), a standard test of lipid peroxidation used by workers in the field.

The TBA assay then is probably an assay of peroxy radical cyclization in lipid peroxidation.

*Membrane Oxidation*

Membrane destruction by free radical autoxidation has been the subject of extensive investigation by biochemists over the past several years. While considerable effort has been directed toward developing an understanding of the steps involved in initiation of the free radical process and the inhibition of the reaction, until very recently little attention has been given to the target of oxidation, the phospholipid.

Polyunsaturated diacylglycerophosphatidylcholines (diacyl-GPC, lecithin) readily undergo autoxidation as multilamellar or unilamellar vesicles. A variety of initiators of peroxidation may be used (Fenton's reagent, xanthine oxidase, adventitious initiation) but the products of oxidation are identical regardless of the method of initiation. High-pressure liquid chromatography has been used to good advantage in the study of lecithin oxidation.[14,15] For example, autoxidation of dilinoleoyl lecithin (**19**) can be monitored directly by reverse-phase high-pressure liquid chromatography. Two series of products are formed, one series containing monooxidation products the other compounds resulting from dioxidation.

The first series contains products of autoxidation of **19** in which one acyl chain (at either C-1 or C-2 of the glycerol moiety) has been oxidized.

**19**

In fact, 16 different monooxidation products are formed from **19**. Eight of the products derive from oxidation at the C-1 acyl chain and eight from the C-2 acyl linoleate. Four of the eight products in each of the acyl chains are trans,cis-diene hydroperoxide products while the other four products have trans,trans-diene stereochemistry. Within the set of trans,cis or trans,trans products, half are formed at C-9 of linoleate with a 10,12-conjugated diene unit, the other half are formed with hydroperoxide substitution at C-13 with a 9,11-diene functionality. Finally, hydroperoxide at C-9 or C-13 may have either R or S configuration. While these stereoiso-

[14] C. G. Crawford, R. D. Plattner, D. J. Sessa, and J. J. Ruckis, *Lipids* **15**, 91 (1980).
[15] N. A. Porter, R. A. Wolf, and H. Weenen, *Lipids* **15**, 163 (1980).

mers would be enantiomers in autoxidation of simple linoleate esters, this is not the case in the oxidation of phospholipid esters. Since natural lecithins have only the R configuration at the C-2 glycerol carbon, the possibility of having R and S configuration in a particular oxidation product leads to a mixture of diastereomers, not enantiomers.

The 16 monooxidation products of **19** do not separate by reverse-phase HPLC. The products may be analyzed, however, by the sequence outlined here for one of the sixteen products, **20**. Treatment of mixtures containing **20** with Ph$_3$P leads to conversion to **21** by simple reduction of **20**. Exhaustive methanolysis (KOH, MeOH) converts all hydroxylinoleate glycerolphosphatidylcholines such as **21** to simple methyl esters,

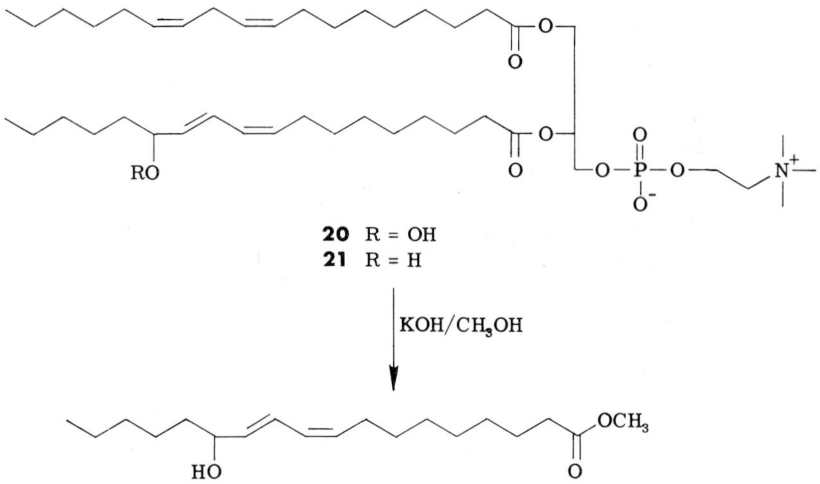

**20**  R = OH
**21**  R = H

KOH/CH$_3$OH

**22**

**22,** and these methyl esters may be separated and analyzed by normal-phase HPLC. In this way, details of oxidation within the membrane bilayer may be assessed. Similar analytical methods may be used for a variety of lecithin molecular species containing linoleate, linolenate, arachidonate, palmitate, and stearate functional groups. In each case, oxidation products are analyzed by normal-phase HPLC analysis of simple methyl esters.

Products of oxidation of phospholipids in model membranes such as multilamellar or unilamellar vesicles are thus similar to products formed from free fatty acid or methyl ester oxidation in the bulk phase. Details of the mechanism of oxidation in bulk phase and membrane phase also appear to be the same. The ratio of trans,cis/trans,trans-diene products formed reflects the hydrogen donating ability in the bilayer much as it does in bulk phase. As an example, consider oxidation of dilinoleate lecithin (di-LinGPC) vs 1-palmitic-2-linoleic lecithin (1-P,2-LGPC)[16] in multilamellar vesicles. At 37°, the ratio of trans,cis/trans,trans-diene hydroperoxide products is 1.28 ± 0.05 for Di-LinGPC while the same ratio is 0.63 ± 0.04 for oxidation of 1-P,2-LGPC. The concentration of hydrogen atom donors (linoleates) in the bilayer is relatively high for Di-LinGPC (two linoleates/lecithin) while 1-P,2-LGPC has half the number of good donors in the bilayer. According to the mechanism presented in Scheme III, the trans,cis/trans,trans-diene ratio should depend directly on the number of H-atom donors available in the oxidation medium and the results of oxidation of Di-LinGPC and 1-P,2-L-GPC described here are consistent with that view. It thus appears that the mechanism for oxidation in a bilayer is qualitatively the same as oxidation in bulk phase. Simple diene hydroperoxides are formed from diene phospholipid esters and the mechanism for their formation in bulk phase and bilayers is the same. The phospholipid bilayer seems to provide a medium where autoxidation occurs in the normal way. It should also be noted that the distribution of phospholipid autoxidation products does not depend on method of initiation. For example, oxidation of linoleate phospholipid bilayers initiated by xanthine oxidase/acetaldehyde at pH 7 gave a product distribution identical to that obtained from adventitious initiation.[16]

Autoxidation of arachidonate phospholipid esters has also been investigated[17] and it appears that here also bilayer and bulk phase autoxidations take a similar course. In multilamellar vesicle oxidations of 1-stearic-2-arachidonic-GPC (1-S,2-AGPC), the same six hydroperoxide products are formed as are found in oxidations of arachidonic acid in benzene. Furthermore, in bilayer, the distribution of products is the same as that found in bulk phase. Cyclic peroxides like those found in bulk phase autoxidations are also formed in the bilayer.

*Summary*

The free radical chemistry of lipid peroxidation is complex. The classical mechanism of autoxidation involving a peroxy radical abstracting hydrogen atom from lipid and oxygen addition to the carbon radical thus formed must be modified to include (1) peroxy radical β fragmentation and (2) peroxy radical cyclization. A host of diene hydroperoxides, cyclic peroxides, bicyclic peroxides and epoxy alcohols may be formed in free fatty acid or phospholipid autoxidation. The distribution of products and the effects of hydrogen atom donors on product distribution are understandable by referring to a general scheme for autoxidation described in Scheme III and in Ref. 10.

[16] N. A. Porter and L. S. Lehman, *J. Am. Chem. Soc.* **104**, 4731 (1982).
[17] H. Weenen and N. A. Porter, *J. Am. Chem. Soc.* **104**, 5216 (1982).

# [33] Overview of Methods Used for Detecting Lipid Peroxidation

*By* T. F. SLATER

Lipid peroxidation is a complex process whereby unsaturated lipid material undergoes reaction with molecular oxygen to yield lipid hydroperoxides; in most situations involving biological samples the lipid hydroperoxides are degraded to a variety of products including alkanals, alkenals, hydroxyalkenals, ketones, alkanes, etc.[1-3] Although attack by singlet oxygen on unsaturated lipid has been shown to give lipid hydroperoxide by a nonradical, nonchain process,[4] the vast majority of situations involving lipid peroxidation proceed through a free radical-mediated chain reaction initiated by the abstraction of a hydrogen atom from the unsaturated lipid by a reactive free radical, followed by a complex sequence of propagative reactions. In the following discussion the unsaturated lipid used for illustrative purposes will be a polyunsaturated fatty acid such as arachidonic acid, $C_{20:4}$; in point of fact, most biological studies on lipid peroxidation in biomembranes have involved the peroxidative breakdown of such materials.

Equation (1) shows the initiation of a chain reaction through the abstraction of a hydrogen atom from a polyunsaturated fatty acid [abbreviated in this equation only as PUFA(H), where the (H) indicates easily abstractable hydrogen atoms] by a reactive free radical, R·:

$$\text{PUFA(H)} + \text{R·} \rightarrow \text{PUFA} + \text{RH} \tag{1}$$

This reaction is followed by the addition of oxygen to give a lipid peroxy free radical and, in the case of polyunsaturated fatty acids (PUFAs) containing unconjugated 1,4-dienes, by double-bond rearrangement to yield conjugated dienes with characteristic UV absorption around 233 nm. The lipid peroxy free radical can itself abstract a hydrogen atom from a neighboring molecule (XH), which may be another polyunsaturated fatty acid (PUFA) substrate molecule to give the corresponding lipid hydroperoxide; the hydroperoxide can break down, as already mentioned, to a

---

[1] N. A. Porter, this volume [32].
[2] J. F Mead, *Free Radicals Biol.* **I**, 51 (1976).
[3] H. Esterbauer, K. H. Cheeseman, M. U. Dianzani, G. Poli, and T. F. Slater, *Biochem. J.* **208**, 129 (1982).
[4] W. A. Pryor and L. Castle, this volume [34].

Copyright © 1984 by Academic Press, Inc.
All rights of reproduction in any form reserved.
ISBN 0-12-182005-X

variety of products, especially when in the presence of transitional metal ions[5]:

$$PUFA\cdot + O_2 \rightarrow PUFAO_2^{\cdot} \qquad (2)$$
$$PUFAO_2^{\cdot} + XH \rightarrow PUFAO_2H + X\cdot \qquad (3)$$
$$PUFAO_2H \rightarrow products \qquad (4)$$

The peroxidation of polyunsaturated fatty acids (PUFAs) can proceed through nonenzymic autoxidative pathways, or through processes that are enzymically catalyzed. The importance of autoxidation in the deterioration of foods, and in the oil industry, has long been recognized and authoritative reviews are available for such aspects.[6,7]

A relatively new growth point for studies on lipid peroxidation has been the realization that many toxic agents can be metabolically activated within cells to free radical intermediates that can initiate lipid peroxidation and result in cell injury.[8,9] Irradiation of tissue, cells, or cell organelles may likewise produce reactive free radicals that can result in similar consequences. Moreover, depletion of normal cellular protective mechanisms, such as the level of endogenous antioxidants or glutathione, may also facilitate a significant level of lipid peroxidation[10,11] (but see also Ref. 12). A very large number of such studies on lipid peroxidation in biological systems have demonstrated the degradation of membrane PUFAs, with a subsequent disorganization of membrane structure and disturbance of membrane function. This overview of the methods used for studying lipid peroxidation will concentrate on such aspects of lipid peroxidation in relation to biomembrane disturbance. First, it will be useful to summarize the biological systems most frequently used in such studies: this initial summary will be divided into a section concerning investigations made *in vitro,* and a section where the lipid peroxidation occurred *in vivo.*

*Commonly Used Model Procedures for Studying Lipid Peroxidation in Biological Systems in Vitro*

Studies *in vitro* have generally used whole cells or homogenates, or suspensions of intracellular organelles such as nuclei, mitochondria, microsomes, lysosomes, plasma membrane, or liposomes. Such investiga-

[5] H. W. Gardner, *J. Agric. Food Chem.* **23**, 129 (1975).
[6] W. O. Lundberg, ed., "Autoxidation and Antioxidants," Vols. 1 and 2. Wiley (Interscience), New York, 1961.
[7] G. Scott, "Atmospheric Oxidation and Antoxidants." Elsevier, Amsterdam, 1965.
[8] T. F. Slater, *in* "Free Radical Mechanisms in Tissue Injury." Pion, London, 1972.
[9] R. Snyder, D. V. Parke, J. J. Kocsis, D. J. Jollow, G. G. Gibson, and C. M. Wittmer, eds., "Biological Reactive Intermediates." Plenum, New York, 1982.
[10] J. Hogburg, S. Orrenius, and R. E. Larson, *Eur. J. Biochem.* **50**, 595 (1975).
[11] M. Younes and C. P. Siegers, *Res. Commun. Chem. Pathol. Pharmacol.* **27**, 119 (1980).
[12] R. Reiter and A. Wendel, *Chem. Biol. Interact.* **40**, 365 (1982).

tions have used a variety of nonenzymic and enzymic mechanisms to stimulate lipid peroxidation.

*Nonenzymic Mechanisms.* Although some studies with nonenzymic mechanisms have used radiation (ionizing,[13] or ultraviolet,[13] or visible radiation in the presence of a suitable photosensitizer[14]) to stimulate lipid peroxidation, most investigations have relied on the addition of transitional metal salts or chelates, in particular iron salts or chelates.[15,16] Moreover, in many cases a reducing agent has been included to enable $Fe^{3+}$ to be converted back to $Fe^{2+}$; for example, ascorbate[17] or cysteine.[18] In some instances, ascorbate itself has been used at an appropriate concentration[15]; there is usually enough iron contamination of normal buffers and other components to ensure a significant rate of lipid peroxidation.[19] A nonenzymic procedure that has also been used on several occasions is based on the cytochrome *P*-450-mediated decomposition of cumene hydroperoxide.[20]

*Enzymic Mechanisms.* Many enzymic mechanisms share common features with the nonenzymic processes summarized above; however, in such cases, an enzyme-catalyzed reaction serves as the reducing source either to oxidise $Fe^{3+}$ to $Fe^{2+}$, or to enable metabolic activation of an added substance (e.g., $CCl_4$) to a free radical intermediate that can initiate peroxidation. As mentioned above, biological suspensions usually contain significant quantities of iron salts; the addition of NADPH to a microsomal suspension then permits a recycling of $Fe^{3+}$ and $Fe^{2+}$ through the reducing power of NADPH–cytochrome *P*-450 reductase. A frequently used model procedure is that based on a supplementation of the NADPH-mediated lipid peroxidation with an iron chelate such as $Fe^{2+}/ADP$.[15]

Another commonly used model procedure is dependent on the metabolic activation of $CCl_4$ to $CCl_3^-$ (and $CCl_3O_2^-$) through the action of the NADPH–cytochrome *P*-450 electron transport chain in microsomal suspensions.[21] Many other substances are metabolically activated by the cytochrome *P*-450 system to free radical intermediates.

[13] I. D. Desai, P. L. Sawant, and A. L. Tappel, *Biochim. Biophys. Acta* **86,** 277 (1964).

[14] T. F. Slater and P. A. Riley, *Nature (London)* **209,** 151 (1966).

[15] P. Hochstein and L. Ernster, *Biochem. Biophys. Res. Commun.* **12,** 388 (1963).

[16] M. Tien, L. A. Morehouse, J. R. Bucher, and S. D. Aust, *Arch. Biochem. Biophys.* **218,** 450 (1982).

[17] P. H. Beswick, K. Cheeseman, G. Poli, and T. F. Slater, *in* "Recent Advances in Lipid Peroxidation and Tissue Injury" (T. F. Slater and A. Garner, eds.), p. 156. Brunel Univ., Dept. of Biochemistry, Uxbridge, Middx. U.K., 1981.

[18] A. J. F. Searle and A. Tomasi, *J. Inorg. Biochem.* **17,** 161 (1982).

[19] S. F. Wong, B. Halliwell, R. Richmond, and W. R. Skowroneck, *J. Inorg. Biochem.* **14,** 127 (1981).

[20] A. P. Kulkarni and E. Hodgson, *Int. J. Biochem.* **13,** 811 (1981).

[21] T. F. Slater and B. C. Sawyer, *Biochem. J.* **123,** 805 (1971).

A classical example of an enzyme catalyzed lipid peroxidation is that of lipoxygenase-mediated reactions.[22] Although much work has been done with soya bean lipoxygenase there is increasing attention paid to lipoxygenases present in mammalian cells such as platelets,[23] reticulocytes,[24] leukocytes,[25] etc.

## Model Procedures for Studying Lipid Peroxidation in Vivo

Studies *in vivo* in this sense refer to situations where lipid peroxidation is proceeding *in vivo*, and where its occurrence is followed either by noninvasive procedures or by invasive techniques that involve surgical operation and/or tissue sampling.

*Noninvasive Techniques.* The main procedure available in the category of noninvasive techniques is the estimation of exhaled alkanes; references to this procedure will be given later.

*Invasive Techniques.* One interesting development concerning invasive techniques is that based on the measurement of an increased chemiluminescence associated with lipid peroxidation by sensitive photon capture techniques applied to exposed organs (e.g., liver) in an anesthetized animal.[26]

Estimations of lipid peroxidation in blood samples have been based on analysis of lipid hydroperoxide content.[27] Similar procedures have been applied to the analysis of tissue fractions, as have techniques for measuring diene conjugation absorption and malonaldehyde content.

In order to minimize artifactual changes in, or production of, such products of lipid peroxidation during and after sampling, it is best to inhibit enzymic and peroxidative processes as quickly as possible at the time of sampling.[28] In this respect, freeze clamping of tissue at the temperature of liquid nitrogen is sometimes useful; otherwise, the rapid inhibition of enzymic reactions by mixing with acid or chloroform/methanol, and inhibition of free radical chain processes by the inclusion of an excess of a suitable free radical scavenger such as butylated hydroxytoluene is often necessary.

With complex biological systems, such as whole-tissue samples, iso-

[22] D. M. Bailey and L. W. Chakrin, *Annu. Rep. Med. Chem.* **16,** 213 (1981).
[23] D. H. Nugteren, *Biochim. Biophys. Acta* **380,** 299 (1975).
[24] S. Rapoport, T. Schewe, R. Wiesner, W. Halangk, P. Ludgwig, M. Janicke-Hohne, C. Tannert, C. Hiebsch, and D. Klatt, *Eur. J. Biochem.* **96,** 545 (1979).
[25] S. Narumiya, J. A. Salmon, F. H. Cottee, B. C. Weatherley, and R. J. Flower, *J. Biol. Chem.* **256,** 9583 (1981).
[26] A. Boveris, E. Cadenas, and B. Chance, *Fed. Proc. Fed. Am. Exp. Biol.* **40,** 195 (1981).
[27] K. Yagi, *Biochem. Med.* **15,** 212 (1976).
[28] P. J. Jose and T. F. Slater, *Biochem. J.* **128,** 141P (1972).

lated cells, etc., it is important to remember that the concentrations of products of lipid peroxidation (e.g., diene conjugation, lipid hydroperoxides, malonaldehyde) at any one instant of time, reflect a balance between the rate of production and the rates of degradation and metabolism. Failure to see a significant increase in a particular product of lipid peroxidation (for example, lipid hydroperoxide, diene conjugates, or malonaldehyde) may hide a stimulation in lipid peroxidation rate above the normal range but not sufficient to exceed the capacity for degrading or metabolising of such a product. In such cases a change in the amount (or turnover) of the parent PUFAs may be detectable.

In the following overview of methods used for detecting lipid peroxidation emphasis will be placed on the general applicability to the model procedures outlined above.

## Overview of Methods Used for Detecting Lipid Peroxidation

*Loss of Lipid Substrate.* Early studies *in vitro* in relation to decreases in the PUFA content of intracellular membranes in response to lipid peroxidation are well illustrated by the work of Schneider *et al.*[29] on mitochondria and of May and McCay[30] on microsomes. The use of gas chromatography to monitor changes in PUFAs in peroxidizing biomembrane suspensions has subsequently been extended by numerous workers to studies on tissue samples, whole cells, and various intracellular organelles.[31] It is necessary to bear in mind, however, that with studies involving lipid peroxidation *in vivo* there is a dynamic basis to PUFA levels, and small changes resulting from lipid peroxidation may be masked by systemic effects of an applied toxic agent, by local influx of PUFAs, and by changes in PUFA metabolism.

*Oxygen Uptake.* This approach is suited only to studies *in vitro* where background respiration is small compared to the contribution of lipid peroxidation to oxygen uptake. Good examples of this procedure are (1) the rapid increase in oxygen uptake by liver microsomal suspensions with NADPH when a solution of ADP/$Fe^{2+}$ is added[15] and (2) the effect of adding $Fe^{2+}$ to mitochondrial suspensions in the absence of added substrate.[29]

*Diene Conjugation.* This method has been much used for estimations of lipid peroxidation after experiments *in vivo* and *in vitro*. A general protocol for this method is that described by Recknagel and Ghoshal,[32]

---

[29] A. K. Schneider, E. E. Smith, and F. E. Hunter, *Biochemistry* **3**, 1470 (1964).
[30] H. E. May and P. B. McCay, *J. Biol. Chem.* **243**, 2288 (1968).
[31] R. A. Jordan and J. B. Schenkman, *Biochem. Pharmacol.* **31**, 1393 (1982).
[32] R. O. Recknagel and A. K. Ghoshal, *Exp. Mol. Pathol.* **5**, 413 (1966).

but many relatively minor modifications of this are in use in individual laboratories.

The difficulty that occurs in attempting to estimate a small amount of diene conjugates in extracts of biological material, which arises from the diene absorption appearing as a small shoulder on a high background absorption, can be significantly improved by the use of double-derivative spectroscopy.[33]

A modification of the "simple" diene absorption method has been proposed by Waller and Recknagel[34] in which the diene is reacted with [$^{14}$C]tetracyanoethylene followed by measurement of the $^{14}$C-labeled Diels–Alder adduct.

There is some indication[35] that diene conjugates are metabolized, at least *in vivo*, as witnessed by the time course of diene conjugation absorption in liver microsomal fractions prepared from rats intoxicated with $CCl_4$.

*Lipid Hydroperoxides.* In principle this should be a reasonable and direct approach to the measurement of lipid peroxidation. In practice, however, the measurement of lipid hydroperoxides in extracts of biological material is complicated by the rather easy breakdown of the hydroperoxides catalyzed by metal ions, by the reduction of hydroperoxides by thiols, and by the metabolism of the hydroperoxides by peroxidases such as GSH-peroxidase.[4] It is therefore desirable to protect the sample against such effects in the period following sampling by decreasing the concentration of transitional metal ions where possible (for example, by adding a strong chelating agent such as Desferal), and by inhibiting enzyme action.

Pryor and Castle, in this volume,[4] describe chemical methods for measuring lipid hydroperoxides that are based on iodometric titration and on peroxidase-catalyzed reactions; generally, however, the measurement of lipid hydroperoxide content as an indication of lipid peroxidation has not been so widely used as other methods such as diene conjugation or malonaldehyde content. This usage may change considerably in the near future with the development of sensitive high-performance liquid chromatography (HPLC) methods for lipid hydroperoxides.

*Fluorescent Analysis of Products of Lipid Peroxidation.* Amino acids react with malonaldehyde to give conjugated Schiff bases that *have* strong optical absorptions, and which also have fluorescent features that may be used for analysis.[36] Fluorescent analysis of biological extracts has been

---

[33] F. Corongiu and A. Milia, *Chem. Biol. Interact.* **44**, 289 (1983).
[34] R. L. Waller and R. O. Recknagel, *Lipids* **12**, 914 (1977).
[35] C. D. Klaassen and G. L. Plaa, *Biochem. Pharmacol.* **18**, 2019 (1969).
[36] C. J. Dillard and A. L. Tappel, this volume [41].

developed particularly by Tappel's laboratory,[37] and applications have been made to clinical situations by several other groups.[38]

*Chemiluminescence.* Lipid peroxidation of tissue fractions is associated with the emission of light, and a number of studies have shown reasonable correlation between this chemiluminescence and other measures of lipid peroxidation. For example, in liver microsomes a good relationship was found between chemiluminescence and malonaldehyde production during short periods of incubation at 37°; with incubation times longer than approximately 15 min the chemiluminescence decreased relative to malonaldehyde values.[39] With isolated hepatocytes, however, Smith *et al.*[40] have reported that the relationship between chemiluminescence and malonaldehyde production was best at longer incubation times, perhaps reflecting the metabolism of malonaldehyde that is known to occur in such organized systems.

An analysis of chemiluminescence linked to oxidative reactions in the intact liver has been presented by Boveris *et al.*[41]; an extension of their work has allowed the investigation of lipid peroxidation in liver *in situ*.

Light emission during lipid peroxidation may arise largely from the formation of singlet oxygen followed by decay back to the triplet ground state [Eq. (5)]

$$^1O_2 \rightarrow {}^3O_2 + h\bar{\nu} \tag{5}$$

The wavelengths of this emission are in the red and infrared regions.[42] Most studies on the chemiluminescence associated with lipid peroxidation of biological samples have not established unequivocally that the emission results from singlet oxygen decay.

One route to the formation of singlet oxygen in such systems is through the tetroxide[43] followed by decomposition [Eq. (6)].

$$2RO_2^{\cdot} \rightarrow RO_2O_2R \rightarrow 2RO\cdot + {}^1O_2 \tag{6}$$

[37] B. L. Fletcher, C. J. Dillard, and A. C. Tappel, *Anal. Biochem.* **52**, 1 (1973).
[38] D. Wickens, M. H. Wilkins, J. Lunec, G. Ball, and T. L. Dormandy, *Ann. Clin. Biochem.* **18**, 158 (1981).
[39] J. R. Wright, R. C. Rumbaugh, H. D. Colby, and P. R. Miles, *Arch. Biochem. Biophys.* **192**, 344 (1979).
[40] M. T. Smith, H. Thor, P. Hartzell, and S. Orrenius, *Biochem. Pharmacol.* **31**, 19 (1982).
[41] A. Boveris, E. Cadenas, R. Reiter, M. Filipkowski, Y. Nakase, and B. Chance, *Proc. Natl. Acad. Sci. U.S.A.* **77**, 347 (1980).
[42] N. I. Krinsky, *in* "Singlet Oxygen" (H. H. Wasserman and R. W. Murray, eds.), p. 597. Academic Press, New York, 1979.
[43] J. A. Howard and K. U. Ingold, *J. Am. Chem. Soc.* **90**, 1056 (1968).

Another possibility is in a reversal of the initial reaction of a lipid free radical with triplet oxygen [Eq. (7)].

$$PUFA \cdot + {}^3O_2 \rightarrow PUFA \; O_2^{\cdot} \rightarrow PUFA \cdot + {}^1O_2 \tag{7}$$

With increasingly sensitive photon detectors, and more detailed studies of the chemical origin of the chemiluminescence, it seems probable that this technique will be rather widely used in studies on isolated cells and on organs *in situ*.

*The Thiobarbituric Acid Reaction.* The most generally used procedure for this reaction utilizes tissue samples that have been mixed with a relatively strong acid precipitant [e.g., an equal volume of 10% (w/v) trichloroacetic acid]; the mixture is then centrifuged and the supernatant solution is heated with thiobarbituric acid. After cooling, absorption is measured at 532–535 nm. There are many minor modifications on this overall procedure, which has been used as a measure of lipid peroxidation for many years in a wide variety of biological situations.[8] The reaction is usually described as measuring malonaldehyde *or* "malonaldehyde-like substances." Overall, the method is very easy to use, and is quite sensitive; it may be calibrated against malonaldehyde bisdimethyl acetal. The molecular extinction coefficient of the product relative to malonaldehyde is ~150,000.[3]

It has been known for a long time[8] that numerous substances other than malonaldehyde react with thiobarbituric acid under the typical acid conditions described above; fortunately, many of these potentially disturbing substances do not occur in appropriately high concentrations in tissue extracts under normal conditions. Where such interfering substances do occur in moderate concentrations then correction can often be applied where necessary by appropriate blanks. Obviously, however, each experimental system must be looked at critically in relation to the suitability of using the simple thiobarbituric acid reaction as a measure of lipid peroxidation: commonly occurring situations that affect the procedure are the presence of high sucrose concentrations[44] and of added iron salts[45] in excess of ~100 $\mu M$.

Although open to interference by many substances[46] the thiobarbituric acid reaction has remained extremely popular with investigations studying lipid peroxidation in biological systems *in vitro*. The thiobarbituric acid reaction has been directly compared with other methods of measuring lipid peroxidation, such as diene conjugation, chemiluminescence,

[44] G. P. Plaisance, *J. Biol. Chem.* **29,** 207 (1917).

[45] E. D. Wills, *Biochim. Biophys. Acta* **84,** 475 (1964).

[46] J. M. C. Gutteridge, J. Stocks, and T. L. Dormandy, *Anal. Chim. Acta* **70,** 107 (1974).

oxygen uptake, PUFA loss, and lipid hydroperoxide content, and in liver microsomes[17,31] and in isolated hepatocytes[40] the comparisons have been satisfactory enough to warrant continued usage of this method.

The objection that the thiobarbituric acid reaction does not accurately reflect the content of malonaldehyde has been tested by comparing the results of the thiobarbituric acid reaction with a direct measurement of malonaldehyde by HPLC. In the system tested, liver microsomes in which lipid peroxidation had been stimulated by ADP/$Fe^{2+}$, there was a good relationship between the thiobarbituric acid reaction and direct malonaldehyde determination.[47]

An important aspect of the thiobarbituric acid reaction in relation to lipid peroxidation is that thiobarbituric acid does not react[3] efficiently with some products of the peroxidative breakdown of PUFAs, such as 4-hydroxyalkenals, in terms of molar extinction coefficients at 535 nm. Some of these products, for example, 4-hydroxynonenal, have powerful biological effects.[48] In consequence, the thiobarbituric acid reaction may not adequately reflect the biological consequences of lipid peroxidation in situations where the ratio of the production rate of malonaldehyde/4-hydroxyalkenals varies.[3]

Clearly, the thiobarbituric acid reaction has several advantages in terms of simplicity and of sensitivity. However, disadvantages also exist as outlined above. It seems prudent in each particular case where the thiobarbituric acid reaction is to be used to cross-check its relationship with lipid peroxidation by reference to as many of the other methods (e.g., PUFA loss; lipid hydroperoxide formation; diene conjugation) as is feasible. Moreover, if the results of the thiobarbituric acid reaction are to be equated with malonaldehyde content then a direct measurement of malonaldehyde should be done for comparison.

*Hydroxyaldehydes.* As already mentioned, among the products of lipid peroxidation of microsomal suspensions and other biological samples are a variety of carbonyl-containing substances including hydroxyalkenals.[3] The latter, in particular, have striking biological activities and can be considered to be important with respect to the damaging consequences of lipid peroxidation.[49]

It seems unlikely, however, that these products of lipid peroxidation will be used as routine measures of lipid peroxidation since the methodology involved in their separation and estimation [derivatization with 2,4-

[47] H. Esterbauer and T. F. Slater, *IRCS Med. Sci.* **9**, 749 (1981).
[48] M. U. Dianzani, *in* "Free Radicals, Lipid Peroxidation and Cancer" (D. C. H. McBrien and T. F. Slater, eds.), p. 129. Academic Press, New York, 1982.
[49] A. Benedetti, M. Comporti, and H. Esterbauer, *Biochim. Biophys. Acta* **620**, 281 (1980).

dinitrophenylhydrazine, thin-layer chromatography (TLC), and HPLC] is time consuming and complex.

*Hydroxy Fatty Acids.* The PUFA hydroperoxides formed in the early stages of the peroxidation of PUFAs are converted to the hydroxy derivatives by enzymic pathways in biological material. The hydroxy fatty acids can be separated by HPLC as described by Capdevila *et al.*[50] However, for reasons similar to those described for the hydroxyaldehydes, it seems unlikely that estimation of hydroxy fatty acids will be used routinely in the near future to follow the course of lipid peroxidation.

*Alkanes.* The formation of hydrocarbon gases such as ethane and pentane during lipid peroxidation has been recognised for about 10 years.[51] Procedures have been elaborated for measuring this gas production *in vitro*[52] and for whole animals *in vivo.*[53] The use with intact animals is the only genuinely noninvasive method available for studying lipid peroxidation under conditions *in vivo*.

There are, however, some complications that relate to interpretation of the data obtained from whole animals. For instance, the site(s) of formation of the exhaled gas within the animal, the influence of microbial flora, and the effects of a particular toxic agent under study on the production and metabolism of these gases through nonperoxidative routes. Nevertheless, with appropriate sageguards the method has been a useful addition to the procedures available for studying lipid peroxidation.

## Concluding Remarks

The methods most commonly used for estimating the extent of lipid peroxidation in biological systems have been loss of lipid substrate, oxygen uptake, diene conjugation, the thiobarbituric acid reaction, and alkane production. In a number of cases where a variety of methods have been checked against each other in different biological systems there has been generally good correlation between them in relation to the relative extents of lipid peroxidation with times of incubation. The stoichiometries of the methods are not the same, however; the production of malonaldehyde, for example, may be only a small proportion of the total PUFA loss.

For following the general change in lipid peroxidation with time in biological systems any one of the above listed methods is capable of giving significant information; the formation of lipid hydroperoxides, fluo-

[50] J. Capdevila, L. J. Marnett, N. Chacos, R. A. Prough, and R. W. Estabrook, *Proc. Natl. Acad. Sci. U.S.A.* **79,** 767 (1982).

[51] C. Riely, G. Cohen, and M. Lieberman, *Science* **183,** 208 (1974).

[52] H. Kappus and H. Muliawan, *Biochem. Pharmacol.* **31,** 597 (1982).

[53] C. J. Dillard and A. L. Tappel, *Lipids* **14,** 989 (1979).

rescent analysis, and chemiluminescence should also be considered seriously in such respects. As emphasized in the preceding discussion, however, each method has its own drawbacks and limitations and, as already emphasized, it is safest with each particular situation under study to use a variety of methods for cross-checking purposes.

## [34] Chemical Methods for the Detection of Lipid Hydroperoxides

By WILLIAM A. PRYOR and LAURENCE CASTLE

The autoxidation of biological materials occurs by mechanisms that are well known;[1] oxidation by oxygen leads to the formation of hydroperoxides[2] in a radical chain reaction.[3] The net process is shown in Eq. (1), where RH represents an unsaturated biomolecule such as a membrane phospholipid or polyunsaturated fatty acid (PUFA).

$$RH + O_2 \rightarrow ROOH \tag{1}$$

Clearly, the most direct and meaningful quantitation of peroxidation would be achieved from measurements of either loss of substrate or uptake of oxygen. Though these two measures have been used with considerable success by, for example, rubber chemists and food scientists,[4,5] they are often not relevant in the biochemical context where the concept of a single, definable, substrate may not hold and where there are other oxygen-utilizing processes, such as respiration, that could mask any extra demand for oxygen for autoxidative degradation. The major initial reaction products of lipid peroxidation are hydroperoxides,[6] as shown in Eq. (1), and their assay offers the next most direct measure of peroxidation. These rather labile species can undergo both enzymic and nonenzymic decomposition to give products that include volatile hydrocarbons, malonaldehyde and malonaldehyde precursors, and carbon monoxide. Analysis of these secondary products forms the basis of several tests for peroxidation that will be described in subsequent sections of this chapter. In spite of

[1] W. A. Pryor, ed. "Free Radicals in Biology," Vol. I, p. 1. Academic Press, New York, 1976.
[2] R. Criegee, H. Pilz, and H. Flygare, Chem. Ber. 72, 1799 (1939).
[3] J. L. Bolland, Quart. Revs. 3, 1 (1949).
[4] J. W. Hamilton and A. L. Tappel, J. Am. Oil. Chem. Soc. 40, 52 (1963).
[5] E. E. Dumelin and A. L. Tappel, Lipids 12, 894 (1977).
[6] Sometimes incorrectly called lipid peroxides or lipoperoxides.

METHODS IN ENZYMOLOGY, VOL. 105
Copyright © 1984 by Academic Press, Inc.
All rights of reproduction in any form reserved.
ISBN 0-12-182005-X

this lability, the quantitative determination of hydroperoxides provides a valuable and popular index of peroxidation. This is especially true for model studies employing emulsions and liposomes, where the decomposition of hydroperoxides by transition-series metals can be minimized.

The most sensitive chemical assays for lipid hydroperoxides that are of use to biochemists are (1) measurement of conjugated dienes, (2) iodometry, (3) peroxidase-catalyzed reduction of hydroperoxide, and (4) oxidation of a "marker substrate." These will be discussed in turn.

### The Measurement of Conjugated Dienes

In the formation of a hydroperoxide from a 1,4-nonconjugated diene during autoxidation, the diene moiety rearranges into conjugation. Four isomeric conjugated diene hydroperoxides can be formed on rearrangement. Chan and Levett have used high-pressure liquid chromatography to separate the approximately equimolar mixture of the four isomers that are formed in the air-oxidation of methyl linoleate.[7] Both cis,trans and trans,-trans isomers were isolated and showed absorption maxima at 236 and 233 nm with extinction coefficients of 26,000 and 28,000 $M^{-1}$ cm$^{-1}$, respectively. Thus, an average extinction coefficient of 27,000 $M^{-1}$ cm$^{-1}$ at 234 nm can be used in the analysis of mixtures of isomers.

It is not possible to describe a single experimental procedure for this method, since the details vary with the nature of the biological sample. However, a typical procedure to measure conjugated dienes would be to extract the lipids from the sample to be tested using 2 : 1 (v/v) chloroform/ methanol and evaporate this extract to dryness under a stream of nitrogen at ambient temperature.[8] The dried extract is then redissolved in hexane or cyclohexane of spectroscopic quality, and its absorbance at 234 nm measured against a solvent blank or by difference spectroscopy versus nonperoxidized lipid.[9] The latter is preferable since the technique of difference spectroscopy removes part of the error associated with measurement of the conjugated diene absorbance which, by normal absorption spectroscopy, appears as a rather imprecisely defined shoulder on the strong end absorption of the nonconjugated lipid itself.[10]

This sensitive assay can be used to measure lipid peroxidation both *in vivo* and *in vitro*. When studying model systems comprising aqueous suspensions of purified lipid, it is often possible to analyze for conjugated diene directly without the need to extract the lipid into organic solution. Studies with PUFA have shown that the ratio of formation of lipid hydro-

[7] H. W.-S. Chan and G. Levett, *Lipids* **12**, 99 (1977).

[8] J. A. Buege and S. D. Aust, this series, Vol. 52, p. 302.

[9] R. O. Recknagel and A. K. Ghoshal, *Exp. Mol. Pathol.* **5**, 413 (1966).

[10] H. V. Thomas, P. K. Mueller, and R. L. Lyman, *Science* **159**, 532 (1968).

peroxide and conjugated diene is not always strictly $1:1$.[11] (This is, of course, especially true for oxidations involving singlet oxygen, where a nonradical, nonchain process leads to the formation of approximately equimolar amounts of conjugated and nonconjugated diene hydroperoxides.[12]) However, even though in some cases the concentrations of the conjugated diene and hydroperoxide functionalities do not always agree, the two tend to be proportional to one another, especially at early stages of the reaction, and the spectrophotometric determination of the diene function at 234 nm gives an index of lipid peroxidation that can be performed quickly and with a minimum of apparatus.

*Iodometric Assay*

The oxidation of iodide by hydroperoxides proceeds according to Eq. (2). The classical method for the determination of liberated iodine, titration with thiosulfate, has been superceded by its spectroscopic estimation as the triiodide anion [Eq. (3)]. A major disadvantage of iodometric assay is its susceptibility toward interference by molecular oxygen, and major efforts have been directed in recent years to minimize this difficulty and to increase the sensitivity of the technique. The reduction of lipid hydroperoxides must employ a large excess of iodide, for two reasons. First, to allow the reduction to proceed at a reasonable rate and, second, to provide sufficient unreacted iodide to force equilibrium (3) overwhelmingly to the right. The key to applying this reduction to the analysis of minute levels of hydroperoxides in biological samples has been to afford protection of the excess iodide against reaction with molecular oxygen. Without this precaution, high blanks and poor sensitivity result.

$$ROOH + 2H^+ + 2I^- \rightarrow ROH + H_2O + I_2 \tag{2}$$
$$I^- + I_2 \rightleftharpoons I_3^- \tag{3}$$

Two experimental approaches to this problem have been made. In the first, described originally by Takagi *et al.*[13] and in a much improved form by Buege and Aust,[8] the reduction of hydroperoxide is performed under anaerobic conditions and unreacted iodide is then protected against oxidation by complexation with cadmium. The sample is then transferred to the spectrophotometric cuvette for estimation of triiodide at 353 nm. The second protocol, by Hicks and Gibicki[14] and by Vilsen and Nielsen,[15]

[11] W. A. Pryor, J. P. Stanley, E. Blair, and G. B. Cullen, *Arch. Environ. Health* **31**, 201 (1976).
[12] M. J. Thomas and W. A. Pryor, *Lipids* **15**, 544 (1980).
[13] T. Takagi, Y. Mitsuno, and M. Masumura, *Lipids* **13**, 147 (1978).
[14] M. Hicks and J. M. Gebicki, *Anal. Biochem.* **99**, 249 (1979).
[15] B. Vilsen and H. Nielsen, submitted, (1983).

involves performing the assay under anaerobic conditions in the optical cuvette and with the cuvette in place in the spectrophotometer. This has two important advantages in that it is possible to ensure that reduction of the hydroperoxide is complete, and the rate of formation of iodine due to adventitious oxygen can be measured both before the test sample is added and after the hydroperoxide has reacted. Thus, one can compensate for the effect of adventitious oxygen without the need for separate controls.

These last-mentioned reports represent the "state of the art" in the iodometric determination of lipid hydroperoxides. They display high sensitivity, exact stoichiometry, and can be performed with a minimum of apparatus. With reaction in 2–3 ml HOAc/MeOH or HOAc/CHCl$_3$ under subdued light, and with estimation of $I_3^-$ at 290 or 360 nm, 1 nmole hydroperoxide can be assayed, although this is the lower limit. The extinction coefficients for $I_3^-$ reported by Hicks et al.[14] are 4.41 × 10$^4$ and 2.80 × 10$^4$ $M^{-1}$ cm$^{-1}$ at 290 and 360 nm, respectively, in HOAc/MeOH and agree fairly well with values for $I_3^-$ in aqueous solution.[16] Other values for the extinction coefficients have been reported,[17] and it is recommended that a calibration curve be constructed using a standard hydroperoxide such as t-butyl or cumyl hydroperoxide. It has been shown that $I_3^-$ obeys Beers' law within the useful range of spectrophotometers.[14]

The assay will not tolerate substances that promote the decomposition of hydroperoxides, such as transition metal ions, easily oxidized substances (such as mercaptans), or compounds that react with iodine (such as acetone). Hydrogen peroxide is detected by this assay but endoperoxides are not. Dialkyl peroxides are generally reduced too slowly to react with acidified iodide under the assay conditions.[18] Colored materials absorbing in the region 290–360 nm will increase the background absorbance of samples and so reduce the sensitivity of the assay. The majority of interfering substances can be eliminated by the precaution of extracting the materials to be tested using chloroform/methanol, evaporation of this extract to dryness under a stream of nitrogen at ambient temperature, and submission of this material to the assay. If "whole" samples must be tested, controls spiked with a standard organic hydroperoxide should be included.

### Peroxidase-Catalyzed Reduction of Hydroperoxide

Aerobic organisms have evolved with a variety of peroxidases to prevent excessive accumulation of lipid hydroperoxides and hydrogen perox-

[16] A. D. Awtrey and R. E. Connick, J. Am. Chem. Soc. 73, 1842 (1951).
[17] E. Lovaas and F. Palmieri, FEBS Meet., 12th Dresden Abstr. 3302, report 3.21 × 10$^4$ and 1.83 × 10$^4$ $M^{-1}$ cm$^{-1}$ as the $I_3^-$ extinction coefficient in methanolic solution.
[18] R. D. Mair and A. J. Graupner, Anal. Chem. 36, 194 (1964).

ide. The function of this class of enzyme is to reduce hydroperoxides to the corresponding alcohol with the aid of a donor of reducing equivalents, according to Eq. (4).

$$ROOH + DH_2 \xrightarrow{\text{(peroxidase)}} ROH + H_2O + D \qquad (4)$$

The utility of peroxidases in the estimation of lipid hydroperoxides *in vitro* was first appreciated by Heath and Tappel, who ingeniously coupled two enzymic systems to give us the first glutathione peroxidase-based assay.[19] In their procedure, reduced glutathione (GSH) is first oxidized by the hydroperoxides on catalysis by glutathione peroxidase [Eq. (5)]. The oxidized glutathione (GSSG) is in turn rereduced by the action of glutathione reductase with NADPH [Eq. (6)]. Since both peroxidase and reductase are specific for glutathione alone, the overall reaction is as shown in Eq. (7) and the amount of hydroperoxide submitted to the test is found simply from the loss of NADPH as measured at 340 nm ($\varepsilon = 6200$ $M^{-1}$ cm$^{-1}$).

$$ROOH + 2\ GSH \rightarrow ROH + GSSG + H_2O \qquad (5)$$
$$GSSG + 2\ NADPH \rightarrow 2\ GSH + 2\ NADP^+ \qquad (6)$$

overall;

$$ROOH + 2\ NADPH \rightarrow ROH + H_2O + 2\ NADP^+ \qquad (7)$$

Several variations of this basic method followed quickly, most notably from research groups in Japan where glutathione peroxidase was available commercially at the time. Miura and co-workers described an alternative technique for the estimation of GSSG in which excess, unreacted, GSH is removed by treatment with the sulfhydryl reagent *N*-ethylmaleimide. GSSG is then reacted with *o*-phthalaldehyde in base to give a fluorescent product that is determined with excitation at 343 nm and emission at 420 nm.[20] In a more recent paper, this same research group describes a procedure that is essentially identical except that unreacted GSH is not removed using an SH reagent but, rather, is assayed using *o*-phthalaldehyde under conditions in which GSSG does not form the fluorescent product.[21]

Horseradish peroxidase (HRP) has been used as an alternative to glutathione peroxidase. HRP will accept reducing equivalents from quite a variety of hydrogen donors and the judicious choice of donor can form the basis of a sensitive spectrophotometric assay for hydroperoxide. Thus, Yamaguchi and Misaki have estimated lipid hydroperoxides in a 50-$\mu$l

[19] R. L. Heath and A. L. Tappel, *Anal. Biochem.* **76**, 184 (1976).
[20] T. Miura, J. Hiraizumi, and M. Kimura, Koen Yoshishu-Seitai Seibun no Bunseki Kagaku Shinpojumu, 4th, 49 (1979). (*Chem. Abstr.* **92**, 159957p).
[21] T. Miura, J. Hiraizumi, and M. Kimura, *J. Pharmacobio-Dyn.* **3**, S-8 (1980).

sample of blood serum by the HRP-catalyzed oxidation of homovanillic acid to a fluorescent product.[22] They also applied HRP to the catalytic oxidation of 4-aminoantipyrine-N,N-dimethylaniline by lipid hydroperoxide from the autoxidation of methyl linoleate and estimated the colored oxidation product at 565 nm.

These assays offer both sensitivity and specificity. The measurement of nanoequivalents of hydroperoxides is feasible and only hydroperoxides and hydrogen peroxide are measured; dialkyl peroxides and endoperoxides are not substrates for the peroxidase and, therefore, are not reduced. If $H_2O_2$ is present, it can be eliminated by prior treatment of the sample to be tested with catalase. These assays are, however, considerably more involved than, for example, the iodometric procedures. The use of enzymic catalysis confers specificity but brings with it a susceptibility to interference from inhibitors of the enzymic process such as detergents and organic solvents. For example, Kohda et al. have concluded that the coupled system of Heath and Tappel[19] cannot be used for the determination of lipid hydroperoxides in human serum because of the presence of many kinds of interfering substances which cannot be removed by the various pretreatments applicable to clinical tests. These workers have suggested an alternative procedure for such samples that employs a peroxygenase from pea seeds, N-methylindole as the substrate, and fluorometric detection.[23] With the exception of the coupled system of Heath and Tappel,[19] these peroxidase procedures are not stoichiometric and the assay of comprehensive controls and standards is obligatory. This is especially true with HRP, where the rate of reaction with certain hydroperoxides is too slow to allow the reduction to proceed to completion in a reasonable time.[24] It is probably fair to say that a universal method has not yet emerged that can be confidently used without modification to suit a particular application.

### Oxidation of a "Marker Substrate"

There exist in the literature a number of procedures for the determination of hydroperoxides that are based on their decomposition, promoted by a transition metal, with concomitant oxidation of a "marker substrate" to form a product that can be detected by fluorimetry,[24] absorbance spectroscopy,[25] or chemiluminescence.[26] The nature of oxidation has not been

[22] T. Yamaguchi and H. Misaki, Fr. Demande FR 2,476, 677 (Chem. Abstr. **96**, 65263q).

[23] K. Kohda, K. Arisue, A. Maki, and C. Hayashi, Jpn. J. Clin. Chem. **11**, 306 (1982).

[24] T. Miura, personal communication.

[25] A. Miike, Y. Shimizu, T. Tatano, and K. Watanabe, Eur. Pat. Appl. EP 38,205 (Chem. Abstr. **96**, 65272s).

[26] P. J. O'Brien and L. G. Hulett, Prostaglandins **19**, 683 (1980).

elucidated in most cases, though in some a peroxidase mechanism seems to operate while in others the involvement of free oxy-radicals has been demonstrated. The transition metal is typically iron in the form of a heme compound such as hematin or microperoxidase[24] (a degradation product of cytochrome $c$ that retains the heme and 7-11 amino acids of the original enzyme). Microperoxidase has been used for some time for the estimation of hydrogen peroxide.[27] These assays have many of the features offered by those employing peroxidase enzymes, but are still largely in a period of development and are as yet confined largely to the patent literature.

Acknowledgment

Our research on peroxides is supported by NIH, NSF, and the NFCR.

[27] N. Feder, *J. Cell Biol.* **51**, 339 (1971).

# [35] Comparative Studies on Different Methods of Malonaldehyde Determination

*By* R. P. Bird and H. H. Draper

Malonaldehyde (MA) is of interest primarily as a product of lipid peroxidation *in vivo* and as an index of oxidative rancidity in foods. In biological materials it exists in its free form and as a complex with various tissue constituents. It has also been identified among the products of the oxidative decomposition of amino acids, complex carbohydrates, pentoses, and hexoses formed in the presence of a metal catalyst, as a product of free radicals generated by ionizing radiation *in vivo*, and as a byproduct of prostaglandin biosynthesis. However, peroxidation of fatty acids with three or more double bonds (notably arachidonic acid) is believed to be its major source. Because of its interest as an indicator of lipid peroxidation, various methods have been proposed for its estimation.

## Spectrophotometric Methods

### *Determination of MA in the Free Form*

In its free form MA can be estimated by UV absorptiometry,[1] by polarography,[2] or by high-performance liquid chromatography.[3,4] Deter-

[1] T. W. Kwon and B. M. Watts, *Anal. Chem.* **35**, 733 (1963).

Copyright © 1984 by Academic Press, Inc.
All rights of reproduction in any form reserved.
ISBN 0-12-182005-X

mination of free MA by UV absorbance is made possible by the fact that, below pH 4.65, it exists largely in its undissociated cyclic form ($e_{245} = 30,000$). To separate free MA from interfering materials in the sample, it is first distilled at an acid pH and then the UV absorbance of the distillate at 245 nm is determined. However, recovery of MA is poor (50–60%) and some samples may contain other distillable compounds with UV absorbance characteristics which interfere with the absorbance of MA in the free form. Distillation is also tedious when used to analyze large numbers of samples.

### Determination of MA Derivatives

Malonaldehyde reacts with a variety of compounds to form derivatives which can be estimated spectrophotometrically. These include aniline, 4-hexylresorcinol, $N$-methylpyrrole, indole, 4-aminoacetophenone, ethyl $p$-amino benzoate, 4,4′-sulfonyldianiline, $p$-nitroaniline, and azulene.[5] However, other aldehydes react with most of these compounds to form yellow or orange complexes which may interfere in the spectrophotometric determination of MA.

The most widely employed method for the determination of MA in biological materials is based on its reaction with thiobarbituric acid (TBA). One molecule of MA reacts with two molecules of TBA with the elimination of two molecules of water to yield a pink crystalline pigment with an absorption maximum at 532–535 nm and secondary maxima at 245 nm and 305 nm.[6]

Various modifications have been introduced into the TBA method to adapt it to the determination of MA in different biological substances. Although heating the sample with TBA at an acid pH is common to all procedures, the conditions used to carry out the reaction differ from one laboratory to another with respect to sample preparation, type of acid used, pH of the reaction mixture, reaction time, use of metal catalysts, and composition of the TBA reagent. Several of these factors have been shown to influence the results obtained.

Acid can affect the TBA–MA reaction in various ways. For maximum complex formation, the reaction should be carried out at a pH of 2 to 3 (Fig. 1).

[2] A. M. Bond, P. P. Deprez, R. D. Jones, G. G. Wallace, and M. H. Briggs, *Anal. Chem.* **52**, 2211 (1980).
[3] H. Esterbauer and T. F. Slater, *IRCS Med. Sci.* 749 (1981).
[4] A. Saari Csallany, Ming Der Guan, and B. Addis, personal communication.
[5] E. Sawicki, T. W. Stanley, and H. Johnson, *Anal. Chem.* **35**, 199 (1963).
[6] R. O. Sinnhuber, I. C. Yu, and Te. C. Yu, *Food Res.* **23**, 620 (1958).

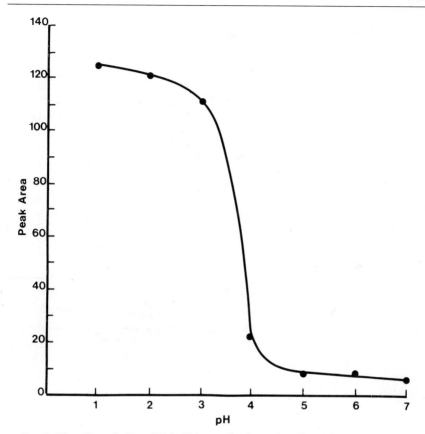

FIG. 1. The effect of pH on TBA–MA complex formation. Free MA was reacted with TBA at 100° for 30 min.[6] Complex formation was determined by HPLC.[19]

As shown by Tarlagis et al.,[7] the reaction requires acid for catalysis. Acid extraction of biological samples (usually carried out with TCA) yields higher TBA values than neutral saline extraction. Acidification of human and rat urine to pH 2–3 with HCl increases the yield of TBA–MA complex several-fold (unpublished results). This may be due to release of bound MA and/or to acid catalysis of the TBA reaction.

Although acidic conditions are essential for TBA–MA complex formation, an excess of strong acid also may inhibit color development. The TBA reagent gives lower MA values when prepared in strong acid than when dissolved in water.[8] Heating with strong acid alters the structure of

[7] B. G. Tarladgis, A. M. Pearson, and L. R. Dugan, Jr., *J. Sci. Food. Agric.* **15,** 602 (1964).
[8] B. G. Tarladgis, A. M. Pearson, and L. R. Dugan, Jr., *J. Am. Oil Chem. Soc.* **39,** 34 (1962).

TBA and leads to degradation products which absorb at about the same wavelength as the TBA–MA complex.[8]

Various acids have been used in the determination of MA by the TBA reaction and some have been reported to give higher values than others.[9,10] In our experience, however, the pH of the reaction mixture is more important than the selection of the acidifying agent.

Elevated TBA values have been reported for samples which contain inorganic iron or to which inorganic iron is added.[11,12] Various inorganic ions, including $Co^{2+}$, $Ni^{2+}$, $Mn^{2+}$, $Zn^{2+}$, $Ca^{2+}$, $Cr^{2+}$, $Ce^{2+}$, $Fe^{2+}$, and $Fe^{3+}$ increase the TBA value.[13] Iron salts catalyze the breakdown of linoleic and linolenic acid hydroperoxides to form MA.[14] Metals, especially $Fe^{2+}$, and $Fe^{3+}$, catalyze the degradation of amino acids, sugars (deoxyribose, hexoses, pentoses) and DNA in the presence of air to yield MA. This oxidative degradation is not inhibited by antioxidants or by EDTA[15,16]; in fact, these compounds have been reported to increase TBA–MA formation.[17]

The specificity of the TBA test for MA has been a subject of protracted controversy. TBA breakdown products, as well as yellow to orange products of reactions between TBA and various other compounds, have absorbance maxima in the region 440–460 nm and significant absorbance at 532 nm.[9,18,19] Acetaldehyde–sucrose mixtures react with TBA to form a pigment which has a visible absorption spectrum similar to that of the TBA–MA complex. However, on a molar basis the fluorescence intensity of the TBA–MA complex is substantially greater than that of other TBA reaction products.[9]

MA reacts with compounds which contain primary amino groups, and hence the presence of these derivatives in biological materials may affect the TBA value. Under the usual conditions of the TBA test, open chain mono- and disubstituted MA addition products (e.g., R—N=CH—CH=CHOH and R—N=CH—CH=CH—NH—R) give 100% recovery of MA on a molar basis.[20] However, when MA is incorporated into cyclic

[9] J. M. C. Gutteridge and T. R. Tickner, *Anal. Biochem.* **9**, 250 (1978).

[10] L. W. Ohkawa, N. Ohishi, and K. Yagi, *J. Lipid Res.* **19**, 1053 (1978).

[11] F. Bernheim, M. L. C. Bernheim, and K. M. Wilbur, *J. Biol. Chem.* **174**, 257 (1948).

[12] C. H. Castell, B. Moore, and W. Neal, *J. Fish Res. Board Can.* **23**, 737 (1966).

[13] C. H. Castell and G. A. Boyce, *J. Fish Res. Board Can.* **23**, 1587 (1966).

[14] T. Asakawa and S. Matsushita, *Lipids* **14**, 401 (1980).

[15] J. M. C. Gutteridge, *FEBS Lett.* **105**, 278 (1979).

[16] J. M. C. Gutteridge, *FEBS Lett.* **128**, 343 (1981).

[17] J. M. C. Gutteridge and F. Xiaochang, *FEBS Lett.* **123**, 71 (1981).

[18] V. S. Waravdekar and L. D. Saslaw, *J. Biol. Chem.* **234**, 1945 (1959).

[19] R. P. Bird, S. S. O. Hung, M. Hadley, and H. H. Draper, *Anal. Biochem.* **128**, 240 (1983).

[20] H. Buttkus and R. J. Bose, *J. Am. Oil Chem. Soc.* **49**, 440 (1972).

products (e.g., between the ureido or guanidino substituents of $\alpha$-amino acids such as critrulline or arginine), the recovery of MA as indicated by the TBA reaction is much reduced.[21] The products of imine amine type reactions, such as those between MA and pyrazoline or pyrazole ring systems, release very little MA under the conditions of the TBA test.[21] The reactions of MA with secondary amines such as dimethylamine, diethylamine, piperidine, pyrrolidine, and morpholine yield $\beta$-dialkyl aminoacroleins of trans-s-trans conformation. These acroleins decompose under acidic or alkaline conditions to yield a colored reaction product with TBA.[2] The TBA reaction measures total MA present in the free form under the conditions of the TBA reaction.

From the foregoing, it is apparent that the results of the spectrophotometric determination of MA in biological samples by the TBA method are strongly influenced by the conditions employed. In addition to considerations pertaining to the conditions used to carry out the TBA reaction, MA may be generated as an artifact if the sample is overheated or if metal catalysts are used. Acid conditions and heat decompose peroxides, as well as MA addition products, to form free MA. Another source of error arises from the presence of pigments, particularly in plant materials, which absorb in the 532 nm region and thereby contribute to spuriously high estimates of MA content.[19] For this reason, a distillation step has been commonly used for the analysis of food samples.[22] Although TBA has been shown to react with a number of compounds to produce pigments which could interfere in the spectrophotometric determination of the TBA–MA complex, many of these compounds are not normal constituents of biological materials or are not formed under the conditions of the TBA assay. For some materials, the TBA method yields results which are similar to those obtained by a more specific HPLC procedure, but for others (especially pigmented samples) it gives substantially higher values.[19]

Fluorometric Methods

Some compounds, including 4,4-sulfonyldianiline, ethyl $p$-aminobenzoate, $p$-aminobenzoic acid and 4-aminoacetophenone, react with MA to form complexes which can be measured spectrofluorometrically. The TBA–MA complex also has characteristic fluorescence excitation and emission maxima (at 532 and 553 nm, respectively) which theoretically could be utilized as a basis for the determination of MA in natural

[21] K. Kikugawa, K. Tsukuda, and T. Kurechi, *Chem. Pharm. Bull.* **28**, 3323 (1980).
[22] B. G. Tarladgis, B. M. Watts, M. T. Younathan, and L. R. Dugan, Jr., *J. Am. Oil Chem. Soc.* **37**, 44 (1960).

materials.[5] However, fluorometric methods have not been widely used for the determination of MA in biological samples owing to the presence of naturally occurring fluorescent compounds or the formation of fluorescent derivatives by compounds other than MA.

HPLC Methods

HPLC procedures recently have been described for the determination of MA in biological samples as the free aldehyde[3,4] and as the TBA–MA complex.[19] The latter procedure, described in brief below, has been found satisfactory for the analysis of food samples, animal tissues, and (with some modifications) human and rat urine.

*TBA Reagent.* One percent TBA in distilled water.

*HPLC Conditions.* A 0.39 × 30-cm $\mu$Bondapak $C_{18}$ stainless-steel analytical column attached to a 3 × 22-mm guard column packed with $C_{18}$/Corasil (Waters) is used. The ultraviolet absorbance detector is equipped with a 546-nm interference filter and is attached to an electronic integrator. The monitor is set at 0.01 AUPS. The eluting solvent is 11% HPLC grade methanol in double-distilled water degassed by filtering through a 0.45-$\mu$m filter (Millipore) under vacuum with constant stirring. The flow rate is 2 ml/min.

*Standard Curve.* A standard curve is prepared using TBA–MA complex[6] which has been checked for purity by HPLC, NMR, and elemental analysis. Instrument response is plotted against the weight of the complex injected. Instrument sensitivity is 1 ng.

*Procedure.* The samples (0.5–1.0 g) are homogenized in 5 ml of 5% TCA solution. After centrifuging at 1000 g for 10 min, an aliquot of the supernate is reacted with an equal volume of TBA reagent in a boiling water bath for 30 min. The pH is adjusted to 1.5 with 4 N HCl. At the end of the incubation any precipitate is promptly removed by centrifugation to prevent adsorption of the TBA–MA complex. An aliquot of the reaction mixture is adjusted to pH 6–7 with 0.15 N NAOH in distilled water and methanol (1 : 1, v/v), and an appropriate aliquot (usually 10–20 $\mu$l) is injected into the instrument. After 15–20 determinations, the column is washed with methanol.

The method has been found to be specific for MA and to have satisfactory reproducibility when applied to a variety of food and tissue samples (coefficient of variability 7.0% for samples containing 1–2 $\mu$g MA/g). For most samples, it yields lower values than the spectrophotometric TBA procedure. It is uncertain, however, what portion of the MA determined was originally present in the sample in the free state, what portion may have been "bound" to other compounds, or what fraction may have been generated from precursors such as endoperoxides. It has been observed, for example, that there is negligible free MA in rat urine, but the procedure yields significant amounts of MA formed by hydrolysis of small molecular weight compounds (L. G. McGirr, personal communication). For urine, maximum yield is obtained following hydrolysis at pH 3 rather than at pH 1.5. HCl (4 $N$) is used for pH adjustment, followed by addition of an equal volume of TBA solution. The yield of MA in rat urine rises in vitamin E deficiency, indicating that urinary MA may be a useful index of lipid peroxidation *in vivo* (L. Polensek and H. H. Draper, unpublished results).

## [36] Concentrating Ethane from Breath to Monitor Lipid Peroxidation *in Vivo*

*By* Glen D. Lawrence and Gerald Cohen

### Background

The measurement of exhaled hydrocarbons has evoked widespread interest in recent years as a noninvasive method for monitoring *in vivo* lipid peroxidation. The exhalation of ethane by mice was first demonstrated in this laboratory[1] in a study of $CCl_4$-induced lipid peroxidation in liver. Pretreatment of mice with $\alpha$-tocopherol (vitamin E), a lipid antioxidant which protects against $CCl_4$ hepatotoxicity, effectively suppressed ethane production. Subsequent studies detected low levels of spontaneous alkane production by rodents and showed that dietary vitamin E and selenium suppressed spontaneous ethane[2-4] and pentane[5-6] production. In the $CCl_4$-toxicity model, a good correlation has been observed[7] between ethane exhalation and hepatic levels of conjugated dienes, a standard indicator of lipid peroxidation.

In the early experiments,[1] a small desiccator (2.4 liters) was used as the experimental chamber, and it was necessary to pool five $CCl_4$-treated mice in a single chamber in order to accumulate sufficient amounts of

---

[1] C. A. Riely, G. Cohen, and M. Lieberman, *Science* **183**, 208 (1974).
[2] G. Lawrence, G. Cohen, and L. Machlin, *Ann. N.Y. Acad. Sci.* **393**, 227 (1982).
[3] D. G. Hafeman and W. G. Hoekstra, *J. Nutr.* **107**, 656 (1977).
[4] D. G. Hafeman and W. G. Hoekstra, *J. Nutr.* **107**, 666 (1977).
[5] C. J. Dillard, E. E. Dumelin, and A. L. Tappel, *Lipids* **12**, 109 (1977).
[6] C. J. Dillard, R. E. Litov, and A. L. Tappel, *Lipids* **13**, 396 (1978).
[7] T. P. Lindstrom and M. W. Anders, *Biochem. Pharmacol.* **27**, 563 (1978).

Copyright © 1984 by Academic Press, Inc.
All rights of reproduction in any form reserved.
ISBN 0-12-182005-X

ethane for detection by direct injection of 0.5 ml of the chamber air into the gas chromatograph. Hafeman and Hoekstra[4] worked with larger animals (rats) and waited several hours in order to detect significant differences in spontaneous ethane production between vitamin E-deficient and control rats.

Tappel and co-workers[5] did not work in an enclosed rebreathing system and utilized an activated alumina trap immersed in an ethanol/liquid nitrogen slush to concentrate hydrocarbons from exhaled breath. Their method traps pentane efficiently, but ethane either was not produced by the animals or was not adsorbed by the alumina trap. We have found that ethane is not efficiently adsorbed on alumina, even at low temperature.[8]

We describe here a simple method[8] for concentrating ethane, ethylene, and longer chain hydrocarbons from larger volumes of air for subsequent analysis on a gas chromatograph. This method allows geater than a 100-fold increase in the sensitivity for ethane analysis over earlier and currently used procedures. We also include a description of materials that can be used to fabricate an animal chamber, a discussion of the advantages and disadvantages of measuring exhaled hydrocarbons compared to other techniques for assessing peroxidative damage, and some discussion of the relative merits of ethane vs pentane measurements.

### Breath Collection Chamber

An air circulation chamber is modified from earlier studies[1,3] and is shown in Fig. 1. The system works well for long-term breath collection (multiple hours). The system shown in Fig. 1 was designed for a mouse, but the chamber size can be increased to accommodate larger animals. We have used modified chambers to study rats up to 500 g in body weight. The air from the chamber is circulated through a series of three traps: 5% $H_2SO_4$ to trap exhaled $NH_3$, 10–15% KOH to trap respiratory $CO_2$, and a coldfinger immersed in dry ice/2-propanol to condense water vapor. Thiazole Yellow is used as an indicator dye in the KOH trap to indicate when the base is expended. Gas inlet and outlet tubes are sealed into either the walls or top of the chamber in an arrangement which allows efficient mixing of the chamber air. An inexpensive Hypalon and Teflon oscillating pump (Fisher Scientific, N.Y.) is used to circulate the air through the traps and back to the chamber. The rate of flow is regulated with a powerstat; a flow rate of 100–200 ml/min is convenient for a 0.5-liter chamber. An oxygen reservoir is attached to an inlet to the chamber to allow pure oxygen to enter as respiratory $CO_2$ is scrubbed from the chamber air. The pump is turned off and the oxygen inlet (stopcock B in

[8] G. D. Lawrence and G. Cohen, *Anal. Biochem.* **122**, 283 (1982).

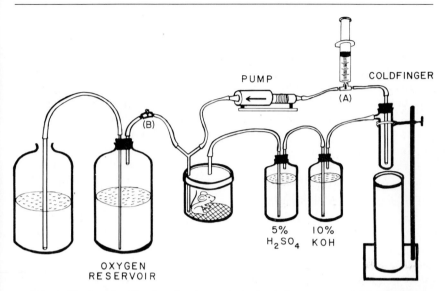

Fig. 1. The breath collection chamber and air circulating system are described in the text. Air flow is in the direction shown by the arrow on the pump. (For clarity, the coldfinger is shown above the Dewar flask.) The pump is turned off and the oxygen supply is closed at stopcock (B) when a sample is removed from stopcock (A) for analysis. An equal volume of purified air is added back to the chamber via stopcock (B) to maintain constant oxygen tension.[8]

Fig. 1) is closed when a sample (generally 50 ml) is removed from the system at stopcock A. The sample is removed with a clean, ungreased, gas-tight syringe. An identical volume of purified air is added to the system as replacement before the oxygen inlet is reopened. If this is not done, oxygen will enter the chamber from the reservoir after each sample is removed and give rise to increasing $P_{O_2}$ values. Recent studies have shown that the oxygen tension can affect the yield of hydrocarbon gas.[9]

An equation for calculating the total amount ($X_t$) of ethane produced by the animal at time, $t$, is:

$$X_t = X_n[(V_c - V_a)/V_s] + \Sigma X_i,$$

where $X_n$ is the moles in the $n$th sample (determined by gas chromatography), $V_c$ is the total volume of the chamber and circulating system (calculated by determining the dilution of an ethane standard injected into the system), $V_a$ is the volume of the animal in the chamber (determined by weighing the animal and assuming an average density of 1.0 g/ml), $V_s$ is

[9] G. Cohen, in "Role of Lipid Peroxides in Biology and Medicine" (K. Yagi, ed.). Academic Press, New York, 1982.

the volume of air removed from the chamber for analysis, and $\Sigma X_i$ is the summation of the moles of ethane removed from the chamber in previous samples.

Chambers can be fabricated from a number of materials. Glass has the advantage of being inert and impermeable to gases. Large, airtight, glass chambers are usually expensive, however, and commercial suppliers may be difficult to find. Acrylics,[8] or other durable plastics, are generally much less costly and they are relatively inert and impermeable to gases. Flexible connections are achieved with Teflon or polyvinyl tubing, which are relatively impermeable to ethane.[8] The entire operational system (Fig. 1) is tested for leaks by injecting a sample of ethane and analyzing aliquots of chamber air for several hours.

## Concentrating Hydrocarbons from Air Samples

A Hewlett–Packard Model 5750 gas chromatograph (or other suitable gas chromatograph) is equipped with a Chemical Data Systems (Oxford, PA) Model 310 Concentrator. The concentrator consists of two 3.2 mm × 60-cm traps and a pneumatic valve encased in an insulated valve oven; the valve diverts gas flow over either of the two traps. Alkanes are bound by adsorbents in the traps and, subsequently, they are rapidly desorbed directly into the gas chromatograph. The valve configuration is such that the gas flow through the traps is reversed when switching from the trap "load" to the trap "desorb" position. This configuration backflushes the desorbed gas and avoids having the substances pass over the entire trap bed, which would result in peak broadening on the gas chromatograph. A programmable console is interfaced with the concentrator unit to control the valve position and heating of the traps. A pulse heater inside a fiberglass insulating jacket heats the traps at a rate of 400°/min during the desorption phase. We pack the leading half of each stainless-steel trap with Tenax (Chemical Data Systems, Oxford, PA) or activated alumina, and the succeeding half with activated charcoal (type SK-4, 60/80 mesh, Applied Science Labs, State College, PA). Activated charcoal at room temperature is an excellent adsorbent for ethane and ethylene,[8] but it binds pentane tenaciously. A systematic evaluation of various adsorbents showed that the adsorption characteristics of pentane on Tenax are similar to ethane on activated charcoal. Accordingly, both hydrocarbons can be adsorbed sequentially (pentane on the Tenax and ethane on the charcoal) at room temperature in a single, mixed trap, as described above, and both can be desorbed at or above 220°. Packing the leading end of the traps with Tenax or alumina (or other selected adsorbents) is recommended even when pentane is not analyzed; this prevents other, less volatile substances from reaching the charcoal adsorbent and fouling the ethane trap.

We perform routine studies by sampling 50-ml aliquots of chamber air. This provides a 100-fold increase in sensitivity compared to the earlier sampling of only 0.5 ml. We have also tested up to 1 liter samples with a slightly modified trapping procedure with essentially 100% recovery.[8] The analysis of large gas volumes is limited by the presence of "trace" alkane gases in room air and/or in high purity commercial sources of air, oxygen, and other gases. It should be noted that the charcoal traps in the concentrator are exposed to continuous carrier gas flow (e.g., zero grade helium), which may contain low levels of ethane and other alkanes. It is prudent to carefully regulate the time interval between samples (e.g., every 10 min over the alternate traps) to avoid changes which could contribute to the baseline (or zero time) values. In addition, we thoroughly desorb the traps by prolonged heating at the beginning of each working day and, as needed, during the day.

*Gas Chromatography*

A 3.2-mm × 2-m stainless-steel column packed with Poropak N (60/80 mesh, Applied Science Labs., State College, PA) gives excellent separation of methane, ethane, and ethylene at 60°. An elevated temperature (e.g., 140°) is required to elute pentane from this column. The resolution of the lower molecular weight hydrocarbons on Poropak N far outweighs any inconvenience due to a need to program a temperature ramp to elute pentane and other higher molecular weight hydrocarbons. The Poropak N column can be operated isothermally at 60° for the study of ethane. It should be emphasized that the study of pentane production is not well suited for rebreathing systems because pentane is lost from the chamber due to metabolism by the animal (see next section).

When the isothermal determination of ethane and pentane is desired, this can be accomplished at 50° on a 3.2-mm × 2-m stainless steel column packed with Porasil B (80/100 mesh, Applied Science Labs., State College, PA). However, there may be a problem resolving the methane and ethane peaks, particularly when larger amounts of methane are produced by flora in the gut of the animal. Improved resolution can be achieved on a longer column (e.g., 4–6 m). The methane peak can also be substantially reduced or eliminated by waiting several minutes before desorbing the bound hydrocarbons with the pulse heater; methane is selectively lost from the activated charcoal trap and passed to the external environment by continuous carrier gas flow at ambient temperature. However, ethane, ethylene, and other hydrocarbons may exhibit some peak broadening when exposed to continuous carrier flow for more than 15 min.

### Advantages and Disadvantages of Measuring Exhaled Hydrocarbons

The main advantages to measuring exhaled ethane as an index of lipid peroxidation are (1) the method is noninvasive; (2) it does not give rise to handling artifacts, such as induced lipid peroxidation on manipulation of biopsied or autopsied tissue; and (3) it permits observation of the same animal over the course of hours or, sequentially, over days or months. The main disadvantages are (1) alkane gases are relatively minor products, and do not, as yet, permit direct quantitative correlation with lipid peroxidation; (2) a number of areas, such as the role of diet and bacterial peroxidation in the intestines (see below), remain to be further clarified and/or controlled; and (3) the organ or tissue of origin cannot be immediately pinpointed without ancillary information, such as tissue analyses for other lipid peroxidation products.

Recently, Smith and co-workers[10] compared five methods for measuring lipid peroxidation in isolated hepatocytes. $CCl_4$-stimulated lipid peroxidation was assessed by production of malondialdehyde, fluorescent products, chemiluminescence, ethane, and pentane, and the time course for formation of each product was followed. Ethane was the most sensitive of the techniques studied (a 200-fold increase over zero time levels in 4 hr, compared to a 50-fold increase in chemiluminescence and less than a 25-fold increase for the other three products). Ethane production was also the most immediate (about a 30 min lag for significant increases in ethane production, compared to 90 min for significant increases in chemiluminescence, malondialdehyde, or fluorescent products).

The measurement of ethane has advantages over pentane measurements. Ethylene and $C_3$–$C_5$ straight-chain hydrocarbons are eliminated from the air in an enclosed chamber by untreated mice[8] or rats,[11] whereas ethane is not. Aliphatic hydrocarbons are hydroxylated by the microsomal monooxygenase system.[12] In mice injected (ip) with a hydrocarbon mixture, we observed increased recoveries of exhaled ethylene and $C_3$–$C_5$ alkanes after treatment with an anesthetic agent that contains a competitive substrate for the microsomal enzymes, whereas no significant change occurred in the recovery of ethane.[8] The lower volatility of pen-

[10] M. T. Smith, H. Thor, P. Hartzell, and S. Orrenius, *Biochem. Pharmacol.* **31**, 19 (1982).
[11] H. Frank, T. Hintze, D. Bimboes, and H. Remmer, *Toxicol. Appl. Pharmacol.* **56**, 337 (1980).
[12] U. Frommer, V. Ullrich, and H. Staudinger, *Hoppe-Seyler's Z. Physiol. Chem.* **351**, 903 (1970).

tane (bp 36°) poses an additional problem in the air circulating chamber described above, even if the coldfinger is removed. Even in nonrebreathing systems, the recovery of generated pentane will depend upon the effects of coadministered drugs on the metabolic disposition of pentane.

The problem of bacterial peroxidation of dietary polyunsaturated fats in the gut, giving rise to increased hydrocarbon exhalation, has recently been addressed.[2,13] We have observed that high-fat diets (e.g., 20% corn oil) and vitamin E-deficient diets each give rise to high background levels of ethane, which can be suppressed by maintaining the animals on a fat-free diet for 24 hr,[14] or by starving the animals for 24 hr. We recommend feeding the animals a fat-free diet for at least 36 hr to ensure complete (or nearly complete) elimination of dietary lipids from the gut.

*Conclusions*

The measurement of exhaled hydrocarbons has the advantage of being noninvasive, i.e., does not require extraction of tissue or internal probing. The immediacy and sensitivity of ethane production, combined with the ability to concentrate ethane from large volumes of breath, make this technique valuable for studying pathological conditions in which lipid peroxidation is a factor. The described method is directly applicable to the study of small animals and, with modification, can be adapted to larger animals, and human subjects.

[13] D. Gelmont, R. A. Stein, and J. F. Mead, *Biochem. Biophys. Res. Commun.* **102**, 932 (1981).
[14] G. Lawrence and G. Cohen, unpublished observation.

# [37] Assay of Ethane and Pentane from Isolated Organs and Cells

*By* Armin Müller and Helmut Sies

Volatile hydrocarbons such as ethane and pentane are known to originate from the peroxidation of polyunsaturated fatty acids present in membrane lipids, and the formation of these hydrocarbon gases has been proposed as a sensitive index for lipid peroxidation in toxicological studies.[1-4]

[1] C. Riely, G. Cohen, and M. Lieberman, *Science* **183**, 208 (1974).
[2] A. Wendel and E. E. Dumelin, *Methods Enzymol.* **77**, 10 (1981).
[3] G. Lawrence and G. Cohen, this volume [36].
[4] C. J. Dillard and A. L. Tappel, *Lipids* **14**, 989 (1979).

METHODS IN ENZYMOLOGY, VOL. 105
Copyright © 1984 by Academic Press, Inc.
All rights of reproduction in any form reserved.
ISBN 0-12-182005-X

Possible alternative sources for these alkanes have not yet been characterized. Upon acute administration of xenobiotics increased amounts of alkanes were detected in the exhaled air of animals. Increased hepatic levels of malondialdehyde[1] and diene conjugation[5] indicated that the liver is a major site of alkane production. In order to eliminate contributions by extrahepatic tissues in more detailed analysis, experimental models such as isolated liver microsomes,[6,7] isolated hepatocytes,[8–10] and the isolated perfused rat liver[11,12] have been employed. These afford the possibility for the direct study of metabolic influences and mechanistic aspects with respect to hepatic alkane formation and lipid peroxidation. It is noted that pentane may represent a risky indicator of *in vivo* lipid peroxidation due to its hepatic metabolism. Because ethane is not metabolized, its measurement as an index of lipid peroxidation may be more reliable.[13]

### Isolated Perfused Liver

#### Alkane Collection Chamber[11]

The rat liver is isolated for perfusion[14] and is placed on a special perfusion table equipped with a cylindrical Plexiglas chamber (Fig. 1). The wall of the chamber contains a cleft for the portal vein cannula, and the top plate contains a perforation for the caval vein cannula. The effluent perfusate from this cannula passes the $O_2$ and pH electrodes for continuous monitoring. The chamber has two ports in the side wall, one for sampling, the other for pressure equilibration (stopcocks 1 and 2 in Fig. 1). When a sample is being drawn after opening stopcock 1, stopcock 2 is also opened to maintain isobaric conditions in the collection chamber. The connected balloon is filled with a reservoir of nitrogen or synthetic air of purest grade. In order to avoid contamination with alkanes contained in the ambient air, the balloon is flushed in 20 min intervals. Caval effluent perfusate amounts to 97–99% of the influent perfusate. The remainder consists of the so-called drip-off and bile which drip through a sieve. During the experiments care is taken to maintain a closed system by

[5] T. Lindstrom and M. Anders, *Biochem. Pharmacol.* **27,** 563 (1978).
[6] H. Kappus and H. Muliawan, *Biochem. Pharmacol.* **31,** 597 (1982).
[7] R. Reiter and A. Wendel, *Biochem. Pharmacol.* **32,** 665 (1983).
[8] N. De Ruiter, H. Ottenwälder, H. Muliawan, and H. Kappus, *Toxicol. Lett.* **8,** 265 (1981).
[9] N. H. Stacey, H. Ottenwälder, and H. Kappus, *Toxicol. Appl. Pharmacol.* **62,** 421 (1982).
[10] M. T. Smith, H. Thor, P. Hartzell, and S. Orrenius, *Biochem. Pharmacol.* **31,** 19 (1982).
[11] A. Müller, P. Graf, A. Wendel, and H. Sies, *FEBS Lett.* **126,** 241 (1981).
[12] A. Müller and H. Sies, *Biochem. J.* **206,** 153 (1982).
[13] G. D. Lawrence and G. Cohen, *Anal. Biochem.* **122,** 283 (1982).
[14] H. Sies, *Methods Enzymol.* **52,** 48 (1978).

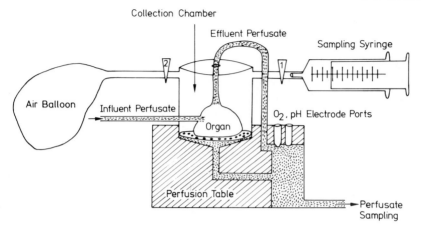

FIG. 1. System used for sampling of alkanes produced by the isolated perfused liver (nonrecirculating perfusion).

adjusting the height of the effluent perfusate. The chamber is sealed with a silicone rubber (Elastosil A 33, Wacker-Chemie, Munich, West Germany) containing no alkanes in the series from ethane to *n*-pentane. Control experiments in which alkanes were present in the chamber without the perfused liver showed that the amount of alkanes remains constant with time, even in the picomole range which is of interest for experiments with perfused organs.[11]

*Sampling*

Samples (5 ml) from the gas space around the liver are taken at appropriate intervals (5 to 15 min) using Hamilton gas-tight syringes containing no lubricants. The gas inside the chamber is mixed with the syringe several times before withdrawal. Samples can be stored in the syringe for several hours if required. The ethane produced by the organ is partitioned, in the steady state, into a portion dissolved in the effluent perfusate, and another portion diffusing into the gas space of the collection chamber. For determining the amount of dissolved gas in the perfusate, about 500 ml is collected into a vessel supplemented with about 2 mmol EDTA and 2 mmol 2,6-di-*t*-butyl-4-methylphenol as an antioxidant. Subsequently the perfusate is heated to 60° and the amount of ethane is determined in the gas space of the vessel.[11] Two vessels in sequence will allow for complete assessment of the amounts present.

In the case of pentane the experimental situation is complicated somewhat by the fact that pentane is metabolized whereas ethane is not, as shown by the time course of these hydrocarbons when added to the collection chamber.[11,15] Pentane dissolved in the perfusion medium (about 1 pmol/ml) is completely taken up by the liver, so that no pentane is detectable in the effluent perfusate.

## Gas Chromatographic Analysis

Samples of 5 ml are injected into a sampling loop and 3 ml is introduced to the Porasil C column (siliceous material, 80-100 mesh) of a gas chromatograph (e.g., Carlo Erba Fractovap Model 2151 AC with flame ionization detection).[2] Nitrogen of purest grade is used as carrier gas with a flow of 40 ml/min. In the case of measuring only ethane (as usual), the analyses are run isothermally at 50°. This condition was found to be favorable for separating the ethane peak from the inert gas peak leaving the column at the beginning. For pentane alone, the run is carried out at 90° to shorten the time required for analysis. The flow rate through the flame ionization detector is 450 ml synthetic air per minute and 35 ml hydrogen per minute, both of purest grade. Electrometer recording is routinely performed at 0.1 pA/mV. The column is flushed and cleaned overnight at 140°.

## Calculation

The system is calibrated with calibration gas (Messer Griesheim, Duisburg, West Gemany) ranging from 0.1 to 0.8 ppm alkanes in nitrogen. For the analysis of the alkane peaks automatic integration or a graphic approximation method is used. For calculation of the amount of alkanes the following equation is employed:

$$\text{Alkane production rate} \left( \frac{\text{pmol}}{\text{g liver wet wt} \times \text{min}} \right)$$

$$= \frac{(a_{t_n} - a_{t_{n-1}}) \times (V_C - V_L) + a_{t_{n-1}} \times V_S}{W_L \times \Delta t} \times PF$$

with $a_t$, alkane concentration in samples as calculated from the calibration peaks (pmol/ml) at a given time; $n$, sample number; $V_C$, volume of the collection chamber (ml); $V_L$, volume of the liver (ml); $V_S$, sample volume drawn from the collection chamber (ml); $W_L$, wet weight of the liver (g); $\Delta t$, sample interval (min); and $PF$, partitioning factor allowing for the partitioning of release into the collection chamber and perfusate.

[15] M. Thelen and A. Wendel, *Biochem. Pharmacol.* **32**, 1701 (1983).

The partitioning of ethane between the evolution from the liver surface and the release into the effluent perfusate was found to be remarkably constant at perfusate flow rates from 25 to 34 ml/min. The ethane release observed in the collection chamber amounted to $7 \pm 1\%$ ($n = 7$) of the total release of ethane from the organ.[11] This was found to pertain not only for controls, but also for experiments with $t$-butyl hydroperoxide or ethanol or paraquat in which increased rates of ethane production are observed. However, this percentage may be subject to variations depending on the biological system or the perfusion conditions and should be checked in each specific case. For example with mouse livers, perfused at a rate of 5 ml/min, the release of ethane into the collection chamber was found to be 25% of the total hepatic ethane release.[15] Assuming a constant percentage of partitioning implies a homogeneous response of the organ with respect to ethane production.

For calculation of cumulative ethane production against time, the correction for alkane removal from the chamber is done according to the equation in Ref. 2.

*Comments*

For the measurements in the picomole range it is important to realize that ambient air contains alkanes. If their concentration amounts to <10 ppb as determined prior to the experiment, the contamination of the perfusion medium can be considered insignificant in relation to the alkane concentration in the collection chamber. In the case of a higher concentration in the air the concentration of alkanes in the perfusate can be high, and subsequently they may be released into the collection chamber, probably favored by the shift in the perfusate temperature to 37°. Since this effect may distort results, such contamination has to be taken into account or somehow lessened; e.g., the water of the perfusion medium can be preheated and stored under alkane-free nitrogen or synthetic air atmosphere.

Isolated Hepatocytes and Subcellular Organelles

Isolated hepatocytes[16] are incubated aerobically (alkane-free synthetic air) at 37° in gas-tight cell culture flasks (20 or 62 ml).[17-19] Following an

[16] P. Moldeus, J. Högberg, and S. Orrenius, this series, Vol. 52, p. 60.
[17] D. Köster-Albrecht, H. Kappus, and H. Remmer, *Toxicol. Appl. Pharmacol.* **46,** 499 (1978).
[18] N. H. Stacey and H. Kappus, *J. Toxicol. Environ. Health* **9,** 277 (1982).
[19] N. De Ruiter, H. Ottenwälder, H. Muliawan, and H. Kappus, *Arch. Toxicol.* **49,** 265 (1982).

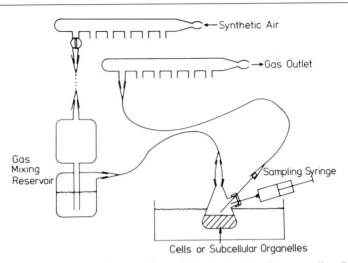

Fig. 2. System used with microsomal or cells incubation for alkane sampling. Synthetic air is used prior to experiment for flushing the system. Multiple (six) incubation vessels can be adapted to gas supply as indicated on top. Reservoir is used for mixing when taking sample with syringe. Modified from Ref. 6.

incubation period an 8-ml sample is withdrawn from the gas phase, and 5 ml is injected into the column. The amounts of alkane are calculated using calibration gas (see above) and can be expressed either in 1 pmol of alkane per $10^8$ cells or per milligram of protein or per gram of wet weight. For sequential sampling to follow a time course, a system is used as shown in Fig. 2.[6] Prior to the incubation the flask is flushed with synthetic air (alkane-free), in order to remove all contaminants. Eight-milliliter samples are taken from the head space above the incubation mixture, replaced by synthetic air. The 8 ml gas present in the syringe is pumped backward and forward several times, in order to achieve homogeneous distribution. For calculation, the dilutions and the removals must be taken into account.

For simultaneous recording of oxygen concentration, a syringe equipped with an $O_2$ electrode and a headspace for alkane sampling was employed in experiments with mouse liver microsomes.[7]

Concluding Remarks

In the table the rates of ethane formation from *in vitro* (perfused liver, hepatocytes and microsomal fractions) and *in vivo* systems are compared. The results show that ethane production ranges in the same order of

ETHANE PRODUCTION FROM *in Vitro* AND *in Vivo* MODELS[a]

| Model | Ethane production (pmol/min/g liver wet wt) |
|---|---|
| Isolated perfused liver (rat) | |
| Control[b,c]; mouse[d,e] | 1.0–1.4 |
| Ethanol (30 m$M$)[c] | 3.2 |
| Acetaldehyde (1 m$M$)[c] | 2.5 |
| Paraquat (3 m$M$)[b] | 4.2 |
| Paraquat (3 m$M$), selenium deficient[f] | 10.5 |
| $FeCl_2$ (0.26 m$M$)[b] | 10.0 |
| Linolenate (0.1 m$M$)[e] | 48.0 |
| Phenobarbital pretreated[b] | 0.9 |
| Selenium deficient[f] | 0.6 |
| Isolated hepatocytes (rat) | |
| Control[g–k] | 0.3–1.6 |
| $FeCl_2$ (0.5 m$M$)[g] | 4.2 |
| ADP-$Fe^{2+}$ (0.1 m$M$)[i] | 15 |
| $CCl_4$ (2.5[h] or 5 m$M$[k]) | 12–14 |
| $Hg^{2+}$ (0.1 m$M$)[j] | 2.5 |
| *In vivo* (rat) | |
| Control,[l–r] mouse[s] | 0.2–0.8 |
| Paraquat (78 $\mu$mol/kg)[q] | 0.5 |
| Selenium deficient[q] | 2.2 |
| Vitamin E deficient[q] | 2.5 |
| Ethanol (5 g/kg)[m,n] | 0.8–1.9 |
| $CBrCl_3$ (100 mg/kg)[l] | 4.1 |
| $CCl_4$ (1 g/kg)[m] | 6.0 |
| Selenium deficient[q,r] | 0.2 |
| Vitamin E deficient[q,r] | 0.8–1.0 |
| Paracetamol (500 mg/kg), mouse[l] | 42 |
| | (pmol/min/mg protein) |
| Microsomes (rat liver) | |
| NADPH[u] | 0.1 |
| NADPH + $FeCl_2$ (24 $\mu$M)[u] | 2.0 |
| NADPH + $CCl_4$ (10 m$M$)[u] | 0.8 |
| NADPH + ADP-$Fe^{2+}$ (244 $\mu$M)[n] | 3.1 |
| Peritoneal macrophages (mouse) | |
| Control[v] | 0.7 |
| $FeCl_2$ (1 m$M$)[v] | 1.1 |

[a] *In vivo* data were calculated assuming 4 g liver wet weight per 100 g body weight. Isolated hepatocyte data were calculated assuming 1 g liver wet weight per $10^8$ cells. Hepatic microsomal protein is ~50 mg/g wet wt. Ethane production rates were calculated for the first 1–2 hr of measurement.

[b] A. Müller, P. Graf, A. Wendel, and H. Sies, *FEBS Lett.* **126,** 241 (1981).

[c] A. Müller and H. Sies, *Biochem. J.* **206,** 153 (1982).

Footnotes to table (*continued*)

[d] M. Thelen and A. Wendel, *Biochem. Pharmacol.*, **32,** 1701 (1983).
[e] A. Müller and H. Sies, unpublished results (high linolenate blank corrected for).
[f] E. Cadenas, H. Wefers, A. Müller, R. Brigelius, and H. Sies, *Agents Actions Suppl.* 203 (1982).
[g] N. H. Stacey and H. Kappus, *J. Toxicol. Environ. Health* **9,** 277 (1982).
[h] N. H. Stacey, H. Ottenwälder, and H. Kappus, *Toxicol. Appl. Pharmacol.* **62,** 421 (1982).
[i] N. De Ruiter, H. Ottenwälder, H. Muliawan, and H. Kappus, *Arch. Toxicol.* **49,** 265 (1982).
[j] N. H. Stacey and H. Kappus, *Toxicol. Appl. Pharmacol.* **63,** 29 (1982).
[k] M. T. Smith, H. Thor, P. Hartzell, and S. Orrenius, *Biochem. Pharmacol.* **31,** 19 (1982).
[l] D. Köster-Albrecht, U. Köster, H. Kappus, and H. Remmer, *Toxicol. Lett.* **3,** 363 (1979).
[m] U. Köster, D. Albrecht, and H. Kappus, *Toxicol. Appl. Pharmacol.* **41,** 639 (1977).
[n] H. Frank, T. Hintze, D. Bimboes, and H. Remmer, *Toxicol. Appl. Phamacol.* **56,** 337 (1980).
[o] M. Sagai and T. Ichinose, *Life Sci.* **27,** 731 (1980).
[p] C. J. Dillard, E. E. Dumelin, and A. L. Tappel, *Lipids* **12,** 109 (1977).
[q] R. F. Burk, R. A. Lawrence, and L. M. Lane, *J. Clin. Invest.* **65,** 1024 (1980).
[r] R. F. Burk and J. M. Lane, *Toxicol. Appl. Pharmacol.* **50,** 467 (1979).
[s] A. Wendel, S. Feuerstein, and K.-H. Konz, *Biochem. Pharmacol.* **28,** 2051 (1979).
[t] A. Wendel and S. Feuerstein, *Biochem. Pharmacol.* **30,** 2513 (1981).
[u] H. Kappus and H. Muliawan, *Biochem. Pharmacol.* **31,** 597 (1982).
[v] N. De Ruiter, H. Muliawan, and H. Kappus, *Toxicology* **17,** 265 (1980).

magnitude for liver *in vitro* and *in vivo,* indicating that the liver may be a major contributory organ for the ethane released from the animal. In fact, the *in vivo* model shows slightly lower values than the *in vitro* models. This could be due to ethane trapping by the body lipids of the animals[13] or to systematic differences arising from the incubation conditions *in vitro.* Ethane release is considered a more reliable index of lipid peroxidation than pentane. Unlike ethane, pentane is metabolized by the perfused liver at a rate of 0.6 pmol/min/g wet weight from rat[11] or mouse,[15] in agreement with work on the intact animal.[13,20]

Ethane is produced as one of several end products of lipid peroxidation,[21] and therefore changes in the pathway of lipid breakdown may influence the amount of ethane that is released. This should be considered when analytical data are interpreted, so that caution is required in making statements on the overall process of lipid peroxidation.

### Acknowledgments

Fruitful discussions with Profs. H. Kappus and A. Wendel are gratefully acknowledged. Supported by Deutsche Forschungsgemeinschaft, Schwerpunktsprogramm "Mechanismen toxischer Wirkungen von Fremdstoffen."

[20] H. Frank, T. Hintze, D. Bimboes, and H. Remmer, *Toxicol. Appl. Pharmacol.* **56**, 337 (1980).
[21] C. K. Chow, *Am. J. Clin. Nutr.* **32**, 1066 (1979).

## [38] Detection of Malonaldehyde by High-Performance Liquid Chromatography

*By* HERMANN ESTERBAUER, JOHANNA LANG, SYLVIA ZADRAVEC, and TREVOR F. SLATER

Malonaldehyde, or substances with properties very similar to malonaldehyde ("malonaldehyde-like" substances), occurs in many biological samples such as peroxidized tissues, cells and cell fractions, foods from plant and animal sources, fats, and dairy products.[1-4] Malonaldehyde can react with thiols[5] and amino acids[6-8] and introduce cross-link-

[1] W. A. Pryor, *Free Radicals Biol.* **1–5** (1977–1982).
[2] T. F. Slater, *in* "Free Radical Mechanisms in Tissue Injury" (J. R. Lagnodo, ed.), p. 38. Pion, London, 1972.
[3] E. Schauenstein, H. Esterbauer, and H. Zollner, *in* "Aldehydes in Biological Systems" (J. R. Lagnodo, ed.; P. H. Gore, trans.), p. 133. Pion, London, 1977.
[4] G. M. Siu and H. H. Draper, *J. Food Sci.* **43**, 1147 (1978).
[5] H. Buttkus, *J. Am. Oil Chem. Soc.* **46**, 88 (1968).
[6] B. C. Shin, J. W. Huggins, K. L. Carraway, *Lipids* **7**, 229 (1972).
[7] K. S. Chio and A. L. Tappel, *Biochemistry* **8**, 2827 (1969).
[8] V. Nair, D. E. Vietti, and C. S. Cooper, *J. Am. Chem. Soc.* **103**, 3030 (1981).

Copyright © 1984 by Academic Press, Inc.
All rights of reproduction in any form reserved.
ISBN 0-12-182005-X

ages into proteins[7] and nucleic acids.[9,10] The aldehyde appears to be involved in the formation of age pigments,[11] and there have been reports that it has mutagenic properties[12,13] and may be a chemical carcinogen.[14]

Malonaldehyde is generally measured by the 2-thiobarbituric acid (TBA) method, originally developed by Ottolenghi.[15] The TBA method, however, is not specific for free malonaldehyde, since many other substances that can occur in biological material give positive reactions with TBA.[2] For example, it has been shown that autoxidized linolenic acid contains at least nine breakdown products that give positive responses in the TBA reaction when studied on TLC plates.[16] The TBA method is certainly easy to use, and with experiments on tissue samples *in vitro* gives good kinetic information in relation to the extent of lipid peroxidation. The validity of equating the TBA reaction with the content of malonaldehyde itself in the sample, however, has been questioned many times.

There is no doubt that free malonaldehyde reacts stoichiometrically with TBA reagent in acidic solution, but it has not been made clear, in our view, whether the TBA reaction performed on biological samples is demonstrating principally malonaldehyde, or "malonaldehyde-like" substances, or decomposition products that arise during the heating stage of the reaction of the tissue sample with TBA.[2]

Poyer and McCay[17] have used Sephadex G-10 chromatography to demonstrate the presence of free malonaldehyde in peroxidized liver microsomal suspensions. Their method is time consuming and requires rather large sample volumes. A high-performance liquid chromatography (HPLC) method using an octadecyl silyl (ODS) column has also been proposed[18] but requires a steam distillation step before the HPLC separation; moreover, on an ODS column, malonaldehyde is eluted shortly after or with the void volume, and many compounds present in biological samples can therefore interfere with the malonaldehyde peak.

A new HPLC procedure for the analysis of free malonaldehyde in biological samples such as cells and cell fractions and autoxidized fatty acids, originally developed for studying malonaldehyde formation by rat

[9] B. R. Brooks and O. L. Klammert, *Eur. J. Biochem.* **5**, 178 (1968).
[10] U. Reiss, A. L. Tappel, and K. S. Chio, *Biochem. Biophys. Res. Commun.* **48**, 921 (1972).
[11] C. J. Dillard and A. L. Tappel, *Lipids* **8**, 183 (1973).
[12] K. D. Munkres, *Mech. Ageing Dev.* **10**, 249 (1979).
[13] T. M. Yau, *Mech. Ageing Dev.* **11**, 37 (1979).
[14] D. J. Shamberger, T. L. Andreone, and C. E. Willis, *J. Natl. Cancer Inst.* **53**, 1771 (1974).
[15] A. Ottolenghi, *Arch. Biochem. Biophys.* **77**, 355 (1959).
[16] J. M. C. Gutteridge, J. Stocks, and T. L. Dormandy, *Anal. Chim. Acta* **70**, 107 (1974).
[17] L. Poyer and P. B. McCay, *J. Biol. Chem.* **246**, 263 (1971).
[18] Y. Kukuda, D. W. Stanley, and F. R. van de Voort, *J. Am. Oil Chem. Soc.* **58**, 773 (1981).

liver microsomes,[19] is described in this chapter. The analysis time is 10–15 min; the detection limit is ~5 pmol in the injected sample of 20 $\mu$l volume. Even though the determination of malonaldehyde by HPLC is rather quick and very sensitive it should not be considered as a substitute for the common TBA assay, when a large number of samples must be analyzed within a short time period. In such cases HPLC analysis of a few selected samples will generally provide sufficient information about the existence or absence of free malonaldehyde and the quantitative relationship between the TBA number and the amount of free malonaldehyde.

## Principle of the Method

An aqueous solution or suspension containing malonaldehyde is adjusted to pH 6.5–8 and a sample (20 $\mu$l) is separated by HPLC on an aminophase column with acetonitrile/0.03 $M$ Tris buffer, pH 7.4 (1 : 9, v/v). The effluent is monitored at 270 nm, which is the absorption maximum of free malonaldehyde in the enolate anionic form.[20] The malonaldehyde peak in the chromatogram is identified by comparison with a reference chromatogram of freshly prepared free malonaldehyde. The concentration of malonaldehyde is calculated from the peak height or area, based on a calibration chromatogram performed with a standard solution (1–100 $\mu M$) of malonaldehyde. The smallest concentration of malonaldehyde in the parent aqueous solution that can be quantified by this method is 0.25 $\mu M$. A clean-up of the aqueous solution prior to HPLC is necessary with some samples (oxidized free fatty acids, complete cells, nuclei) in order to remove substances interfering with the spectrophotometric scan of the HPLC peaks.

## Measurement of Malonaldehyde in Peroxidizing Liver Microsomes

### Basic Procedure

A standard procedure, which has been successfully used for following the rate and extent of malonaldehyde formation by liver microsomes, in which peroxidation is stimulated by ADP/$Fe^{2+}$, is as follows. An S-5 Spherisorb-$NH_2$ column is connected to a standard HPLC instrument equipped with a high-pressure pump set to an upper pressure limit of 2500 psi, a Rheodyne injection valve with a 20 $\mu$l loop, a UV detector set at 270 nm, and a recorder. The column is equilibrated at the beginning of

[19] H. Esterbauer and T. F. Slater, *IRCS Med. Sci.* **9**, 749 (1981).
[20] T. W. Kwon, D. B. Menzel, and H. S. Okott, *J. Food Sci.* **30**, 808 (1965).

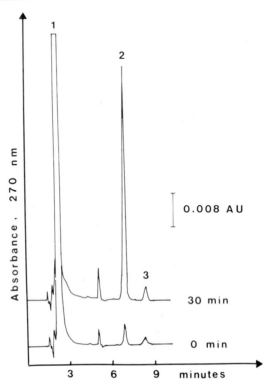

FIG. 1. High-performance liquid chromatographic separation of malonaldehyde in microsomal suspension (1 mg protein/ml) incubated in the presence of an NADPH-regenerating system and ADP/Fe at 37° for 0 and 30 min. 1, Nicotinamide, 2, malonaldehyde; 3, NADP.

each daily series of measurements with at least 50 ml of the eluent prior to analysis. The eluent used for equilibration and separation is acetonitrile/0.03 $M$ Tris, pH 7.4 (1 : 9, v/v). A sample of at least 20 $\mu$l of the microsomal suspension is withdrawn from the incubation medium with a Hamilton 50-$\mu$l syringe, injected into the HPLC, and separated in the usual way with a flow of the eluent of 1 ml/min. A typical chromatogram with peak identification is shown as an example in Fig. 1.

If no variable-wavelength detector is available, a single-wavelength detector with a setting in the range of 250 and 280 nm can also be used to detect the malonaldehyde peak. Disadvantages to the measurement at the absorption maximum (270 nm) are a lower detector response and possible interferences with substances eluted near malonaldehyde, and having high absorption at the wavelength setting used.

*Calibration and Preparation of the Malonaldehyde Standard*

A calibration chromatogram of an accurately prepared standard malonaldehyde solution (the recommended concentration is 20 $\mu M$) adjusted with Tris buffer to pH 7.4 is also run every time for peak identification and quantification. The concentration of the malonaldehyde in the experimental sample is $h_{exp}c_{st}/h_{st}$ = nanomoles of malonaldehyde/milliliter of microsomal suspension where $h_{exp}$ and $h_{st}$ are the heights (or areas) of the malonaldehyde peaks in the experimental sample and the standard sample, respectively, and $c_{st}$ is the concentration of the standard in nmole/ml.

A malonaldehyde stock solution (10 m$M$) is prepared from malonaldehyde-bisdiethylacetal (Merck, Darmstadt, F.R.G.) which has been distilled and stored at 4°. To 220 mg of the acetal (1 mmol) 100 ml of 1% (v/v) sulfuric acid is added. After 2 hr standing at ambient temperature, 1 ml of the solution is brought to a volume of 100 ml with 1% (v/v) sulfuric acid, and the malonaldehyde concentration is checked by measuring the UV absorbance in 1-cm cuvettes at 245 nm ($\varepsilon$ = 13,700) or by performing the TBA assay ($\varepsilon$ = 153,000 at 535 nm). The $\varepsilon$ value for the malonaldehyde-TBA reaction product is the average of four slightly differing values reported in the literature.[21] The malonaldehyde concentrations obtained by the three different methods (weight, UV, TBA) agreed with each other within ±2%. For the HPLC calibration curve aliquots of the 10 m$M$ stock solution are appropriately diluted with 0.1 $M$ Tris buffer (pH 7.4) to give a final malonaldehyde concentration in the range of that present in the peroxidized microsomal suspension, preferably 20 $\mu M$. It is not necessary to establish a complete calibration curve daily, since an excellent linearity exists over a range of 1–100 nmol/ml independent of detector attenuation, flow rate, temperature, and so on (Fig. 2). The attenuation of the detector should be adjusted according to the malonaldehyde concentration and to the sensitivity of the detector used. The range usually employed is 0.04–0.32 absorbance units for full recorder scale.

*Kinetic Experiment*

In kinetic experiments, where malonaldehyde formation of microsomes should be measured by HPLC in short time intervals, it is possible to start the subsequent sample analysis before the malonaldehyde of the preceding injection is eluted; in such cases the next sample can be injected, when the first major peak (see Fig. 1) is completely eluted. Alter-

[21] H. Esterbauer, K. H. Cheeseman, M. U. Dianzani, G. Poli, and T. F. Slater, *Biochem. J.* **208**, 129 (1982).

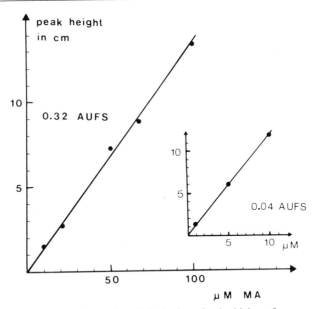

FIG. 2. Calibration curve for malonaldehyde detection by high-performance liquid chromatography.

natively, it is also possible to stop the lipid peroxidation process in aliquots of the suspension by addition of Desferal (9 ml suspension, 1 ml aqueous solution of Desferal 0.5 mg/ml) or by precipitation of the protein with an equal volume of acetonitrile (1 ml suspension, 1 ml acetonitrile). The Desferal-treated samples are kept on ice and analyzed by HPLC by injecting either the complete suspension or the clear supernatant obtained after centrifugation at 100,000 $g$. In samples treated with acetonitrile the protein precipitate is immediately removed by centrifugation at 3000 $g$ and the clear supernatant is used for HPLC. With microsomal suspensions, in which lipid peroxidation has been stimulated by ADP/Fe$^{2+}$, no significant difference in the kinetic of malonaldehyde formation was found between the different methods of sample processing.[19] This may be true, however, only for this particular system which obviously forms only free malonaldehyde and no significant amounts of other "malonaldehyde-like" substances. The close correlation between the different methods used on microsomal suspensions peroxidized for 60 min in the presence of ADP/Fe$^{2+}$ is shown in the table. It has been found in our experiments that neither Desferal addition nor protein precipitation leads to a complete arrest of the malonaldehyde formation (measured by TBA or HPLC)

ANALYSIS OF MALONALDEHYDE BY HPLC AND TBA IN
MICROSOMAL SUSPENSION[a]

| Method | Malonaldehyde (nmol/ml incubation mixture) |
|---|---|
| HPLC, complete mixture | 68.5 |
| HPLC, 100,000 $g$ supernatant after Desferal addition | 69.0 |
| HPLC, 3000 $g$ supernatant after acetonitrile addition | 70.0 |
| Thiobarbituric acid,[b] complete mixture | 67.0 |

[a] Microsomes were incubated in the presence of ADP/Fe$^{2+}$.
[b] The TBA assay was performed as described by T. F. Slater and B. C. Sawyer, *Biochem. J.* **123**, 805 (1971).

when used for suspensions in the early phase of peroxidation, where a high percentage of polyunsaturated fatty acids is still available. It is therefore advisable for all kinetic experiments, to keep the samples cold throughout the storage time and to perform the HPLC separations as soon as possible.

## Measurement of Malonaldehyde in Other Biological Materials

The method described above for peroxidizing liver microsomes can be used in principle for all other biological samples in which malonaldehyde production during a peroxidation process stimulated by Fe$^{2+}$, CCl$_4$, or other agents is under study. A simple sample pretreatment prior to HPLC separation may be necessary in some cases to avoid severe column contamination and disturbance of the separation profile through other eluting substances. We have performed a number of preliminary studies with samples other than liver microsomes in order to demonstrate the broad applicability of the method and make suggestions for investigators who wish to utilize the method for other systems.

### Microsomes and Mitochondria

Microsomes derived from rat kidney, brain, and lung as well as rat liver mitochondria incubated in the presence of an NADPH-regenerating system and ADP/Fe$^{2+}$, inorganic pyrophosphate/Fe$^{2+}$, or CCl$_4$ can be studied without any alteration of the basic procedure.

*Total Homogenates, Nuclei, Erythrocytes*

A repeated direct injection of such protein-rich suspensions into the HPLC would lead to an obstruction of the injection valve and to a severe contamination of the column head. Such samples therefore should be separated only after an appropriate sample clean up which removes most of the protein. Based on our experience, we recommend a precipitation with acetonitrile. The suspension is diluted with an equal volume of acetonitrile, mixed on a Vortex mixer for a short time, and then centrifuged at 3000 $g$ for 5 min. The clear supernatant is injected into the HPLC. This simple sample pretreatment was successfully used for demonstrating malonaldehyde formation by total homogenates of liver, kidney, lung, heart, skeleton muscle, brain, and spleen as well as liver nuclei, when incubated in the presence of ADP/$Fe^{2+}$ and an NADPH-regenerating system. Similarly erythrocyte suspensions, in which lipid peroxidation was induced by hydrogen peroxide,[22] can be studied for malonaldehyde production.

*Systems with Ascorbate—Iron*

Ascorbate-iron is frequently used to study the "nonenzymic" lipid peroxidation. Ascorbate, if present in concentrations of 100 $\mu M$ or higher, leads to an interference with the malonaldehyde peak in the HPLC separation since it exhibits a rather strong absorption at 270 nm and elutes from the column with a retention time close to that of malonaldehyde. This problem may be overcome by addition of an oxidant, such as hydrogen peroxide, to the sample prior to HPLC separation or by including hydrogen peroxide in the eluent. This leads to the conversion of ascorbate to dehydroascorbate, which does not disturb the HPLC detection of malonaldehyde. For suspensions of liver microsomes, in which lipid peroxidation was stimulated by 500 $\mu M$ ascorbate/20 $\mu M$ $Fe^{2+}$ a concentration of 0.1 m$M$ hydrogen peroxide in the eluent, i.e., 0.03 $M$ Tris pH 7.4–acetonitrile 9:1 (v/v) was sufficient to suppress any interference from ascorbate, without affecting the column performance or the malonaldehyde concentration.

*Autoxidized Free Fatty Acids*

Solutions of partly autoxidized free fatty acids or fatty acid derivatives should not be injected directly into the HPLC, because these lipophilic materials can lead to severe column contamination, which badly affects the quality of the malonaldehyde separation. The disturbing lipophilic materials can be removed readily, however, by filtering the sample prior

[22] T. G. Bidder and P. D. Jaeger, *Life Sci.* **30**, 1021 (1982).

to HPLC through a small ODS column (for example Waters RP18 Sep-pak cartridge); alternatively an ODS pre-column can be used. The method of using a mobile phase containing hydrogen peroxide was used by us for measuring free malonaldehyde contained in ascorbate/Fe autoxidized samples of linoleic, linolenic, and arachidonic acid.

*Systems Where HPLC Is Not Applicable*

Direct HPLC determination of malonaldehyde is not applicable to whole blood and EDTA-treated or heparinized serum, plasma, and urine. These samples give on HPLC separation with the aminophase column a number of peaks that elute near to or at the position of malonaldehyde. For the same reason the method is unsuitable for trichloroacetic acid or perchloric acid extracts of tissues. None of the above described clean-up procedures successfully removes from such samples all substances that interfere with a clear HPLC elution profile of malonaldehyde. In blood, serum, and plasma the interfering peaks approximately correspond to a 5–10 $\mu M$ malonaldehyde peak; taking this as background, we used direct HPLC also to study the reaction of malonaldehyde with human serum.[23]

Column Selection and Column Life Performance

HPLC columns suitable for direct determination of malonaldehyde are the 200 × 4.5 mm S5 Spherisorb amino column (Phase Separations Ltd.) and the 300 × 2.9 mm Carbohydrate Analysis Column (Waters Associates). The material of the latter column is also an aminophase. Other columns tested include Spherisorb ODS, Zorbax ODS, Lichrosorb RP-18, Spherisorb-Phenyl, Spherisorb-Nirile, and Lichrosorb-Diol. None of these columns gave a separation comparable to the amino phases. With the ODS columns, malonaldehyde was not retarded, while on the other columns malonaldehyde gave a broad peak. A slurry packed Nucleosil 10 Dimethylaminophase gave good separations; columns prefilled with this material are, however, not yet commercially available.

The Spherisorb-NH$_2$ column is delivered in hexane/1% acetonitrile. For solvent change it is washed with 15 ml of chloroform, 15 ml of methanol, 15 ml water, followed by the malonaldehyde eluent, acetonitrile/0.03 $M$ Tris, pH 7.4 (1 : 9, v/v). Washing with the latter eluent should be continued, until no drifting of the baseline is observed at 0.04 AUFS. The Carbohydrate Analysis column is shipped in methanol–water, 7 : 3. Prior to use, the column is washed with 300 ml water, followed by 100 ml of

[23] H. Esterbauer, *in* "Free Radicals, Lipid Peroxidation and Cancer" (D. C. H. McBrien and T. F. Slater, eds.), p. 101. Academic Press, New York, 1982.

acetonitrile/tris buffer, pH 7.4 (1:9, v/v). Both aminophase columns tested showed decreasing retention times for malonaldehyde during the first weeks of use. Malonaldehyde is generally eluted on new columns at a retention time of ~12–15 min. If the column is continuously used, the retention time decreases to about 6–8 min within 2 weeks. Afterward only a slight further decrease of the retention time occurs. The use of an aminophase or ODS precolumn did not eliminate this phenomenon, and it seems unlikely, therefore, that the problem is related to avoidable column contaminations. In our opinion the decrease of column performance is due to the fact that a part of the bound aminophase is hydrolyzed and washed out from the column on prolonged use. The rapid progress in designing new and more stable materials for HPLC columns will certainly also lead to more stable aminophases.

### Acknowledgments

We are grateful to the National Foundation for Cancer Research for financial support. This chapter is dedicated by the authors to Professor Dr. E. Schauenstein on the occasion of his 65th birthday.

# [39] Assay for Blood Plasma or Serum

*By* KUNIO YAGI

## Principle

Since the amounts of lipid peroxides in the blood are rather small, a sensitive assay method is needed. For this purpose, the most appropriate among several reactions for detecting lipid peroxides is the thiobarbituric acid (TBA) reaction, because of its sensitivity.[1,2] TBA reaction with lipid peroxides gives a red-colored pigment, the structure of which is shown in Fig. 1.[3] Malondialdehyde also gives the same product upon reaction with TBA. Since this product is fluorescent,[4] a sensitive assay can be made by fluorometry. The excitation and emission spectra of this red pigment are shown in Fig. 2. When tetraethoxypropane or tetramethoxypropane,

[1] F. Bernheim, M. L. C. Bernheim, and K. M. Wilbur, *J. Biol. Chem.* **174**, 257 (1948).
[2] K. M. Wilbur, F. Bernheim, and O. W. Shapiro, *Arch. Biochem.* **24**, 305 (1949).
[3] R. O. Sinnhuber, T. C. Yu, and T. C. Yu, *Food Res.* **23**, 626 (1958).
[4] K. Yagi, *Biochem. Med.* **15**, 212 (1976).

Copyright © 1984 by Academic Press, Inc.
All rights of reproduction in any form reserved.
ISBN 0-12-182005-X

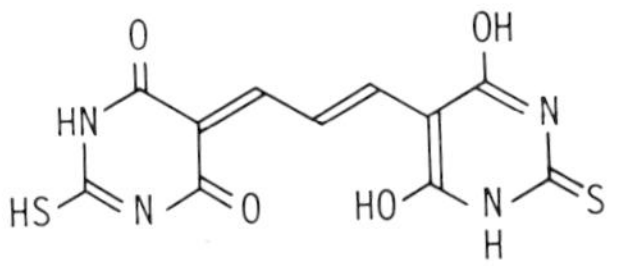

FIG. 1. Structure of the product from the reaction of lipid peroxides with TBA.

which converts quantitatively to malondialdehyde during the reaction procedure, is reacted with TBA, the fluorescence intensity of the reaction product runs parallel with its concentration. These results show that lipid peroxides can be measured by TBA reaction with fluorometry.

The maximum formation of the reaction product from TBA and hydroperoxide of linoleic, linolenic, or arachidonic acid is attained at around pH 3.5,[5] and the reaction rate becomes higher upon elevating temperature. When 95° is adopted for the reaction, 60 min is required for the maximum degree of reaction.[5]

Although it is clear that lipid peroxides can be measured fluorometrically using TBA reaction under the above conditions, elimination of TBA-reacting substances other than lipid peroxides is necessary for the measurement of lipid peroxides in serum. The elimination procedure must be simple to avoid artifact due to the peroxidation during the procedure. One of the best procedures is to isolate lipids by precipitating them along with serum protein with the phosphotungstic acid–sulfuric acid system. By this procedure, water-soluble substances, which react with TBA to yield the same product as lipid peroxides, are removed.[4]

Another substance that reacts with TBA is sialic acid. It was found that sialic acid cannot react with TBA in acetic acid solution,[6,7] though it reacts strongly with TBA in trichloroacetic acid solution. This indicates that trichloroacetic acid adopted by some researchers[8,9] is not suitable for samples which contain sialic acid.

Bilirubin was found to react with TBA, but the fluorescence of this product was different from the fluorescence of the reaction product of lipid peroxides with TBA. Accordingly, bilirubin does not disturb the analysis of lipid peroxides by TBA reaction, if the fluorescence intensity of the product is determined at 553 nm.[6,7]

It was anticipated that platelet aggregation, if it occurs during the

[5] H. Ohkawa, N. Ohishi, and K. Yagi, *J. Lipid Res.* **19**, 1053 (1978).
[6] H. Ohkawa, N. Ohishi, and K. Yagi, *Anal. Biochem.* **95**, 351 (1979).
[7] K. Yagi, *in* "Lipid Peroxides in Biology and Medicine" (K. Yagi, ed.), p. 223. Academic Press, New York, 1982.
[8] H. I. Kohn and M. Liversedge, *J. Pharmacol. Exp. Ther.* **82**, 292 (1944).
[9] T. F. Slater, *Biochem. J.* **106**, 155 (1968).

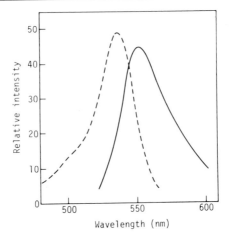

FIG. 2. Excitation and emission spectra of the product from the reaction of lipid peroxides with TBA. Dotted line: excitation spectrum (monitored at 565 nm); solid line: emission spectrum (excited at 515 nm).

drawing of the blood, would liberate the TBA-reacting substances, and the effect of the aggregation was found to be eliminated by the treatment with phosphotungstic acid–sulfuric acid system.[7]

### Reagents

0.9% NaCl aqueous solution

N/12 $H_2SO_4$

10% phosphotungstic acid aqueous solution

TBA reagent, a mixture of equal volumes of 0.67% TBA aqueous solution and glacial acetic acid

n-Butanol

### Procedure[10]

As a standard method for the microdetermination of the lipid peroxide level in blood plasma or serum, the following procedure is recommended.

1. Using a pipet for determination of blood cells, 0.05 ml of the blood is taken (e.g., from the ear lobe).

2. The blood is put into 1.0 ml of physiological saline in a centrifuge tube, and shaken gently.

[10] When only a small amount of the blood is available (e.g., small animals or infants), the lipid peroxide level in blood plasma is more easily measured. When 20 $\mu$l of serum is available, steps 4–10 can be adopted.

3. After centrifugation at 3000 rpm for 10 min, 0.5 ml of the supernatant is transferred to another centrifuge tube. In the case of the serum, 20 $\mu$l of the specimen is taken.

4. To this solution, 4.0 ml of N/12 $H_2SO_4$ is added and the mixture is shaken gently.

5. Then, 0.5 ml of 10% phosphotungstic acid is added and mixed. After standing at room temperature for 5 min, the mixture is centrifuged at 3000 rpm for 10 min.

6. The supernatant is discarded, and the sediment is mixed with 2.0 ml of N/12 $H_2SO_4$ and 0.3 ml of 10% phosphotungstic acid. The mixture is centrifuged at 3000 rpm for 10 min.

7. The sediment is suspended in 4.0 ml of distilled water, and 1.0 ml of TBA reagent is added. The reaction mixture is heated for 60 min at 95° in an oil bath.

8. After cooling with tap water, 5.0 ml of $n$-butanol is added and the mixture is shaken vigorously.

9. After centrifugation at 3000 rpm for 15 min, the $n$-butanol layer is taken for fluorometric measurement at 553 nm with 515 nm excitation.

10. Taking the fluorescence intensity of the standard solution, which is obtained by reacting 0.5 nmol of tetramethoxypropane with TBA by steps 7–9, as $F$ and that of the sample as $f$, the lipid peroxide level ($Lp$) can be expressed in terms of malondialdehyde:

$$\text{Plasma } Lp = 0.5 \times \frac{f}{F} \times \frac{1.05}{0.05} \times \frac{1.0}{0.5} = \frac{f}{F} \times 21 \qquad \text{(nmol/ml of blood)}$$

$$\text{Serum } Lp = 0.5 \times \frac{f}{F} \times \frac{1.0}{0.02} = \frac{f}{F} \times 25 \qquad \text{(nmol/ml of serum)}$$

# [40] Spectrophotometric Detection of Lipid Conjugated Dienes

*By* Richard O. Recknagel and Eric A. Glende, Jr.

*Principles of the Method*

Like many other substances, naturally occurring lipids exhibit simple end absorption in ultraviolet light as the wavelength is lowered toward 200 nm (Figs. 1 and 2). The spectra of a variety of organic molecules contain-

Copyright © 1984 by Academic Press, Inc.
All rights of reproduction in any form reserved.
ISBN 0-12-182005-X

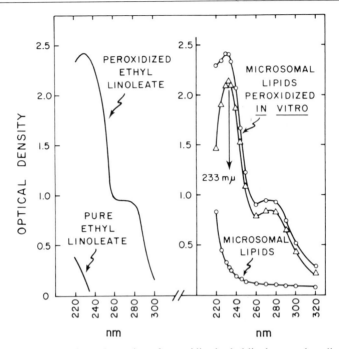

FIG. 1. Conjugated diene absorption of peroxidized ethyl linoleate and rat liver microsomal lipids peroxidized *in vitro*. The lower trace of the right panel is the end absorption of nonperoxidized microsomal lipids. The upper trace of the right panel is the absorption spectrum of peroxidized microsomal lipids. The middle trace is the difference spectrum. Left panel: from J. L. Bolland and H. P. Koch, *J. Chem. Soc.* 445 (1945). From R. O. Recknagel and A. K. Ghoshal, *Lab. Invest.* **15**, 132 (1966); reproduced with permission.

ing conjugated dienes, however, are characterized by intense absorption, the so-called K band, which may range, with respect to peak absorption, from 215 to 250 nm, depending on nearby substituent groups.[1,2] The spectra of peroxidized lipids are characterized by an intense K band near 233 nm, with a lesser secondary absorption maximum due to ketone dienes, in the region 260–280 nm. The appearance of conjugated dienes in peroxidized unsaturated fatty acids is due to resonance following free radical attack on hydrogens of methylene groups separating double bonds in these compounds.[3] Since the absorption at 233 nm due to conjugated dienes is superimposed on the end absorption of the nonperoxidized lipid,

[1] R. B. Woodward, *J. Am. Chem. Soc.* **63**, 1123 (1941).

[2] R. B. Woodward, *J. Am. Chem. Soc.* **64**, 72 (1942).

[3] R. O. Recknagel, E. A. Glende, Jr., R. L. Waller, and K. Lowrey, *in* "Toxicology of the Liver" (G. Plaa and W. R. Hewitt, eds.), p. 213. Raven, New York, 1982.

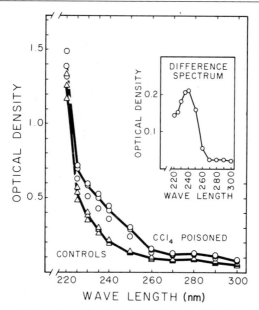

FIG. 2. Conjugated diene absorption in rat liver microsomal lipids 5 min after intragastric administration of CCl₄. The lower trace is end absorption of nonperoxidized microsomal lipids from normal untreated rats. From K. S. Rao and R. O. Recknagel, *Exp. Mol. Pathol.* **9**, 271 (1968); reproduced with permission.

one principle of the method involves determination of the difference between the 233 nm absorption of the peroxidized lipid and the 233 nm absorption of a corresponding sample of nonperoxidized lipid serving as control (Figs. 1 and 2).

Ultraviolet spectrophometric detection of conjugated dienes has been used for many years in the food industry for detection of autoxidized lipids.[4] The method appears to have been applied for the first time to the problem of liver cell lipid peroxidation of toxigenic origin in 1966[5] and has been widely used since. Increasingly, for a variety of pathological processes, the question has been raised whether peroxidative decomposition of membrane lipids has occurred *in vivo*. A second principle of the method recognizes that for whole-animal studies involving possible lipid peroxidation, the fraction of endogenous lipids actually peroxidized may not only be low, but the process of lipid peroxidation may be confined to a particular subcellular structure, e.g., the endoplasmic reticulum (ER) of

[4] R. T. Holman, *in* "Progress in the Chemistry of Fats and Other Lipids" (R. T. Holman, W. O. Lundberg, and T. Malkin, eds.), Vol. 2, p. 51. Pergamon, Oxford, 1954.
[5] R. O. Recknagel and A. K. Ghoshal, *Lab. Invest.* **15**, 132 (1966).

rat liver cells after $CCl_4$ intoxication. The end absorption of nonperoxidized lipids from cellular components unaffected by the pathological process under investigation may mask the absorption at 233 nm. It is for this reason that the starting point for detection of lipid conjugated dienes is homogenization of the tissue followed by differential centrifugation to isolate the major subcellular fractions. All aqueous phase procedures are conducted in the presence of ethylenediaminetetraacetate (EDTA) or other $Fe^{2+}$-chelating agents to avoid possible lipid peroxidation during homogenization and centrifugation procedures. Total lipids of a particular isolated subcellular fraction can be extracted conveniently with chloroform–methanol 2 : 1, according to Folch et al.[6] with modifications introduced by Bligh and Dyer.[7] Extracted lipids in chloroform are eventually dried down by removing the chloroform under a stream of oxygen-free nitrogen, or in suitable evaporation devices. Absolutely oxygen-free nitrogen can be obtained by passing commercial "oxygen-free" nitrogen over hot reduced copper mesh. The dried-down and chloroform-free lipid is dissolved in spectrophotometric grade cyclohexane for spectrophotometric analysis. The final concentration of lipid in cyclohexane can be determined according to Chiang et al.[8] All absorbance measurements are then corrected to a uniform base of 1 mg lipid per ml of cyclohexane.

According to this method, estimation of lipid peroxidation in vivo depends on determination of the difference spectrum at 233 nm for lipids from control animals and lipids from treated animals. Since the difference spectrum depends on two measured quantities, error in the difference spectrum will be greater than for each of the measured quantities alone. Because of this tendency for greater error in the difference spectrum, we routinely obtain a mean difference spectrum, with at least three animals in the control group and three animals in the experimental group. Finally, since the spectrophotometric readings are adjusted to a final concentration of 0.1 g% at the time of assay, the mean difference spectrum is multiplied by 10 to obtain the mean difference spectrum at a concentration of 1%. This yields the so-called mean delta $E_{1\,cm}^{1\%}$ value.[9]

The shift of the double bond which occurs as a result of radical attack on polyenoic fatty acids can take place under special circumstances in the absence of molecular $O_2$, and hence in the absence of peroxidation reactions per se. For example, conjugated dienes appear in the lipids of rat liver microsomes that have been incubated anaerobically in the presence

[6] J. Folch, M. Lees, and G. H. Sloane-Stanley, J. Biol. Chem. **226**, 497 (1957).
[7] E. G. Bligh and W. J. Dyer, Can. J. Biochem. Physiol. **37**, 911 (1959).
[8] S. P. Chiang, C. F. Gessert, and O. H. Lowry, in Research Report 56-113, Air University School of Aviation Medicine, p. 1, USAF, Texas, 1957.
[9] R. O. Recknagel and A. K. Ghoshal, Exp. Mol. Pathol. **5**, 413 (1966).

of NADPH and halothane[10] or NADPH and $CCl_4$.[10] Evidently, generation of radical products from halocarbon substrates can lead to the shift in the double bond. This initial part of the overall process does not require $O_2$. However, the rapid reaction of lipid organic radicals with $O_2$[11] will ensure that peroxidation reactions, in the strict sense of formation of peroxy functions, will occur in aerobic systems where polyenoic fatty acids are attacked by radicals. It is for this reason that the shift of the double bond, which does not in and of itself involve peroxide formation, is nevertheless a clear sign that lipid peroxidation has occurred, if molecular $O_2$ is present.

### Reagents

Chloroform
Methanol
Chloroform–methanol, 2 : 1, v/v. We use commercially available reagent grades of these solvents without further purification
Cyclohexane, spectrophotometric grade
$O_2$-free nitrogen. Commercial "oxygen-free" nitrogen is passed over hot, reduced copper mesh to remove the last traces of oxygen. The hot copper mesh is reduced with hydrogen gas. Alternatively, several brands of oxygen traps, originally designed for purifying carrier gases in gas–liquid chromatography, are available from chromatography supply houses. Some of these do not incorporate an indicator of exhaustion.

### Details of Procedure

Although the following procedure is written for detection of lipid conjugated dienes in the microsome fraction obtained from rat liver, it is equally applicable to any isolated subcellular component, tissue, etc.

1. The microsome fraction is obtained by centrifugation of liver homogenates prepared in media (isotonic sucrose, saline-phosphate buffer, etc.) containing at least 0.001 $M$ EDTA. Final sedimentation of the microsomes is carried out in centrifuge tubes of a type which resists methanol and thereby facilitates the next step. The final supernatant fraction and any fat clinging to the walls of the centrifuge tube are discarded. The sedimented microsomes (usually from 2 to 3 g of whole liver) are transferred quantitatively with 6–7 ml methanol to a suitable graduated container, e.g., a graduated, 40-ml heavy-walled, stoppered centrifuge tube (Corning No. 8144). The volume is adjusted to 7 ml with methanol.

[10] C. L. Wood, A. J. Gandolfi, and R. A. Van Dyke, *Drug Metab. Dispos.* **4**, 305 (1976).
[11] G. R. McMillan and J. G. Calvert, *Oxid. Combust. Rev.* **1**, 84 (1965).

2. Add 14 ml chloroform, stopper, and mix thoroughly. Allow the mixture to stand at room temperature for about 10 min with occasional mixing.

3. Centrifuge the mixture at approximately 260 $g$ for 10 min at room temperature to sediment the insoluble material. Decant the supernatant fraction into another 40-ml graduated, stoppered centrifuge tube. Adjust the final volume of the lipid extract to 30 ml by the addition of $CHCl_3 : CH_3OH$ (2 : 1, v : v). Alternatively, insoluble material in the original $CHCl_3 : CH_3OH$ extract can be removed by filtration, e.g., through a medium-porosity sintered glass funnel.

4. Add 10 ml $H_2O$. Mix gently by inversion only. Too vigorous mixing at this point results in emulsions which sometimes are difficult to break.

5. Centrifuge 10 min at approximately 260 $g$.

6. By means of aspiration using a water vacuum or house vacuum remove the upper $CH_3OH$–water phase and any fluffy material at the interface between the two phases. Note: At this state some workers place the tubes in an ice bath for 5 min, recentrifuge, and aspirate off any remaining $CH_3OH$–water droplets.

7. Transfer a 2-ml aliquot of the chloroform phase into a clean tube, place in a water bath at 40–50°, and remove the chloroform under a stream of oxygen-free nitrogen. All traces of chloroform must be removed. At this step, some workers dissolve the dried-down lipid in several milliliters of fresh chloroform, transfer to a clean tube, and dry down again.

8. Dissolve the extracted, chloroform-free lipid in 3 ml of cyclohexane and record optical density (1 cm light path) from 300 to 220 nm, against a cyclohexane blank.

9. After the optical density measurements remove 0.25-ml aliquots from each sample for analysis of total lipid content, according to Chiang et al.[8]

10. Correct all optical density measurements to a uniform base of 1 mg lipid per ml of cyclohexane.

*General note:* Once the microsome fraction is obtained, we usually carry out the procedure through step 8 without overnight delay. If necessary, the lipid in cyclohexane can be kept frozen for subsequent analysis.

## Detection of Lipid Conjugated Dienes with [$^{14}C$]Tetracyanoethylene

Quantitative determination of conjugated diene unsaturation can be carried out with [$^{14}C$]tetracyanoethylene in a Diels–Adler condensation.[12]

[12] R. L. Waller and R. O. Recknagel, *Lipids* **12**, 914 (1977).

This method was used for quantitative analysis of conjugated dienes in triglycerides, phospholipids, and peroxidized tissue lipids. The amount of $^{14}C$ found in the Diels–Adler adducts was shown to be a measure of conjugated diene content. A disadvantage of this method is that it is much more difficult and time consuming than spectrophotometric determination of conjugated dienes in suitably extracted lipids. Further, [$^{14}C$]tetra-cyanoethylene is not commercially available, and the method is no more sensitive than the spectrophotometric method. However, the data obtained with this method demonstrated rigorously that the increase in absorption in the 230 to 235 nm region observed in rat liver microsomal lipids after $CCl_4$ or $BrCCl_3$ administration is not some nonspecific artifact, but is actually due to appearance of conjugated dienes in the fatty acid side chains of the tissue lipids. The work with [$^{14}C$]tetracyanoethylene permitted drawing of the important conclusion that the fast and relatively simple method of UV spectrophotometric scanning of extracted lipids, as herein described, is a reliable method for detection of double-bond isomerization in the fatty acid side chains of membrane lipids peroxidized *in vivo*.

# [41] Fluorescent Damage Products of Lipid Peroxidation

*By* Cora J. Dillard and Al L. Tappel

Free radical chain reactions, which take place during lipid peroxidation,[1] lead to the formation of lipid hydroperoxides that decompose to many secondary products. Among the products is a three-carbon dialdehyde, malonaldehyde. Malonaldehyde readily reacts with free amino groups on biological compounds such as amino acids, proteins, amino phospholipids, nucleic acids, etc., to yield $N,N'$-disubstituted 1-amino-3-iminopropenes, which are fluorescent conjugated Schiff bases.[2] The chromorphic system responsible for the fluorescence is N—C=C—C=N. When ribonuclease reacts with peroxidizing polyunsaturated lipid and with malonaldehyde, it becomes inactivated and yellow fluorescent products are formed.[3] The spectral properties of these chromophoric systems are nearly identical with those in chloroform–methanol extracts of age

[1] E. Sawiki, T. W. Stanley, and H. Johnson, *Anal. Biochem.* **35**, 199 (1963).
[2] K. S. Chio and A. L. Tappel, *Biochemistry* **8**, 2821 (1969).
[3] K. S. Chio and A. L. Tappel, *Biochemistry* **8**, 2827 (1969).

Copyright © 1984 by Academic Press, Inc.
All rights of reproduction in any form reserved.
ISBN 0-12-182005-X

pigment (lipofuscin).[4,5] Fluorescent product formation also occurs in peroxidizing subcellular particles,[6,7] amino acid and phospholipid systems,[8,9] peroxidizing arachidonic acid and DNA,[10] antioxidant-deficient mouse[11] and rat tissues,[12] etc. A major portion of fluorescent molecular damage in biological tissues is found in the lipid-soluble phase of chloroform–methanol extracts. The simple method described here is for the extraction and measurement of fluorescent molecular damage products of lipid peroxidation from animal tissues. As recently reviewed,[13] extraction of fluorophores is probably the most accurate method for determining the total amount of lipofuscin in a specific tissue. This method was initially described by Fletcher et al.[14]

*Reagents.* The reagents required are spectral grade choloroform and methanol and reagent grade quinine sulfate and sulfuric acid.

*Steps of Procedure.* (1) Extract tissue with chloroform-methanol, 2 : 1. (2) Record fluorescence spectra. (3) Calculate relative fluorescence.

*Extraction Procedure.* The surfaces of freshly excised tissues are blotted on filter paper to remove excess moisture, and samples of approximately 0.2 g are weighed to the nearest milligram. The minced tissue sample is placed into a glass and Teflon tissue homogenizer, and chloroform–methanol,[3] 2 : 1 (v : v), in a volume-to-weight ratio of 20 : 1 is added. Tissues are homogenized for 1 min at approximately 1300 rpm in a 45° water bath using a three-eighths in., 1/4 hp variable-speed drill to drive the pestle. Fibrous tissues, such as muscle and heart, are homogenized for 1.5 min. After homogenization, an equal volume of water is added, and the contents are thoroughly mixed on a Vortex mixer and then transferred to a 12-ml centrifuge tube and centrifuged at about 3000 rpm for 1–2 min or until the phases are separated. The chloroform-rich layer is removed to a small tube with a disposable Pasteur pipet rinsed with methanol. Add 1 ml of the chloroform layer and 0.1 ml of methanol to a 1-$cm^2$ quartz cuvette and record the fluorescence spectra. A similar procedure is used to extract fluorescent products from *in vitro* peroxidizing systems.[5]

[4] B. Strehler, D. Mark, A. Mildvan, and M. Gee, *J. Gerontol.* **14**, 430 (1959).
[5] H. Shimasaki, O. S. Privett, and I. Hara, *J. Am. Oil Chem. Soc.* **54**, 119 (1977).
[6] C. J. Dillard and A. L. Tappel, *Lipids* **6**, 715 (1971).
[7] W. R. Bidlack and A. L. Tappel, *Lipids* **8**, 177 (1973).
[8] C. J. Dillard and A. L. Tappel, *Lipids* **8**, 183 (1973).
[9] W. R. Bidlack and A. L. Tappel, *Lipids* **8**, 203 (1973).
[10] U. Reiss and A. L. Tappel, *Lipids* **8**, 199 (1973).
[11] A. Tappel, B. Fletcher, and D. Deamer, *J. Gerontol.* **28**, 415 (1973).
[12] K. Reddy, B. Fletcher, A. Tappel, and A. Tappel, *J. Nutr.* **103**, 908 (1973).
[13] B. M. Zuckerman and M. A. Geist, *in* "Age Pigments" (R. S. Sohal, ed.), p. 283. Elsevier, Amsterdam, 1981.
[14] B. L. Fletcher, C. J. Dillard, and A. L. Tappel, *Anal. Biochem.* **52**, 1 (1973).

*Spectrophotofluorometric Analyses.* Any number of different commercially available spectrophotofluorometers and x-y recorders can be used to measure fluorescence or record fluorescence spectra. This section describes the use of an Aminco–Bowman spectrophotofluorometer fitted with a ratio photometer (American Instrument Co., Inc., Silver Spring, MD) to measure the excitation and emission spectra. Slit positions 3, 4, and 6 of the spectrophotofluorometer are set at 3, 1, and 3 mm. The excitation light source used is a Hanovia xenon lamp; the photomultiplier tube is an RCA 1P 21. The sensitivity vernier setting is usually 100, where maximum amplification of the signal is obtained. The sensitivity setting is adjusted as required by the fluorescence intensity of the sample.

*Calculation of Relative Fluorescence.* The standard used for fluorescence intensity and wavelength calibration is 1 $\mu$g of quinine sulfate/ml of 0.1 $N$ $H_2SO_4$. The relative fluorescence intensities of samples are always compared to and expressed in terms of the fluorescence intensity of the standard by multiplying each percentage fluorescence by the sensitivity setting used. For quantitative comparisons of replicate and linearity determinations, the fluorescence of the standard is recorded just before the spectra of the sample are recorded, and corrections are made for the variation in light intensity. Further calculations can be made to express the relative fluorescence on a per gram or per tissue basis; e.g., if the relative fluorescence of a 0.200-g sample is 100 then (0.200 g/100) × (1 g/x), and x = relative fluorescence of 1 g of tissue.

*Removal of Interfering Fluorescent Compounds from Lipid Extracts.* The chloroform–methanol extracts from membranes and tissues contain

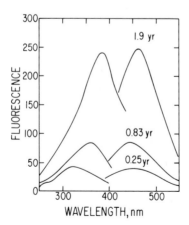

FIG. 1. Fluorescence spectra of mouse testes at three ages. Extraction of 0.2-g samples was done as described in the text. A quinine sulfate standard (1 $\mu$g/ml of 0.1 $N$ $H_2SO_4$) measured 1950 relative fluorescence units.

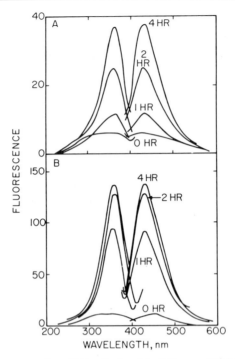

FIG. 2. Fluorescence spectra of (A) mitochondria (1.9 mg protein) and (B) microsomes (1.6 mg protein) peroxidized over a 4-hr time period. The 0.5-ml sample was extracted into 3 ml of chloroform–methanol, 2 : 1. The sensitivity setting was 0.03; 1 $\mu$g of quinine sulfate/ml 0.1 $N$ $H_2SO_4$ had a relative fluorescence intensity of 60 at a sensitivity setting of 0.3.

flavins as interfering fluorescent compounds. The flavins have 520-nm emission and 280-, 350-, and 450-nm excitation. The solvent system of 2 : 1 chloroform–methanol is polar enough to extract flavin compounds. These flavins are easily removed from the chloroform–methanol by the water-wash step of the extraction procedure. Retinol is a fat-soluble compound that has fluorescence excitation at 325–340 nm and emission in the 475 nm region. Retinol is not removed by the water wash. Retinol fluorescence can be removed from the chloroform-rich extract by exposing it to high-intensity ultraviolet light for 30 sec. Photocatalytic degradation of retinol in chloroform is very rapid. For this purpose, exposure to ultraviolet light from a Pen-Ray quartz lamp (Ultraviolet Products, Inc., San Gabriel, CA) when the sample is in a quartz cuvette will remove the fluorescence of retinol. Other conjugated polyene compounds should be similarly photooxidized by this procedure to leave the fluorescent product of peroxidation damage in solution for fluorescence measurement.

*Lipid Soluble Fluorophores from Tissues and Peroxidizing Systems.* Lipid-soluble fluorescent damage products with excitation maxima in the 340–370 nm region and emission maxima in the region 420–470 nm accumulate in animal tissues as a function of *in vivo* lipid peroxidation and of the aging processes. This is shown for extracts of mouse testes as a function of age in Fig. 1. This procedure has also been used to extract fluorescent damage products from human testes for further characterization.[15] Fluorescent damage products also form during peroxidation of biological tissues or subcellular fractions *in vitro*, as shown for mitochondria and microsomes in Fig. 2. This methodology has been applied in studies of aging in nematodes[16] and *Drosophila melanogaster*,[17,18] as a measure of *in vivo* lipid peroxidation in red blood cells of humans treated with the drug diaminodiphenylsulfone,[19] and in a number of other recently reviewed studies.[20]

[15] R. Trombly, A. L. Tappel, J. G. Coniglio, W. M. Grogan, Jr., and R. K. Rhamy, *Lipids* **10**, 591 (1975).
[16] M. R. Klass, *Mech. Ageing Dev.* **6**, 413 (1977).
[17] J. A. Sheldahl and A. L. Tappel, *Exp. Gerontol.* **9**, 33 (1974).
[18] J. Miquel, A. L. Tappel, C. J. Dillard, M. M. Herman, and K. G. Bensch, *J. Gerontol.* **29**, 622 (1974).
[19] B. D. Goldstein and E. M. McDonagh, *J. Clin. Invest.* **57**, 1302 (1976).
[20] R. S. Sohal, ed., "Age Pigments." Elsevier, Amsterdam, 1981.

## [42] Calibration of Microspectrophotometers as It Applies to the Detection of Lipofuscin and the Blue- and Yellow-Emitting Fluorophores *in Situ*[1]

*By* William S. Stark, Gregory V. Miller, and Kendall A. Itoku

The purpose of this communication is to emphasize the importance of careful calibration in spectrofluorometric work. We have studied lipofuscin,[2–5] the "wear and tear pigment"[6] of classical pathology, which accu-

[1] Supported by NIH Grant EY 03408. Equipment purchased in part on NSF Grant BNS 76-11921.
[2] G. E. Eldred, G. V. Miller, W. S. Stark, and L. Feeney-Burns, *Science* **216**, 757 (1982).
[3] W. S. Stark and K. E. W. P. Tan, *Photochem. Photobiol.* **36**, 371 (1982).
[4] G. V. Miller, "Fluorescence Studies of Substances in Visual Systems." M.A. Thesis, University of Missouri-Columbia, 1982.
[5] G. E. Eldred, W. S. Stark, G. V. Miller, and L. Feeney-Burns, *Invest. Ophthalmol. Visual Sci. Suppl.* **20**, 162 (1981).

Copyright © 1984 by Academic Press, Inc.
All rights of reproduction in any form reserved.
ISBN 0-12-182005-X

mulates with age. Metabolically active postmitotic cells of numerous tissues have the greatest accumulations of these heterogeneous lipophilic granules.[6,7] These are lysosomal residual bodies which are thought to contain indigestable remnants of damaged cellular membranes that were autophagized or phagocytized.[8,9]

Fluorescence microscopists identify lipofuscin granules by their yellow-orange autofluorescence[10,11] under near-UV excitation. A major contradiction to this observation exists however. Reports on the fluorescence properties of lipid extracts of lipfuscin-laden cells show peak emission in the blue (415–490 nm) region of the visible spectrum.[12–16]

## Methods

To determine whether the UV-excited fluorescence is yellow or blue we have obtained *in situ* corrected emission and excitation spectra using microspectrofluorometry. To obtain accurate excitation and emission spectra, it is necessary to calibrate and correct for (1) the spectra sensitivity of the photomultiplier, (2) the excitation intensities, (3) the spectral characteristics of excitation and emission "monochromators," and (4) the transmission properties of the light path through the microscope.

1. We calibrated the photomultiplier and the excitation intensities with a photodiode (EG&G, Inc., HUV 4000B). This device is sensitive, stable, convenient, and inexpensive. At present, we have calibrated our photodiode against six other devices. All calibrations averaged agree quite well with data we purchased from the company, differing in absolute sensitivity by at most 0.05 log units from 350 to 600 nm. Neutral density filters are often used in conjunction with the photodiode to be certain that light intensities are always within the diode's linear range. Neutral density

---

[6] P. S. Timiras, "Developmental Physiology and Aging," p. 429. Macmillan, New York, 1972.

[7] W. Reichel, *J. Gerontol.* **23**, 145 (1968).

[8] E. Essner and A. B. Novikoff, *J. Ultrastruct. Res.* **3**, 374 (1960).

[9] C. Biava and M. West, *Am. J. Pathol.* **47**, 287 (1965).

[10] S. Bommer, *Acta Derm. Venereol.* **10**, 391 (1929).

[11] E. A. Porta and W. S. Hartroft, *in* "Pigments in Pathology" (N. Wolman ed.), p. 191. Academic Press, New York, 1969.

[12] A. S. Csallany and K. L. Ayaz, *Lipids* **11**, 412 (1976).

[13] R. Trombly and A. Tappel, *Lipids* **10**, 441 (1975).

[14] A. N. Siakotos, I. Watanabe, A. Saito, and S. Fleischer, *Biochem. Med.* **4**, 361 (1970).

[15] B. L. Fletcher, C. J. Dillard, and A. L. Tappel, *Anal. Biochem.* **52**, 1 (1973).

[16] L. Feeney-Burns, E. R. Berman, and H. Rothman, *Am. J. Ophthalmol.* **90**, 783 (1980).

filters are rarely absolutely neutral, and nominal densities are rarely accurate. Thus, their transmission must be carefully determined in a spectrophotometer.

2. We calibrated excitation intensities with our photodiode, using one additional refinement. We focus the light onto a calibrated pinhole aperture to accurately determine area and to control for chromatic aberration. Excitation intensities are then accurately measured with the aperture placed at the focus of the sample.

3. To obtain monochromatic stimuli, we used interference filters placed in the excitation beam, or for emission spectra, in front of the photomultiplier. Alternatively, monochromators can be used, though calibration to a standard lamp can be technically and conceptually complicated. Interference filters were calibrated for spectral maxima and for the integrated area under the transmission curve using a Cary 210 spectrophotometer. The wavelength accuracy of the spectrophotometer was verified by calibrating holmium oxide which has precise sharp absorbance peaks. When a mercury (Hg) arc lamp is used for excitation, the sharp mercury lines must be compared with the filter transmission curves to know the exact excitation wavelengths.

4. The transmission through everything between the sample (microscope slide) and analyzer (monochromator or interference filters plus photomultiplier) must be determined. The nonneutrality of this transmission is dominated by the filter systems used in epiillumination. Light passes through a dichromatic beam splitter and an emission barrier filter. For emission spectra, we calibrated spectral lights just above the level of the microscope slide and just below the level of the analyzer. This accounted for any changes as light passed through the objective, the filter system, and the microphotometer (which delimits measuring area) with its aperture set at its experimental setting. When we determined excitation spectra, we used several different filter systems at different wavelengths and measured emission above 570 nm. In this situation, it is important to determine each filter system's relative transmission integrated through the entire emission wavelength range studied.

Spectrophotometry customarily involves factoring a blank from the sample. In our work on lipofuscin, we substracted the emission spectrum of the retinal pigment epithelium averaged from human donors under 40 years of age from the emission from older donors. This is because lipofuscin accumulates with age. For the excitation spectrum, factoring "background" (young) from "sample plus background" (old) is different. In this case, the excitation value at each wavelength is proportional to $(I_o^{-1} - I_y^{-1})$ where the $I$ values are intensities needed to obtain criterion photo-

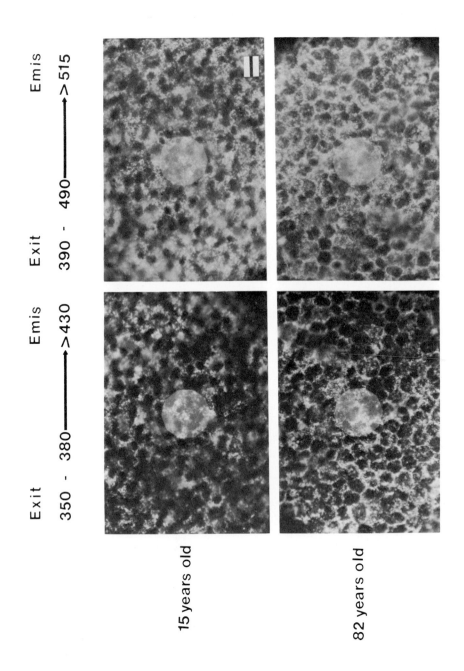

multiplier readings from old (o) and young (y) donor samples, respectively.[17]

In situ fluorescence measurements were taken from small sheets of retinal pigment epithelium dissected from human eyes.[18] These sheets were from human eyes fixed in a cacodylate-buffered mixture of paraformaldehyde and gluteraldehyde,[19] dried onto microscope slides, and then overlaid with a cover slip using very-low-fluorescence immersion oil (Cargille, type A). One report[20] suggested that fixation might alter our spectra. Thus, unfixed cells were also examined for comparison. While no difference in fluorescence was detected, the fixed cells did retain cellular morphology much better and were therefore used in these experiments.

Results and Discussion

The photographs of human RPE in Fig. 1 are representative of their appearance in our microspectrofluorometer. While lipofuscin granules are present in this young tissue it is important to note that it required longer exposure times for young tissue than old to obtain pictures of similar tonal quality.

The in situ excitation spectra, Fig. 2 (for emission >570 nm) of young and old tissues peak in the UV. These findings support our use of the 365 nm mercury line as the excitation stimulus for emission measurements. However, we should add that, if, as expected, "lipofuscin" is a heterogeneous collection of chromophores, different excitation wavelengths will stimulate different proportions of the many chromophores; thus, exciting wavelength is by definition, arbitrarily chosen. The in situ emission spectra of pooled measurements from young and old tissues (Fig. 2) are similar

[17] W. S. Stark, D. G. Stavenga, and B. Kruizinga, Nature (London) 270, 581 (1979).
[18] Eyes were from the Lions Eye Bank of Missouri. Drs. Graig Eldred and Lynette Feeney-Burns dissected the retinal pigment epithelium for this work.
[19] L. Feeney and S. Wissig, Tissue Cell 3, 9 (1971).
[20] J. S. Collins and T. H. Goldsmith, J. Histochem. Cytochem. 29, 411 (1981).

FIG. 1. Photomicrographs of human retinal pigment epithelium (bar = 20 $\mu$m). Tissues from old (82 years) and young (15 years) donors are shown as they appear with ultraviolet (UV) excitation (Exit 350–380 nm) and with the emission (Emis > 430 nm); also with blue excitation (Exit 390–490 nm) with the emission (Emis > 515 nm) as shown. Exposure times for the young were 5–6× exposure for the old to obtain these photographs of similar brightness (ASA 400 daylight Kodacolor). This black and white photograph shows where our aperture delimited ~six cells (center circle of light). Lipofuscin granules show up as small dots. The same photograph presented in color,[4] shows the broad-band yellow emission to UV excitation and the bright goldenrod yellow to blue excitation with different emission filtering.

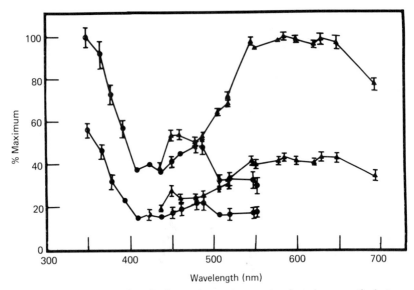

FIG. 2. *In situ* corrected excitation and emission spectra from human retinal pigment epithelium. Upper lines represent data from old tissue. Lower lines represent data from young tissue. Excitation measurements were limited to wavelengths >350 nm by the glass optics in the apparatus. The excitation spectra are for emission >570 nm. For excitation, "old" means 82.5 ± 5.8 SD years, $n$ = 15 experiments, "young" is 21.9 ± 8.7, $n$ = 15. The exciting stimulus was 365 nm for emission spectra. For emission, "old" is 83.5 ± 5.5, $n$ = 11, and "young" is 24.0 ± 9.3, $n$ = 11. The difference spectra for these data, presented elsewhere,[2-4] show the fluorescence properties of the substance accumulating with age, i.e., lipofuscin. Most notably lipofuscin has a broad band yellow (540–620 nm) emission maximum to UV excitation.

in shape, differing primarily in magnitude. This suggests that "background" from younger tissue is largely a smaller amount of the same fluorescence as in older tissue. In both cases a broad band peak exists with maximum emission occurring in the yellow-orange region (550–620 nm). The difference spectrum representing emission of age-related fluorophores[2-4] has similar properties. The major emission of lipofuscin is a broad band of yellow-orange peak with an additional inflection at about 460 nm.

We demonstrated that lipofuscin, the age pigment, fluoresces yellow when excited with ultraviolet *in vitro* as well as *in situ*.[2] The fact that there had been numerous reports that the emission was blue dramatizes the need for careful calibrations. The emission properties of lipofuscin have remained unclear for some time. An early report using other techniques had shown two emission peaks 440–460 and 530–560.[21] Strehler[22] reported lipofuscin emission as a broad band from 500 to 630 nm. Moreover, Hendly et al.[23] noted that just correcting for the spectral sensitivity of their phototube had the effect of broadening the peak and shifting the maximum 50–70 nm toward longer wavelengths. These, however, were interspersed by the numerous other reports[12-16] of a blue emission peak in tissue extracts.

The consequence of the reported blue emission from what was considered to be a lipofuscin extract has been to divert attention toward a number of blue-emitting fluorophores as possible contributors to *in situ* lipofuscin fluorescence, lipid peroxidation products which may cross-link with primary amines to form blue-emitting Schiff base molecules.[13-15]

[21] H. Hyden and B. Lindstrom, *Discuss. Faraday Soc.* **9**, 436 (1950).
[22] B. L. Strehler, *in* "Advances in Gerontological Research" (B. L. Strehler, ed.). Academic Press, New York, 1964.
[23] D. D. Hendley, A. S. Mildvan, M. C. Reporter, and B. L. Strehler, *J. Gerontol.* **18**, 144 (1963).

# [43] Detection of the Metabolism of Polycyclic Aromatic Hydrocarbon Derivatives to Ultimate Carcinogens during Lipid Peroxidation

*By* Thomas A. Dix and Lawrence J. Marnett

NADPH initiates the mixed-function oxidase-dependent metabolism of polycyclic hydrocarbons to carcinogenic derivatives by donating electrons to cytochrome *P*-450 via NADPH–cytochrome *P*-450 reductase. Certain dihydrodiol metabolites of polycyclic hydrocarbons are further oxidized to dihydrodiolepoxides that are extremely mutagenic and appear to represent the ultimate carcinogenic forms of the parent hydrocarbon. For example, 7,8-dihydroxy-7,8-dihydrobenzo[*a*]pyrene (BP-7,8-diol), a metabolite of the environmental pollutant benzo[*a*]pyrene, is converted by microsomal mixed-function oxidases to diastereomeric diol epoxides (Scheme 1). The (+)-*anti* enantiomer,[1] derived from (−)-BP-7,8-diol, is the most mutagenic and carcinogenic derivative of benzo[*a*]pyrene.

[1] The *anti*-diolepoxide has the oxirane ring on the opposite face of the tetrahydrobenzo ring from the 7-hydroxyl group, whereas the *syn*-diolepoxide has both groups on the same face.

METHODS IN ENZYMOLOGY, VOL. 105
Copyright © 1984 by Academic Press, Inc.
All rights of reproduction in any form reserved.
ISBN 0-12-182005-X

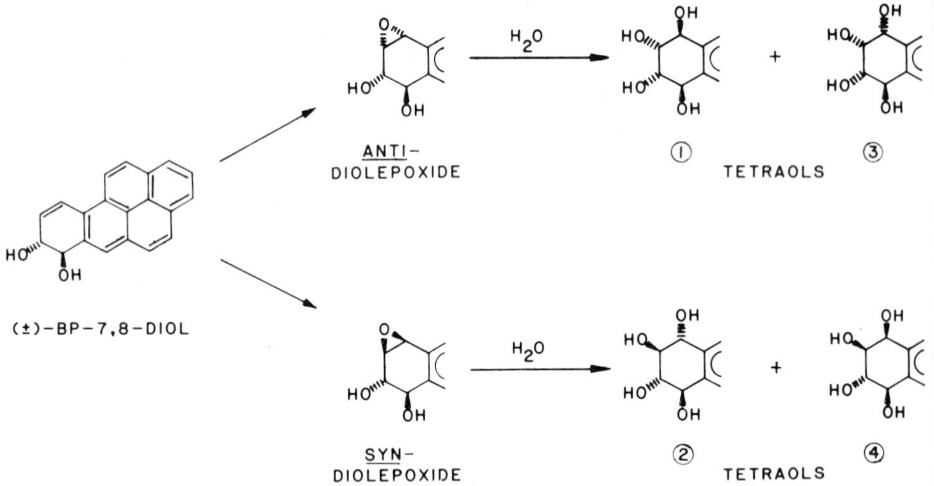

SCHEME 1. Products of BP-7,8-diol epoxidation.

NADPH also serves as the source of electrons for lipid peroxidation. This laboratory has recently reported that NADPH-dependent lipid peroxidation leads to the epoxidation of BP-7,8-diol.[2] We utilized the stereochemistry of oxygen insertion to distinguish between cytochrome $P$-450- and lipid peroxidation-dependent epoxidation. Scheme 1 illustrates that there are two enantiomeric pairs of diastereomeric diolepoxides formed from the racemic BP-7,8-diol. As seen in the table, mixed-function oxidase-dependent epoxidation of ($\pm$)-BP-7,8-diol by microsomes from uninduced rats results in approximately equal formation of *anti*- and *syn*-diolepoxides. In contrast, peroxidative epoxidation results in an excess of *anti*-diolepoxides. Other peroxide-dependent epoxidations of BP-7,8-diol yield predominantly the *anti*-diolepoxide.[3–5] The diolepoxides are unstable in aqueous media, so their formation is quantitated from the yields of their stable tetraol hydrolysis products, separable by reverse-phase high-performance liquid chromatography.

Resolved (+)-BP-7,8-diol is an especially useful probe for distinguishing mixed-function oxidase- and lipid peroxidation-dependent metabolism. Microsomal and purified cytochromes $P$-450 metabolize this enantiomer exclusively to the ($-$)-*syn*-diolepoxide,[6] whereas peroxidative

[2] T. A. Dix and L. J. Marnett, *Science* **221,** 77 (1983).
[3] T. A. Dix and L. J. Marnett, *J. Am. Chem. Soc.* **103,** 6744 (1981).
[4] L. J. Marnett, J. T. Johnson, and M. J. Bienkowski, *FEBS Lett.* **106,** 13 (1979).
[5] K. Sivarajah, H. Mukhtar, and T. Eling, *FEBS Lett.* **106,** 17 (1979).
[6] D. R. Thakker, H. Yagi, H. Akagi, M. Koreeda, A. Y. H. Lu, W. Levin, A. W. Wood, A. H. Conney, and D. M. Jerina, *Chem. Biol. Interact.* **16,** 281 (1977).

STEREOCHEMISTRY OF (±)-BP-7,8-DIOL EPOXIDATION

| System[a] | anti/syn ratio[b] |
|---|---|
| Mixed-function oxidase epoxidation | 1.1 ± 0.3 (1.1)[c] |
| Peroxidative epoxidation | 2.5 ± 0.4 |

[a] All data from uninduced rat liver microsomes.
[b] Data from Dix and Marnett.[2]
[c] Data from Thakker et al.[6]

epoxidation gives predominantly (−)-*anti*-diolepoxide.[2] The formation of the *anti*-diol epoxide from (+)-BP-7,8-diol is, therefore, diagnostic of peroxidative diol epoxidation.

## Materials and Methods

### Substrates

Racemic and (+)-BP-7,8-diol, as well as (±)-*anti*- and *syn*-diolepoxides are available from the National Cancer Institute Chemical Carcinogen Reference Standard Repository, NIH, Bethesda, MD. (±)-BP-7,8-diol can also be synthesized from 9,10-dihydrobenzo[a]pyren-7(8H)-one (Aldrich), as described by McCaustland et al.[7] The enantiomers of BP-7,8-diol can be resolved as their di(−)-menthoxyacetates, as described by Yang et al.[8] *Trans*- and *cis*-tetraol standards are prepared by solvolysis of the *anti*- and *syn*-diolepoxides, and characterized by UV spectroscopy and mass spectrometry of their tetraacetate derivatives, as described by Yagi et al.[9]

### Reagents

Tris–HCl buffer, 0.05 M pH 7.5
1.0 mM butylated hydroxyanisole in ethyl acetate
Rat liver microsomes, in 0.2 M KCl buffer prepared from Long–Evans rats as described by Schelin et al.[10]
NADPH (0.1 M), in buffer

[7] D. J. McCaustland, D. L. Fischer, K. C. Kolwyck, W. P. Duncan, J. C. Wiley, C. S. Menon, J. F. Engel, J. K. Selkirk, and P. P. Roller, in "Carcinogenesis" (R. I. Freudenthal and P. W. Jones, eds.), Vol. 1, p. 349. Raven, New York, 1976.

[8] S. K. Yang, H. V. Gelboin, J. D. Weber, V. Sankaran, D. L. Fischer, and J. F. Engel, *Anal. Biochem.* **78,** 520 (1977).

[9] H. Yagi, D. R. Thakker, O. Hernandez, M. Koreeda, and D. M. Jerina, *J. Am. Chem. Soc.* **99,** 1604 (1977).

[10] C. Schelin, A. Tunek, B. Jernström, and B. Jergil, *Mol. Pharmacol.* **18,** 529 (1980).

ADP (0.4 $M$)/$Fe^{3+}$ (1.5 m$M$), in buffer
EDTA (10 m$M$)/$Fe^{2+}$ (10 m$M$), in deoxygenated buffer
[$^{14}$C]BP-7,8-diol (1.2 m$M$, specific activity 5.7 mCi/mmol), in methanol
Thiobarbituric acid solution, prepared as described by Buege and Aust.[11]

*Incubation Procedure*

*Mixed-Function Oxidase-Dependent Epoxidation.* Microsomes (0.5 mg/ml) and 30 $\mu$l of the [$^{14}$C]BP-7,8-diol solution are added to the Tris buffer to give a final volume of 0.99 ml. Following a 3 min preincubation, reaction is initiated by the addition of 10 $\mu$l of the NADPH solution and allowed to continue for 20 min at 37°. Reaction is terminated by adding half of the incubation mixture to 1.0 ml of the thiobarbituric acid solution, while the remainder is extracted with 3 × 1.0 ml ethyl acetate containing butylated hydroxyanisole. The thiobarbituric acid mixture is vortexed, heated in a boiling water bath for 15 min, and analyzed for thiobarbituric acid-reactive material (indicative of lipid peroxidation), as described by Buege and Aust.[11] The protein residue that remains after the ethyl acetate extraction may be analyzed for protein-bound metabolites (indicative of reactive intermediates), as described by Tunek *et al.*[12]

*Lipid Peroxidation-Dependent Epoxidation.* Microsomes (0.5 mg/ml) and 30 $\mu$l of the [$^{14}$C]BP-7,8-diol solution are added to incubation buffer to give a final volume of 0.97 ml. After preincubation, 10 $\mu$l of the NADPH, ADP/$Fe^{3+}$, and EDTA/$Fe^{2+}$ solutions are added simultaneously to initiate the reaction. The termination and workup are as described above.

*HPLC Analysis*

HPLC separations are performed on a Varian Model 5060 instrument using a Dupont Zorbax reverse-phase column. A Varian Vari-Chrome detector is set at 344 nm (an absorbance maximum of the tetraol products) and used to monitor the column effluent. A Romac Instrument Flo-One HP radioactivity flow detector attached to the exit port of the variable-wavelength detector is used to monitor radioactivity.
The following gradient separation program is used:
Reservoir:   B = 55/45 water/methanol
             A = 100% methanol
Flow:        1.5 ml/min

[11] J. A. Buege and S. D. Aust, this series, Vol. 52, p. 302.
[12] A. Tunek, K. L. Platt, P. Bentley, and F. Oesch, *Mol. Pharmacol.* **14,** 920 (1978).

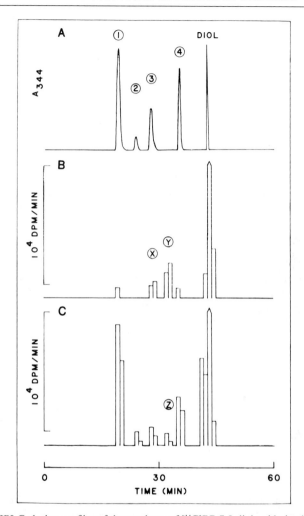

FIG. 1. HPLC elution profiles of the products of [¹⁴C]BP-7,8-diol oxidation by uninduced rat liver microsomes. (A) Absorbance profile of authentic tetraol and BP-7,8-diol standards, monitored at 344 nm. Numbered peaks are identified in Scheme 1. (B) Radioactivity profile of mixed-function oxidase-dependent metabolism of [¹⁴C]BP-7,8-diol. Peaks x and y are unknown metabolites that are not diolepoxide derived. (C) Radioactivity profile of lipid peroxidation-dependent metabolism of [¹⁴C]BP-7,8-diol. Peak z cochromatographs with an authentic standard of 7,8,9-trihydroxy-10-methoxybenzo[a]pyrene, which arises from the methanolysis of the *anti*-diolepoxide. Incubation and HPLC analysis conditions are given in the text.

Gradient:     0–20 min, 100% B isocratic
              20–40 min, 100–80% B
              40–50 min, 80–0% B
              50–60 min, 0% B isocratic

Unlabeled tetraol standards are added to the ethyl acetate extract, solvent removed, and the residue dissolved in a minimal volume of methanol for HPLC injection. Tetraol products are quantitated by the cochromatography of radiolabeled peaks with the unlabeled standards. Alternatively, unlabeled BP-7,8-diol can be used for the incubation and the tetraol formation estimated by the peak height ratios. Figure 1 shows the HPLC separation of the tetraol and BP-7,8-diol standards, as well as radioactivity profiles of mixed-function oxidase- and lipid peroxidation-dependent metabolism of (±)-BP-7,8-diol. Resolved (+)-BP-7,8-diol can be used to detect lipid peroxidation-dependent epoxidation by microsomes that exhibit a high rate of mixed-function oxidase-dependent metabolism. The formation of tetraols derived from the (−)-*anti*-diolepoxide is largely due to lipid peroxidation since this diolepoxide is not formed from (+)-BP-7,8-diol by mixed-function oxidases to an appreciable extent.

### Acknowledgment

This work was supported by the National Institutes of Health (GM 23642).

## [44] Detection of Picomole Levels of Lipid Hydroperoxides Using a Dichlorofluorescein Fluorescent Assay

*By* RICHARD CATHCART, ELIZABETH SCHWIERS, and BRUCE N. AMES

Numerous methods currently exist for the determination of lipid hydroperoxides. The thiobarbituric (TBA) method is based on the acid-catalyzed decomposition of the lipid hydroperoxide to malondialdehyde (MDA), which reacts with thiobarbituric acid to form a red chromogen.[1–5] Although as low as 100 pmol of malondialdehyde can be detected,[6] not all

[1] F. Bernheim, M. L. C. Bernheim, and K. M. Wilbur, *J. Biol. Chem.* **174**, 257 (1947).
[2] S. Patton and G. W. Kurtz, *J. Dairy Sci.* **34**, 669 (1951).
[3] I. K. Dahle, E. G. Hill, and R. T. Holman, *Anal. Biochem. Biophys.* **98**, 253 (1962).
[4] W. A. Pryor and J. P. Stanley, *J. Org. Chem.* **40**, 3615 (1975).
[5] W. A. Pryor, J. P. Stanley, and E. Blair, *Lipids* **11**, 370 (1976).
[6] M. Hicks and J. M. Bebicki, *Anal. Biochem.* **99**, 249 (1979).

Copyright © 1984 by Academic Press, Inc.
All rights of reproduction in any form reserved.
ISBN 0-12-182005-X

lipid hydroperoxides yield MDA in the acid reaction. As a consequence, many potentially significant lipid hydroperoxides such as cholesterol hydroperoxide are not detected with this method.

The iodine method, based on the oxidation of $I^-$ by the peroxide to the $I_3^-$ chromophore,[6] has an advantage over the TBA test in its specificity toward the functional ROOH group. Therefore, a wider range of organic hydroperoxides can be used. The major disadvantage in this method is the severe interference by dissolved oxygen and the lower inherent sensitivity to organic peroxides.

We have reported a fluorescent assay for lipid hydroperoxides which is highly sensitive and very simple to perform.[7] The assay is based on the original work of Keston and Brandt, in which dichlorofluorescin is oxidized to the fluorescent dichlorofluorescein (DCF) by hydrogen peroxide and peroxidase.[8,9] By substituting hematin for peroxidase, and optimizing the reaction conditions, we are able to detect as low as 25 pmol of peroxide. Furthermore, we obtain the same response from a wide variety of organic hydroperoxides. There are limitations to this method, but it is offered as an attractive alternative to the current methods.

*Assay Method*

*Reaction Conditions.* Of 25 m$M$ sodium phosphate buffer, pH 7.2, 100 ml was mixed with 14 ml of hematin solution (0.01 mg/ml in sodium phosphate buffer, pH 7.2) and boiled for 15 min. While purging with argon this solution was cooled on ice and 2 ml dichlorofluorescin was added. The dichlorofluorescin was prepared from 2′,7′-dichlorofluorescin diacetate (DCFDA) by mixing 0.5 ml 1 m$M$ DCFDA in ethanol with 2.0 ml 0.01 $N$ NaOH. This was allowed to stand at room temperature for 30 min, then neutralized with 10 ml 25 m$M$ sodium phosphate buffer, pH 7.2. This solution was stored on ice and discarded each day. Of the hematin dichlorofluorescin solution 2.9 ml was mixed with 200 $\mu$l of the hydroperoxide (in methanol, acetonitrile, or water) and incubated under an argon atmosphere in a Teflon sealed vial at 50° for 50 min. The reaction was cooled for 2 min in water (room temperature) and the fluorescence determined with a Kratos FS 950 fluorometer equipped with a 4-ml fluorescence cell. A 400- to 470-nm lamp was used for excitation, and the emission was measured using a 550-nm filter. The range on the FS 950 was set at 1, the sensitivity 0.6 kV, low suppression, and the output monitored with a strip chart recorder set to 10 mV full scale. The fluorescent yield in

[7] R. Cathcart, E. Schwiers, and B. N. Ames, *Anal. Biochem.*, submitted (1983).
[8] A. S. Keston and R. B. Brandt, *Anal. Biochem.* **11**, 1 (1965).
[9] M. J. Black and R. B. Brandt, *Anal. Biochem.* **58**, 246 (1974).

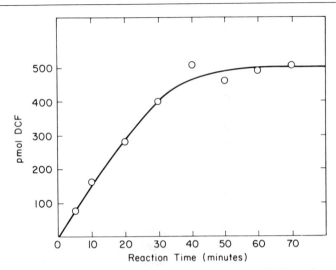

FIG. 1. Reaction time course. Each reaction contained 0.5 nmol $H_2O_2$, and was incubated at 50° for the indicated time. Blank reactions were run in parallel without $H_2O_2$ and subtracted from those reactions with $H_2O_2$.

millivolts was proportional to the amount of DCF in the reaction from 10 pmol to 2 nmol.

*Requirements.* Hematin was required for the reaction. Amounts of hematin greater than 1.2 $\mu g/ml$ only increased the blank. It has been observed by others that hematin has peroxidase activity.[10,11] The most likely mechanism of hematin catalysis is one in which hematin forms a hematin–peroxide complex, which degrades to a ferryl-oxo compound and hydroxyl radical, both of which are capable of oxidizing dichlorofluorescin.[12,13]

Although deaeration of the reaction was not absolutely necessary, this procedure did lower the blank by about 50% and increased both sensitivity and reproducibility. The reaction with peroxide was found to be linear for 50 min at 50°, and reached completion at 500 pmol DCF produced for 500 pmol $H_2O_2$ (Fig. 1). Longer reaction times only increased the blank. The assay volume of 3 ml is convenient for a standard fluorometer cell and

[10] B. N. Ames, R. Cathcart, E. Schwiers, and P. Hochstein, *Proc. Natl. Acad. Sci. U.S.A.* **78,** 6858 (1981).
[11] R. R. Howell and J. B. Wyngaarden, *J. Biol. Chem.* **235,** 3544 (1960).
[12] H. B. Dunford, T. Araiso, D. Job, J. Ricard, R. Rutter, L. P. Hager, R. Wever, W. M. Kast, R. Boelens, N. Ellfolk, and M. Ronnberg, eds., "The Biological Chemistry of Iron," p. 337. Reidel, New York, 1982.
[13] H. B. Dunford, *Adv. Inorg. Biochem.* **4,** 41 (1982).

keeping the solution anaerobic. The assay, however, has the potential of being adapted for postcolumn characterization for HPLC which might increase sensitivity considerably.

*Peroxide Specificity.* Several peroxides and lipid hydroperoxides representing a fairly wide range of structures were tested in the assay. As shown in Fig. 2, the assay is linear in the picomole range and each hydroperoxide generated a limit of approximately 1 mol of DCF per mole of peroxide. This was true for both the water-soluble and insoluble hydroperoxides. Although, as demonstrated, nonpolar lipid hydroperoxides such as linoleic and cholesterol hydroperoxide were active in the aqueous assay, methanol can be substituted as a solvent for the reaction. No reaction was observed with cholesterol, linoleic acid, cumenol or *t*-butanol up to 2000 pmol.

*Interference.* In the process of testing various commercial solvents as vehicles for organic hydroperoxides, it was found that many were unsuitable because they apparently contain various peroxides. For example, when freshly opened diethyl ether was used, no activity was seen above the blank. However, after standing at room temperature for 2 days, 10 μl was sufficient to cause a large positive response. Reagent grade acetone

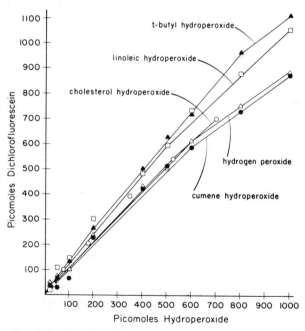

Fig. 2. DCF produced in response to various hydroperoxides.

TABLE I
CONCENTRATION OF ANTIOXIDANTS REQUIRED TO
SUPPRESS THE REACTION BY 50%[a]

| Antioxidant | C50% INH ($\mu M$) | Molar excess over dichlorofluorescin |
|---|---|---|
| Uric acid | 10 | 15.6 |
| Ascorbic acid | 6.5 | 10.1 |
| Vitamin E | 14.5 | 22.7 |

[a] Each reaction contained 500 pmol $t$-butyl hydroperoxide and 2000 pmol dichlorofluorescin (0.64 $\mu M$ final concentration). The concentration of each compound required to suppress the yield of DCF 50% was determined and is expressed as concentration for 50% inhibition (C50% INH).

also resulted in a large positive response, however glass-distilled acetone (Burdick and Jackson) gave only a very slight increase above the blank. Tetrahydrofuran, which is capable of peroxide formation, was also found to be positive. On the other hand, many solvents were found to quench the reaction severely. Ethanol, $n$-butyl alcohol, isopropanol, dimethyl sulfoxide, and chloroform were all found to decrease the yield of DCF in the assay. This is not unexpected, as many of these solvents react with hydroxyl radicals and iron-oxo complexes. Methanol was found to be the best organic solvent of those tried, and in fact the test can be run in 100% methanol (same time and temperature), but the yield of fluorescence is only 10% of that in water, presumably because of oxidant scavenging.

The biologically significant antioxidants urate, ascorbate, and vitamin E also interfered in the reaction. The concentration of each required to suppress the reaction by 50% is presented in Table I. Vitamin E, which would be present in crude lipid extracts, does interfere with the reaction. In preliminary experiments using plasma lipid extracts, we have found it necessary to use an initial lipid fractionation on silica to remove the vitamin E inhibitory activity.

*Reactivity of Other Peroxides.* Table II lists a number of both diacyl peroxides and dialkyl peroxides tested in the assay. The activity of each is expressed relative to that of hydrogen peroxide. The acyl peroxides are considerably more reactive than the alkyl peroxides, which exhibit virtually no reactivity in the test. The acyl peroxide reactivity is probably a consequence of the carboxyl carbon center which tends to destabilize the oxygen–oxygen bond. Two steroid endoperoxides showed no activity, as was also the case for prostaglandin $H_2$ (0%). The latter finding offers the

TABLE II

### TABLE II
#### REACTIVITY OF VARIOUS PEROXIDES RELATIVE TO HYDROGEN PEROXIDE[a]

| Peroxide | | Percentage |
|---|---|---|
| t-Butyl hydroperoxide | | 100 |
| Linoleic hydroperoxide (9- and 13-) | | 100 |
| Cholesterol 5-α-hydroperoxide | | 100 |
| Cumene hydroperoxide | | 100 |
| t-Butyl peroxymaleic acid | | 71 |
| Benzoyl peroxide | | 66 |
| Lauroyl peroxide | | 43 |
| trans-1-Hydroperoxy-5-phenyl-4-pentene | $Ph-CH=CH-(CH_2)_3-OOH$ | 24 |
| Di-t-butyl peroxide | | 0 |
| Di-cumyl peroxide | | 0 |
| 9(11),22-Cholestadien-24β-methyl-5,8-peroxy-3β-ol-acetate | | 0 |
| 6,9(11),22-Cholestatrien-24β-methyl-5,8-peroxy-3β-ol-acetate | | 0 |

[a] The values were determined using 500 pmol of peroxide and standard assay conditions.

possibility of quantifying the hydroperoxide of prostaglandin $G_2$ without interference from the endoperoxide group.

The primary peroxide 1-hydroperoxy-5-phenyl-4-pentene, demonstrates low reactivity, and may be fairly typical of other primary peroxides in this assay. In this case, there appears to be a competing side reaction, in which the hematin, acting like a Lewis acid, reacts with the primary peroxide group to form an aldehyde (Paul Weller and L. J. Marnett, unpublished results).

In summary, the classes of peroxides which this assay either does not detect, or detects only poorly, have been defined. However, peroxides of biological significance, such as $H_2O_2$ and the secondary and tertiary peroxides such as lipid hydroperoxides are detectable at picomole levels.

### Acknowledgments

This work was supported by Department of Energy Contract DE-AT03-76EV70156 to B.N.A. and by National Institute of Environmental Health Sciences Center Grant ES01896. We thank L. J. Marnett for many helpful discussions.

## [45] Measurement of $O_2^-$ Production by Human Neutrophils. The Preparation and Assay of NADPH Oxidase-Containing Particles from Human Neutrophils

By MICHÈLE MARKERT, PATRICIA C. ANDREWS, and BERNARD M. BABIOR

Neutrophils are phagocytic cells whose principal function is the destruction of invading bacteria. Bacterial killing is largely accomplished by means of an oxygen-dependent microbicidal system that employs as antimicrobial agents a group of lethal oxidants which the neutrophils generate when stimulated by bacterial targets.[1] The precursor of these lethal oxidants is $O_2^-$. This substance is manufactured from oxygen by stimulated neutrophils and is subsequently converted into highly reactive lethal oxidants such as $OH\cdot$ and $HOCl$ by a complicated series of secondary reactions. $O_2^-$ is therefore the key intermediate in the generation of microbicidal oxidants by neutrophils.

[1] B. M. Babior and C. A. Crowley, in "Metabolic Basis of Inherited Disease" (J. B. Stanbury, J. B. Wyngaarden, D. S. Fredrickson, J. L. Goldstein, and M. S. Brown, eds.), p. 1956. McGraw-Hill, New York, 1983.

Copyright © 1984 by Academic Press, Inc.
All rights of reproduction in any form reserved.
ISBN 0-12-182005-X

$O_2^-$ is produced by neutrophils through the action of an NADPH oxidase.[2] This oxidase, which is dormant in resting neutrophils but is activated when the cells are exposed to bacterial targets (or other suitable stimuli), is a membrane-bound flavoprotein that catalyzes the following reaction:

$$NAD(P)H + 2O_2 \rightarrow NAD(P)^+ + H^+ + 2O_2^-$$

Either reduced pyridine nucleotide can be used as substrate, but the enzyme prefers NADPH over NADH by a factor of 10–30, as judged by the Michaelis constants ($K_m$ values for NADPH and NADH are around 30 $\mu M$ and 0.5 m$M$, respectively).

The oxidase can be assayed by measuring $O_2^-$ production by neutrophil preparations. The measurement can be made using either intact neutrophils or broken cell preparations.

Isolation of Human Neutrophils

The purification of neutrophils from freshly drawn whole blood involves three basic steps: dextran sedimentation, lysis of contaminating red blood cells, and Ficoll–Hypaque density gradient centrifugation to separate neutrophils from platelets and mononuclear cells.

*Procedure*

*Dextran Sedimentation.* Into a 50-ml plastic syringe is placed 6 ml special acid–citrate–dextrose (ACD) anticoagulant (0.20 $M$ sodium citrate/0.14 $M$ citric acid/0.22 $M$ glucose). Thirty milliliters of blood is then drawn into the syringe, followed by 15 ml of 6% Dextran 70 in 0.154 $M$ (0.9%) saline (Macrodex from Pharmacia, or an equivalent preparation). After mixing its contents, the syringe is taped to the bench by its plunger and allowed to stand vertically at room temperature for an hour. By the end of this period, most of the red cells will have settled to the bottom half of the syringe, leaving a turbid straw-colored upper layer which contains most of the white cells. This layer is expressed from the syringe into a 50-ml conical centrifuge tube through a length of plastic tubing fitted to the nipple of the syringe (Pharmaseal K50L extension tube, or equivalent). The white cells are then pelleted by centrifugation at 4° for 12 min at 100 $g$.

Dextran sedimentation may also be carried out in a graduated cylinder, provided it is made of plastic or is siliconized before use.

*Hypotonic Lysis of Red Cells.* To lyse the residual contaminating red

---

[2] T. G. Gabig and B. M. Babior, *J. Biol. Chem.* **254**, 9070 (1979).

cells, the white cell pellet is carefully but thoroughly suspended in 3 ml of ice-cold distilled water by gentle agitation on a Vortex mixer. After exactly 30 sec, tonicity is restored by the addition of 1 ml of ice-cold 0.6 $M$ KCl. Cell suspensions from multiple tubes may be pooled at this time. The white cells are pelleted by centrifugation at 4° for 4 min at 160 $g$, and hypotonic lysis is repeated, still using 3 ml of water and 1 ml KCl regardless of whether or not the cells have been pooled. The white cells are again pelleted by centrifugation, then suspended at the desired concentration in ice-cold Dulbecco's phosphate-buffered saline (PBS) or Hanks' balanced salt solution (HBSS). The preparation at this point generally contains about 90% neutrophils. The remainder of the cells consist of monocytes, lymphocytes, and eosinophils.

For some experimental applications, neutrophils of this degree of purity are satisfactory. For other applications, the cells must be further purified by Ficoll–Hypaque density gradient centrifugation.

*Ficoll–Hypaque Density Gradient Centrifugation.* For Ficoll–Hypaque density gradient centrifugation, the cells should be suspended in 20 ml of *Ca- and Mg-free* PBS at 0–4°. The suspension is carefully layered onto 10 ml of Ficoll–Hypaque (Pharmacia) in a 50-ml conical centrifuge tube and spun at 250 $g$ for 20 min at 4°. The granulocytes (neutrophils and eosinophils) are denser than the Ficoll–Hypaque layer, and sediment to the bottom of the tube, leaving the monocytes and lymphocytes at the interface between the two fluid layers. The supernatant is drawn off, taking care to avoid mixing the mononuclear cells with the Ficoll–Hypaque layer. The neutrophil pellet is then washed with complete PBS or HBSS and finally suspended in the same buffer at the desired concentration.

The preparation at this point contains >99% granulocytes (90–95% neutrophils, and the remainder eosinophils). There should be fewer than 1 platelet per 1000 white cells. The yield from 100 ml blood is about $10^8$ cells.

## Remarks

Neutrophils are fragile cells, and should be handled gently. Once begun, their isolation should be carried through to completion without interruption. Following the centrifugation steps, the cell pellets should be resuspended as soon as possible by gentle agitation on the Vortex mixer. Plasticware or siliconized glassware should be used throughout. Glassware and plasticware should be scrupulously clean; by far the most common cause of clumped, useless neutrophil preparations is dirty glassware.

Viability can be checked by mixing the neutrophil suspension with an equal volume of 0.5% Trypan blue in 0.154 $M$ (0.9%) saline; dye should not enter more than 5% of the cells.

## Methods for Measuring $O_2^-$ Production by Intact Neutrophils

In these assays, the output of $O_2^-$ by neutrophils which have been stimulated with an activating agent is determined by measuring superoxide dismutase-inhibitable cytochrome $c$ reduction. Two methods are described: a discontinuous method, in which $O_2^-$ production is measured during a fixed time interval, and a continuous method, in which $O_2^-$ production is monitored constantly over time.

### Activating Agents

Neutrophils can be stimulated to manufacture $O_2^-$ by a large number of activating agents, including among others C5a (a complement component), A23187 (an ionophore), $N$-formyl-Met-Leu-Phe (a chemotactic peptide), digitonin, and serum-treated bacteria.[1] The agents most frequently used in assays of $O_2^-$ production, however, are opsonized zymosan and phorbol myristate acetate.

*Opsonized Zymosan.* Zymosan is a yeast cell wall preparation which is readily opsonized (coated by complement component C3b) by incubation in serum. To prepare zymosan for opsonization, boil 45 mg of commercial zymosan for 10 min in 10 ml 1 $M$ NaOH, then isolate the alkali-treated particles by centrifugation (250 $g$ for 10 min) and wash three times in buffer (PBS or HBSS). Opsonize by suspending the washed particles in 5–10 ml fresh autologous serum (i.e., serum obtained from the same person who donated the neutrophils) and incubating them for 30 min at 37°. The time of incubation is fairly critical. Wash the now opsonized zymosan twice with buffer, then suspend in buffer at the desired concentration. Opsonized zymosan may be stored frozen if desired, but for best results it should be prepared fresh each day.

*Phorbol Myristate Acetate.* Phorbol myristate acetate is stored as a stock solution in dimethyl sulfoxide (2 mg/ml) at −20° or below under desiccation. The agent is stable indefinitely when stored in this manner. For use, the stock solution is diluted in buffer to the desired concentration.

Phorbol myristate acetate is an irritant and a cocarcinogen, so care should be taken to avoid skin contact with this agent.

*The Assays*

*Reagents.* All reagents are dissolved or suspended in buffer except superoxide dismutase, which is dissolved in water; see the table.

*Discontinuous (Fixed Time) Assay.* This assay is most useful for experiments in which $O_2^-$ production by neutrophils must be measured in a large number of samples.

Place 0.7 ml of neutrophil suspension in each of two siliconized or plastic test tubes. Add 10 $\mu$l superoxide dismutase to one of the test tubes, and 10 $\mu$l of water to the other. Incubate at 37° for 2 min, then add to each test tube 0.05 ml cytochrome *c* followed directly by 0.75 ml prewarmed (37°) opsonized zymosan or phorbol myristate acetate. Incubate at 37° for the desired time (10–15 min is generally satisfactory); agitate occasionally, or carry out the incubations in a shaking water bath. Then stop the reactions by placing the tubes in melting ice, and remove cells by centrifuging at 1500 *g* for 5 min at 4°. Finally, measure cytochrome *c* reduction in a double-beam spectrophotometer by scanning between 530 and 570 nm, using the dismutase-free supernatant as the sample and the dismutase-containing supernatant as the reference. The height of the peak at 550 nm represents the absorbance due to superoxide-dependent cytochrome *c* reduction ($A_{superoxide}$). The amount of $O_2^-$ generated in 1 ml of the reaction mixture can then be calculated from the formula:

$$O_2^- \text{ (nmol)} = 47.7 \times A_{superoxide}$$

*Continuous Assay.*[3] This assay is especially suited for measuring the rate of $O_2^-$ production early in the course of the reaction.

Into each of two 1-ml cuvettes place 50 $\mu$l of neutrophil suspension, 50 $\mu$l of cytochrome *c* and 0.85 ml buffer. Add 10 $\mu$l superoxide dismutase to the reference cuvette and 10 $\mu$l water to the sample cuvette. Place the cuvettes in the thermostatted cell compartment of a double-beam spectrophotometer and allow them to come to 37°. After they have reached this temperature, add 50 $\mu$l of activating agent, mix quickly, and follow the reduction of cytochrome *c* at 550 nm.

*Remarks*

In carrying out these assays, attention must be paid to two points. First, the cytochrome *c* concentration must be high enough to trap the $O_2^-$ before it has a chance to dismute to oxygen and $H_2O_2$. This is a particular problem at pH values below neutrality, because the rate of

[3] H. J. Cohen and M. E. Chovaniec, *J. Clin. Invest.* **61,** 1081 (1978).

REAGENTS USED IN MEASURING $O_2^-$ PRODUCTION BY DISCONTINUOUS AND
CONTINUOUS ASSAYS

| | Concentration | |
|---|---|---|
| Reagent | Discontinuous assay | Continuous assay |
| Neutrophil suspension | $1–5 \times 10^6$ cells/ml | $10–50 \times 10^6$ cells/ml |
| Horse heart cytochrome $c$, superoxide dismutase-free (Sigma type III or VI) | 30 mg/ml | 12.5 mg/ml |
| Superoxide dismutase (bovine erythrocyte) | 3 mg/ml | 1 mg/ml |
| Opsonized zymosan | 8 mg/ml | 15 mg/ml |
| or | | |
| Phorbol myristate acetate | 2 $\mu$g/ml | 20 $\mu$g/ml |

spontaneous dismutation increases as the pH drops, reaching a maximum around pH 4.8.[4] Trapping problems can be ruled out if necessary by showing that the amount of cytochrome $c$ reduced in the assay does not change when the concentration of cytochrome $c$ in the assay mixture is doubled. Second, there is the possibility that some of the reduced cytochrome $c$ could be reoxidized by the $H_2O_2$ that accumulates in the reaction mixture over the course of time, leading to falsely low values for $O_2^-$ production by neutrophils. Reoxidation generally causes no difficulty, but if it should become a problem it can be prevented by adding catalase (10 units/ml) to the assay mixture.

Typical values for $O_2^-$ production by human neutrophils range between 1 and 3 nmol/min/$10^6$ cells.

### The Particulate NADPH Oxidase from Human Neutrophils

The oxidase is obtained from zymosan-activated neutrophils by a modification of the method of Hohn and Lehrer.[5] $O_2^-$ production by this enzyme is assayed as superoxide dismutase-inhibitable cytochrome $c$ reduction using NADPH as electron donor.

### Preparation of the Oxidase

Opsonized zymosan is suspended in complete HBSS containing 2 m$M$ NaN$_3$ at a concentration of 22.5 mg/ml. Two ml of zymosan suspension prewarmed to 37° is mixed with 2 ml of a prewarmed suspension of neu-

[4] B. H. J. Bielski, *Photochem. Photobiol.* **28,** 645 (1978).
[5] D. C. Hohn and R. I. Lehrer, *J. Clin. Invest.* **55,** 707 (1975).

trophils in Ca- and Mg-free HBSS (0.5–1.5 × 10$^8$ cells/ml). The mixture is incubated for 7 min at 37° with gentle shaking. The incubation is stopped with 4 ml ice-cold HBSS, and the cells are pelleted by centrifugation at 160 g for 5 min at 4°. The pellet is resuspended in 6 ml of ice-cold 0.34 M sucrose containing 0.5 mM phenylmethyl sulfonyl fluoride, and the cells are disrupted at 0° by homogenization (Potter–Elvehjem homogenizer; homogenize to 90% cell rupture) or sonication (Heat Systems sonifier *fitted with a cup horn;* two 20-sec bursts at full power with a 1-min interval between them). After disrupting the cells, the preparation is centrifuged at 160 g for 5 min at 4° to remove zymosan, nuclei and unbroken cells. The supernatant is then centrifuged at 27,000 g for 30 min at 4°. The pellet, which contains the $O_2^-$-forming activity, is resuspended in sucrose at 1–2 mg protein/ml.

The $O_2^-$-forming activity is very sensitive to temperature, so the particle suspension should always be kept on ice while experiments are being performed. The preparation may be stored for months at −70° without loss of activity; to prevent inactivation at the time of assay, stored preparations should be thawed in the cold (in ice or in a cold room).

*Assay of the Oxidase*

*Reagents*

Flavin adenine dinucleotide (FAD), 0.2 mM in water; store in the dark

NADPH, 1.0 mM in water; prepare fresh daily

Triton X-100, 0.2% (w/v) in water

Horse heart cytochrome *c*, superoxide dismutase-free (Sigma type III or VI), 10 mg/ml in water

Superoxide dismutase, bovine erythrocyte, 10 mg/ml in water

*The Assay.* To a tube containing 1.0 ml of HBSS or 0.1 M potassium phosphate buffer pH 7.4 is added 0.2 ml FAD, 0.2 ml NADPH, 0.2 ml Triton X-100 solution, and 0.2 ml cytochrome *c*. Let this mixture come to room temperature, then add to it 0.2 ml of $O_2^-$-forming particles. Immediately transfer 1 ml of the reaction mixture to a tube containing 10 μl superoxide dismutase. Then place both tubes in the dark, and incubate them for the desired time (5–10 min is usually satisfactory). Stop the incubation by adding 10 μl superoxide dismutase to the dismutase-free tube (the tube in which the reaction mixture was originally prepared). Finally, measure cytochrome *c* reduction spectrophotometrically as described above, using the portion to which superoxide dismutase was added before incubation as the reference and the other portion as the sample. The amount of $O_2^-$ produced during the incubation can be calculated from the height of the 550 nm peak ($A_{superoxide}$) by using the equation given above.

$O_2^-$-forming activity in a good preparation of NADPH oxidase should exceed 20 nmol/min/mg protein.

*Remarks*

The enzyme is assayed at room temperature instead of 37° because of its susceptibility to inactivation at high temperatures, a property discussed above. It is this property that accounts for the rapid decline in the reaction rate over time that occurs during the course of the assay even at room temperature. The enzyme is also sensitive to salts, and for this reason it is isolated in sucrose; its salt sensitivity, however, is not so great as to interfere with the assay, even though the assay is carried out in a fairly concentrated salt solution. Hard sonication also seems to inactivate the enzyme, so gentle methods employing a cup sonicator or a Potter–Elvehjem homogenizer are used to disrupt the cells.

Some workers have obtained the NADPH oxidase from neutrophils activated with phorbol myristate acetate, but in our view opsonized zymosan is a more reliable activating agent for this purpose. $NaN_3$ is used to prevent oxidative destruction of the enzyme during the neutrophil activation step[6]; Triton X-100 is used in the assay because about 60% of the oxidase activity is latent and is only expressed when the plasma membrane vesicles which contain the enzyme are disrupted with detergent.[7]

---

[6] R. C. Jandl, J. André-Schwartz, L. Borges-DuBois, R. S. Kipnes, B. J. McMurrich, and B. M. Babior, *J. Clin. Invest.* **61**, 1176 (1978).

[7] G. L. Babior, R. E. Rosin, B. J. McMurrich, W. A. Peters, and B. M. Babior, *J. Clin. Invest.* **67**, 1724 (1981).

# [46] Measurement of $O_2^-$ Secreted by Monocytes and Macrophages[1]

*By* RICHARD B. JOHNSTON, JR.

Superoxide anion ($O_2^-$) produced during phagocytosis or surface perturbation of phagocytes is largely released to the outside of the cell. There it can be detected by its ability to reduce chemically an electron-accepting compound. Ferricytochrome $c$ has been particularly useful in this regard.

[1] This work was supported in part by United States Public Health Service Grant AI 14148.

Copyright © 1984 by Academic Press, Inc.
All rights of reproduction in any form reserved.
ISBN 0-12-182005-X

Its reduction by $O_2^-$ provided the basis for the first assay of superoxide dismutase (SOD) activity,[2] and modifications of the same assay have been used to detect generation of $O_2^-$ by neutrophils,[3] monocytes fresh from the blood[4] or in culture,[5] or macrophages.[5] The reduction of ferricytochrome $c$ is not, of course, an assay specific for $O_2^-$. The required specificity is achieved by the use of SOD, for which $O_2^-$ is the only known substrate. Accordingly, the assay with cells is run with and without SOD, and only SOD-inhibitable reduction of cytochrome $c$ is used to calculate the amount of $O_2^-$ released.

## Reagents

Hanks' balanced salt solution (HBSS) without phenol red (Grand Island Biological Co. (GIBCO), Grand Island, New York, or Microbiological Associates, Los Angeles, California). Indicator is absent, and the pH should be checked if reused.

Krebs–Ringer phosphate buffer, pH 7.35, containing dextrose, 2 mg/ml (KRPD).

Ferricytochrome $c$, horse heart, type III (Sigma Chemical Co., St. Louis, Missouri). Dissolve in HBSS to a stock concentration of 1.2 m$M$. Filter through a micropore membrane (Millipore Corp., Bedford, Massachusetts) and store at $-20°$ in an airtight container in volumes sufficient for a single experiment.

Superoxide dismutase (SOD). Preparations with high specific activity and purity can be purchased from Diagnostic Data, Inc., Mountain View, California (bovine) or from Diagnostic Materials, Ltd., Oxford, England (bovine or human). The product of Miles Laboratories, Elkhart, Indiana is satisfactory. Dissolve in water for a stock concentration of 1–5 mg/ml, and store at $-20°$. This can be refrozen a few times without loss of activity.

Dimethyl sulfoxide (DMSO; Sigma).

Phorbol myristate acetate (PMA; Consolidated Midland Corp., Brewster, New York). Dissolve in DMSO at 2 mg/ml and store at $-70°$ in airtight polypropylene or glass tubes. Avoid contact with aqueous solutions or vapor condensation because PMA loses its activity rapidly in water.

Fresh (complement-preserved) human serum. Venous blood from normal adults should be allowed to clot for about 45 min at room tempera-

[2] J. M. McCord and I. Fridovich, *J. Biol. Chem.* **244**, 6049 (1969).
[3] B. M. Babior, R. S. Kipnes, and J. T. Curnutte, *J. Clin. Invest.* **52**, 741 (1973).
[4] R. B. Johnston, Jr., J. E. Lehmeyer, and L. A. Guthrie, *J. Exp. Med.* **143**, 1551 (1976).
[5] R. B. Johnston, Jr., C. A. Godzik, and Z. A. Cohn, *J. Exp. Med.* **148**, 115 (1978).

ture, then centrifuged at 4°. The supernatant serum should be pooled, then frozen at −70°.

Zymosan (ICN Pharmaceuticals, Cleveland, Ohio). Suspend in a glass tube to a concentration of 12 mg/ml in saline and heat in boiling water bath, while mixing about every 10 min, for 1 hr. Centrifuge, wash once, and resuspend in physiologic saline or KRPD to a stock concentration of 50 mg/ml. Opsonize by incubating 1 vol zymosan with 3 vol fresh human serum in water bath at 37° for 20 min, with agitation. Centrifuge at 6000–8000 $g$ for 15 min, wash once with KRPD, and resuspend in KRPD to a concentration of 10 mg/ml.

Macrophage culture medium. Results in this assay should not be affected significantly by culture of the macrophages in any of the different standard culture media.

*Procedure*

Macrophages and human monocytes cultured for this assay should be plated at a density sufficient to yield at the time of assay 40–100 $\mu$g of cell protein on a 35-mm-diameter plate.[5] This will require in the neighborhood of 2–6 × $10^6$ macrophages or monocytes per dish. These are added in a volume of 1 ml. The cell suspension should be mixed well between each pipetting to ensure that a consistent number of cells is plated.

On the basis of surface area the appropriate density will require approximately one-quarter as many cells for a 16-mm-diameter culture dish and approximately three times more cells for a 60-mm-diameter dish. The need for concern about cell density stems from the observation that although greater numbers of plated cells will release more $O_2^-$ when stimulated, the extent of release does not increase in proportion to the increase in cell number.[5] Therefore, when expressed as specific activity (nmol/mg cell protein), $O_2^-$ release, in general, declines with increasing cell number.

Release of $O_2^-$ may be quantitated with macrophages cultured for 2 hr, overnight, or for days. The macrophages are prepared by washing quickly twice with KRPD (at room temperature) to remove nonadherent cells. Washing is accomplished here by vigorous swirling of the dish. Immediately after removal of the second wash, the reaction mixture is added, and the reaction is begun by placing the dishes in an incubator at 37° with 100% air or with 95% air–5% $CO_2$. In order to avoid having the macrophages remain long without a full cover of medium, washing and addition of the incubation mixture are usually performed with groups of four to six plates each.

The reaction mixture, prepared in one large tube before the cells are washed, should contain KRPD and cytochrome $c$ to give a final concentration of 80 $\mu M$ in the 1.5 ml volume added to each 35-mm-diameter dish. If the stimulus is to be opsonized zymosan, it should be added to give a concentration of 1 mg/ml in the final reaction mixture. If the stimulus is PMA, a volume of KRPD with cytochrome $c$ sufficient for about six culture dishes (10 ml for 35-mm-diameter dishes) should be placed in a separate tube, and PMA should be added to give a final concentration of 0.5 $\mu g$/ml. On addition of PMA, the reaction mixture should be mixed and immediately added to plated cells to avoid inactivation of the PMA before contact is made with the macrophages. The volume of DMSO used to deliver the PMA (1 part in 4000) does not by itself stimulate $O_2^-$ or inhibit zymosan-stimulated $O_2^-$ release.

With each stimulus, an additional reaction mixture should be prepared that contains SOD at a final concentration of 40 $\mu g$/ml. In most cases, SOD at this concentration eliminates all cytochrome $c$ reduction by stimulated macrophages; autoclaving the SOD removes at least 90% of this inhibitory activity. In each experiment, "blanks" are prepared by incubating each type of reaction mixture in tissue culture dishes without macrophages. Each experimental determination should be run with duplicate or triplicate cultures.

The reduction of cytochrome $c$ by stimulated "activated" mouse macrophages occurs at a linear rate for about 10 min; with resident macrophages the rate is linear for 30–60 min (Sasada and Johnston, unpublished). The rate is nearly linear for 60–90 min with any macrophage type,[5] however, and we have generally used a 60-min or 90-min incubation period with mouse macrophages and cultured human monocytes. The incubation period should depend on the kinetics of the reaction obtained with the cell density, cell type, stimulus, and other conditions employed.

The reaction is stopped by transfer of the incubation mixture by Pasteur pipet to centrifuge tubes in an ice bath, followed promptly by centrifugation at 1200 $g$ for 10 min, or at 8000 $g$ for 2 min in a microcentrifuge (Eppendorf), using 1.5 ml tubes. One or two milliliters of HBSS is added to the dishes to prevent drying of the cells. The supernatant is transferred to separate tubes, and absorbance of the supernatant at 550 nm is determined in a spectrophotometer. Reaction mixtures from dishes that did not contain macrophages are used as blanks, after absorbance of the blanks at 550 nm is compared to that of water. If the cytochrome concentration is proper and the reagents clean, the $OD_{550}$ of the blanks should be 0.55 to 0.65. Spectrophotometers utilizing single-prism monochrometers are not suitable for quantitative measurement of reduced cytochrome $c$. The spectral band width should not exceed 1 nm or, at the most, 2 nm. If the actual wavelength deviates from 550 by as much as 5 nm, light absorption by reduced cytochrome $c$ will be lost almost completely.

The cells remaining in the dish are washed three time with HBSS. Copper tartrate reagent[6] is added directly to the dish, and the protein content of the dish is determined by the method of Lowry et al.[6] using bovine serum albumin as standard. The dishes used as blanks give a significant Lowry reaction, and this value must be subtracted from that of the dishes containing cells. The extent of $O_2^-$ release can be corrected for the protein content of each individual dish if one is certain that cells are not dislodged during the incubation with cytochrome or subsequent washes. Alternatively, if macrophage adherence is uniform from dish to dish, the mean of the protein content of five dishes washed free of non-adherent cells but not carried through the assay can be used for expression of the extent of $O_2^-$ release in individual dishes. Release of $O_2^-$ can also be expressed on the basis of cell number, using a conversion factor that equates cell protein and number[5] or, preferably, an actual determination of cell number by counting nuclei of disrupted cells.[7]

The release of $O_2^-$ by freshly prepared monocytes in suspension can be measured by the assay described for neutrophils given elsewhere in this volume.[8] With some stimuli the $O_2^-$ response is not as rapid as that of the neutrophil,[4] and a slightly longer incubation time may be desirable.

## Calculation of Data

The $OD_{550}$ of the reaction mixtures is converted to nanomoles of cytochrome c reduced using the extinction coefficient $\Delta E_{550} = 21.0 \times 10^3 \ M^{-1}$ $cm^{-1}$.[9] Thus, with a 1-cm light path and a 1.5-ml reaction mixture, the observed $OD_{550}$ should be multiplied by 71.4 to yield the number of nanomoles of $O_2^-$ measured, since 1 mol of $O_2^-$ reduces 1 mol of ferricytochrome c. For a reaction mixture of 1 ml, the conversion factor is 47.6.

This conversion depends upon the assumption that the cytochrome c in the blank is fully oxidized and, therefore, that the observed OD represents the absorbance of only the reduced product (a $\Delta OD$, reduced − oxidized). This assumption can be tested by fully oxidizing the reagent cytochrome c in solution with a few milligrams of potassium ferricyanide, by fully reducing the cytochrome c with a few milligrams of sodium dithionite, and by comparing the $OD_{550}$ of the untreated, oxidized, and reduced solutions against that of water. We found that 98–99% of the ferricytochrome c in fresh reagent solutions is oxidized and, therefore, for routine purposes, do not adjust for the state of oxidation of our reagent material.

[6] O. H. Lowry, N. J. Rosebrough, A. L. Farr, and R. J. Randall, J. Biol. Chem. **193**, 265 (1951).
[7] G. M. Shaw, P. C. Levy, and A. F. Lobuglio, J. Immunol. **121**, 573 (1978).
[8] M. Markert, P. C. Andrews, and B. M. Babior, this volume [45].
[9] V. Massey, Biochim. Biophys. Acta **34**, 255 (1959).

# [47] Superoxide Production

By PETER J. O'BRIEN

The methodologies currently employed to measure superoxide generation in biological systems involve oxidation, reduction, or binding of superoxide to an indicator to form a stable product.

## Superoxide Production by Leukocytes

Superoxide radicals have been shown to play an important role in host defenses against microorganisms and contribute to phagocytic bactericidal activity. Polymorphonuclear leukocytes (PMN) and other phagocytic cells ingest opsonized particles by encompassing them within phagocytic vesicles formed from the plasma membrane. A rapid cyanide-insensitive respiratory burst ensues forming superoxide, $H_2O_2$ and hydroxyl radicals.[1] The enzyme system responsible seems to be a reduced pyridine nucleotide oxidase located in the plasma membrane[2,3] and consequently in phagocytic vesicles.[4] The system is absent from PMNs and vesicles made from PMNs, obtained from patients with chronic granulomatous disease, which fail to generate sufficient concentrations of activated oxygen species to kill susceptible bacteria.[1] The human PMN oxidase is believed to include a flavoprotein[5] and a $b$-type cytochrome[6] and prefers NADPH as the cofactor. Superoxide formation by intact leukocytes is increased up to 50-fold by activation of the leukocyte. This system can also be activated by a wide variety of soluble agents without involving phagocytosis. These include surface-active agents, e.g., digitonin,[3] hypertonic medium,[3] fluoride,[7] tumor promoter agents,[8] a calcium ionophore,[9] and small synthetic N-formylmethionyl peptides which induce chemotaxis in phagocytic cells.[10] The latter effect is unusual

[1] B. M. Babior, N. Engl. J. Med. 298, 659 (1978).
[2] K. Takanaka and P. J. O'Brien, FEBS Lett. 110, 283 (1980).
[3] K. Takanaka and P. J. O'Brien, Arch. Biochem. Biophys. 169, 428 (1975).
[4] H. J. Cohen, P. E. Newburger, and M. E. Chovaniec, J. Biol. Chem. 255, 6584 (1980).
[5] B. M. Babior and R. S. Kipnes, Blood 50, 517 (1977).
[6] A. W. Segal and O. T. G. Jones, Nature (London) 276, 515 (1978).
[7] J. G. R. Elferink, Biochem. Pharmacol. 30, 1981 (1981).
[8] I. R. DeChatelet, P. S. Shirley, and R. B. Johnston, Blood 47, 545 (1976).
[9] G. Z. Zabucchi and D. Romeo, Biochem. J. 156, 209 (1976).
[10] J. E. Lehmeyer, R. Snyderman, and R. B. Johnston, Blood 54, 35 (1979).

METHODS IN ENZYMOLOGY, VOL. 105
Copyright © 1984 by Academic Press, Inc.
All rights of reproduction in any form reserved.
ISBN 0-12-182005-X

in being enhanced by $Ca^{2+}$, $Na^{2+}$, or cytochalasin. Digitonin activation also requires external $Ca^{2+}$.[11]

Two principal methods have been used to measure superoxide formation, one involving the superoxide dismutase inhibition of cytochrome $c$ reduction and the other involving the trapping of the superoxide by 5,5-dimethylpyrroline $N$-oxide (DMPO). The superoxide DMPO adduct is unstable and generates hydroxyl radicals[12] so that it is difficult to use for measuring hydroxyl radicals when the level of hydroxyl radical is less than 3% of the rate of superoxide generation.[13] However cysteine converts the unstable adduct to the stable hydroxyl adduct. DMPO has been used to assay superoxide formation by polymorphonuclear leukocytes following activation by phorbol myristate[14] or by a NADPH oxidase preparation solubilized from granules.[15] By far the most popular method for following superoxide production by leukocytes has been the cytochrome $c$ assay. This method was first reported in 1969 as a method for following superoxide formation by xanthine oxidase.[16] Cytochrome $c$ has the advantage over DMPO of having a 60,000-fold higher trapping rate constant. Using this method, polymorphonuclear leukocytes have been shown to form superoxide at a rate far higher than any other biological system. Rates of 35–80 nmol/mg protein/min have been reported for vesicle fractions[4,17] and recently an extract from a phagocytic vesicle fraction from pig blood polymorphonuclear leukocytes has been reported to form 1 $\mu$mol/mg protein/min.[17] With intact leukocytes rates reported are 0.5–95 nmol/min/$10^7$ cells[18] depending on the method of activation. This method may not be applicable for studying superoxide formation by leukocytes activated with surface active agents, which results in the release of reductases from the cells, or formation by leukocyte homogenates or granules which contain cytochrome $c$ reductase and oxidase activity.[19] In these cases the use of acetylated or succinoylated cytochrome $c$ (described later) is a more suitable method. It is however applicable to studies on

[11] H. J. Cohen and M. E. Chovaniec, *J. Clin. Invest.* **61,** 1088 (1978).
[12] E. Finkelstein, G. M. Rosen, and E. J. Rauckman, *Mol. Pharmacol.* **21,** 262 (1982).
[13] H. Rosen and S. J. Klebanoff, *J. Clin. Invest.* **64,** 1725 (1979).
[14] M. R. Green, H. A. O. Hill, M. J. Okalow-Zublowska, and A. W. Segal, *FEBS Lett.* **100,** 23 (1979).
[15] J. V. Bannister, P. Bellavite, M. J. Serra, P. J. Thomalley, and F. Rossi, *FEBS Lett.* **145,** 323 (1982).
[16] J. M. McCord and I. Fridovich, *J. Biol. Chem.* **244,** 6049 (1969).
[17] H. Wakeyama, K. Takeshige, R. Takayanagi, and S. Minakami, *Biochem. J.* **205,** 593 (1982).
[18] D. R. Light, C. Walsh, A. O'Callaghan, E. J. Goetzl, and A. I. Tauber, *Biochemistry* **20,** 1468 (1981).
[19] K. Kakinuma and S. Minakami, *Biochim. Biophys. Acta* **538,** 50 (1978).

intact leukocytes, leukocyte phagocytic vesicles, or superoxide-forming oxidase preparations.

### Isolation of Human Neutrophils

Human peripheral neutrophils are obtained from heparinized blood. The neutrophils are isolated by sequential dextran sedimentation (Sigma Chemical Company) and Ficoll–Hypaque (Winthrop Laboratories, New York) gradient centrifugation.[20] Contaminating erythrocytes are lysed by treatment with 0.85% ammonium chloride for 10 min at 37°. The neutrophils are then washed three times and finally resuspended in 0.01 $M$ phosphate-buffered saline, pH 7.2, to a final concentration of 2.5 $\times$ $10^7$ cells/mi ascertained by counting with a Coulter counter.

### Preparation of Guinea Pig Neutrophils

Neutrophils can be obtained from guinea pigs injected intraperitoneally with 1% sterile sodium caseinate. Cells are harvested from the peritoneal activity of guinea pigs 18 hr later. The cells are washed and suspended at 2.5 $\times$ $10^6$/ml in Krebs–Ringer phosphate buffer, pH 7.4, containing 5 m$M$ glucose and 0.5 m$M$ CaCl$_2$.

### Cytochrome c Reduction Assay for following Superoxide Formation by Activated Neutrophils

Superoxide is assayed spectrophotometrically by following the reduction of ferricytochrome $c$ at 550 nm and measuring its inhibition by superoxide dismutase.

#### Reagents

1.2 m$M$ ferricytochrome $c$ (type III Sigma Chemicals)
Leukocytes 2.5 $\times$ $10^6$ cells/ml
Krebs–Ringer phosphate buffer containing 5 m$M$ glucose and 0.5 m$M$ CaCl$_2$
Phorbol myristate acetate (Sigma Chemicals) dissolved in dimethyl sulfoxide at a concentration of 1 mg/ml
Superoxide dismutase (Sigma Chemicals)

All incubations are carried out in plastic centrifuge tubes. One milliliter of cells (preincubated at 37°) is added to each assay tube containing 1 $\mu$l of phorbal myristate acetate and 0.4 ml of buffer. After 90 sec at 37°, 0.1 ml ferricytochrome $c$ (final concentration of 76 $\mu M$). As a comparison, a mixture of the same composition, but with 20 $\mu$g superoxide dismutase

[20] A. Boyum, *Scand. J. Clin. Lab. Invest.* **21** (Suppl. 97), 31 (1968).

per milliliter, is treated in the same way. The tubes are incubated for 15 min at 37°. The reaction is terminated by placing the tubes in ice. The tubes are then centrifuged at 8000 $g$ for 30 sec in a microcentrifuge. Cell separation occurs in less than 5 sec. The supernatants are decanted and the amount of reduced cytochrome $c$ is assayed as follows: 0.2 ml of supernatant is added to 2.8 ml of 0.1 $M$ phosphate buffer pH 7.4 and the absorbance at 550 nm is measured. Utilizing potassium ferricyanide and sodium dithionite the amount of reduced cytochrome $c$ and the total amount of cytochrome $c$ present are calculated by using a reduced–oxidized extinction coefficient of 21.1 m$M^{-1}$ cm$^{-1}$. Superoxide dismutase inhibits the reduction of ferricytochrome $c$ by more than 90%, indicating that the reaction is specific for superoxide. Superoxide generation is expressed as nanomoles of ferricytochrome $c$ reduced per $10^6$ neutrophils.

## Spin Trapping of Superoxide Formed by Activated Neutrophils

The spin-trapping method has the advantage over the above of being direct and less ambiguous. However the rate constants for the reaction of superoxide with cytochrome $c$ are considerably higher than those with the following spin traps. Superoxide reacts with the spin trap 5,5-dimethyl-1-pyrroline $N$-oxide (DMPO) to form an unstable superoxide adduct (DMPO-OOH) which in the presence of thiols converts the adduct to the stable nitroxide 5,5-dimethyl-2-hydroxylpyrrolidoxyl DMPO-OH which can be readily measured by EPR spectroscopy. The presence of thiols thereby increases the sensitivity of this reaction to superoxide. DMPO-OH is also formed by the reaction of OH· and DMPO. The above assay conditions are used but instead of cytochrome $c$, 0.18 $M$ DMPO and 0.5 m$M$ cysteine are added. Incubation is for 20 min at 37° with shaking. Superoxide dismutase (1 $\mu$g/ml) is added to the controls. Samples for EPR spectroscopy are injected into 6-in. glass capillary tubes and placed in the cavity of an EPR spectrometer. EPR spectra are recorded with a modulation frequency of 100 kHz, modulation amplitude 1.0 G, microwave power ~1 mW, and X-band frequency 9.5 GHz. The field strength (~3200 G) is set at the midfield peak and scanned with time. Results are expressed as nanomoles of superoxide formed per minute, per $10^6$ cells.

### Superoxide Production by Liver Cell Fractions

Intracellular formation of $H_2O_2$ in the liver cell has been attributed to peroxisomes, mitochondria and endoplasmic reticulum.[21,22] Oxidases

---

[21] D. P. Jones, L. Eklow, H. Thor, and S. Orrenius, *Arch. Biochem. Biophys.* **210**, 505 (1981).

[22] B. Chance, H. Sies, and A. Boveris, *Physiol. Rev.* **59**, 527 (1979).

seem to be responsible for $H_2O_2$ formation in peroxisomes but superoxide formed by autoxidation of reduced CoQ in antimycin A-blocked mitochondrial membranes[22] could explain $H_2O_2$ formation in mitochondria in the presence of certain substrates.[23,24] Dissociation of the oxygenated ferrous complex of cytochrome $P$-450 and autoxidation of reduced NADPH : cytochrome $P$-450 could explain $H_2O_2$ formation by endoplasmic reticulum. Superoxide formation by microsomes is induced eightfold when microsomes from phenobarbital-treated rats are used.[25] The uncoupling of the microsomal cytochrome $P$-450 monooxygenase system by a variety of drugs also results in an increased production of $O_2^-$ and $H_2O_2$. In the presence of chelated $Fe^{2+}$ complexes, hydroxyl radicals are formed as a result of a Fenton-type reaction. The formation of hydroxyl radicals could have important toxicological implications by initiating a free radical chain reaction leading to lipid peroxidation. This may also be enhanced by the intracellular glutathione (GSH) depletion that has been shown to occur following incubation of the cell with these drugs (e.g., benzphetamine, aminopyrine, ethylmorphine). It is also likely that microsomal catalyzed alcohol oxidation is partly mediated by these hydroxyl radicals.[26,27] Superoxide oxidizes epinephrine to the colored adrenochrome (a six-electron process). Unfortunately the intermediate semiquinone autoxidizes to form superoxide. The microsomal NADPH–cytochrome $c$ reductase reduces the quinone and results in an overestimation of the quantity of superoxide generated. Furthermore epinephrine is also readily oxidized to adrenochrome by autoxidation or cooxidation.

Another method widely used is the reduction of nitro blue tetrazolium (NBT). Unfortunately reductases in subcellular fractions readily reduce NBT. The cytochrome $c$ method is also difficult when applied to subcellular fractions because of the highly active cytochrome $c$ reductases and oxidases. Cytochrome $c$ is also rapidly reduced by thiols or ascorbate. Acetylation of 60% of lysine residues of ferricytochrome $c$ results in a decrease of more than 95% in its reduction by mitochondrial and microsomal reductases and its oxidation by cytochrome oxidase.[28] Reduced acetylated cytochrome $c$ is more susceptible to autoxidation than cyto-

[23] G. Powis and I. Jansson, *Pharmacol. Ther.* **7**, 297 (1979).

[24] H. Kuthan, H. Tsuji, H. Graf, V. Ullrich, J. Werringloer, and R. W. Estabrook, *FEBS Lett.* **91**, 343 (1978).

[25] C. Auclair, D. DeProst, and J. Hakim, *Biochem. Pharmacol.* **27**, 355 (1978).

[26] A. I. Cederbaum, A. Quereshi, and P. Messenger, *Biochem. Pharmacol.* **30**, 825 (1981).

[27] M. Ingelman-Sundberg and I. Johannson, *J. Biol. Chem.* **256**, 6321 (1981).

[28] A. Azzi, C. Montecucco, and C. Richter, *Biochem. Biophys. Res. Commun.* **65**, 597 (1975).

chrome $c$. However the ability of acetylated ferricytochrome $c$ to be reduced by $O_2^-$ radicals was maintained and mitochondrial membranes were found to form superoxide at a rate of 0.5 nmol $min^{-1}$ $mg^{-1}$. Intact mitochondria form little superoxide presumably because of their superoxide dismutase content. Recently succinoylation of 45% of the lysine residues was found to be more effective in decreasing the reduction of cytochrome $c$ by NADPH–cytochrome $P$-450 reductase or reduced $b_5$ and the oxidation by cytochrome $c$ oxidase. The bimolecular rate constant for the reduction by superoxide however was estimated to be 10% of that with cytochrome $c$. A rate of superoxide ion formation by liver microsomes from phenobarbital-treated rats was reported to be 9.6 nmol/min/mg protein.[29] Hexobarbitone and coumarin increased the rate twofold.[29] It is expected that this method underestimated the rate of superoxide formation as both cytochrome $c$ and superoxide dismutase are unable to penetrate the microsomal membrane or the lumen of the microsomal vesicles. The similar rate of $H_2O_2$ generation may reflect this. The high lipid solubility of the spin traps 2-ethyl-1-hydroxy-2,5,5-trimethyl-3-oxazolidine (OXANOH) or DMPO should enable them to trap superoxide formed by membranes more effectively. However, rates of superoxide formation were less when these spin traps were used.[30] Possibly the charges on these traps due to their weakly amine character prevent their interaction with cytochrome $P$-450.

### Superoxide Production by Rat Liver Microsomes Using Partially Succinoylated Cytochrome c

*Preparation of Partially Succinoylated Ferricytochrome c.*[29] Succinoylation of 45% of the lysine groups of cytochrome $c$ leaves the 695 nm band intact and while preventing its reduction by reductases still enables it to react with superoxide at 10% of its normal rate. This modified cytochrome $c$ can be prepared as follows. Finely grained succinic anhydride (0.42 mmol) is added to a vigorously stirred ice-cold solution of ferricytochrome $c$ (8 $\mu$mol) in 40 ml of potassium phosphate buffer (0.03 $M$) over a period of 30 min. KOH (2 $M$) is added to keep the pH of the solution at 7.6. The solution is then stirred for another 20 min, transferred to a dialysis bag, and dialyzed against double-distilled water, containing 0.1 m$M$ EDTA, at 4° overnight. The modified cytochrome $c$ preparations are concentrated threefold and stored at $-20°$.

---

[29] H. Kuthan, V. Ullrich, and R. W. Estabrook, *Biochem. J.* **203**, 551 (1982).
[30] G. M. Rosen, E. Finkelstein, and E. J. Rauckman, *Arch. Biochem. Biophys.* **215**, 367 (1982).

*Assay Conditions*

Liver microsomes from phenobarbital-treated male Sprague–Dawley rats are prepared in 0.25 *M* sucrose. The microsomes were diluted to a final concentration of 0.4 mg/ml in 0.1 *M* Tris–HCl buffer, pH 7.7, containing 0.1 m*M* EDTA and 30 $\mu M$ succinoylated ferricytochrome *c* in semimicrocuvettes of 10 mm light path. The reaction is followed continuously as 550 nm with a recording spectrophotometer with the cuvette compartment maintained at 25°. After 1 min 0.4 m*M* NADPH is added, and the rate of reduction of cytochrome is measured for about 2 min (the optical density increase is approximately 0.02). At this time 0.5 $\mu M$ superoxide dismutase is added and the percentage inhibition of the rate of reduction is measured. The rate of superoxide formation is expressed as nanomoles of cytochrome *c* reduced per minute per milligram.

*Superoxide Production by Mitochondrial Membranes Using*
  *Partially Acetylated Cytochrome c*

*Preparation of Acetylated Cytochrome c.*[19] Acetylation of 60% of lysine residues of horse heart ferricytochrome *c* results in a 95% decrease of its ability to be reduced by reductases and oxidized by mitochondrial oxidase. Its ability to be reduced by superoxide is maintained.

At 0°, 300 mg of ferricytochrome *c* is dissolved in 10 ml of sodium acetate (50% saturated). Acetic anhydride (100 mol/mol cytochrome *c*) is slowly added by stirring at 0° into the solution of cytochrome *c* over a period of 30 min. The reaction mixture is dialyzed at 0° for 24 hr against a 0.1 *M* sodium phosphate buffer of pH 7.0. The acetylated cytochrome *c* is laid on top of a column packed with Amberlite IRC-50 equilibrated with 0.1 *M* sodium phosphate buffer pH 7.0. Acetylated cytochrome *c* is further eluted with the same buffer, while unmodified and slightly acetylated cytochrome *c* is absorbed by the Amberlite IRC 50. The unabsorbed acetylated cytochrome *c* is further fractionated on a DEAE-cellulose column equilibrated with 0.02 *M* sodium phosphate buffer of pH 7.0. The extent of acetylation is determined by analyzing the remaining amino groups by the ninhydrin method.[31] The percentage modification is given by percentage acetylation = 100 (1 − *S* acetylated/*S* native) where *S* is the slope of the plot of absorbance at 570 nm versus the concentration of native or acetylated cytochrome *c*.

[31] C. H. W. Hirs, this series, Vol. 11, p. 325.

*Assay Conditions*

Rat liver mitochondrial membrane fragments are prepared by sonication and washed by centrifugation. Mitochondrial fragments (180 $\mu$g/ml) are incubated in 1 ml of 50 m$M$ phosphate buffer, pH 7.4, containing 10 m$M$ KCl, 2 $\mu M$ antimycin, 6 $\mu M$ acetylated cytochrome $c$ in semimicro-cuvettes of 10 mm light path. The reaction is monitored continuously at 550 nm and is initiated by addition of 3 m$M$ succinate. After 1 min 0.5 $\mu M$ superoxide dismutase is added.

*A Spin-Trapping Assay for Superoxide Formation by*
*Rat Liver Microsomes*

Superoxide oxidizes OXANOH to its corresponding nitroxide 2-ethyl-2,5,5-trimethyl-3-oxazolidinoxyl (OXANO) and can be readily determined with electron paramagnetic resonance.[30] In the presence of $Cu^{2+}$ ions this hydroxylamine readily autoxidizes so that chelating agents (e.g., diethylenetriaminepentaacetic acid; DETAPAC) need to be added to OXANOH solutions to prevent autoxidation. This method has been successfully used to monitor superoxide formation by microsomal mixed function oxidase, NADPH–cytochrome $P$-450 reductase and microsomes. In the case of pig liver NADPH–cytochrome $P$-450 reductase, rates of superoxide formation was as high as 64 nmol $O_2^-$/mg/min.[30] OXANOH is much more effective than DMPO in trapping superoxide.[30]

*Preparation of OXANOH*

OXANO is prepared and stored in a desiccator. In contrast OXANOH needs to be prepared fresh just prior to use because of its ease of autoxidation.[32] The nitroxide is reduced to the corresponding hydroxylamine (OXANOH) by bubbling a 10 m$M$ solution of the nitroxide with hydrogen in the presence of the catalyst platinum for 45 min. The purity of the hydroxylamine can be demonstrated using silica gel thin-layer plates eluted with benzene : ethyl acetate 3 : 1. Liver microsomes are prepared by differential centrifugation from a homogenate prepared by homogenizing a liver from a 125 g male Sprague–Dawley rat in 50 m$M$ chelated phosphate buffer at pH 7.4 containing 0.1 m$M$ DETAPAC and 0.25 $M$ sucrose. The microsomes are washed several times by resuspending the pellet in 0.15 $M$ KCl, 0.1 m$M$ DETAPAC at pH 7.4 and centrifuging this

[32] G. M. Rosen, E. J. Rauckman, and K. W. Hanck, *Toxicol. Lett.* **1,** 71 (1977).

mixture at 100,000 $g$ for 40 min. They are then suspended in the 50 m$M$ chelexed phosphate buffer at pH 7.4 containing 0.15 $M$ KCl and 0.1 m$M$ DETAPAC at a concentration of 1 mg/ml. To a 1 ml suspension is added 1 m$M$ OXANOH and 150 $\mu$g/ml NADPH. The EPR machine is set at the midfield peak of the EPR signal of OXANO and scanned with time. Rates of superoxide production of 0.73–3.9 nmol/mg/min have been reported using this method.[30] Unfortunately superoxide dismutase only inhibits 18% possibly because of problems in diffusing to the sites where superoxide is generated. Microsomes from phenobarbital-induced rats should give much higher rates of superoxide formation. The efficiency of superoxide detection is determined by dividing the increase in the rate of OXANOH oxidation caused by adding xanthine oxidase to the microsomal system, by the rate of OXANOH oxidation caused by xanthine and xanthine oxidase alone. The efficiency decreases as the microsomal protein increases.

Reduced quinones reduce acetylated cytochrome $c$.[33] Recently the spin trapping method has proved especially useful in following superoxide formation that accompanies the reductive activation of quinones, anthracycline, or aminoquinone anticancer drugs by microsomal reductases.[34]

---

[33] H. Thor, M. T. Smith, P. Hartzell, G. Bellomo, S. A. Jewell, and S. Orrenius, *J. Biol. Chem.* **257**, 12419 (1982).

[34] B. Kalyanaraman, E. Perez-Reyes, and R. P. Mason, *Biochim. Biophys. Acta* **630**, 119 (1980).

---

## [48] Endothelial Culture, Neutrophil or Enzymic Generation of Free Radicals: *In Vitro* Methods for the Study of Endothelial Injury

*By* CHARLES F. MOLDOW and HARRY S. JACOB

*Principle*

*In vitro* culture of endothelial cells from a variety of anatomic sites and various species permits the study of the inflammatory response at a cellular level. The interactions of endothelial cells with polymorphonuclear leukocytes (PMN)—particularly those "activated" by inflammatory mediators such as certain complement components—seem critical to the vasodilatation and permeability alterations which characterize acute inflammation. The metabolic products of activated PMN, especially reac-

Copyright © 1984 by Academic Press, Inc.
All rights of reproduction in any form reserved.
ISBN 0-12-182005-X

tive oxygen species (ROS) and lysosomal proteases, contribute to the inflammatory response, and have been implicated in many clinical syndromes.[1-3] Our assay procedure (*vide infra*) describes a method to determine toxic effects of PMN products upon endothelial cells *in vitro*. It uses cultured human umbilical vein endothelial cells that are mixed with activated PMN to induce release of endothelial bound radiolabel and measures both cellular injury and detachment. Specific inhibitors may be used to determine the contribution of ROS (or other substances) produced by the neutrophil, while PMN production of these products can be measured directly and correlated with the degree of endothelial injury. ROS may also be generated enzymically in the absence of PMN to further explore their effects upon the endothelial cells.

*Preparation of Endothelial Cells*

Human umbilical vein endothelial cells are grown in microtiter wells (Linbro, No. 76-003-05, Flow Laboratories, Inc., Hamden, Connecticut) using established procedures.[4] Umbilical cords obtained following normal vaginal delivery can be stored at 4° in cord buffer for up to 48 hr. The cords are gently washed free of adherent blood and placed upon a sterile disposable drape. Short cords (<20 cm) and those with clamp marks are discarded. The end of each cord is trimmed with a sterile surgical scapel blade, and the umbilical vein cannulated with a sterile plastic or metal syringe adaptor which is clamped securely with a plastic tie (Cable Ties, Tyton Corp., Milwaukee, Wisconsin). The venous lumen is gently flushed with 20–30 ml of cord buffer (37°) to remove clots and the other end of the cord then cannulated with a sterile adaptor connected to a small piece (3–4 mm) of tubing that can be clamped. The adaptor is securely fastened as above and the venous lumen is gently flushed several times. After several umbilical cords have been similarly prepared, they are held vertically and gently infused with 10 ml of sterile cord buffer containing 2 mg/ml collagenase (Worthington Biochemical Corp., Freehold, New Jersey). The collagenase-containing veins (infusion syringe in place) are put on the sterile drape for 20 min. The cellular yield may be increased by permitting the collagenase to digest at 37°, but overly vigorous handling of the cords

[1] H. S. Jacob, P. R. Craddock, D. E. Hammerschmidt, and C. F. Moldow, *N. Engl. J. Med.* **302**, 789 (1980).

[2] T. Sacks, C. F. Moldow, P. R. Craddock, T. Bowers, and H. S. Jacob, *J. Clin. Invest.* **61**, 1161 (1978).

[3] O. Yamada, C. F. Moldow, T. Sacks, P. R. Craddock, and H. S. Jacob, *Inflammation* **5**, 115 (1981).

[4] E. A. Jaffe, "The Biology of Endothelial Cells" (E. A. Jaffe, ed.). Nijhoff, The Hague, in press, 1983.

will increase contamination by smooth muscle cells and fibroblasts. Therefore, our laboratory permits collagenase action at room temperature, sacrificing cellular yield for purer preparations. Following the 20-min incubation, each cord is held vertically and unclamped over a sterile plastic tube containing 10 ml cord media. The effluent solution is collected and the veins are rinsed with 20 ml cord buffer (37°). The cells are collected by centrifugation (200 $g$, 10 min, 25°), washed once with cord media, resuspended by mixing in cord media, and finally plated at a density of $2 \times 10^4$ cells per square centimeter of culture surface.

*Comments*

Approximately 50% of the cells harvested are viable by trypan blue exclusion, but residual blood cells may obscure the endothelial cells making counting difficult. In general the cells from two cords are pooled to prepare one microtiter plate. After 12 hr in a $CO_2$ incubator (5% $CO_2$, 37°) 20–50% of the cells are adherent, and the microtiter wells should be covered by polygonal endothelial cells 4 to 8 days postculture. The adherent endothelial cells may be washed as early as 12–18 hr after culture; however, this will remove endogenous growth factors and is not necessary. Some investigators pretreat the culture plate surface with agar, fibronectin, or collagen solutions to enhance adherence; if cells are plated at high density, this is not necessary. Cells are usually used when newly confluent, and contamination with nonendothelial cells is minimal at this time. Endothelial cells may be identified by growing aliquots upon glass coverslips and determining the cellular reactivity with commercial factor VIII antibody (Calbiochem-Behring, San Diego, California) using indirect immunofluorescence.[5] In general 90–95% of the confluent cells will be reactive with the factor VIII antibody.

*Preparation of PMN*

Peripheral blood is drawn into a heparinized syringe (6–8 U/ml final concentration) containing 0.5 ml 6% hetastarch (Hespan, American Hospital Supply Corp.) for each milliliter of blood drawn. The contents of this syringe are mixed and left in an inverted upright position at room temperature for 30 min; the well-separated, and less dense, plasma buffy-coat layer is then gently pushed through a bent needle into a sterile plastic tube. The cells are sedimented (200 $g$ for 10 min) and resuspended in 0.5 ml of Hanks'–albumin solution. Cold distilled water (12 ml) is added to lyse contaminating red cells and mixed for 20–25 sec; thereafter, 4 ml of

[5] E. A. Jaffe, L. W. Hoyer, and R. L. Nachman, *J. Clin. Invest.* **52**, 2757 (1973).

3.6% (w/v) NaCl solution is added with mixing to restore tonicity and the mixture resedimented at 200 $g$ for 10 min. The cells are resuspended in 5.0 ml Hanks'–albumin solution and carefully layered upon 5.0 ml Ficoll–Hypaque solution. This mixture is centrifuged for 20 min (400 $g$) at 4°. The interface contains mononuclear cells, and the pellet contains PMNs which are resuspended in 10 ml Hanks'–albumin solution, counted and adjusted to $1-5 \times 10^7$/ml.

### Cell-Mediated Cytotoxicity

Newly confluent endothelial cells are labeled with 0.1 ml sodium ($^{51}$Cr) chromate (10 $\mu$Ci/ml) in serum-free cord medium for 4–6 hr at 37°. This supernatant fluid is discarded and the cells washed twice with Hanks'–albumin solution. The washed cells should manifest approximately $2 \times 10^4$ cpm per $5 \times 10^4$ endothelial cells. To assess cytotoxicity—for example, that induced by activated complement plus PMN—reagents can then be added to the microtiter wells. Six wells are assayed in each group, and a representative study is illustrated below.

Nos. 1–6    Buffer control—200 $\mu$l Hanks'–albumin solution

Nos. 7–12    Serum control—200 $\mu$l Hanks'–albumin solution containing 10% (v/v) heated human serum (56° $\times$ 30 min)

Nos. 13–18    Complement control—200 $\mu$l Hanks'–albumin solution containing 10% complement-activated human serum (zymosan-activated serum)

Nos. 19–24    Neutrophil-activating agent—200 $\mu$l Hanks'–albumin solution containing stimulators such as phorbol myristate acetate (PMA) (1–100 ng), serum-treated zymosan particles (1 mg), or other neutrophil-activating material

Nos. 25–30    Resting neutrophil control—$1-50 \times 10^5$ neutrophils in 10% heat-decomplemented serum and Hanks'–albumin solution media in a final volume of 200 $\mu$l

Nos. 31–36    $H_2O_2$ control—200 $\mu$l Hanks'–albumin solution containing $H_2O_2$ diluted to give greater than 50% chromium release

Nos. 36–96    Activated PMN—neutrophils plus stimulating agent(s) in a final 200 $\mu$l volume of Hanks'–albumin solution

Nos. 36–96    Same as above plus specific inhibitor(s) to be studied

The microtiter plate is spun at 200 $g$ for 5 min (room temperature) and incubated at 37° for 30–180 min. The plate is then recentrifuged at 200 $g$ for 5 min and 150 $\mu$l is removed from each well with a multichannel

REPRESENTATIVE CYTOTOXICITY ASSAY[a]

| Ratio PMN : endothelial cell | PMN plus heat-decomplemented sera | PMN plus complement-activated sera |
|---|---|---|
| 10 : 1 | 22 ± 3(9 ± 0.6) | 42.1 ± 8(12 ± 1) |
| 20 : 1 | 27 ± 5(11 ± 0.9) | 47 ± 4(17 ± 1) |
| 50 : 1 | 20 ± 1(10 ± 1) | 54.1 ± 8(22 ± 3) |
| 100 : 1 | 26 ± 4(15 ± 4) | 56 ± 4(26 ± 5) |
| 15 $\mu$mol $H_2O_2$ | — | 90 ± 1(67 ± 4) |

[a] The values refer to the mean and SD from six separate microtiter wells; values outside parentheses represent total $^{51}$Cr release (cytotoxicity plus detached cells), while those within parentheses refer to soluble $^{51}$Cr release (cytotoxicity alone). Buffer, heated serum, or complement-activated serum in the absence of cells induced 6 to 11% soluble $^{51}$Cr release and 20–25% total $^{51}$Cr release. Serum complement was activated with zymosan.

micropipet. This "soluble fraction" is placed in individual disposable tubes (6 × 50 mm). The endothelial cells are washed twice by additions of Hanks'–albumin solution (200 $\mu$l) to each well followed by gentle aspiration. The two washes are pooled in a second group of disposable tubes (6 × 50 mm) and designated the "nonadherent fraction." Finally 1 $M$ NaOH (200 $\mu$l) is added to each well for 30 min, and collected in a third disposable tube, designated the "adherent fraction." All tubes are counted, and $^{51}$Cr release calculated in two ways.

$$\text{Percentage cytotoxicity} = \frac{\text{cpm soluble fraction}}{\text{cpm(soluble + nonadherent + adherent fractions)}} \times 100 \quad (1)$$

$$\text{Percentage cytotoxicity plus detached intact endothelial cells} = \frac{\text{cpm(soluble + nonadherent fractions)}}{\text{cpm(soluble + nonadherent + adherent fractions)}} \times 100 \quad (2)$$

Total counts per well are calculated so variations in specific activity or cell number per well will not affect the percentage of counts released. The mean and standard deviation of each group of six wells are calculated. Specific $^{51}$Cr release (cytotoxicity or cytotoxicity plus cellular detachment) reflects the difference between radiolabel release induced by the activated PMN and the most complete control (usually the resting cells). A representative experiment demonstrating endothelial injury induced by PMN and activated complement is shown in the table.

Using this assay we have examined the toxic role of relevant reactive oxygen species known to be produced by activated PMNs as follows: (1) using ROS inhibitors; (2) correlating PMN production of ROS with $^{51}Cr$ release from endothelial cells; and (3) determining endothelial $^{51}Cr$ release in response to enzymically generated ROS.

*Inhibitors.* The following commercially available scavengers have been tested in this[1,2] and related assays.[6] Only catalase (1–20 µg/ml; use thymol-free preparations) has consistently inhibited endothelial cell injury induced by neutrophils. Other compounds investigated include superoxide dismutase (1–20 µg/ml), mannitol (20 m$M$), ethanol (20 m$M$), L-amino acids (20 m$M$), azide (1 m$M$), $CN^-$ (1 m$M$), and thiourea (20 m$M$).[1,6] These scavengers do not themselves induce $^{51}Cr$ release from endothelial cells, but specific controls for such an artifactual effect should be included in each microtiter assay. The various ROS scavengers may be titrated in these assays to select the minimal effective concentrations. In addition PMNs from patients with specific defects in leukocyte metabolism (e.g., neutrophil myeloperoxidase deficiency and chronic granulomatous disease) may also be utilized, if available.

*The Production of ROS.* The production of ROS by neutrophils may be monitored directly using established techniques. Superoxide anion is determined by the capacity of $O_2^-$ to reduce ferricytochrome $c$, as described by Goldstein *et al.*[7] Release of $H_2O_2$ by granulocytes is measured by following the decrease in fluorescence intensity of scopoletin due to its peroxidase-mediated oxidation by $H_2O_2$.[3,8] Hexose monophosphate shunt activity is determined in PMNs by incubation with 1 µCi of D-[1-$^{14}C$]glucose (New England Nuclear, Boston, Massachusetts) as described by DeChatelet *et al.*[9]

*Enzymic Generation of Oxygen Intermediates.* Labeled endothelial cells have been treated with enzyme–substrate combinations to generate different ROS. The microtiter wells may be arranged as indicated previously using buffer, enzyme, and substrate alone as controls. The enzyme–substrate combinations (and appropriate scavengers) are then used to induce (or prevent) $^{51}Cr$ leakage which may be correlated with cellular injury and detachment. Successful systems in the laboratory have utilized purine–xanthine oxidase[2,6] and glucose–glucose oxidase.[2,6] The assay is

[6] S. J. Weiss, J. Young, A. F. LoBuglio, A. Slivka, and N. F. Nimeh, *J. Clin. Invest.* **68**, 714 (1981).
[7] I. M. Goldstein, D. Roos, H. B. Kaplan, and G. Weissmann, *J. Clin. Invest.* **56**, 1155 (1975).
[8] R. K. Root, B. Metcalf, N. Ochino, and B. Chance, *J. Clin. Invest.* **55**, 945 (1975).
[9] R. L. DeChatelet, P. Wang, and C. E. McCall, *Proc. Soc. Exp. Med. Biol.* **140**, 1434 (1972).

generally initiated with addition of enzyme, and carried out for 90–180 min at 37°. This assay is performed in the absence of serum and contains no cells other than endothelium, so the degree of cellular injury may be assessed by measuring release of cytoplasmic lactate dehydrogenase (LDH)[10] into the supernatant fraction.

## General Remarks

Assay of [51]Cr leakage is not ideal, as this assay usually requires very significant cellular injury, so that sublethal effects may be underestimated. Assays utilizing endothelial cells labeled with amino acids, thymidine, or uridine have been tried but have provided no advantage in our laboratory. Assaying release of cytoplasmic contents such as LDH is generally not practical, since the endothelial cytoplasmic proteins must be differentiated from that of PMNs. Our laboratory has had some success utilizing fluorescein diacetate[11,12] to label viable cells. This assay can be automated using a fluorescence-activated cell sorter, but analysis may be complicated when several cell types are present simultaneously; thus, stimulated effector cells such as PMNs generally adhere to target endothelial cells so that nonfluorescent cells reflect contaminating PMNs as well as nonviable endothelial cells. In sum, therefore, the most convenient assay continues to be [51]Cr release. Spontaneous leakage is usually less than 10% (in a 90- to 180-min incubation), and 85–95% of this label is released with complete cellular disruption. Rarely, commercial reagents will induce significant "spontaneous" [51]Cr leakage without neutrophils. In this instance, heat-stable contaminants appear responsible, and reagents must be changed. In addition, endothelial cytotoxic sera have been encountered by others in patients with scleroderma[13] and thrombotic thrombocytopenic purpura[14]; however, our laboratory has not yet encountered consistently cytotoxic serum samples. Studies of macrophage-mediated tumor cell lysis[15] have suggested that intracellular glutathione metabolism may be critical for cellular resistance to the effects of inflammatory cell ROS. Recent, unpublished reports suggest a similar mechanism for endothelial cell resistance to oxidant injury, since inhibition of endothelial glutathione reductase with 1,3-bis(chloroethyl)nitrosourea (BCNU) increases the sensitivity of the endothelial cells to the cytotoxic

[10] R. D. Glass and D. Doyle, *Science* **176**, 180 (1972).
[11] B. Rotman and B. X. Papermaster, *Proc. Natl. Acad. Sci. U.S.A.* **55**, 134 (1966).
[12] F. Celada and B. Rotman, *Proc. Natl. Acad. Sci. U.S.A.* **57**, 630 (1967).
[13] M. B. Kahaleh, G. K. Sherer, and E. C. LeRoy, *J. Exp. Med.* **149**, 1326 (1979).
[14] E. R. Burns and D. Zucker-Franklin, *Blood* **60**, 1030 (1982).
[15] C. F. Nathan, B. A. Arrick, H. W. Murray, N. M. DeSantis, and Z. A. Cohn, *J. Exp. Med.* **153**, 766 (1980).

effects of activated PMNs.[16] This suggests that various cultures of endothelial cells may manifest differing propensities to PMN-induced damage based on intrinsic antioxidant defenses.

## Reagents

*"Cord buffer."* 0.14 $M$ NaCl, 0.0004 $M$ KCl, 0.001 $M$ NaPO$_4$ buffer pH 7.4, 0.01 $M$ glucose

*Hanks'–albumin.* Hanks' balanced salt solution (GIBCO, Grand Island, New York) containing 0.5% (v/v) human albumin (Travenol Labs Inc., Glendale, California)

*"Cord media."* Media 199, containing 20% calf serum, penicillin (200 U)/ml, sodium bicarbonate (1.12 mg/ml), streptomycin (200 mg/ml), and L-glutamine (2 m$M$) (GIBCO, Grand Island, New York) is prepared weekly, stored at 4°, and warmed to 37° before use

*Zymosan-activated serum.* Zymosan (Sigma Chemical Corp., St. Louis, Missouri) (10 mg/ml) is mixed with human serum for 30 min at 37°, and the mixture then spun at 400 $g$ for 30 min. The supernatant fluid is decanted and may be stored (−70°) for 2 weeks. The pellet can be resuspended in Hanks' balanced salt solution, washed twice, and resuspended (8–10 mg/ml) in this same solution to provide opsonized zymosan particles which, when phagocytosed, activate PMNs.

*Ficoll–Hypaque solution.* Mix 50 ml of 75% (v/v) Hypaque M (Winthrop Labs, New York, New York) with 61.8 ml distilled water plus 268.4 ml Ficoll (Sigma, St. Louis, Missouri—27 g in 300 ml distilled water) and add 0.8 ml sodium heparin (1000 U/ml). Autoclave with slow pressure exhaust for 15 min and store at 4°.

*Phorbol myristate acetate (PMA)* (Sigma, St. Louis, Missouri). Dissolve 5 mg (well-ventilated hood) in 5 ml DMSO and store in 0.5-ml aliquots at −70°. Prior to use dilute in Hanks'–albumin solution and use at final concentration of 1–100 ng per microtiter well.

---

[16] J. D. Levine and J. M. Harlan, personal communication.

# [49] Systemic Consequences of $O_2^-$ Production

## By ROLANDO DEL MAESTRO

Free radical reactions have been implicated in a wide variety of diseases including inflammation, irradiation-induced injury, and ischemia. The role played by superoxide anion radical ($O_2^-$) in these conditions is being elucidated using a number of model systems. This discussion is an attempt to provide an approach to the understanding of the systemic consequences of $O_2^-$ generation.

A valuable conceptual model has been to categorize disease states by the major site of pathological $O_2^-$ generation. Increased free radical generation may occur predominantly (1) intracellular, (2) extracellularly, and (3) both intracellularly and extracellularly. Decreased $O_2^-$ generation may occur (1) intracellularly and (2) extracellularly.

### Increased Intracellular Generation

The essential factor in this group of conditions appears to be an alteration of intracellular protective and control mechanisms and/or an overwhelming of these mechanisms. The intracellular components of the cell are therefore exposed to an increased flux of radical species. The table gives a partial list of the large number of diseases in which an increased intracellular flux of $O_2^-$ has been suggested to play a pathological role. The mitochondria have been suggested as the major site of $O_2^-$ production intracellularly,[1,2] and the exposure of lung submitochondrial particles to hyperoxia results in increased $O_2^-$ production.[3] Cellular damage in these conditions may be related to peroxidative injury to the biomembranes of intracellular organelles mediated not by $O_2^-$ itself but by other activated oxygen species derived from increased $O_2^-$ generation.[4]

### Increased Extracellular Generation

Acute and chronic inflammatory conditions are characterized by an increased $O_2^-$ generation predominately in the extracellular space. Four

---

[1] G. Loschen, L. Flohe, and B. Chance, *FEBS Lett.* **18**, 261 (1971).

[2] A. Boveris and B. Chance, *Biochem. J.* **134**, 707 (1973).

[3] J. F. Turrens, B. A. Freeman, J. G. Levitt, and J. D. Crapo, *Arch. Biochem. Biophys.* **217**, 401 (1982).

[4] K. Brawn and I. Fridovich, *Acta Physiol. Scand.* **492**, 9 (1980).

Copyright © 1984 by Academic Press, Inc.
All rights of reproduction in any form reserved.
ISBN 0-12-182005-X

DISEASES IN WHICH INCREASED INTRACELLULAR GENERATION MAY
PLAY A ROLE

Hyperoxygenation syndromes
  Hyperbaric oxygen
  Hyperoxygenation—respirator pulmonary disorders
  Retrolental hyperplasia
  Pulmonary dysplasia (neonatal)
  Complete ischemia—reperfusion disorders (cardiac arrest, transplantation)
Hypooxygenation syndromes
  Incomplete ischemia—hypoxia (shock, cerebral or myocardial infarction)
Chemicals
  Paraquat
  Carbon tetrachloride
  Chemotherapeutic drugs (adriamycin, streptonigrin, etc.)
  Carcinogens (benzopyrene)
Hemolytic anemias—drug induced (i.e., phenylhydrazine)
Aging

prerequisites appear to be necessary before suggesting a role for $O_2^-$ in inflammation.

1. The presence of inflammatory cells capable of releasing $O_2^-$ such as polymorphonuclear leukocytes, macrophages, or monocytes.[5]

2. The appropriate stimulus (bacteria, fungi, immune complexes, etc.) capable of inducing the respiratory burst with the generation of $O_2^-$.

3. Low levels of superoxide dismutase and catalase in the extracellular space which results in the increased susceptibility of the interstitial space components to an $O_2^-$ flux.

4. The availability of metal complexes chelated such that they are available for the generation of hydroxyl radical (OH·) and possibly other oxidizing species.

These prerequisites appear to be fulfilled in the extracellular space during inflammatory conditions and free radical induced alterations have been invoked as mechanisms of bactericidal killing[5] and tissue injury. Extracellular macromolecules such as hyaluronic acid[6] and collagen[7] have been degraded *in vitro* by an enzymatic generation of $O_2^-$. Major consequences to cellular integrity may result from the initiation of propagating free radical chain reactions in plasmalemmal membranes. The release of the resulting lipid hydroperoxide products into the extracellular space

[5] B. M. Babior, *N. Engl. J. Med.* **298,** 659 (1978).
[6] J. M. McCord, *Science* **185,** 529 (1974).
[7] R. A. Greenwald and W. W. Moy, *Arthritis Rheum.* **22,** 251 (1979).

may increase microvascular permeability[8,9] and alter leukocyte chemo-taxis.[10] It would appear reasonable to suggest that during inflammatory conditions free radical products may result in significant alterations in both the permeability and structural characteristics of inflamed tissue.

### Increased Extracellular and Intracellular Generation

Increased $O_2^-$ flux in both the intracellular and extracellular space occurs during the irradiation of tissue and possibly by certain chemicals. Irradiation results in the disruption of water molecules resulting in the generation of a group of chemical species including hydrogen atoms, hy-drated electrons and $OH\cdot$. If $O_2$ is present, both $O_2^-$ and $H_2O_2$ are formed. The subsequent generation of $OH\cdot$ by any one of a group of different mechanisms may be essential for the DNA damage[4] which results in cellular death. The generation of these reactive species in the extracellu-lar space may result in alterations of the components of the extracellular space and plasmalemmal peroxidation. This extracellular $O_2^-$ flux may result in the formation of chemotactic factors which result in the emigra-tion of granulocytes into the irradiated region and possibly the activation of these cells by other products or irradiation.

### Decreased Intracellular Generation

In humans, the gene for Cu,Zn superoxide dismutase (EC 1.15.1.1), an enzymatic scavenger of $O_2^-$, is located on chromosome 21. Patients suf-fering from Trisomy 21 (mongolism) have 50% more intracellular Cu,Zn superoxide dismutase.[11] Patients suffering from a variety of psychiatric disorders also have been shown to have increased Cu,Zn superoxide dismutase levels and it has been suggested that a portion of the mental aberrations seen in these patients may be related to increased $O_2^-$ scav-enging by the excess superoxide dismutase available.[11] Superoxide anion radical has also been shown to participate in the intracellular activity of enzymes such as indoleamine 2,3-dioxygenase,[12] galactose oxidase,[13] and

[8] R. F. Del Maestro, J. Björk, and K.-E. Arfors, *Microvasc. Res.* **22**, 239 (1981).
[9] R. F. Del Maestro, J. Björk, and K.-E. Arfors, *Microvasc. Res.* **22**, 255 (1981).
[10] R. F. Del Maestro, M. Planker, and K.-E. Arfors, *Int. J. Microcirc. Clin. Exp.* **1**, 105 (1982).
[11] A. M. Michelson, K. Puget, P. Durosay, and J. C. Bonneau, *in* "Superoxide and Superox-ide Dismutases" (A. M. Michelson, J. M. McCord, and I. Fridovich, eds.), p. 467. Academic Press, New York, 1977.
[12] F. Hirata and O. Hayaishi, *J. Biol. Chem.* **250**, 5960 (1975).
[13] G. A. Hamilton and R. D. Libby, *Biochem. Biophys. Res. Commun.* **55**, 333 (1973).

2-nitropropane dioxygenase[14]; therefore, its more efficient scavenging could be detrimental.

### Decreased Extracellular Generation

If one or more of the prerequisites for suggesting a role for $O_2^-$ in inflammation is unfulfilled, then decreased extracellular generation of $O_2^-$ may occur, resulting in deficient bactericidal killing. Conditions resulting in leukopenia or alterations in the inflammatory cell's ability to release $O_2^-$ (chronic granulomatous disease)[15] result in an increased susceptibility to infection. Interestingly chronic renal disease is associated with both increased extracellular superoxide dismutase[16] and increased infection rates.

### Concluding Remarks

The consequences of the systemic generation of $O_2^-$ clearly present a challenge to those concerned with biological phenomena. The study and interpretation of $O_2^-$ interactions is fraught with difficulties. However, only through a more complete understanding of this complicated and reactive species will we be able to assess its role in disease processes.

### Acknowledgments

These studies were supported by the Canadian Medical Research Council. The author is a recipient of a Canadian Life Insurance Medical Scholarship.

[14] T. Kido, K. Soda, T. Suzuki, and K. Asada, *J. Biol. Chem.* **251**, 6994 (1976).
[15] B. M. Babior, *N. Engl. J. Med.* **298**, 721 (1978).
[16] S. L. Marklund, E. Holme, and L. Hellner, *Clin. Chem. Acta* **126**, 41 (1982).

## [50] Calcium and Calmodulin in Neutrophil Activation

*By* HAROLD P. JONES and JOE M. McCORD

The importance of calcium in neutrophil activation has been appreciated for more than 20 years. Nevertheless, the biochemical mechanisms linking elevations in cytosolic calcium to neutrophil responses are poorly understood. Approximately 4 years ago, we first proposed that calmodulin might play a role in the calcium-mediated activation of the neutrophil.

METHODS IN ENZYMOLOGY, VOL. 105
Copyright © 1984 by Academic Press, Inc.
All rights of reproduction in any form reserved.
ISBN 0-12-182005-X

Since that time, numerous laboratories in addition to ours have implicated or suggested a role for calmodulin in the various components of bacterial killing. In the brief discussion that follows, we attempt to define criteria that can be used to establish calmodulin involvement in neutrophil responses and to discuss the approaches that may be utilized to satisfy these criteria.

In order to implicate calmodulin in any physiological response, at least four criteria must be satisfied. First of all, calmodulin must be present in the cell. Second, the response being studied must be sensitive to calcium over a concentration range consistent with calmodulin-dependent stimulation ($10^{-7}$–$10^{-5}$ $M$). Third, potent inhibitors of calmodulin-dependent activation such as trifluoperazine and W-7 [$N$-(6-aminohexyl)-5-chloro-1-naphthalene sulfonamide] should prevent the cellular responses. Finally, the addition of calmodulin to cell-free preparations should, in the presence of calcium, stimulate the activity of an enzyme crucial in mediation of the physiological process.

## Assay and Localization of Calmodulin in Neutrophils

While the presence of calmodulin in the neutrophil has been established by at least three groups,[1-3] the desire to quantify calmodulin may be important in future studies. There are two basic approaches for the quantification of calmodulin in neutrophil homogenates. One of these depends upon the biological activity of the regulator while the other depends upon its immunological recognition.

Activity measurements of calmodulin in the neutrophil depend upon the measurement of calmodulin-sensitive stimulation of bovine or porcine brain phosphodiesterase. Since this assay has been presented in detail in a previous volume in this series,[4] it will not be discussed here except in regard to specific modifications required for application to neutrophil extracts. We and others have observed the rapid loss of biological activity of calmodulin in neutrophil extracts following homogenization. This inactivation is probably due to the high amount of proteolytic activity present in these extracts. To circumvent this problem, one may take advantage of the thermal stability of calmodulin. Intact neutrophils suspended in an EDTA-containing buffer are rapidly heated to 90° and maintained at that temperature for 5 min. The cell suspension is then cooled, sonicated to

[1] H. P. Jones, G. Ghai, W. F. Petrone, and J. M. McCord, *Biochim. Biophys. Acta* **714**, 1522 (1982).

[2] K. Takeshige and S. Minakami, *Biochem. Biophys. Res. Commun.* **99**, 484 (1981).

[3] J. G. Chafouleas, J. R. Dedman, R. P. Munjaal, and A. R. Means, *J. Biol. Chem.* **254**, 10262 (1979).

[4] J. C. Matthews and M. J. Cormier, this series, Vol. 57, p. 107.

ensure complete cellular disruption, and the resultant homogenate centrifuged to remove precipitated proteins and cellular debris. Biologically active calmodulin, stable to these procedures, remains in the supernate and can be assayed as previously described.[4] This procedure provides a mechanism for rapidly eliminating proteases which inactivate calmodulin. It also eliminates other calmodulin-binding proteins which might interfere, in a competitive fashion, with the measurement of calmodulin-dependent activation of the phosphodiesterase.

Alternatively one may measure the amount of calmodulin in a cellular extract using the radioimmunoassay procedure described by Chafouleas et al.[3] This procedure is advantageous due to the enhanced sensitivity of the technique and due to the fact that when conducted in the presence of a chelating agent, the calmodulin-binding proteins do not interfere. On the other hand, since the assay measures antigenically similar material and not biological activity one may detect inactive or degraded forms of calmodulin. Indeed, the radioimmunoassay of calmodulin in neutrophil extracts by Chafouleas et al.[3] indicated higher quantities of calmodulin than they could detect by an activity assay.

### Calcium Dependence of the Physiological Response

In order to establish that calmodulin may play a role in the mediation of a neutrophil function it is important to demonstrate the calcium dependency of the event. In intact cells, one should be able to show that changes in intracellular calcium concentrations are required for the physiological response. In broken cell preparations one should demonstrate that a change in calcium concentration from $10^{-7}$ to $10^{-5}$ $M$ will result in the activation of an enzyme system implicated in the biological event.

In intact cells, the calcium dependency of a neutrophil function can best be shown by demonstrating that calcium antagonists block the event. Since many neutrophil functions appear to be triggered by the release of intracellular calcium, the preferred choice of antagonist is the intracellular calcium antagonist, TMB-8 [8-[(N,N-dimethylamino)octyl-3,4,5-trimethoxybenzoate hydrochloride]. This compound at a concentration of 0.5 m$M$ antagonizes the calcium-dependent events of neutrophil activation[5,6] and its inhibitory effect upon a neutrophil function may be considered a necessary prerequisite for the consideration of calmodulin as a control element for any single activation process.

In broken cell preparations one should demonstrate that an enzyme or a series of enzymatic events, linked to the response of interest, is sensi-

[5] J. E. Smolen, H. M. Korchak, and G. Weissmann, *Biochim. Biophys. Acta* **677**, 512 (1981).
[6] R. J. Smith and S. S. Iden, *Biochem. Biophys. Res. Commun.* **91**, 263 (1979).

tive to free calcium concentrations ranging from $10^{-7}$ to $10^{-5}$ $M$ since calmodulin is converted from its inactive to active form over this concentration range. Homogenates of known total calcium concentration (determined by specific ion electrodes or atomic absorption spectroscopy) or subcellular fractions resuspended in solutions of known total calcium concentration should be titrated with aliquots of EGTA to produce environments of varying free calcium concentration. Free calcium concentrations can be determined using a modified version of the program of Perrin and Sayce as described by Anderson et al.[7] Using this technique one can determine the calcium sensitivity of the enzymatic reaction. Due to the large dilution of intracellular calmodulin which occurs upon cellular disruption one may find it necessary to augment the reaction mixture with exogenous calmodulin (100 $\mu$g/ml). In addition, care should be taken that sufficient $Mg^{2+}$ is present in all assays to ensure that the observed effects are due to perturbations in free calcium concentration.

*Drug Inhibition Studies*

Several potent calmodulin antagonists exist which should block specific neutrophil responses if calmodulin is involved in mediating their activation. The two inhibitors most commonly chosen are trifluoperazine, a phenothiazine, and W-7. In intact neutrophils physiological responses which depend upon calmodulin should be inhibited at concentrations of these drugs consistent with their action as calmodulin inhibitors.[8] Full inhibition of activity should be observed at trifluoperazine and W-7 concentrations of 10 $\mu M$ or less. Concentrations greater than this may cause inhibition, but that inhibition may be through calmodulin-independent processes. Indeed, at trifluoperazine and W-7 concentrations of 20 and 50 $\mu M$, respectively, cellular lysis occurs.[5] However, at concentrations below 10 $\mu M$, neither of these drugs inhibits either the changes in membrane permeability or calcium influx associated with neutrophil activation proving that their stabilizing effects are distal to the initial events occurring at the membrane surface.[9]

*Activation of Specific Enzymes with Calmodulin*

While the calcium dependence and drug sensitivity of a neutrophil response may strongly suggest calmodulin involvement, final and definite

[7] J. M. Anderson, H. C. Charbonneau, H. P. Jones, R. O. McCann, and M. J. Cormier, *Biochemistry* **19**, 3113 (1980).
[8] B. Weiss and R. M. Levin, *Adv. Cyclic Nucleotide Res.* **9**, 285 (1978).
[9] P. H. Naccache, T. F. P. Molski, T. Alobaidi, E. L. Becker, H. J. Showell, and R. I. Sha'afi, *Biochem. Biophys. Res. Commun.* **97**, 62 (1980).

proof rests upon the ability to show the calmodulin-dependent stimulation of an enzyme involved in the biochemical pathway linking the stimulus to the response. This is by far the most difficult criterion to meet since it requires a detailed knowledge of the biochemistry of the response being studied. In the neutrophil, in most cases this is not known. In time, as these pathways are delineated, we will be in a better position to study the role of calmodulin in activation.

In summary, once an enzyme is targeted for study in regard to calmodulin activation, the following experimentation should be conducted to determine calmodulin involvement. First of all, the ability of exogenously added calmodulin (100 $\mu$g/ml) to stimulate enzymatic activity should be assessed. Activation by calmodulin should occur only in the presence of calcium and over a range of calcium concentrations ($10^{-7}$–$10^{-5}$ $M$) consistent with calmodulin activation. Second, activation of the enzyme should be blocked by the addition to the incubation mixture of antibodies directed against calmodulin or by the addition of micromolar levels of trifluoperazine or W-7. And, finally, specificity of the calmodulin-dependent activation should be demonstrated by showing that other calcium-binding proteins such as troponin C and parvalbumin are ineffective in enzyme activation. Positive results from these studies provide convincing evidence for calmodulin involvement in specific neutrophil processes.

# [51] Measurement of Oxidizing Radicals by Polymorphonuclear Leukocytes

*By* DENNIS P. CLIFFORD and JOHN E. REPINE

Recent evidence has suggested that highly charged oxygen metabolites (commonly called $O_2$ radicals) may play an important role in disease, making detection of these molecules important in research and health care. In particular, release of $O_2$ metabolites from polymorphonuclear leukocytes (PMN) appears to hold a preeminent position in inflammation, bactericidal activity, and tissue injury. The important $O_2$-derived metabolites from PMN are most likely superoxide anion ($O_2^-$), hydroxyl radical ($\cdot$OH), singlet oxygen ($^1O_2$), hypochlorous acid (HOCl), and hydrogen peroxide ($H_2O_2$). The following is a summary of the commonly used methods available to researchers to identify these oxygen radicals in systems using PMN. Because of space limitations, detailed methods to perform

Copyright © 1984 by Academic Press, Inc.
All rights of reproduction in any form reserved.
ISBN 0-12-182005-X

each test cannot be included but are documented in the references and elsewhere in this volume.

### Detection of Superoxide Anion $O_2^-$

Following phagocytosis of particles or stimulation in other ways, PMN undergo a marked increase in their consumption of $O_2$ which is called the "respiratory burst." The main product of this $O_2$ uptake by PMN is $O_2^-$. $O_2^-$ is measured in several ways—nitro blue tetrazolium reduction, ferricytochrome $c$ reduction, and epinephrine oxidation.

*Nitro Blue Tetrazolium Reduction (NBT).*[1] In the unreduced state, NBT is an artificial electron receptor that is soluble and yellow in color. Upon reduction by $O_2^-$, NBT becomes a relatively insoluble blue-black colored substance called formazan as shown:

$$O_2^- + NBT \text{ (yellow)} \rightarrow O_2 + \text{formazan (blue-black)}$$

This blue-black color of reduced NBT may be directly visualized in the phagocytic vesicles of PMN microscopically. It may also be extracted with an appropriate solvent and quantitated spectrophotometrically. When reduction of NBT is inhibited by addition of superoxide dismutase (SOD), which scavenges free $O_2^-$, the specificity of the reaction is improved. NBT reduction by PMN may be increased by adding phorbol myristate acetate (PMA) or other agents which stimulate the respiratory burst and cause PMN to produce more $O_2^-$. This maneuver frequently increases the ability of this assay to distinguish abnormalities in NBT reduction such as those seen in PMN from patients and carriers of chronic granulomatous disease (CGD).

*Cytochrome c Reduction.*[2] Reduction of ferricytochrome $c$ may occur from $O_2^-$ made by intact PMN. Again, supporting evidence for specificity is provided if addition of SOD inhibits this reaction. SOD inhibitable reduction of cytochrome $c$ has become the standard assay for measuring $O_2^-$ from PMN. The test is simply performed. Usually, PMN are incubated with opsonized particles, such as zymosan, or chemical agents, such as PMA, to induce the respiratory burst. Superoxide anion-mediated reduction of cytochrome $c$ is then measured by increases in absorbancy at 550 nm spectrophotometrically. The reaction is maximal after about 8 min and continues for approximately 20 min before diminishing. The reaction rate is constant between 10 and 20 min of incubation, so the slope of the line from the data points can usually be used to estimate the maximum rate of $O_2^-$ production.

[1] C. Beauchamp and I. Fridovich, *Anal. Biochem.* **44**, 276 (1971).
[2] E. J. Land and A. J. Swallow, *Arch. Biochem. Biophys.* **14S**, 365 (1971).

*Oxidation of Epinephrine to Adrenochrome.*[3] While the previous two methods above use $O_2^-$ reductant capabilities, $O_2^-$ can also act as an oxidant on certain substrates. An example of this type of reaction is oxidation of epinephrine to adrenochrome. In this assay, when $O_2^-$ gains an electron to form $H_2O_2$, oxidation of epinephrine to adrenochrome occurs and can be quantitated spectrophotometrically.

### Measurement of Hydroxyl Radical (·OH) Production

·OH is perhaps the most reactive $O_2$ metabolite produced by PMN. ·OH is most likely derived from the interaction of $O_2^-$, $H_2O_2$, and iron salts. In PMN, this most likely occurs through modified Haber–Weiss and/or possibly Fenton type reactions. In general, detection of ·OH uses reactions that take advantage of the unusual reactivity of ·OH and its ability to generate easily measured products from various substrates.

*Ethylene ($C_2H_4$) Production from Methional and Ketomethylthiobutyric Acid (KMB).*[4] When ·OH reacts with methional or KMB, ethylene is liberated. $C_2H_4$ may then be quantitated by gas chromatography. Indeed, stimulated PMN incubated with either KMB or methional in stoppered test tubes under a variety of experimental conditions produce amounts of $C_2H_2$ in head-space gas from the test tube which can be easily sampled and measured by gas chromatography.

*Methane ($CH_4$) Production from Dimethyl Sulfoxide (DMSO).*[5] Analogous to ethylene production, ·OH reacts with DMSO to release $CH_4$. As above, DMSO may be incubated with PMN under conditions stimulating phagocytosis and test tube head-space gas analyzed for $CH_4$ by gas chromatography. The specificity of the production of $CH_4$ from the reaction of ·OH with DMSO appears to be greater than reactions producing $C_2H_4$ from KMB or methianol.

*Electron Spin Resonance (ESR).*[6] Another method of detecting ·OH radical formation in PMN uses electron spin resonance spectroscopy and a "spin trap." A spin trap is a compound that is converted to relatively long-lived free radicals by less stable free radicals so that they may be identified by their ESR spectrum. The most commonly used trap is 5,5-dimethyl-1-pyrroline *N*-oxide (DMPO). When combined with ·OH, DMPO forms a relatively stable adduct of DMPO/·OH which may be readily identified by its ESR characteristics. This reaction is dependent on phagocytosis, normal respiratory burst activity, and $O_2^-$ formation.

[3] J. M. McCord and I. Fridovich, *J. Biol. Chem.* **244**, 6049 (1969).

[4] A. I. Tauber and B. M. Babior, *J. Clin. Invest.* **60**, 374 (1977).

[5] J. E. Repine, J. W. Eaton, M. W. Anders, J. R. Hoidal, and R. B. Fox, *J. Clin. Invest.* **64**, 1642 (1979).

[6] G. R. Buettner and L. W. Oberly, *Biochem. Biophys. Res. Commun.* **83**, 69 (1978).

*Measurement of Singlet Oxygen ($^1O_2$)*

It is not yet possible to rank oxidizing radicals according to their importance in biologic systems. However, many researchers have suggested that $^1O_2$ is one of the more potent $O_2$ metabolites. The most common technique for identifying $^1O_2$ is to use a reagent which reacts with $^1O_2$ to give an isolable product. The isolable product is then quantitated giving an indirect measure of the formation of $^1O_2$. However, since there are no highly specific $^1O_2$ "traps," these methods are inaccurate at best. Among the substances used are the following.

*Furans.* Furans react with $^1O_2$ to form diketones. This is one of the least specific tests for measurement of $^1O_2$ since probably any strong oxidant will produce the same product.[7]

*Dienes.* Dienes react with $^1O_2$ to form endoperoxides and the corresponding diepoxide. Either of these products can be quantitated by gas chromatography. This reaction may be more specific than the reaction of $^1O_2$ with furans.[8]

*Olefins.* The most important reaction in this group is the conversion of cholesterol to its $5\alpha$-hydroperoxide. While more specific, this reaction is inefficient, trapping only one in a thousand of the $^1O_2$ formed. Therefore, this reaction is better suited to detect the presence of $^1O_2$, but not the quantity.[9]

While all of the aforementioned reactions are nonspecific, specificity may be heightened by performing experiments in $D_2O$ which prolongs the half-life of $^1O_2$ and, thus, intensifies reactions mediated by $^1O_2$. In addition, the reactions may also be inhibited by "quenchers" such as $\beta$-carotene, bilirubin, and DABCO (triethylenediamine) which also provide some increased confidence regarding measurement of $^1O_2$.

*Detection of Hypochlorous Acid (HOCl)*[10]

The $H_2O_2$–myeloperoxidase–halide system has been proposed as a major bactericidal system in PMN. $Cl^-$ is thought to be the preferred halide because it is commonly found *in vivo*. The toxic reactive product of this system has been identified as hypochlorous acid. However, it is extremely difficult to measure HOCl because it has such a short half-life in biological systems with multiple substrates. Recently, advantage has been taken of the ability of HOCl to convert taurine to a stable product, taurine

---

[7] C. S. Foote, *Acc. Chem. Res.* **1**, 103 (1968).
[8] C. S. Foote, "Free Radicals in Biology" (W. A. Pryor, ed.), p. 85. Academic Press, New York, 1968.
[9] C. S. Foote, *Science* **162**, 963 (1968).
[10] S. J. Weiss, R. Klein, A. Slivka, and M. Wei, *J. Clin. Invest.* **70**, 598 (1982).

chloramine. This can be quantitated by its ability to oxidize iodide to iodine or the sulfhydryl compound 5-thio-2-nitrobenzoic acid to the disulfide 5,5-dithiobis(2-nitrobenzoic acid). Specificity is achieved by attempts at selective scavenging with catalase, or myeloperoxidase inhibitors, azide, cyanide, or aminotriazole.

## Measurement of Hydrogen Peroxide ($H_2O_2$)

Hydrogen peroxide ($H_2O_2$) is the least reactive of the $O_2$ metabolites in PMN but appears to be the central $O_2$ metabolite around which other $O_2$ species interact. Quantifying production of $H_2O_2$ is complicated, since whenever $O_2^-$ is formed, it will dismutate to form $H_2O_2$. In addition, reaction of $H_2O_2$ and $Fe^{2+}$ forms $\cdot OH$ (Fenton reaction and reaction of $H_2O_2$ and $O_2^-$, when catalyzed by $Fe^{2+}$ salts, also leads to $\cdot OH$ production (Haber–Weiss reaction). Among the many methods devised to detect $H_2O_2$, the following are in common usage.

*Release of $^{14}CO_2$ from Carboxylated Compounds.*[11] One method for the detection of $H_2O_2$ in PMN is based on the ability of $H_2O_2$ to oxidize compounds with carboxylate groups causing liberation of $CO_2$ as a product. Formate, for example, may be labeled with $^{14}C$ and the $^{14}CO_2$ readily collected and measured:

$$H^{14}COOH + H_2O_2 \text{ (catalase)} \rightarrow {}^{14}CO_2 + 2H_2O$$

In fact, any compound with a COOH group may be so used. Indeed, benzoate, palmitate, and other free fatty acids have all been used.

*Oxidation of Scopoletin.*[12] Scopoletin may be used to detect $H_2O_2$. The method is based on assay of decreases in scopoletin fluorescence following its oxidation by $H_2O_2$ and horseradish peroxidase. More specifically, when activated by light at 350 nm, scopoletin fluoresces with a peak at 460 nm. When oxidized by $H_2O_2$ and horseradish peroxidase, scopoletin loses its fluorescence and in direct proportion to $H_2O_2$ concentration. Oxidation of leucodiacetyl-2,7-dichlorofluorescin to its fluorescent oxidation product in the presence of horseradish peroxidase may also be used, but is limited by measuring only the excess of $H_2O_2$ over the amount consumed by the PMN during a period of time. One limitation of this method is that it only measures extracellular $H_2O_2$.[13]

*Release of $O_2$ by Catalase.*[14] Since catalase degrades $H_2O_2$ to $O_2$ and water, released $O_2$ can be measured manometrically or with an oxygen

[11] S. J. Klebanoff and S. H. Pincus, *J. Clin. Invest.* **50,** 2226 (1971).
[12] R. K. Root, J. Metcalf, N. Oshino, and B. Chance, *J. Clin. Invest.* **55,** 945 (1975).
[13] A. S. Keston and R. Brandt, *Anal. Biochem.* **11,** 1 (1965).
[14] M. Zatti, F. Rossi, and P. Patriarca, *Experientia* **24,** 669 (1968).

electrode. Catalase must be added to the system, and unless cyanide or azide is added to inhibit $H_2O_2$ breakdown by competing systems, this method will detect little $H_2O_2$.

### Chemiluminescence (CL)[15]

Under conditions stimulating phagocytosis, PMN emit photons of light (chemiluminescence) which can be measured in a scintillation counter operated in the singlet mode. The specificity of this phenomenon is very low. Nearly every $O_2$ metabolite has been implicated in the generation of CL and attempts at adding scavengers have not improved the specificity of the nature of light produced. The difficulties with specificity probably reflect, at least in part, complexities involved in transfer of energy from generating and accepting molecules which are then converted to an excited state. This procedure is useful as an overall evaluation of the respiratory burst in intact PMN and in chemical systems, but its lack of specificity is a drawback. Care should be taken to perform CL at 37°.

### Summary

$O_2$ radicals are important in health and disease. The most commonly used ways of identifying $O_2$ radicals in PMN are described above. Several shortcomings exist in these methods reflecting the unusual, complex nature of $O_2$ radical biochemistry. Some general principles include (1) $O_2$ radicals are very short-lived, reacting with many other compounds and each other quickly. (2) There are no highly specific assays for $O_2$ radicals. (3) Highly specific scavengers of $O_2$ radicals also do not exist. (4) No methods have been found to detect and quantitate $O_2$ radicals *in vivo*. (5) Solubility and membrane permeability of various scavengers and/or test reagents may affect the measurement of $O_2$ radicals in PMN and other biological systems.

In general, the best approach to measurement of $O_2$ radicals involves using the best assay available and showing that the reaction is inhibited by scavengers in proportion to their reactivity with the specific $O_2$ radical being assayed.

### Acknowledgments

Supported in part by grants from the American Heart Association of Colorado, National Institutes of Health (HL 24248 and HL 21248), American Lung Association of Colorado, Council for Tobacco Research, and the Kroc, Hill, Swan and Kleberg Foundations.

[15] R. C. Allen, R. L. Stjernholm, and R. H. Steele, *Biochem. Biophys. Res. Commun.* **47**, 679 (1972).

# [52] Antimicrobial Activity of Myeloperoxidase

*By* SEYMOUR J. KLEBANOFF, ANN M. WALTERSDORPH,
and HENRY ROSEN

Peroxidases when combined with $H_2O_2$ and a halide (chloride, bromide, iodide, and the pseudohalide thiocyanate) form a potent cytotoxic system which contributes to the host defense against invading microorganisms and possibly tumor cells.[1–5] Neutrophils and monocytes contain the same peroxidase (myeloperoxidase, MPO), and eosinophils a different peroxidase (eosinophil peroxidase, EPO), in cytoplasmic granules and these enzymes are discharged into the phagosome following particle ingestion. Phagocytosis also is associated with a respiratory burst and much of the added oxygen consumed is converted to $H_2O_2$. Peroxidase, $H_2O_2$, and a halide interact in the phagosome to destroy the ingested organism. The components of the peroxidase system can also be released extracellularly where they may attack adjacent normal or malignant cells, uningested organisms, or soluble mediators. Lactoperoxidase (LPO) which differs from MPO and EPO, may contribute to the antimicrobial activity of milk and saliva. In these biological fluids, $H_2O_2$-generating microorganisms (e.g., lactic acid bacteria such as lactobacilli or streptococci in saliva), enzymes (e.g., xanthine oxidase in milk), or phagocytes may serve as the source of $H_2O_2$, and the pseudohalide thiocyanate, which is present in high concentration, may be the important cofactor for the peroxidase system. The peroxidase reacts with $H_2O_2$ to form an enzyme–substrate complex which oxidizes the halide to a toxic agent or agents. At least for some halides, the reactive species is the hypohalous acid, and the toxicity results from a combination of halogenation and oxidation of essential components at or near the cell surface.

[1] S. J. Klebanoff and R. A. Clark, "The Neutrophil: Function and Clinical Disorders." North-Holland Publ., Amsterdam, 1978.

[2] J. Schultz, *in* "The Reticuloendothelial System. A Comprehensive Treatise 2. Biochemistry and Metabolism" (A. J. Sbarra and R. R. Strauss, eds.), p. 231. Plenum, New York, 1980.

[3] J. M. Zgliczynski, *in* "The Reticuloendothelial System. A Comprehensive Treatise 2. Biochemistry and Metabolism" (A. J. Sbarra and R. R. Strauss, eds.), p. 255. Plenum, New York, 1980.

[4] S. J. Klebanoff, *in* "The Reticuloendothelial System. A Comprehensive Treatise 2. Biochemistry and Metabolism" (A. J. Sbarra and R. R. Strauss, eds.), p. 279. Plenum, New York, 1980.

[5] S. J. Klebanoff, *in* "Advances in Host Defense Mechanisms" (J. I. Gallin and A. S. Fauci, eds.), Vol. 1, p. 111. Raven, New York, 1982.

METHODS IN ENZYMOLOGY, VOL. 105
Copyright © 1984 by Academic Press, Inc.
All rights of reproduction in any form reserved.
ISBN 0-12-182005-X

A variety of methods have been employed for the measurement of the toxicity of the peroxidase system. In general, these methods depend on the nature of the target cell and include measurement of replication in growth medium, $^{51}$Cr release, metabolic activity, and morphologic changes. We will concentrate here on bactericidal activity as measured by decrease in colony-forming units, using *Escherichia coli* as the target, MPO as the peroxidase, and chloride as the halide.

### Reagents

*Escherichia coli*. The bacteria are maintained on blood agar plates and transferred in the late afternoon prior to the day of use to trypticase soy broth (BBL Microbiological Systems, Becton, Dickinson & Co., Cockeysville MD). Overnight (37°) cultures are washed twice in 0.1 *M* sodium sulfate and suspended in the same solution to the required absorbancy at 540 nm. The absorbancy is that which on prior experiments had indicated a concentration of $5 \times 10^7$ organisms per milliliter.

*Myeloperoxidase*. A whole leukocyte or granule extract can be employed as a source of peroxidase. In general, lysis is best achieved with weak acid or with detergent [e.g., 0.05% Triton X-100, 0.02% cetyltrimethylammonium bromide (CTAB)].[6–8] Chloride should be avoided in the extraction medium as this or another halide is a requirement for the microbicidal system, and a concentration of detergent should be employed which is not itself toxic to the test organism. The use of purified MPO, prepared by any of a number of described methods,[2,9–21] is preferred.

[6] P. Patriarca, R. Cramer, M. Marussi, F. Rossi, and D. Romeo, *Biochim. Biophys. Acta* **237**, 335 (1971).

[7] A. Jörg, J.-M. Pasquier, and S. J. Klebanoff, *Biochim. Biophys. Acta* **701**, 185 (1982).

[8] Z. A. Cohn and J. G. Hirsch, *J. Exp. Med.* **112**, 983 (1960).

[9] K. Agner, *Acta Chem. Scand.* **12**, 89 (1958).

[10] J. Schultz and H. W. Schmukler, *Biochemistry* **3**, 1234 (1964).

[11] G. F. Rohrer, J. P. von Wartburg, and H. Aebi, *Biochem. Z.* **344**, 478 (1966).

[12] I. Olsson, T. Olofsson, and H. Odeberg, *Scand. J. Haematol.* **9**, 483 (1972).

[13] S. R. Himmelhoch, W. H. Evans, M. G. Mage, and E. A. Peterson, *Biochemistry* **8**, 914 (1969).

[14] R. K. Desser, S. R. Himmelhoch, W. H. Evans, M. Januska, M. Mage, and E. Shelton, *Arch. Biochem. Biophys.* **148**, 452 (1972).

[15] J. E. Harrison, S. Pabalan, and J. Schultz, *Biochim. Biophys. Acta* **493**, 247 (1977).

[16] A. R. J. Bakkenist, R. Wever, T. Vulsma, H. Plat, and B. F. van Gelder, *Biochim. Biophys. Acta* **524**, 45 (1978).

[17] N. R. Matheson, P. S. Wong, and J. Travis, *Biochemistry* **20**, 325 (1981).

[18] M. Yamada, M. Mori, and T. Sugimura, *Biochemistry* **20**, 766 (1981).

[19] P. C. Andrews and N. I. Krinsky, *J. Biol. Chem.* **256**, 4211 (1981).

[20] M. R. Anderson, C. L. Atkin, and H. J. Eyre, *Arch. Biochem. Biophys.* **214**, 273 (1982).

[21] W. M. Nauseef, R. K. Root, and H. L. Malech, *J. Clin. Invest.* **71**, 1297 (1983).

Stock concentrated solutions in water are stable for long periods at $-20°$. On the day of the experiment, the peroxidase is diluted in water and assayed as follows. Ten microliters of the MPO preparation is added to a 1-cm light path cuvette containing 3.0 ml of the assay solution. This solution is made up daily as follows: $H_2O$, 26.9 ml, 0.1 $M$ sodium phosphate buffer, pH 7.0, 3.0 ml, 0.1 $M$ $H_2O_2$, 0.1 ml, and Guaiacol (Sigma Chemical Co., St. Louis MO), 0.048 ml, and stored in a capped opaque glass flask at room temperature. The cuvette components are rapidly mixed, the absorbancy is determined at 470 nm for 1 min in a recording spectrophotometer, and the optical density change per minute is calculated from the initial rate. One unit of enzyme is the amount that consumes 1 $\mu$mol of $H_2O_2$ per minute. Four moles of $H_2O_2$ are required to produce 1 mol of the tetraguaiacol product which has an extinction coefficient ($E$) of 26.6 m$M^{-1}$ cm$^{-1}$ at 470 nm.[22] Units are calculated as follows:

$$\text{units/ml} = (\Delta OD \times V_t \times 4)/(E \times \Delta_t \times V_s)$$

where $V_t$ = total volume (milliliters), $V_s$ = sample volume (milliliters), $\Delta OD$ = density change, and $\Delta_t$ = time of measurement (minutes). Under the specified conditions the calculation simplifies to

$$\text{units/milliliter} = \Delta OD/\text{minutes} \times 45.1$$

The MPO diluted to 13.5 units/ml is maintained on ice and used within 30 min of dilution.

*$H_2O_2$ ($10^{-4}$ $M$).* On the day of the experiment, the stock 30% $H_2O_2$ is diluted with water to 0.1 $M$ and just prior to the experiment the 0.1 $M$ solution is further diluted to $10^{-4}$ $M$ and maintained on ice.

*Sodium chloride (0.1 M).*

*Sodium phosphate buffer pH 7.0 (0.1 M).*

*Sodium sulfate (0.67 M).*

*Trypticase soy agar.* On the day of the experiment, the agar (BBL), previously dispensed into 500-ml Erlenmeyer flasks, is liquefied in an autoclave or boiling water bath and maintained at 60°.

## Procedure

The reagents are added to 12 × 75-mm polystyrene test tubes (Falcon Labware, No 2052, Becton Dickinson Labware, Oxnard, CA) in the order shown in the table. When sodium chloride is omitted (or its concentration reduced), sodium sulfate is added to maintain tonicity. The tubes are incubated at 37° in a shaking water bath for the required period (generally 15–60 min), 0.1 ml of the reaction mixture is placed in duplicate 100 × 15-

[22] B. Chance and A. C. Maehly, this series, Vol. 2, p. 764.

BACTERICIDAL ACTIVITY OF THE
MYELOPEROXIDASE–H$_2$O$_2$–CHLORIDE SYSTEM[a]

| Component | Tube number | | | | |
|---|---|---|---|---|---|
| | 1 | 2 | 3 | 4 | 5 |
| Water | 0.35 | 0.295 | 0.345 | 0.3 | 0.295 |
| Phosphate buffer pH 7.0 | 0.05 | 0.05 | 0.05 | 0.05 | 0.05 |
| Sodium sulfate | 0.05 | | | | 0.05 |
| Sodium chloride | | 0.05 | 0.05 | 0.05 | |
| Escherichia coli | 0.05 | 0.05 | 0.05 | 0.05 | 0.05 |
| Myeloperoxidase | | 0.005 | 0.005 | | 0.005 |
| H$_2$O$_2$ | | 0.05 | | 0.05 | 0.05 |
| Organisms/ml × 10⁻⁶ | 4.04 | 0.001 | 3.30 | 3.46 | 3.54 |

[a] The final concentration of reagents were sodium phosphate buffer pH 7.0, $10^{-2}$ $M$; sodium sulfate, $6.7 \times 10^{-2}$ $M$; sodium chloride, $0.1$ $M$; $E.\ coli$, $5 \times 10^6$; myeloperoxidase, 0.135 units/ ml; H$_2$O$_2$, $10^{-5}$ $M$. Incubation 60 min.

mm petri dishes (Falcon No. 1029), and 0.1 ml is serially diluted in 0.1 $M$ sodium sulfate. Aliquots (0.1 ml) of the $10^2$ and $10^3$ dilutions are placed in petri dishes and 5–10 ml of liquid agar are immediately added to all plates by pouring. The organisms are thoroughly mixed with the agar by swirling with a figure 8 motion and the agar allowed to solidify at room temperature. The petri dish cover is kept slightly ajar, exposing the agar to air during the solidification period to minimize condensation on the undersurface of the cover. When multiple tubes are employed in the experiment, each reaction is begun at 2-min intervals with the addition of $E.\ coli$, MPO, and H$_2$O$_2$, thus allowing time for sequential dilution and plating of samples at the end of the incubation period. The plates are placed at 37° for 24–48 hr and the viable cell count is determined with a Quebec darkfield colony counter (American Optical Co., Buffalo, NY). The data (see the table) are expressed as organisms per milliliter of original reaction mixture. For statistical analysis, reaction mixtures yielding sterile plates are assigned a value of 10 organisms per milliliter, the limit of the sensitivity of the assay since 0.1-ml aliquots of the reaction mixture are employed. The values are converted to their logarithm, and arithmetic means, standard errors, and $p$ values are determined for the logarithmically transformed data. The data are expressed as the antilog of the arithmetic mean of the logarithms (the geometric mean of the original untransformed values). Statistical differences are determined using Student's two-tailed $t$ test for independent means (not significant, $p > 0.05$).

The table is a typical experiment demonstrating the bactericidal activity of the MPO–$H_2O_2$–chloride system and the requirement for each component. The chloride can be replaced by other halides (bromide, iodide), $H_2O_2$ by a $H_2O_2$ generating system (glucose + glucose oxidase, xanthine + xanthine oxidase), and MPO by EPO (for purification see Refs. 7,14,23,24) or LPO (available commercially). The effectiveness of the halide varies with the peroxidase; MPO for example is much more effective with chloride relative to iodide or bromide than is EPO or LPO. The pH optimum of the MPO-mediated antimicrobial system is generally acid. Under the conditions employed in the table, chloride was effective at concentrations down to $10^{-3}$ $M$ at pH 7.0 whereas in phosphate buffer pH 5.0 bactericidal activity was observed at a chloride concentration of $5 \times 10^{-5}$ $M$. MPO is inhibited by excess $H_2O_2$.[25-27]

The experiment shown in the table was performed in the absence of added protein. It may be necessary for some organisms, or other target cells, to add protein to the reaction mixture to maintain viability in control tubes. Protein can scavenge the reaction products of the peroxidase system and inhibit toxicity.[28] Thus the lowest protein concentration needed to maintain viability should be employed. If required, we generally add gelatin at a concentration of 0.001% or less. Even at 0.001%, gelatin inhibited toxicity at low chloride concentration ($10^{-3}$ $M$) under the conditions employed in the table.

[23] G. T. Archer and M. Jackas, *Nature* (*London*) **205**, 599 (1965).
[24] R. Wever, H. Plat, and M. N. Hamers, *FEBS Lett.* **123**, 327 (1981).
[25] K. Agner, *Acta Physiol. Scand.* **2** (Suppl 8), 1 (1941).
[26] K. Agner, *Acta Chem. Scand.* **17**, 332 (1963).
[27] S. J. Klebanoff, *Infect. Immun.* **25**, 153 (1979).
[28] C. F. Nathan and S. J. Klebanoff, *J. Exp. Med.* **155**, 1291 (1982).

## [53] Determination of Microbial Damage Caused by Oxygen Free Radicals, and the Protective Role of Superoxide Dismutase*

### By HOSNI M. HASSAN

The importance of superoxide dismutase (SOD) as a defense against the cellular damage caused by oxygen free radicals has been extensively demonstrated in microbial systems.[1-3] Most of the studies have been carried out with gram-negative enteric organisms,[1,2] but a few studies have been done using gram-positive staphylococci.[4] In this type of study, a generator of oxygen free radicals and a test organism are required. Cellular damage may be assessed by enumeration of viable cells, loss of some vital cellular function, such as transport of nutrients, structural changes as observed by electron microscopy, or by the release of some intracellular marker enzymes. In this chapter methods for generating oxygen free radicals, for manipulating the cellular concentration of SOD, and for assessing the damage caused by oxygen-free radicals are presented.

### Methods for Generating Oxygen Free Radicals

The superoxide free radical ($O_2^-$) is generated in many biological reactions that reduce molecular oxygen. Systems that generate $O_2^-$ will also form hydrogen peroxide ($H_2O_2$) and the hydroxyl radical ($OH\cdot$). To assure that the system is generating predominately $O_2^-$, catalase and hydroxyl radical scavengers should be added. It should also be noted that the *Escherichia coli* cell envelope is impermeable to the superoxide anion[5]; therefore, it is necessary to choose the generating system that delivers $O_2^-$ to the desired target site.

---

* Paper number 8756 of the Journal Series of the North Carolina Agricultural Research Service, Raleigh, NC 27650. The use of trade names in the publication does not imply endorsement by the North Carolina Agricultural Research Service of the products named, nor criticism of similar ones not mentioned.

[1] H. M. Hassan and I. Fridovich, *J. Bacteriol.* **129**, 1574 (1977).
[2] E. M. Gregory and I. Fridovich, *J. Bacteriol.* **114**, 1193 (1973).
[3] H. M. Hassan and I. Fridovich, *in* "Enzymatic Basis of Detoxication" (W. B. Jakoby, ed.), Vol. 1, p. 311. Academic Press, New York, 1980.
[4] B. M. Babior, J. T. Curnutte, and R. S. Kipnes, *J. Lab. Clin. Med.* **85**, 235 (1975).
[5] H. M. Hassan and I. Fridovich, *J. Biol. Chem.* **254**, 10846 (1979).

Copyright © 1984 by Academic Press, Inc.
All rights of reproduction in any form reserved.
ISBN 0-12-182005-X

*Extracellular Generation of $O_2^-$*

*Enzymic.* $O_2^-$ is generated during the normal catalytic action of several oxidative enzymes, i.e., xanthine oxidase, aldehyde oxidase, and dihydroorotic dehydrogenase. The reaction of xanthine oxidase with xanthine[6] is used most frequently to generate $O_2^-$ at a steady and controlled rate. The rate of $O_2^-$ generation can be determined by measuring the rate of SOD-inhibitable cytochrome *c* reduction, at 550 nm.[7]

MATERIALS. All reagents are prepared in 50 m$M$ phosphate, 0.1 m$M$ EDTA, pH 7.8.

Xanthine oxidase. This enzyme may be prepared from cream[8] or may be obtained commercially. The stock enzyme can be stored frozen in ammonium sulfate. It should be diluted to give the desired rate of $O_2^-$ production. For example, the enzyme stock may be diluted so that a 5-$\mu$l sample will catalyze the reduction of 3 ml of cytochrome *c* $10^{-5}$ $M$) at a rate equal to 0.05 absorbancy units/min at 550 nm. Under these conditions, the rate of $O_2^-$ generation is 2.4 × $10^{-6}$ $M$/min (using the extinction coefficient for the reduction of ferricytochrome *c* as given by Massey,[9] $\Delta\varepsilon_{550} = 21,000$ $M^{-1}$ cm$^{-1}$).

Xanthine, 5 m$M$. Xanthine is dissolved by boiling. The stock solution may be stored frozen or refrigerated; however, it should be reheated before use to redissolve the precipitated xanthine.

PROCEDURE. Pipet 0.8-ml aliquots of a washed cell suspension containing $10^7–10^8$ viable cells into sterile test tubes. Bring the volume to 1 ml by adding 0.2 ml of xanthine or of sterile phosphate buffer. Equilibrate the tubes to 37° in a shaking (200 rpm) water bath. Remove 0.1 ml samples (and save on ice) for determining the initial viable counts. Start the reaction by adding xanthine oxidase (1–10 $\mu$l) to yield the desired rate of $O_2^-$ generation. At the specified time intervals, remove 0.1-ml samples and place into ice-cold 9.9 ml dilution blanks (0.05 $M$ phosphate, and 0.1 m$M$ MgSO$_4$, pH 7.0). This 100-fold dilution and the low temperature are normally enough to stop the reaction; however, allopurinol[10,11] (25–50 $\mu M$) may be used to inhibit the xanthine oxidase reaction. Spread aliquots (0.1 ml) from the appropriate dilutions onto nutrients agar plates [or onto any appropriate medium, i.e., trypticase soy yeast extract agar[1] (TSY) or

---

[6] J. M. McCord and I. Fridovich, *J. Biol. Chem.* **244**, 6079 (1969).
[7] J. D. Crapo, J. M. McCord, and I. Fridovich, this series, Vol. 53, p. 382.
[8] W. R. Waud, F. O. Brady, R. D. Wiley, and K. V. Rajagopalan, *Arch. Biochem. Biophys.* **169**, 695 (1975).
[9] V. Massey, *Biochim. Biophys. Acta* **34**, 255 (1959).
[10] T. Spector and D. G. Johns, *J. Biol. Chem.* **245**, 5079 (1970).
[11] V. Massey, H. Komai, G. Palmer, and G. B. Elion, *J. Biol. Chem.* **245**, 2837 (1970).

glucose minimal agar[12]]. Incubate the plates at 37° for 48–72 hr. Count the plates with 30–300 colonies and estimate viable counts per milliliter of the original sample. Calculate the percentage viable count from the relationship viable count at any time/viable count at zero time × 100. Plot the data as percentage viable count versus time. Tubes without xanthine and without xanthine oxidase should be included as controls. The effects of superoxide dismutase (5–25 µg/ml), catalase (5–50 µg/ml), and different hydroxyl radical scavengers may be incorporated in the experimental design.

*Photochemical.* Electronically excited dyes (via the absorption of a photon) are active oxidants that can readily be reduced to the semiquinone form by a variety of reductants [EDTA, methionine, $N,N,N',N'$-tetramethylethylenediamine (TEMED), NADH, etc.]. The autoxidation of the semiquinone form by oxygen generates $O_2^-$. The photoreduction of flavins by EDTA was first noted by Frisell *et al.*[13] and was used later by Massey *et al.*[14,15] as a convenient source of $O_2^-$.

MATERIALS. All reagents are prepared in 50 m$M$ phosphate, pH 7.8.

Methionine, 0.15 $M$

Riboflavin, 5 m$M$

Light source, 20–200 W "cool white" fluorescent tubes

The rate of $O_2^-$ generation can be estimated by measuring the rate of SOD-inhibitable ferricytochrome $c$ reduction as mentioned above, or by using any other suitable detector (i.e., nitroblue tetrazolium,[16] tetranitromethane,[6] or hydroxylamine[17]).

PROCEDURE. Pipet 0.8-ml aliquots of a washed cell suspension containing $10^7$–$10^8$ viable cells into sterile test tubes. Add 0.1 ml of riboflavin and 0.1 ml of methionine. Place the tubes on a shaker (200 rpm) at room temperature, or in an incubator at the desired temperature. Start the reaction by turning on the light source. The light intensity may be varied by changing the number of light tubes. Handle the samples and determine viable counts and rate of killing as outlined in the section Enzymic.

*Other Methods.* Several other methods have been used to generate extracellular $O_2^-$. These include but are not limited to (1) γ-irradiation of aerobic solutions (approximately 600 rads/min) in the presence of for-

[12] H. J. Vogel and D. M. Bonner, *J. Biol. Chem.* **218,** 97 (1956).
[13] W. R. Frisell, C. W. Chung, and C. G. MacKenzie, *J. Biol. Chem.* **234,** 1297 (1959).
[14] V. Massey, S. Strickland, S. G. Mayhew, L. G. Howell, P. C. Engel, R. G. Matthews, M. Schuman, and P. A. Sullivan, *Biochem. Biophys. Res. Commun.* **36,** 891 (1969).
[15] D. Ballou, G. Palmer, and V. Massey, *Biochem. Biophys. Res. Commun.* **36,** 898 (1969).
[16] C. Beauchamp and I. Fridovich, *Anal. Biochem.* **44,** 276 (1971).
[17] E. F. Elster and A. Heupel, *Anal. Biochem.* **70,** 616 (1976).

mate[18]; (2) use of activated polymorphonuclear leukocytes[19] (PMNs); and (3) use of commercially available potassium superoxide ($KO_2$).[20]

*Intracellular Generation of $O_2^-$*

Most biological electron transport chains end with terminal cytochrome oxidases that reduce dioxygen to water without the release of oxygen free radicals. However, a small but significant fraction of the total oxygen consumed in microorganisms is spontaneously reduced via a univalent pathway which generates $O_2^-$. Furthermore, it has been shown that the intracellular flux of $O_2^-$ may be increased by hyperbaric oxygen[21–24] (HPO) or by the use of certain redox-active compounds.[25,26] In *E. coli*, a positive correlation is seen between the intracellular production of $O_2^-$, the induction of the manganese-containing SOD (MnSOD), and the rate of cyanide resistant respiration, as influenced by several redox-active compounds.[26] It is clear, therefore, that conditions known to restrict the synthesis of the oxygen detoxifying enzymes (i.e., using a nutritionally restricted media, or adding inhibitors of protein biosynthesis) should be used in order to accurately demonstrate the damaging effects of intracellular oxygen free radicals.

*Hyperbaric Oxygen.* Exposure to 4 to 20 atm of oxygen ($O_2$) has been routinely used to assess the toxicity of oxygen.[21–24] It is important to maintain a normal atmospheric $CO_2$ concentration in the pressurizing chamber, since removal of $CO_2$ has been reported to cause cellular death in *E. coli*.[27]

MATERIALS

Gas-tight stainless-steel cylinders with pressure gauges (pressure bombs)
Pressurized gases ($N_2$ and $O_2$)
Test organism suspended in minimal medium[12]
Chloramphenicol or puromycin

[18] C. Monny and A. M. Michelson, *Biochimie* **64**, 451 (1982).
[19] B. M. Babior, *N. Engl. J. Med.* **298**, 659 (1978).
[20] S. Marklund, *J. Biol. Chem.* **251**, 7504 (1976).
[21] R. Gerschman, D. L. Gilbert, S. W. Nye, P. Dwyer, and W. O. Fenn, *Science* **119**, 623 (1954).
[22] S. F. Gottlieb, *Annu. Rev. Microbiol.* **25**, 111 (1971).
[23] E. M. Gregory and I. Fridovich, *J. Bacteriol.* **114**, 543 (1973).
[24] H. M. Hassan and I. Fridovich, *J. Bacteriol.* **130**, 805 (1977).
[25] H. L. White and J. R. White, *Mol. Pharmacol.* **4**, 549 (1968).
[26] H. M. Hassan and I. Fridovich, *Arch. Biochem. Biophys.* **196**, 385 (1979).
[27] O. R. Brown and H. F. Howitt, *Microbios* **3**, 241 (1969).

PROCEDURE. Place 2- to 5-ml aliquots of a washed cell suspension containing $10^6$–$10^7$ viable cells per milliliter into sterile petri plates. Add an inhibitor of protein biosynthesis (i.e., chloramphenicol, 100–200 $\mu g$/ml; or puromycin, 100–500 $\mu g$/ml). Equal numbers of plates without added inhibitor should be included as a control. Divide the plates into two groups. Expose one group to the desired oxygen pressure (4–20 atm $O_2$) and the other group to the same pressure of nitrogen. Do not evacuate the chamber before adding $O_2$ or $N_2$. Incubate the pressurized chambers at 37° for the desired length of time (1–5 hr), after which the pressure should be gradually released to avoid cellular breakage. Survivors should be enumerated after dilution by plating on TSY or on an appropriate medium. Calculate percentage survival as indicated in the section Enzymic.

The effects of extracellularly added protective agents (i.e., SOD, catalase, or OH· scavengers) as well as of intracellularly induced enzymes may be tested in the same fashion.

*Redox-Active Compounds.* The lethality of streptonigrin,[25] paraquat,[28] juglone,[26] pyocyanine,[29] and of several other redox-active compounds[26] is enhanced by oxygen. This has been attributed to their abilities to exacerbate the intracellular production of oxygen free radicals during their redox cycling inside the cells. These compounds increase cyanide-insensitive respiration and cause induction of the natural defensive enzymes[26] (i.e., SODs and catalases). Streptonigrin,[1,24,30] paraquat,[28] and plumbagin[31] are routinely used to assess oxygen toxicity and to determine the different mechanisms of cellular protection. When using these compounds as intracellular generators of oxygen free radicals, certain requirements should be met[26,28]: (1) the cells should be permeable to the test compound; (2) an abundant supply of both electrons and dioxygen should be present; (3) the cells should be able to reduce the test compound via an NAD(P)H-dependent diaphorase or comparable enzyme; and (4) conditions that are known to restrict or impair the ability of the cells to induce the defensive enzymes SOD and catalase should be used.

MATERIALS

  Paraquat (0.1–1.0 m$M$), streptonigrin (1–2 $\mu g$/ml), or any other suitable redox-active compound
  Test organism (from the logarithmic phase of growth)
  Suspending media (nutrient broth or glucose-minimal medium)
  Chloramphenicol (100–200 $\mu g$/ml), or puromycin (100–500 $\mu g$/ml)

PROCEDURE. Suspend the washed cells in the appropriate medium to

[28] H. M. Hassan and I. Fridovich, *J. Biol. Chem.* **253**, 8143 (1978).
[29] H. M. Hassan and I. Fridovich, *J. Bacteriol.* **141**, 156 (1980).
[30] H. M. Hassan and I. Fridovich, *J. Biol. Chem.* **252**, 7667 (1977).
[31] F. S. Archibald and I. Fridovich, *J. Bacteriol.* **145**, 442 (1981).

$10^7$–$10^8$ cells/ml, add inhibitor of protein biosynthesis and 2 $\mu$g/ml of streptonigrin or any other test compound. Incubate at 37° and 200 rpm. At intervals, remove samples, dilute, and plate onto TSY (or any appropriate medium) for assessment of survival, as in section Enzymic. Controls testing for the absence of redox-active compound, inhibitor of protein biosynthesis, and oxygen should also be performed. Different concentrations of the redox-active compound should be tested. Cells containing different concentrations of SOD and catalase may be tested for resistance to the oxygen-enhanced lethality of these redox-active compounds.[28,30]

## Methods for Manipulating the Cellular Concentration of SOD

If oxygen toxicity is due to the generation of $O_2^-$ (and other reactive species formed therefrom), then superoxide dismutases should provide an essential defense. Knowledge of the mechanism(s) of induction and regulation of SOD in *E. coli*[1,3,24,26,30,32] has provided a better understanding and a strong support to the superoxide theory of oxygen toxicity.[3,33] Several methods are available to induce the synthesis of SOD in enteric gramnegative facultative anaerobes.[1,28–30] SOD is also inducible in *Streptococcus faecalis,*[23] *Saccharomyces cerevisiae,*[34] *Euglena gracilis,*[35] adult rats,[36] and in numerous other organisms. In general, SOD is induced by conditions that increase the intracellular flux of $O_2^-$, (i.e., exposure to elevated $p_{O_2}$ or to redox-active compounds). In most cases induction is more pronounced in rich media than in poor media.[28] In a similar vein, reduced levels of oxygen decrease the SOD content of facultative anaerobic organisms.[1] In this section procedures for manipulating SOD in *E. coli* will be presented.

### Induction by High $p_{O_2}$

*Procedure.* Inoculate 100 ml of TSY medium with an overnight culture of *E. coli,* grown in the same medium, to an initial optical density at 600 nm ($OD_{600}$) equal to 0.1. Incubate at 37° on a shaker (200 rpm). When the culture reaches an $OD_{600}$ of 0.4 (i.e., after two generations) bubble the culture with 100% oxygen or any desired concentration of oxygen. Allow the culture to grow for at least two more generations (i.e., $OD_{600} \simeq 1.6$).

[32] H. M. Hassan and I. Fridovich, *J. Bacteriol.* **132,** 505 (1977).
[33] J. M. McCord, B. B. Keele, Jr., and I. Fridovich, *Proc. Natl. Acad. Sci. U.S.A.* **68,** 1024 (1971).
[34] E. M. Gregory, S. A. Goscin, and I. Fridovich, *J. Bacteriol.* **117,** 456 (1974).
[35] K. Asada, S. Kanematsu, M. Takehashi, and Y. Kona, *Adv. Exp. Med. Biol.* **74,** 551 (1976).
[36] J. D. Crapo and D. L. Tierney, *Am. J. Physiol.* **226,** 1401 (1974).

Harvest the cells, prepare dialyzed cell-free extracts, and assay for SOD activity.[7] Dialyzed samples may be subjected to polyacrylamide gel electrophoresis followed by activity staining[1,16] to distinguish the degree of inducibility of the different isozymes.

## Induction by Redox-Active Compounds

*Procedure.* The procedure is essentially the same as that listed above, except that different concentrations of paraquat[30] (0.1–1.0 m$M$) or of other redox-active compounds[26] may be added in place of 100% oxygen. For more details on the induction of MnSOD by paraquat see [69] of this volume.

## Deinduction of MnSOD by Anaerobiosis

*Escherichia coli* cells lacking MnSOD can be obtained by growing the cultures anaerobically.[1]

*Procedure.* Strict anaerobic conditions are required. This is best achieved by using an anaerobic chamber. The inoculum should be grown anaerobically for at least two transfers in order to adapt the culture to anaerobiosis. Inoculate the culture into TSY medium that has equilibrated in the anaerobic chamber for two days. After growth is completed pour the culture onto ice containing chloramphenicol (100–200 $\mu$g/ml) before removing from the chamber. This step is essential in order to stop growth and prevent the induction of MnSOD upon exposure to air. Prepare dialyzed cell-free extracts and determine the specific activity of SOD.[7] Run gel electrophoresis and stain for SOD activity[16] to establish the absence of MnSOD.

## Methods for Assessing Biological Damage Caused by Oxygen Free Radicals

Cellular damage by oxygen free radicals is normally prevented by the presence of superoxide dismutases and hydroperoxidases. It is proposed, therefore, that in order for damage to be manifested, these cellular defense mechanisms have to be overcome.

In the previous sections loss of viability was used for the assessment of the damage. Loss of viability is normally caused by irreparable damage to one or more of the vital cellular components. Therefore, the use of specific repair deficient mutants may further amplify the damage and loss of viability.

There are still other methods available for assessing cellular damage caused by oxy-radicals.

## Structural Damage

Electron microscopy has been used to visualize damage caused by oxygen free radicals.[1,37]

*Procedure.* Expose the culture to the desired flux of oxygen free radicals under conditions that do not permit induction of the protective enzymes. Remove the cells by centrifuging for 5 min at 2000 $g$ at 4°. Suspend the cells in cold 0.05 $M$ veronal acetate (VA) buffer, pH 7.6, and centrifuge again. Suspend the washed cells in the same VA buffer containing 4% glutaraldehyde, and let stand for 2 hr at room temperature. Suspend the glutaraldehyde-fixed cells in warm 4% agar containing 0.5% tryptone and 0.5% NaCl. Spread the agar-trapped cells on glass slides and allow the agar to solidify. Cut the agar into small cubes, and place in test tubes. Wash the agar cubes five times in the VA buffer and then soak in 0.5% uranyl acetate for 2 hr followed by washing five times with the VA buffer. Immerse the agar cubes in 2% osmium tetroxide $OsO_4$ for 2 hr. Remove the $OsO_4$ by five washes with VA buffer. Dehydrate the agar cubes by successive 10-min passages through 20, 30, 40, 50, 70, 80, 90, and 95% ethanol followed by three 5-min washes in 100% ethanol. Soak the dehydrated agar cubes for 15 min in ethanol–propylene oxide (1 : 1 by volume), followed by two 5-min soaks in propylene oxide. Pass the cubes successively through propylene oxide : Epon (3 : 1, 1 : 1, and 1 : 3 by volume). Embed the cubes in pure Epon 812 (Luft). Polymerize the resin at 60° for 48 hr. Prepare thin sections using an ultramicrotome equipped with a diamond knife. Place the sections on carbon-coated copper grids and stain with uranyl acetate[38] for 5 min and with lead citrate[39] for 3 min; then examine with an electron microscope.

## Impairment of Nutrient Uptake

Superoxide anion radicals, generated by the xanthine/xanthine oxidase system, have been reported to inhibit the transport of [[14]C]leucine in *E. coli,*[40] and of 2-amino[[14]C]isobutyric acid (AIB) in thymocytes.[41]

*Procedure.* Expose washed cells to a flux of $O_2^-$ (generated extra- or intracellularly) for varied lengths of time. Make sure that appropriate measures are taken to prevent the induction of the protective enzymes (i.e., absence of carbon source in the suspending medium or presence of

[37] H. M. Hassan and I. Fridovich, *Rev. Infect. Dis.* **1,** 357 (1979).
[38] M. L. Watson, *J. Biophys. Biochem. Cytol.* **4,** 475 (1958).
[39] E. S. Reynolds, *J. Cell Biol.* **17,** 208 (1963).
[40] V. M. E. Marshall and B. Reiter, *Proc. Soc. Gen. Microbiol.* **3,** 189 (1976).
[41] L. Kwock, P. S. Lin, and L. Ciborowshi, *Cell Biol. Int. Rep.* **5,** 83 (1981).

inhibitors of protein biosynthesis). Wash the treated cells and use to study nutrient uptake according to standard procedures.[42,43]

*Genetic Damage*

Generators of oxygen-free radicals (i.e., paraquat,[44] $\gamma$ irradiation,[45] and PMNs[46]) have been shown to cause mutations.

*Procedure.* See [30] this volume on the mutagenicity of oxygen free radicals.

### Acknowledgments

I wish to thank I. Fridovich and Carmella S. Moody for critically reading the manuscript; and I am grateful to Ann Farmer for expert typing. This work was supported, in part, by NIH Biomedical Research Support Grant 88204, and NSF Grant PCM-8213853.

[42] H. R. Kaback, *Biochim. Biophys. Acta* **265**, 367 (1972).
[43] H. M. Hassan and R. A. MacLeod, *J. Bacteriol.* **121**, 160 (1975).
[44] C. S. Moody and H. M. Hassan, *Proc. Natl. Acad. Sci. U.S.A.* **79**, 2855 (1982).
[45] J. F. Ward, *Adv. Radiat. Biol.* **5**, 182 (1975).
[46] S. A. Weitzman and T. P. Stossel, *Science* **212**, 546 (1981).

# [54] Oxidizing Radical Generation by Prostaglandin H Synthase

*By* PAUL H. SIEDLIK and LAWRENCE J. MARNETT

Prostaglandin H (PGH) synthase catalyzes the first stage in the biosynthesis of prostaglandins, thromboxanes, prostacyclin, and malondialdehyde.[1–3] It introduces two molecules of oxygen into arachidonic acid generating the hydroperoxy endoperoxide $PGG_2$ and reduces the hydroperoxy group of $PGG_2$ to an alcohol $(PGH_2)$.[4–6] The fatty acid cyclooxygenase (arachidonate $\rightarrow$ $PGG_2$) and peroxidase ($PGG_2 \rightarrow PGH_2$) activities are components of the same 70,000-dalton subunit protein

[1] T. Miyamoto, N. Ogino, S. Yamamoto, and O. Hayaishi, *J. Biol. Chem.* **251**, 2629 (1976).
[2] F. J. Van der Ouderaa, M. Buytenhek, D. H. Nugteren, and D. A. Van Dorp, *Biochim. Biophys. Acta* **487**, 315 (1977).
[3] M. Hemler, W. E. M. Lands, and W. L. Smith, *J. Biol. Chem.* **251**, 5575 (1976).
[4] S. Ohki, N. Ogino, S. Yamamoto, and O. Hayaishi, *J. Biol. Chem.* **254**, 829 (1979).
[5] M. Hamberg, J. Svensson, T. Wakabayashi, and B. Samuelsson, *Proc. Natl. Acad. Sci. U.S.A.* **71**, 345 (1974).
[6] D. H. Nugteren and E. Hazelhof, *Biochim. Biophys. Acta* **326**, 448 (1973).

Copyright © 1984 by Academic Press, Inc.
All rights of reproduction in any form reserved.
ISBN 0-12-182005-X

termed PGH synthase.[1-4,7] Free radical derivatives of arachidonic acid are generated during catalysis and a free radical chain mechanism has been proposed to explain the conversion of arachidonic acid to PGG$_2$.[8,9] Free radicals are also generated during the reduction of PGG$_2$ to PGH$_2$.[10] They may be derived from the hydroperoxy fatty acid, the peroxidase, or the reducing cofactor.[11]

The specificity of the peroxidase for reducing cofactors is broad.[12] Consequently, numerous endogenous compounds and xenobiotics are oxidized or oxygenated by it.[13] The mechanism of oxidation depends upon the cosubstrate but radical mechanisms appear to predominate.[11] Thus, the cooxidation or cooxygenation of xenobiotics provides a useful assay for the generation of oxidizing radicals by PGH synthase. Two convenient assays are described below. They are complementary in that one detects radicals derived from the hydroperoxide whereas the other detects radicals derived from the enzyme (presumably higher oxidation states of the peroxidase equivalent to iron-bound oxene or hydroxyl radical).

In addition to probing for the production of oxygen radicals or enzyme-derived oxidants, these assays are useful for distinguishing antiinflammatory compounds into different mechanistic classes. For example, aspirin, indomethacin, naproxen, and eicosatetraenoic acid inhibit arachidonic acid-dependent but not PGG$_2$-dependent cooxidations.[14,15] This indicates that these compounds bind (covalently or noncovalently) to sites that block arachidonic acid oxygenation but not peroxidase oxidation. A corollary is that the compounds are not reducing cofactors for the peroxidase. A survey of a number of compounds indicates that a considerable number are both cyclooxygenase inhibitors and peroxidase cofactors.[16] This not only raises interesting questions about whether peroxidase modulation is linked to cyclooxygenase activity but also may provide important clues to the pattern of biological activities exhibited by certain

[7] G. J. Roth, E. Machuga, and P. Strittmatter, *J. Biol. Chem.* **256**, 10018 (1981).
[8] R. P. Mason, B. Kalyanaraman, B. E. Tainer, and T. E. Eling, *J. Biol. Chem.* **255**, 5019 (1980).
[9] M. E. Hemler and W. E. M. Lands, *J. Biol. Chem.* **255**, 6253 (1980).
[10] R. W. Egan, J. Paxton, and F. A. Kuehl, Jr., *J. Biol. Chem.* **251**, 7329 (1976).
[11] L. J. Marnett, *Free Radicals Biol.* **6**, in press (1983).
[12] L. J. Marnett and T. E. Eling, in "Reviews in Biochemical Toxicology" (E. Hodgson, J. R. Bend, and R. M. Philpot, eds.), Vol. 5, p. 135. Elsevier Biomedical, New York, 1983.
[13] L. J. Marnett, P. Wlodawer, and B. Samuelsson, *J. Biol. Chem.* **250**, 8510 (1975).
[14] R. W. Egan, P. H. Gale, W. J. A. VandenHeuvel, E. M. Baptista, and F. A. Kuehl, Jr., *J. Biol. Chem.* **255**, 323 (1980).
[15] L. J. Marnett, P. H. Siedlik, and L. W. M. Fung, *J. Biol. Chem.* **257**, 6957 (1982).
[16] R. W. Egan, P. H. Gale, G. C. Beveridge, L. J. Marnett, and F. A. Keuhl, Jr., in "Advances in Prostaglandin and Thromboxane Research" (B. Samuelsson, P. Ramwell, and R. Paoletti, eds.), Vol. 6, p. 153. Raven, New York, 1980.

compounds. For example, differential action on cyclooxygenase and peroxidase has been proposed to explain the observation that acetaminophen is an analgesic but not an antiinflammatory agent.[17]

### Diphenylisobenzofuran Oxidation

*Principle*

Diphenylisobenzofuran (DPBF) is cooxygenated by PGH synthase to dibenzoylbenzene (DBB) [Eq. (1)].[13] Cooxygenation can be triggered by the addition of arachidonic acid, $PGG_2$, or hydroperoxy fatty acids.[18]

$$\text{(1)}$$

When $PGG_2$ or hydroperoxy fatty acids cause oxidation, the source of the oxygen is molecular oxygen, not hydroperoxide oxygen.[18] The stoichiometry of DPBF oxidized to hydroperoxide added is variable and ranges from 2:1 to 3000:1.[18] The source of the oxygen and the stoichiometry suggest that DPBF cooxygenation proceeds by a radical mechanism. The high reactivity of DPBF toward oxidizing agents suggests that it can detect small quantities of radicals in the presence of competing molecules.[19,20] Indeed, PGH synthase-dependent DPBF cooxygenation has been detected in crude homogenates of tissue *in vitro*.[18] The high sensitivity of DPBF to oxidants is amplified by the chain nature of its oxidation.

*Reagents*

Potassium phosphate buffer, 0.1 $M$, pH 7.8
DPBF, 10 m$M$ in acetone
Arachidonic acid or 15-hydroperoxyarachidonic acid (prepared according to Funk *et al.*[21]), 15 m$M$ in methanol
Enzyme preparation (e.g., ram seminal vesicle microsomes)

---

[17] W. E. M. Lands and A. M. Hanel, *Prostaglandins* **24**, 271 (1982).
[18] L. J. Marnett, M. J. Bienkowski, and W. R. Pagels, *J. Biol. Chem.* **254**, 5077 (1979).
[19] J. A. Howard and G. D. Mendenhall, *Can. J. Chem.* **53**, 2199 (1975).
[20] P. B. Merkel and D. R. Kearns, *J. Am. Chem. Soc.* **97**, 462 (1975).
[21] M. O. Funk, R. Isaac, and N. A. Porter, *Lipids* **11**, 113 (1976).

*Assay Procedure*

DPBF (10–50 $\mu M$) is incubated with ram seminal vesicle microsomes or other PGH synthase preparations for 3 min followed by the addition of arachidonic acid or 15-hydroperoxyarachidonic acid (50 $\mu M$). DPBF oxidation is monitored by the decrease in absorbance at 420 nm. Since the reaction can be very rapid, a recording spectrophotometer operated in the time base mode is used to provide a continuous monitor of the course of the reaction. The reaction rate is not linear with time so the initial velocity is measured from a tangent drawn to points early in the reaction course.

Phenylbutazone Oxidation

*Principle*

Phenylbutazone (PB) is a nonsteroidal antiinflammatory agent that is used for the treatment of rheumatoid disorders. It is cooxygenated by the peroxidase activity of PGH synthase to 4-hydroxy-PB [Eq. (2)].[22] Arachidonic acid, $PGG_2$, 15-hydroperoxyarachidonic acid, or hydrogen peroxide can effect oxidation. The source of the hydroxyl oxygen is molecular

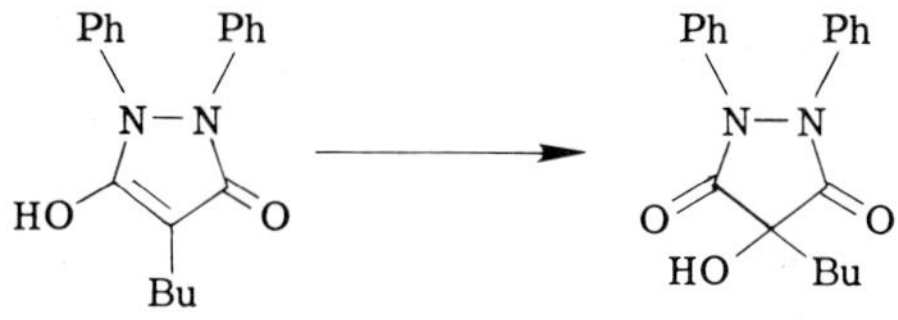

$$(2)$$

oxygen.[22] Although a detailed mechanism of PB oxygenation has not been proposed, it seems likely that PB is oxidized by a higher oxidation state of the peroxidase and the carbon-centered radical formed is scavenged by dioxygen. The resultant peroxy radical is eventually converted to a hydroxyl group. Since the source of the hydroxyl group is molecular oxygen, the uptake of $O_2$ from solution can be used to assay oxidation. The assay is particularly useful for distinguishing antiinflammatory agents since PB itself is a cyclooxygenase inhibitor under certain circumstances. Also, in contrast to DPBF, PB is stable and insensitive to light. This makes it particularly useful for a series of assays that require prolonged periods of time to complete.

[22] L. J. Marnett, M. J. Bienkowski, W. R. Pagels, and G. A. Reed, *in* "Advances in Prostaglandin and Thromboxane Research" (B. Samuelsson, P. W. Ramwell, and R. Paoletti, eds.), Vol. 6, p. 149. Raven, New York, 1980.

*Reagents*

Potassium phosphate buffer, 0.1 *M*, pH 7.8

Phenylbutazone (Sigma), 50 m*M* in methanol

15-Hydroperoxyarachidonic acid or $H_2O_2$, 10 m*M* in methanol or water, respectively

Enzyme preparation (e.g., ram seminal vesicle microsomes)

*Assay Procedure*

PB (500 $\mu M$) is preincubated for 3 min with ram seminal vesicle microsomes or another PGH synthase preparation in the reaction cell of a Gilson oxygraph at 37°. The hydroperoxide (100 $\mu M$) is added and the rate and extent of $O_2$ uptake is recorded. The initial velocity of PB oxidation is calculated by a method analogous to the one described above.

# [55] Assay of *in Situ* Radicals by Electron Spin Resonance

## *By* RONALD P. MASON

Most biological ESR studies of enzymes and their free radical metabolites have used one of two techniques, continuous rapid flow or freeze-quench. The rapid flow technique enables the study of very short-lived free radicals with second-order decay constants near the diffusion limit of $10^{10}$ $M^{-1}$ sec$^{-1}$. Excellent reviews of this technique are available.[1-3] The major limitations of the rapid flow technique are that flow rates as high as 8 ml/sec and a high concentration of enzyme with a large turnover number are needed to achieve the necessary $10^{-8}$–$10^{-7}$ *M* radical metabolite concentration. One laboratory has used a 35 GHz spectrometer, because the fourfold smaller dimensions of the 35 GHz cavity allow a greater than 10-fold decrease in the materials used.[4] With few exceptions, these constraints have resulted in this technique being limited to the study of free radicals that are the one-electron oxidation products of peroxidases, such

[1] H. Beinert and G. Palmer, *Adv. Enzymol.* **27**, 105 (1965).

[2] D. C. Borg, *in* "Biological Applications of Electron Spin Resonance" (H. M. Swartz, J. R. Bolton, and D. C. Borg, eds.), p. 265. Wiley (Interscience), New York, 1972.

[3] I. Yamazaki, *Free Radicals Biol.* **III**, 183 (1977).

[4] D. Borg and J. Elmore, Jr., *in* "Magnetic Resonance in Biological Systems" (A. Ehrenberg, B. Malmstrom, and T. Vanngard, eds.), p. 383. Pergamon, Oxford, 1967.

METHODS IN ENZYMOLOGY, VOL. 105
Copyright © 1984 by Academic Press, Inc.
All rights of reproduction in any form reserved.
ISBN 0-12-182005-X

as the acetaminophen phenoxyl radical.[5] In this and other examples the fast-flow technique is absolutely necessary to detect the primary free radical metabolites.

The second technique, freeze-quench, uses cold isopentane (ca. −140°) to freeze the free radical-containing mixture within a few milliseconds after initiation of the enzymic reaction.[6] This procedure will "freeze" or stabilize many free radicals, and only small volumes of reagents are consumed. Although this technique has been very successful in the study of metalloenzyme redox states,[7,8] the study of free radical metabolites by freeze-quench has been limited almost exclusively to superoxide.[6] Superoxide and many paramagnetic metal ions have linewidths which are so broad at room temperature that ESR signals cannot be detected, and although the use of low temperatures may affect the results, few alternatives are available.

Unfortunately, the freeze-quench technique is of very limited utility in the study of organic free radical metabolites. When a free radical is immobilized in a frozen matrix, a multitude of superimposed spectra (one for each orientation of the free radical in the magnetic field) leads to broad, structureless composite line(s) at $g = 2.00$. Although such a signal proves the formation of a free radical, identification of the free radical metabolite may well be ambiguous in a complex biological system.

### Steady-State Conditions

My interests have concentrated on high-resolution ESR investigations of relatively stable aromatic radical cations and anions with second-order decay constants of $10^5-10^7 M^{-1} \sec^{-1}$. Many classes of free radicals can be detected under physiological steady-state conditions for periods of several minutes to over an hour at room temperature with the use of very simple procedures.[9] In the steady-state condition, the rate of radical formation is equal to the rate of radical decay. Any strategy that will increase the rate of free radical formation or decrease the rate of radical decay will help achieve the necessary $10^{-8}-10^{-7} M$ steady-state radical concentration.

[5] P. R. West, L. S. Harman, P. D. Josephy, and R. P. Mason, submitted for publication, 1983.

[6] D. P. Ballou, this series, Vol. 54, p. 85.

[7] R. C. Bray, in "Biological Magnetic Resonance" (L. J. Berliner and J. Reuben, eds.), Vol. 2, p. 45. Plenum, New York, 1980.

[8] N. J. Blackburn, in "Electron Spin Resonance" (P. B. Ayscough, ed.), Vol. 7, p. 340. The Royal Society of Chemistry, London, 1982.

[9] R. P. Mason, Free Radicals Biol. 5, 161 (1982).

In ESR the upper limit on the sample size is determined by the use of water as the solvent. High concentrations of cells, microsomes, or mitochondria should actually improve the sensitivity of the spectrometer to the extent that protein and membranes replace water. Since optical dispersion is not a limitation in the achievement of optimum signal-to-noise, the use of packed cells or microsomal protein concentrations as high as 40 mg/ml will primarily increase the rate of radical formation and, thus, the steady-state radical concentration. Any other approach that will increase the enzyme activity per unit volume should also be employed.

The rate of radical decay can be decreased by many approaches other than lowering the temperature of the enzymatic incubation as in the freeze-quench technique. Many free radical metabolites are aromatic radical cations and anions with decay constants that are pH dependent. Radical cations, such as benzidine cation radical, are more stable at acidic pH values,[10,11] whereas radical anions, such as metronidazole anion radical, are more stable at basic pH values.[12] Orthosemiquinones, such as the catecholamine anion radicals, are stabilized by complex formation with $Zn^{2+}$.[13]

The spin-trapping technique is a very effective approach to extending the lifetime of the paramagnetic species. Spin traps scavenge many reactive free radicals, even under physiological conditions, to form relatively stable nitroxide adducts.[14-16] The formation of these secondary nitroxides enables the study of superoxide and other free radical metabolites that would otherwise be impossible to study with direct ESR under steady-state conditions.

### Anaerobic Conditions

Most radical anions and some radical cations react with oxygen to form superoxide (and occasionally peroxyl free radicals). Consequently, anaerobic conditions may be required to achieve the necessary $10^{-8}$–$10^{-7}$ M steady-state radical concentrations. The following description of a microsomal-xenobiotic incubation illustrates a simple anaerobic technique.[17]

[10] P. D. Josephy, T. E. Eling, and R. P. Mason, *Mol. Pharmacol.* **23,** 766 (1983).

[11] P. D. Josephy, T. E. Eling, and R. P. Mason, *J. Biol. Chem.* **258,** 5561 (1983).

[12] R. P. Mason and P. D. Josephy, *in* "Toxicity of Nitroaromatic Compounds" (D. Rickert, ed.). Hemisphere, New York, in press, 1984.

[13] B. Kalyanaraman and R. C. Sealy, *Biochem. Biophys. Res. Commun.* **106,** 1119 (1982).

[14] B. Kalyanaraman, *Rev. Biochem. Toxicol.* **4,** 73 (1982).

[15] E. G. Janzen, *Free Radicals Biol.* **4,** 115 (1980).

[16] R. P. Mason, *in* "Spin Labeling in Pharmacology" (J. L. Holtzman, ed.), Academic Press, New York, in press, 1984.

[17] R. P. Mason and J. L. Holtzman, *Biochemistry* **14,** 1626 (1975).

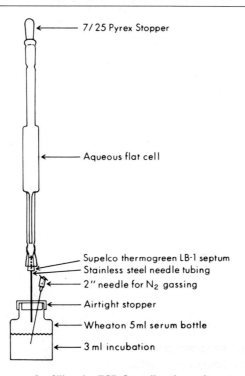

7/25 Pyrex Stopper

Aqueous flat cell

Supelco thermogreen LB-1 septum
Stainless steel needle tubing
2" needle for N₂ gassing
Airtight stopper
Wheaton 5ml serum bottle
3 ml incubation

FIG. 1. Apparatus for filling the ESR flat cell under a nitrogen atmosphere.

The incubation mixture (3 ml) containing substrate (2 m$M$ of furyl-furamide, AF-2) and an NADPH-generating system consisting of NADP$^+$ (0.8 m$M$), glucose 6-phosphate (11 m$M$), and 1.3 units/ml of glucose-6-phosphate dehydrogenase in KCl–Tris–HCl–MgCl$_2$ (150 m$M$, 20 m$M$, pH 7.4, and 5 m$M$) is placed in a rubber-stoppered serum bottle (Fig. 1). Nitrogen gas is then bubbled into the solution for 5 min. The only exit is through the aqueous flat cell, and moderate pressure will allow nitrogen to escape by lifting the ground glass cap. Twelve milligrams of rat hepatic microsomal protein is then added with a syringe through the rubber stop-per and gassing is continued for 30 sec. The stainless-steel needle tubing is then lowered below the surface of the incubation, which forces the solu-tion into the aqueous flat cell. Once the flat cell is filled, the cap can be held closed to stop the flow until the sealed system can be vented by inserting a second needle into the rubber stopper. Once the needle tubing is removed from the force-fitted septum, the flat cell containing the incu-bation under a nitrogen atmosphere can be mounted in the cavity with the

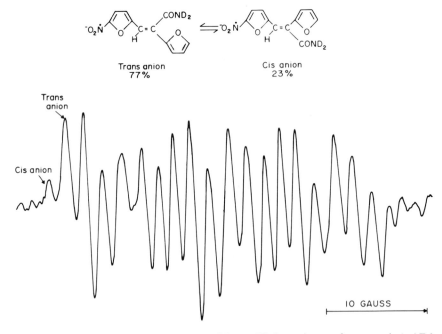

FIG. 2. First-derivative ESR spectrum of the equilibrium mixture of *trans*- and *cis*-AF-2 anion radicals observed under nitrogen in deuterium oxide buffer. The amide hydrogens were exchanged for deuterium to facilitate the analysis of the ESR spectrum. From Ref. 18, with permission.

Varian aqueous cell holders. The ESR spectrum shown in Fig. 2[18] was obtained in this manner.

### Stationary Flat Cell

A more convenient technique is to use aspiration (controlled by a stopcock) or a modified Gilford rapid sampler to fill the flat cell (Fig. 3). This approach has three major advantages. First, the position of the cell in the cavity is unchanged from one incubation to the next, hence the tuning of the spectrometer is unchanged. Therefore, separate incubations can be examined with the spectrometer remaining in the operate mode so that only a small disturbance of the crystal current occurs. With this method, the biochemical controls can be examined under identical instrumental

[18] B. Kalyanaraman, R. P. Mason, R. Rowlett, and L. D. Kispert, *Biochim. Biophys. Acta* **660**, 102 (1981).

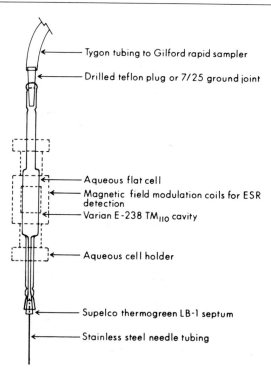

Tygon tubing to Gilford rapid sampler

Drilled teflon plug or 7/25 ground joint

Aqueous flat cell

Magnetic field modulation coils for ESR detection

Varian E-238 TM$_{110}$ cavity

Aqueous cell holder

Supelco thermogreen LB-1 septum

Stainless steel needle tubing

FIG. 3. Apparatus for replacing an incubation in the flat cell *in situ* while still in the operate mode. This technique can be used for anaerobic incubations by combining it with the technique illustrated in Fig. 1.

conditions. Care must be taken not to introduce air bubbles into the flat cell at any time, because the low dielectric constant of air causes a change in the cavity tuning. Air bubbles, as well as the old incubation, can be satisfactorily removed by the passage of quantities of water or buffer through the flat cell. Second, the tedious adjustment of the flat cell in the cavity is done only once. Third, kinetic information with dead times of a few seconds is obtained easily. The reaction can be initiated by the injection of a small volume of the substrate for the reaction into the serum bottle containing the biological material (purified enzyme, cells, or subcellular fractions). The incubation is then aspirated into the microwave cavity. The elapsed time from the initiation of the reaction in a stirred incubation until the sample is within the microwave cavity is less than 4 sec. The radical concentration time dependence can be monitored by repetitively sweeping a small segment of the ESR spectrum. Alternatively, the magnetic field can be adjusted to the position of a particular line in the spectrum of an incubation at steady state, which is then replaced by a fresh incubation. This type of kinetic measurement is best done with the use of a field-frequency locking accessory, but satisfactory results can usually be obtained without one if the line widths are over 0.2 G (see Fig. 5 of Ref. 19). Contamination of the new incubation by the old incubation is not usually a problem. However, in one unpublished case, polymerization products formed a precipitated, but active, horseradish peroxidase coating on the inside of the flat cell, which gave the appearance of a nonenzymic reaction when the controls were done. In such cases, the flat cell must be removed for washing with cleaning solution.

The use of these techniques with high concentrations of unicellular organisms has been most successful,[20,21] but the study of *in vivo* free radicals by ESR is presently limited to organisms with one dimension less than 0.5 mm.

[19] E. Perez-Reyes, B. Kalyanaraman, and R. P. Mason, *Mol. Pharmacol.* **17**, 239 (1980).
[20] R. Docampo, S. N. J. Moreno, R. P. A. Muniz, F. S. Cruz, and R. P. Mason, *Science* **220**, 1292 (1983).
[21] S. N. J. Moreno, R. P. Mason, R. P. A. Muniz, F. S. Cruz, and R. Docampo, *J. Biol. Chem.* **258**, 4051 (1983).

# [56] Chloroplasts: Formation of Active Oxygen and Its Scavenging

*By* KOZI ASADA

Chloroplast thylakoids univalently photoreduce molecular oxygen producing $O_2^-$ through autoxidation of an electron acceptor in photosystem I. In addition, in chloroplasts chlorophyll-photosensitized production of $^1O_2$ is an unavoidable reaction. In chloroplasts, $H_2O_2$ is produced through the superoxide dismutase (SOD)-catalyzed disproportionation of $O_2^-$.[1] Thus, the Mehler reaction[2] is composed of two reaction steps:

$$2O_2 + 2(e^-) \rightarrow 2O_2^-$$
$$2O_2^- + 2H^+ \rightarrow H_2O_2 + O_2$$

where $(e^-)$ is the reduced form of an electron acceptor in photosystem I.

The production of $O_2^-$ is enhanced under conditions where the generation rate of photoreductant in thylakoids exceeds that required for $CO_2$

[1] K. Asada, K. Kiso, and K. Yoshikawa, *J. Biol. Chem.* **249**, 2175 (1974).
[2] A. H. Mehler, *Arch. Biochem. Biophys.* **33**, 65 (1951).

Copyright © 1984 by Academic Press, Inc.
All rights of reproduction in any form reserved.
ISBN 0-12-182005-X

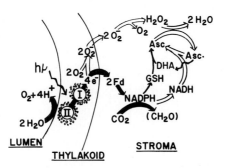

FIG. 1. Formation and scavenging of superoxide and hydrogen peroxide in chloroplasts. When molecular oxygen is reduced by two electrons from photosystem I, one molecule of hydrogen peroxide is produced through the disproportionation of superoxide with SOD in the stroma. This hydrogen peroxide is reduced to water by one molecule of ascorbate (Asc) which has been generated from its oxidation products, monodehydroascorbate (Asc·) and dehydroascorbate (DHA), by the indicated system with two electrons from photosystem I.

reduction; low $CO_2$ concentration and high light intensity. $O_2^-$ is, however, produced even when $CO_2$ is supplied to chloroplasts and the photoproduction of $O_2^-$ appears to be indispensable for prevention of overreduction of electron carriers in the cyclic electron transport pathway.[3] Thus, the photoreduction of $O_2$ is an inevitable reaction in chloroplasts. Scavenging of $O_2^-$ and $H_2O_2$ is essential for chloroplasts to maintain their ability to fix $CO_2$ because several enzymes in the $CO_2$-reduction cycle are sensitive to active oxygen.[4] Production of $O_2^-$ and $H_2O_2$ and their scavenging system are shown in Fig. 1.

## Assay of Superoxide and Hydrogen Peroxide Production in Illuminated Thylakoids

### Photogeneration of Superoxide

*Principle.* The photoreduction of horse heart ferricytochrome *c* by thylakoids is mediated by $O_2^-$ which is produced through the autoxidation of a photoreductant in photosystem I. The rate of cytochrome *c* reduction is measured by the absorbance increase at the $\alpha$ peak.[1]

*Procedure.* To suppress the reoxidation of ferrocytochrome *c* by mitochondrial cytochrome *c* oxidase thylakoids are prepared from intact chloroplasts separated by sucrose or Percoll (modified silica sol) density centrifugation.[1,5,6] Intact chloroplasts are disrupted by osmotic shock in

[3] U. Ziem-Hanck and V. Heber, *Biochim. Biophys. Acta* **591,** 266 (1980).
[4] W. M. Kaiser, *Planta* **145,** 377 (1979).
[5] D. A. Walker, this series, Vol. 69, 94 (1980).
[6] Y. Nakano and K. Asada, *Plant Cell Physiol.* **21,** 1295 (1980).

20 m$M$ potassium phosphate, pH 7.8, and the thylakoids are obtained by centrifugation. To remove stromal SOD the thylakoids suspended in 50 m$M$ potassium phosphate, pH 7.8, containing 1 m$M$ EDTA are kept for an hour at 5° and then centrifuged. This washing procedure repeats twice more.

The reaction mixture (2 ml) contains 50 m$M$ potassium phosphate, pH 7.8, 10 m$M$ NaCl, 20 $\mu M$ ferricytochrome $c$, and thylakoids (10–20 $\mu$g chlorophyll). The reaction rate of cytochrome $c$ is determined using a spectrophotometer modified so that the absorbance increase at 550 nm due to ferrocytochrome $c$ can be monitored while the cuvette is being irradiated from the side by red light at saturating intensity. Actinic light, provided by a projector, is passed through a red filter which cuts off below 650 nm and the photomultiplier is protected from actinic light by a 550-nm interference filter. The reaction rate is determined from the initial absorbance increase 20 sec after illumination.

Under these conditions, cytochrome $c$ photoreduction is inhibited by more than 80% with 1 $\mu M$ SOD. The production rate of $H_2O_2$ in the presence of cytochrome $c$ is suppressed in comparison with that in the absence of cytochrome $c$ or in the presence of SOD and cytochrome $c$. In Tricine, Tris, MES, and HEPES buffers, or in phosphate buffer at low pH, the photoreduction of cytochrome $c$ is only partially inhibited by SOD.[1] An increase of 0.01 absorbance at 550 nm equals the production of 1.05 nmol of $O_2^-$. The difference absorbance coefficient between ferro- and ferricytochrome $c$[7] is taken as 19 m$M^{-1}$ cm$^{-1}$.

In addition to the photoreduction of cytochrome $c$, the following $O_2^-$-induced reactions can be employed to confirm the photoproduction of $O_2^-$ in thylakoids; photooxidation of epinephrine (adrenochrome formation, absorbance increase at 480 nm),[1] sulfite (oxygen uptake),[8] hydroxylamine (nitrite formation, colorimetric assay),[9] manganous pyrophosphate (manganic pyrophosphate, absorbance increase at 258 nm),[10] Tiron (Tiron free radical, EPR),[11] and ascorbate (oxygen uptake).[12] These $O_2^-$-induced reactions are followed by monitoring the respective reaction products shown in parentheses. Since some of these reactions occur without mediation by $O_2^-$, it should be confirmed that the reaction is suppressed under

[7] D. L. Keister and A. San Pietro, *Arch. Biochem. Biophys.* **103**, 45 (1963).
[8] K. Asada and K. Kiso, *Eur. J. Biochem.* **33**, 253 (1973).
[9] E. F. Elstner, C. Stoffer, and A. Henpel, *Z. Naturforsch.* **30C**, 53 (1975).
[10] Y. Kono, M. Takahashi, and K. Asada, *Arch. Biochem. Biophys.* **174**, 454 (1976).
[11] R. W. Miller and F. D. H. MacDowall, *Biochim. Biophys. Acta* **387**, 176 (1975).
[12] E. F. Elstner and R. Kramer, *Biochim. Biophys. Acta* **314**, 340 (1973).

anaerobic conditions and is inhibited by SOD. A spin-trapping method[13] has been also used.[14]

*Properties.* The major photoreductant of $O_2$ is a thylakoid-bound electron acceptor in photosystem I rather than peripheral ferredoxin.[1] Photoreduction of $O_2$ occurs in the thylakoid membranes but not on the surface of the membranes. The $O_2^-$ produced in the membranes appears to diffuse to both sides of thylakoids, lumen and stroma sides.[15] The apparent $K_m$ for $O_2$ is below 10 $\mu M$.[16,17] The $K_m$ is the ratio of the first-order rate constants of photoreduction of the electron acceptor in photosystem I plus the internal back decay of reduced acceptor $(e^-)$ to the second-order rate constant $(k_{e^-})$ for the oxidation of reduced acceptor $(e^-)$ with $O_2$.[18] The reduction rate of $O_2$ in spinach thylakoids is in the range of 10 to 25 $\mu$mol $mg^{-1}$ chlorophyll $hr^{-1}$ (5 to 10% of the electron transport rate) and is enhanced 3- to 5-fold by the addition of paraquat. $k_{e^-}$ is estimated to be about $10^6$ $M^{-1}$ $sec^{-1}$ in "intact" thylakoids, but is increased to $10^8$ $M^{-1}$ $sec^{-1}$ by disrupting the thylakoid membranes with detergent.[18] A high $O_2^-$ production rate (7000 $\mu$mol $mg^{-1}$ chlorophyll $hr^{-1}$) in photosystem I subparticles[16] is partly due to its high $k_{e^-}$.

## Photogeneration of Hydrogen Peroxide

*Principle.* In the presence of peroxidase homovanillic acid (**I**) is converted upon oxidation with $H_2O_2$ to highly fluorescent 2,2'-dihydroxy-3,3'-dimethoxybiphenyl-5,5'-diacetic acid (**II**) which has a $\lambda_{ex}$ of 315 nm and a $\lambda_{em}$ of 425 nm.[19] Monitoring the fluorescence intensity allows continuous measurement of $H_2O_2$ as low as 0.2 nmol in illuminated thylakoids.[1]

[13] This volume [23].
[14] J. R. Harbour and J. R. Boltan, *Biochem. Biophys. Res. Commun.* **64,** 803 (1975).
[15] M. Takahashi, T. Hayakawa, and K. Asada, unpublished.
[16] K. Asada and Y. Nakano, *Photochem. Photobiol.* **28,** 917 (1978).
[17] S. Lien and A. San Pietro, *FEBS Lett.* **99,** 189 (1979).
[18] M. Takahashi and K. Asada, *Plant Cell Physiol.* **23,** 1457 (1982).
[19] G. G. Guilbault, D. N. Kramer, and E. Hackley, *Anal. Chem.* **39,** 271 (1967).

*Procedure.* The reaction mixture (2 ml) contains 60 m$M$ potassium phosphate, pH 7.8, 10 m$M$ NaCl, thylakoids (10–20 μg chlorophyll), 1.5 m$M$ homovanillic acid, and 0.8 μ$M$ horseradish peroxidase.[1] A fluorospectrophotometer is modified so that the fluorescence is monitored while the cuvette is irradiated by red light (>650 nm) at a right angle to the photomultiplier and on the reverse side of the excitation beam. The fluorescence intensity is measured at 425 nm with excitation at 315 nm. Before illumination the fluorescence intensity is recorded and its increased rate is monitored for 1 min after illumination by red light. The fluorescence yield due to $H_2O_2$ is variable depending on chloroplast concentration and compounds having absorbance near 315 and 425 nm. The yield is determined in each sample from the increase in fluorescence intensity by adding an aliquot of $H_2O_2$ (0.5 to 1 nmol in 10 μl) to the reaction mixture after each measurement. $H_2O_2$ is standardized from its absorbance at 230 nm using an absorbance coefficient[20] of 71 $M^{-1}$ cm$^{-1}$.

*Properties.* When ferricytochrome $c$ is added to thylakoids the photoproduction rate of $H_2O_2$ is suppressed by about 70%,[15] but when a compound such as manganous pyrophosphate that is oxidized by $O_2^-$ is added the rate is doubled.[10] The $H_2O_2$ detected in the presence of cytochrome $c$ may be derived from $O_2^-$ ejected to the lumen side from the thylakoid membranes. As $H_2O_2$ is permeable through thylakoid membranes but $O_2^-$ is not,[21] the $O_2^-$ ejected to the lumen is not reactive with cytochrome $c$ in the medium. Therefore the rate estimated by the production of $H_2O_2$ seems to represent the actual photoreduction of $O_2^-$ in thylakoids.

### Scavenging of Superoxide and Hydrogen Peroxide in Chloroplasts

In chloroplasts $O_2^-$ produced in thylakoids is disproportionated by SOD in the stroma and the $H_2O_2$ thus produced is reduced to water by ascorbate peroxidase.[22] Ascorbate is regenerated from its oxidation products, dehydroascorbate (DHA) and monodehydroascorbate, by photoreductants through the system shown in Fig. 1. Ascorbate peroxidase, DHA reductase (EC 1.8.5.1),[22] and NADH-dependent monodehydroascorbate reductase (EC 1.6.5.4)[23a] are localized in the chloroplast stroma, as are the enzymes participating in the generation of GSH and NADH.[22–25]

[20] B. Chance, *Method Biochem. Anal.* **1**, 412 (1956).
[21] M. Takahashi and K. Asada, *Arch. Biochem. Biophys.* **226** (1983).
[22] Y. Nakano and K. Asada, *Plant Cell Physiol.* **22**, 867 (1981).
[23] M. A. Hossain and K. Asada, *Plant Cell Physiol.* **25**, in press (1984).
[23a] M. A. Hossain, Y. Nakano, and K. Asada, *Plant Cell Physiol.* **25**, in press (1984).
[24] P. P. Jablonski and J. W. Anderson, *Plant Physiol.* **67**, 1239–1244 (1981).
[25] P. P. Jablonski and J. W. Anderson, *Plant Physiol.* **69**, 1407–1413 (1982).

## Scavenging of Superoxide, SOD

Spinach chloroplasts contain cyanide-sensitive Cu,Zn-SOD at about 10 $\mu M$ in the stroma[26] and also in the lumen of thylakoids.[27] Stromal SOD activity is assayed by the xanthine–xanthine oxidase–cytochrome $c$ system[28] using a stromal fraction prepared from intact chloroplasts by osmotic shock and centrifugation. The lumen SOD is detected in the supernatant after disruption of the thylakoids by Yeda-pressure cell treatment and centrifugation. The thylakoid membranes (pellet) contain cyanide-insensitive SOD, which is detectable only by 0.1% Triton X-100 containing assay system of SOD.[29] The three types of SOD (Cu,Zn-, Fe-, and Mn-containing enzymes) show a characteristic distribution in photosynthetic organisms at various levels of evolution.[29] Chloroplast components such as cytochrome $f$, plastocyanin, ascorbate, GSH, and manganous ions are reactive with $O_2^-$, but their contribution to the scavenging of $O_2^-$ is below 10% of that with stromal SOD, from their relative concentrations in chloroplasts and reactivities with $O_2^-$.[30]

## Scavenging of Hydrogen Peroxide

### Assay of Ascorbate-Specific Peroxidase

$$\text{Ascorbate} + H_2O_2 \rightarrow \text{dehydroascorbate} + H_2O$$

*Principle.* This assay is based on the spectrophotometric measurement of ascorbate decrease due to the oxidation of ascorbate to DHA by $H_2O_2$.[31,32] Because of the high absorbance of ascorbate at its absorption maximum of 265 nm ($\varepsilon = 14$ m$M^{-1}$ cm$^{-1}$) the absorbance at 290 nm ($\varepsilon = 2.8$ m$M^{-1}$ cm$^{-1}$) is measured.[22]

*Procedure.* To the cuvette 50 m$M$ potassium phosphate, pH 7.0, 0.5 m$M$ ascorbic acid, and enzyme (total volume, 1 ml) are added.[22] Absorbance at 290 nm is followed to determine if the enzyme contains ascorbate oxidase (EC 1.10.3.3). For spinach chloroplasts little activity is found.[22] Then the peroxidase reaction is started by the addition of 20 $\mu$l of 5 m$M$

[26] K. Asada, M. Urano, and M. Takahashi, *Eur. J. Biochem.* **36**, 257 (1973).
[27] T. Hayakawa, S. Kanematsu, and K. Asada, unpublished.
[28] This volume [10].
[29] K. Asada, S. Kanematsu, and T. Hayakawa, *in* "Chemical and Biological Aspects of Superoxide and Superoxide Dismutase" (J. V. Bamniston and H. A. O. Hill, eds.), p. 136. Elsevier, Amsterdam, 1980.
[30] K. Asada, M. Takahashi, K. Tanaka, and Y. Nakano, *in* "Biochemical and Medical Aspects of Active Oxygen" (O. Hayaishi and K. Asada, eds.), p. 45. Univ. Park Press, Baltimore, Maryland, 1977.
[31] G. J. Kelly and E. Latzko, *Naturwissenschaften* **66**, 617 (1979).
[32] S. Shigeoka, Y. Nakano, and S. Kitaoka, *Arch. Biochem. Biophys.* **201**, 121 (1980).

$H_2O_2$, and the absorbance decrease is recorded 10 to 20 sec after this addition. Correction is made for the nonenzymic oxidation of ascorbate by $H_2O_2$ (about 1 nmol $min^{-1}$), and for ascorbate oxidase activity, if any. Under these assay conditions a decrease of 0.01 absorbance corresponds to 3.6 nmol ascorbate oxidized.

*Properties.* In contrast to guaiacol-oxidizable peroxidases such as horseradish peroxidase (HRP), ascorbate peroxidase in spinach chloroplasts and in *Euglena* is very labile[32,33]; in 50 m$M$ potassium phosphate, pH 7.0, half its activity is lost within 10 hr. The activity is stable for several months if the enzyme is kept in 25% sorbitol containing 1 m$M$ ascorbate at 4 or $-20°$.[33] Ascorbate peroxidase is distinguished from HRP by specificity of the electron donor[22,32,33]; the chloroplast peroxidase does not catalyze the oxidation of guaiacol, GSH, NAD(P)H, or cytochrome $c$. Pyrogallol is the only other electron donor but its rate at 10 m$M$ is 26% that of 0.5 m$M$ ascorbate. The apparent $K_m$ values for $H_2O_2$ and ascorbate are 45 and 500 $\mu M$ (spinach)[33] and 56 and 410 $\mu M$ (*Euglena*),[32] respectively, at the optimum pH of 7.0. The reaction is almost completely inhibited by 10 $\mu M$ cyanide, 1 m$M$ azide, 0.1 m$M$ iodoacetate, 10 $\mu M$ *p*-chloromercuribenzoate, and 10 $\mu M$ EDTA, but the inhibition due to EDTA, azide, and iodoacetate is prevented by 0.5 m$M$ ascorbate.[33] The primary oxidation product of ascorbate appears to be monodehydroascorbate and DHA is its disproportionation product.[33] Molecular weights of spinach[33] and *Euglena*[32] enzymes are 45,000 and 76,000, respectively.

*Assay of DHA Reductase*

$$DHA + 2GSH \rightarrow Ascorbate + GSSG$$

*Principle.* The reduction of DHA to ascorbate by GSH is determined by the absorbance increase at 265 nm due to ascorbate formation ($\mathfrak{s} = 14$ m$M^{-1}$ $cm^{-1}$).[22,23]

*Procedure.*[23] To the cuvette 50 m$M$ potassium phosphate, pH 6.5, 5 m$M$ GSH, 0.5 m$M$ DHA, and 0.1 m$M$ EDTA are added (total volume, 1 ml) and the absorbance increase at 265 nm is monitored to determine nonenzymic reduction of DHA by GSH (about 3.5 nmol ascorbate $min^{-1}$). The optimum pH of the reaction is 7.8, but to avoid a high nonenzymic reduction, pH 6.5 is used. Enzymic activity at pH 6.5 is about 40% that at pH 7.8. Production rate of ascorbate is measured by the increase in absorbance at 265 nm 10 to 30 sec after adding the enzyme. An increase of 0.1 absorbance is equal to 7.14 nmol ascorbate formed. The reaction rate is corrected for nonenzymic reaction. As GSSG has an absorbance at 265 nm ($\mathfrak{s} = 0.18$ m$M^{-1}$ $cm^{-1}$ at pH 7.0), the reaction rate is further corrected

---

[33] Y. Nakano and K. Asada, unpublished.

by assuming the formation of 1 mol of GSSG for 1 mol of ascorbate formed, i.e., multiplying a factor of 0.98.

*Properties.*[23,34] The molecular weight of DHA reductase from spinach is 23,000, and no absorbance in the visible region is found. The $K_m$ values[23] for GSH and DHA are 2.5 and 0.07 m$M$, respectively, at pH 7.0. In 50 m$M$ phosphate, pH 7.8, the enzyme is labile and loses half its activity within 2 days. Enzyme activity is stable in 2 m$M$ 2-mercaptoethanol containing 50 m$M$ Tris–Cl, pH 7.8, at $-20°$ for several months.[23] The enzyme inhibited by 1 m$M$ *p*-chloromercuribenzoate, *N*-ethylmaleimide, 5,5′-dithiobis(2-nitrobenzoic acid), and 0.1 m$M$ iodoacetate. Cyanide at 10 m$M$ inhibits its activity by 50%.

[34] C. H. Foyer and B. Halliwell, *Phytochemistry* **16**, 1347 (1977).

# [57] Determination of the Production of Superoxide Radicals and Hydrogen Peroxide in Mitochondria

*By* ALBERTO BOVERIS

## Superoxide Radicals

The mitochondrial respiratory chain produces superoxide radicals ($O_2^-$) at two sites at the flavoprotein NADH dehydrogenase and at the ubiquinone–cytochrome *b* region, likely by autoxidation of ubisemiquinone.[1,2] At pH 7.4, NADH dehydrogenase contributes with about one-third and ubisemiquinone with about two-thirds of total $O_2^-$ production.[2]

The determination of the rate of $O_2^-$ production is based upon the spectrophotometric measurement of oxidation or reduction reactions in which $O_2^-$ is a reactant. The concentration of the spectrophotometric indicator that reacts with $O_2^-$ is adjusted to compete effectively with the spontaneous dismutation of $O_2^-$ so that nearly all $O_2^-$ produced can be detected. The involvement of $O_2^-$ is ascertained by the use of superoxide dismutase which inhibits the reaction rate specifically due to $O_2^-$. Cyanide, which is often used as inhibitor of the mitochondrial cytochrome oxidase ($K_i$ about $3 \times 10^{-5}$ $M$) is also an inhibitor of the often used copper-containing superoxide dismutase ($K_i$ about $3 \times 10^{-4}$ $M$). It is possible to use enough cyanide to partially inhibit cytochrome oxidase

[1] A. Boveris, E. Cadenas, and A. O. M. Stoppani, *Biochem. J.* **156**, 435 (1976).
[2] J. F. Turrens and A. Boveris, *Biochem. J.* **191**, 421 (1980).

Copyright © 1984 by Academic Press, Inc.
All rights of reproduction in any form reserved.
ISBN 0-12-182005-X

without inhibiting superoxide dismutase. Alternatively, bacterial or mitochondrial (manganese-containing) superoxide dismutase can be used.

Mitochondria have Mn-superoxide dismutase in the matrix space, so in order to measure the total production of $O_2^-$ by mitochondrial membranes the dismutase should be inhibited or removed. Since effective inhibitors of Mn-superoxide dismutase are not known, matrical superoxide dismutase is usually removed by repetitive washing of submitochondrial particles obtained by sonication or by other means. Mitochondrial $O_2^-$ production is pH dependent and increases toward the alkaline region.[2–4]

### Preparation of Submitochondrial Particles[5]

Submitochondrial particles are usually prepared by sonication of 5–20 mg mitochondrial protein/ml in 0.23 $M$ mannitol, 0.07 $M$ sucrose, 50 m$M$ Tris–HCl, pH 7.6. Lower protein concentrations during sonication provide submitochondrial particles with lower superoxide dismutase trapped in the intravesicular space but also give more denatured submitochondrial particles. Freezing and thawing (three times) of isolated mitochondria suspended at about 10 mg protein/ml in 140 m$M$ KCl, 20 m$M$ Tris–HCl (pH 7.4) yield acceptable preparations. The number of washings required (usually about two or three) is determined by measuring $O_2^-$ production in the washed submitochondrial particles until a maximal rate is achieved.

### Quantitation of $O_2^-$ Production

*Adrenochrome Formation.* The $O_2^-$-dependent oxidation of epinephrine to adrenochrome[6] is followed spectrophotometrically at 485–575 nm with a dual-wavelength spectrophotometer ($E = 2.96$ m$M^{-1}$ cm$^{-1}$)[7] or at 480 nm with a sensitive (full scale, $A = 0.050$ or 0.100) single-wavelength spectrophotometer ($E = 4.0$ m$M^{-1}$ cm$^{-1}$).[7] Autoxidation of epinephrine is controlled by lowering the pH. Although the specificity of the reaction is rather poor, the superoxide dismutase-sensitive rate of adrenochrome formation gives a useful assay for $O_2^-$ generation at rates of 1–5 $\mu M$/min.

[3] G. Loschen, A. Azzi, C. Richter, and L. Flohé, *FEBS Lett.* **42,** 68 (1974).
[4] A. Boveris and E. Cadenas, *FEBS Lett.* **54,** 311 (1975).
[5] C. T. Gregg, this series, Vol. 10 [33].
[6] H. Misra and I. Fridovich, *J. Biol. Chem.* **247,** 3170 (1972).
[7] S. Green, A. Mazur, and E. Shorr, *J. Biol. Chem.* **220,** 237 (1956).

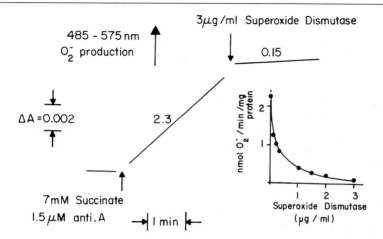

FIG. 1. Production of superoxide radicals by bovine heart submitochondrial particles (0.45 mg protein/ml) suspended in 0.23 $M$ mannitol, 0.07 $M$ sucrose, 30 m$M$ Tris-MOPS buffer [3-($N$-morpholino)propanesulfonic acid], pH 7.6, and supplemented with 1 m$M$ epinephrine and 0.2 $\mu M$ catalase. Numbers near the traces indicate nmol $O_2^-$/min/mg protein. Anti.A: antimycin A. Aminco-Chance dual-wavelength spectrophotometer.

Plant mitochondria contain phenol oxidases that strongly interfere with this assay.[8,9] In the absence of side reactions, the assay detects one adrenochrome formed per one $O_2^-$ produced.[10]

ASSAY. The reaction medium consists of 1 m$M$ epinephrine, 0.23 $M$ mannitol, 0.07 $M$ sucrose, 20 m$M$ Tris–HCl, pH 7.4 (the assayed pH is usually in the range 7.2–8.0). A 60 m$M$ epinephrine bitartrate solution, pH 2.0, kept in ice, is diluted in the reaction medium immediately before use. Supplementation of the reaction medium with 0.1–1 $\mu M$ catalase is convenient.[2,3] Superoxide dismutase is used at 0.1–0.3 $\mu M$ final concentration to give assay specificity (Fig. 1).

Submitochondrial particles (0.3–1.0 mg protein/ml) are necessarily supplemented with substrates and inhibitors in order to obtain maximal $O_2^-$ production. NADH (50 $\mu M$) and succinate (7 m$M$) are used as substrates and rotenone (1 $\mu M$) and antimycin (1 $\mu M$) are the specific inhibitors, respectively, to assay $O_2^-$ production at the NADH dehydrogenase and at the ubiquinone–cytochrome $b$ region. $NADH_2$ and $NADH_2$ bound

[8] P. R. Rich and W. D. Bonner, Jr., *Arch. Biochem. Biophys.* **188,** 206 (1978).
[9] A. Boveris, R. A. Sánchez, and M. T. Beconi, *FEBS Lett.* **92,** 333 (1978).
[10] E. Cadenas, A. Boveris, C. I. Ragan, and A. O. M. Stoppani, *Arch. Biochem. Biophys.* **180,** 248 (1977).

to the dehydrogenase interfere by reacting with $O_2^-$ and generating $O_2^-$ and HO· radicals.[2,11]

*Cytochrome c Reduction.* The rapid reduction of ferricytochrome *c* by $O_2^-$ makes this hemoprotein an excellent quantitative trap for $O_2^-$.[12] However, cytochrome *c* added to submitochondrial particles is more effectively reduced by the electron transport chain (cytochrome *c* reductase activity) than by $O_2^-$. Besides, the ferrocytochrome *c* produced is oxidized by cytochrome oxidase. Acetylated cytochrome *c* (Ac cyt *c*), although not as rapidly reduced by $O_2^-$ as native cytochrome *c*, is much more slowly oxidized by cytochrome oxidase.[13] It is therefore advisable to use the acetylated form of cytochrome *c* for the measurement of mitochondrial $O_2^-$ production.[2,13,14]

$$O_2^- + \text{Ac cyt } c^{3+} \rightarrow O_2 + \text{Ac cyt } c^{2+} \qquad (1)$$

Reduction of acetylated ferricytochrome *c* is followed spectrophotometrically with a dual wavelength spectrophotometer at 550–540 nm ($E = 19$ m$M^{-1}$ cm$^{-1}$) or with a sensitive single wavelength spectrophotometer at 550 nm ($E = 24$ m$M^{-1}$ cm$^{-1}$). Ferricytochrome *c* is acetylated by treatment with acetic anhydride.[13,15] Ferricytochrome *c* (50 mg) is dissolved in 5 ml of a 50% saturated sodium acetate solution. The solution, at 0–2°, is slowly added with 0.07 ml of acetic anhydride (10 molecules acetic anhydride/lysine residue; horse heart ferricytochrome *c* has 19 lysines/molecule), kept in an ice bath with occasional stirring, for 20 min and dialyzed against 50 m$M$ phosphate, pH 7.0.

ASSAY. Acetylated cytochrome *c* is diluted to 10 $\mu M$ in the reaction mixture. Reaction medium, submitochondrial particles, substrates, and inhibitors are used as described for the assay based upon adrenochrome formation.

## Hydrogen Peroxide

Two types of assays, both based on the use of peroxidases, are utilized for the determination of $H_2O_2$ production: (1) the formation of peroxidase–$H_2O_2$ ES (enzyme–substrate) complexes in which accumulation of a measurable intermediate is favored rather than dissociation or conversion to products, and (2) the measurement of the oxidation of hydrogen

[11] P. C. Chan and B. H. J. Bielski, *J. Biol. Chem.* **249**, 1317 (1974).
[12] H. J. Forman and I. Fridovich, *Arch. Biochem. Biophys.* **158**, 396 (1973).
[13] A. Azzi, C. Montecucco, and C. Richter, *Biochem. Biophys. Res. Commun.* **65**, 597 (1975).
[14] K. Takeshige and S. Minakami, *Biochem. J.* **180**, 129 (1979).
[15] S. Minakami, K. Titani, and H. Ishikura, *J. Biochem. (Tokyo)* **64**, 341 (1958).

donors in reactions coupled to $H_2O_2$ generation. Yeast cytochrome $c$ peroxidase and horseradish peroxidase can be used in both types of assays.

*Cytochrome c Peroxidase.* Yeast cytochrome $c$ peroxidase (CCP) forms a stable enzyme–substrate complex with $H_2O_2$ that reacts with ferrocytochrome $c$.[16]

$$CCP + H_2O_2 \rightarrow CCP\text{--}H_2O_2 \tag{2}$$
$$CCP\text{--}H_2O_2 + 2cyt\ c^{2+} + 2H^+ \rightarrow CCP + 2H_2O + 2cyt\ c^{3+} \tag{3}$$

The absorption maxima of the complex and the free enzyme are at 419 and 407 nm, respectively.[16] Formation of the complex is accurately measured with a dual-wavelength spectrophotometer at 419–407 nm ($E = 50$ $mM^{-1}$ $cm^{-1}$) or at 424–400 nm ($E = 60$ $mM^{-1}$ $cm^{-1}$).[17,18] In this assay, ferrocytochrome $c$ must not be allowed to react with the peroxidase–$H_2O_2$ complex, since the spectrophotometric indicator is the intermediate of the aborted reaction. In intact isolated mitochondria, the outer membrane prevents interaction of the complex with the endogenous cytochrome $c$ of the inner membrane. Rates of $H_2O_2$ generation of about 0.1 $\mu M$/min can be accurately measured because of the practical irreversibility of reaction (2). Catalase activity present in the sample, which will cause $H_2O_2$ underestimation, can be inhibited with 10–30 $\mu M$ azide without interference in the assay.[17] Cyanide reacts with cytochrome $c$ peroxidase and the assay cannot be used in the presence of this inhibitor. In some cases, mitochondrial membranes free of cytochrome oxidase activity (by extraction, or supplementation with azide) can be assayed for $H_2O_2$ production in the presence of cytochrome $c$ peroxidase and reduced cytochrome $c$. The samples should not have cytochrome $c$ reductase activity and generation of $O_2^-$, which also reduce cytochrome $c$ (the latter can be eliminated by addition of 0.1–0.3 $\mu M$ superoxide dismutase). Acetylated ferrocytochrome $c$ that acts as hydrogen donor for cytochrome $c$ peroxidase [reaction (3)] can be used with advantages over native cytochrome $c$.

Yeast cytochrome $c$ peroxidase is not commercially available. Yonetani's method[16] of isolating yeast cytochrome $c$ peroxidase yields excellent enzyme preparations. It has been noted that the enzyme isolated by inexperienced hands sometimes does not form a stable $H_2O_2$–peroxidase complex. However, these preparations have a high catalytic activity and can be used to measure acetylated ferrocytochrome $c$ oxidation.

ASSAY. The reaction medium consists usually of 0.23 $M$ mannitol,

---

[16] T. Yonetani, this series, Vol. 10 [60].
[17] A. Boveris, N. Oshino, and B. Chance, *Biochem. J.* **128,** 617 (1972).
[18] A. Boveris and B. Chance, *Biochem. J.* **134,** 707 (1973).

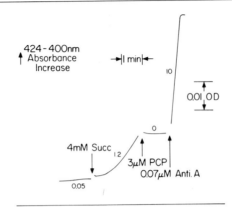

FIG. 2. Production of hydrogen peroxide by pigeon heart mitochondria (0.14 mg protein/ml) suspended in 0.23 $M$ mannitol, 0.07 $M$ sucrose, 30 m$M$ Tris-MOPS buffer, pH 7.4, and supplemented with 1 $\mu M$ yeast cytochrome $c$ peroxidase. Succ.: succinate; PCP: pentachlorophenol; Anti.A: antimycin A. Numbers near the traces indicate nmol $H_2O_2$/min/mg protein. Perkin–Elmer (Model 356) dual-wavelength spectrophotometer.

0.07 $M$ sucrose, 20 m$M$ Tris–HCl, pH 7.4. Cytochrome $c$ peroxidase is added at a 0.5–1 $\mu M$ concentration for the measurement based upon reaction (2) which will give optical changes in the range of $A$ = 0.025–0.060 (Fig. 2).

If the coupled oxidation of acetylated cytochrome $c$ is to be followed, 0.1–0.2 $\mu M$ cytochrome $c$ peroxidase and 5–10 $\mu M$ acetylated cytochrome $c$ are added.

Mitochondrial and submitochondrial particles are usually assayed at 0.1–0.5 mg/ml (0.1–0.3 $\mu M$ superoxide dismutase must be added in the case of submitochondrial particles). Substrates, ADP, and inhibitors are added as required.

*Horseradish Peroxidase and the Coupled Oxidation of Hydrogen Donors.* Horseradish peroxidase oxidizes various hydrogen donors (AH$_2$) in the presence of $H_2O_2$ [reactions (4) and (5)].

$$\text{HRP} + H_2O_2 \rightarrow \text{HRP–}H_2O_2 \qquad (4)$$
$$\text{HRP–}H_2O_2 + AH_2 \rightarrow \text{HRP} + 2H_2O + A \qquad (5)$$

Fluorescent hydrogen donors (AH$_2$) such as scopoletin[19] or fluorescent products (A) such as diacetyldichlorofluorescin[20] are particularly suitable for measuring $H_2O_2$ generation in mitochondrial membranes. $H_2O_2$ production rates of the order of 1 $\mu M$/min are easily detected by this assay. Formation of the HRP–$H_2O_2$ intermediate [reaction (4)] can be

[19] G. Loschen, L. Flohé, and B. Chance, *FEBS Lett.* **18**, 261 (1971).
[20] P. C. Hinkle, R. A. Butow, E. Racker, and B. Chance, *J. Biol. Chem.* **242**, 5169 (1967).

measured at 407–402 nm with a dual wavelength spectrophotometer as an assay for $H_2O_2$ generation.[21] However, the broad specificity of horseradish peroxidase for various hydrogen donors may cause severe underestimation.[21] Controls with cytochrome $c$ peroxidase and calibrations with glucose–glucose oxidase or urate–uricase ($H_2O_2$-generating systems where $O_2$ consumption can be adequately measured with an oxygen electrode) are advisable. Catalase interference can be evaluated by determining the peroxidase-to-catalase heme ratio present in the final reaction mixture.[21] Plant mitochondria contain hydrogen donors (ascorbic acid) that strongly interfere with this method.[21,22]

ASSAY. Reaction medium, mitochondria or submitochondrial particles, substrates, and inhibitors are used as described before. Horseradish peroxidase (0.2–0.4 $\mu M$) (type VI, Sigma Chemical Co.) and 1–2 $\mu M$ scopoletin (Sigma Chemical Co.) are currently used. Scopoletin fluorescence is excited at 365 nm and measured at 450 nm.[19]

[21] A. Boveris, E. Martino, and A. O. M. Stoppani, *Anal. Biochem.* **80,** 145 (1977).
[22] P. R. Rich, A. Boveris, W. D. Bonner, Jr., and A. L. Moore, *Biochem. Biophys. Res. Commun.* **71,** 695 (1976).

# [58] Hydroperoxide Effects on Redox State of Pyridine Nucleotides and $Ca^{2+}$ Retention by Mitochondria[1]

By CHRISTOPH RICHTER

Most of the molecular oxygen consumed during mitochondrial respiration is reduced with four electrons directly to water at the level of cytochrome oxidase while some 1–5% of oxygen is reduced univalently to superoxide. Superoxide radicals are precursors of mitochondrial hydrogen peroxide. Most if not all mitochondrial hydrogen peroxide arises from dismutation of superoxide radicals.[2] While the metabolic significance of intramitochondrial superoxide and hydrogen peroxide remained unknown until recently the presence of glutathione peroxidase in mitochondria suggested an enzymic reduction of hydrogen peroxide and organic peroxides. Experimental evidence for this was first obtained by Oshino and Chance,[3] who showed coupling of the intramitochondrial pyridine nucleotide oxida-

[1] Supported by the Schweizerischer Nationalfonds (Grants 3.229-077 and 3.699.80).
[2] G. Loschen, A. Azzi, C. Richter, and L. Flohé, *FEBS Lett.* **42,** 68 (1974).
[3] N. Oshino and B. Chance, *Biochem. J.* **162,** 509 (1977).

Copyright © 1984 by Academic Press, Inc.
All rights of reproduction in any form reserved.
ISBN 0-12-182005-X

tion to reduction of hydroperoxides. A possible relationship between the redox state of pyridine nucleotides and the transport of $Ca^{2+}$ across the inner mitochondrial membrane was later suggested by Lehninger et al.[4]

The following experiments show a specific oxidation of both mitochondrial NADPH and NADH by small amounts of hydroperoxides through the action of glutathione peroxidase, glutathione reductase, and the energy-linked mitochondrial transhydrogenase. In the presence of succinate the enzymic pyridine nucleotide oxidation is reversible. If mitochondria are loaded with $Ca^{2+}$ before the addition of hydroperoxides the oxidation of pyridine nucleotides is irreversible, and $Ca^{2+}$ is released from mitochondria. The release is accompanied by an intramitochondrial hydrolysis of pyridine nucleotides at the N-glycosidic bond yielding ADPribose and nicotinamide.

*Isolation of Liver Mitochondria*

*Reagents*

MSH buffer, 210 m$M$ mannitol, 70 m$M$ sucrose, 5 m$M$ 4-(2-hydroxyethyl)-1-piperazinesulfonic acid (HEPES), pH 7.4
EDTA
Bovine serum albumin (BSA)

Mitochondria are isolated from female Wistar rats (150–200 g) starved for 16 hr before killing. Livers are washed and homogenized in MSH buffer containing 1 m$M$ EDTA and 0.5 mg BSA/ml. Mitochondria are obtained by differential centrifugation. The first centrifugation is performed at 800 $g$ for 10 min, followed by centrifugation of the resulting supernatant at 9500 $g$ for 10 min. The pellet of the second centrifugation (mitochondrial fraction) is washed twice with MSH buffer.

*Standard Incubation Conditions*

*Reagents*

MSH buffer
Rotenone
$K^+$-succinate
$CaCl_2$
Hydrogen peroxide or organic hydroperoxide (e.g., $t$-butyl hydroperoxide)

[4] A. L. Lehninger, A. Vercesi, and E. A. Bababunmi, *Proc. Natl. Acad. Sci. U.S.A.* **75**, 1690 (1978).

Mitochondria, 2 mg of protein/ml, are incubated under constant stirring at 25° in MSH buffer, 5 $\mu M$ rotenone, and 2.5 m$M$ succinate for 1 min. A fine jet of oxygen is blown onto the surface of the suspension to prevent anaerobiosis (standard incubation). The total volume routinely used is 3 ml. It has to be scaled up about fivefold for extraction and enzymic analysis of pyridine nucleotides.

## Determination of the Redox State of Intramitochondrial Pyridine Nucleotides

The hydroperoxide-induced oxidation of pyridine nucleotides is fully reversible if mitochondria have not been loaded with $Ca^{2+}$ prior to hydroperoxide addition. In contrast, in $Ca^{2+}$-loaded mitochondria the oxidation is irreversible due to intramitochondrial enzymic hydrolysis of pyridine nucleotides (see section below). Oxidation and hydrolysis of pyridine nucleotides are not observed in selenium-deficient mitochondria, i.e., mitochondria lacking active glutathione peroxidase.[5]

*Spectrophotometric Method.* Oxidation and reduction of NAD(P)(H) is continuously measured photometrically[5] by the dual-wavelength technique at 340–370 nm. Mitochondria are incubated under standard conditions. Addition of micromolar amounts of hydrogen peroxide or organic hydroperoxides results in an immediate oxidation of pyridine nucleotides due to the action of glutathione peroxidase, glutathione reductase, and the energy-linked transhydrogenase. (If hydrogen peroxide is used contaminating peroxisomal catalase must be removed by treatment of the mitochondrial preparation with digitonin.[6]) The extent and duration of pyridine nucleotide oxidation depend on the quantity of hydroperoxide offered to mitochondria. To achieve irreversible oxidation of pyridine nucleotides mitochondria are allowed to take up $Ca^{2+}$ (50–100 nmol/mg protein) before addition of hydroperoxide. The amount of pyridine nucleotides oxidized or reduced can be calculated using $\varepsilon_{mM}^{340-370} = 3.5$ for NAD(P)(H).

*Extraction and Enzymic Analysis.* The intramitochondrial levels of NADH, NADPH, NAD$^+$, and NADP$^+$ can be determined individually[7] after alkaline and acid extraction, respectively, and subsequent enzy-

[5] H. R. Lötscher, K. H. Winterhalter, E. Carafoli, and C. Richter, *Proc. Natl. Acad. Sci. U.S.A.* **76**, 4340 (1979).

[6] T. L. Chan, J. W. Greenawalt, and P. L. Pederson, *J. Cell Biol.* **45**, 291 (1970).

[7] W. Hofstetter, T. Mühlebach, H. R. Lötscher, K. H. Winterhalter, and C. Richter, *Eur. J. Biochem.* **117**, 361 (1981).

matic analysis. Here the procedure of Williams and Corkey is closely followed.[8]

*Reagents*

0.25 $N$ KOH in ethanol
1 $M$ triethanolamine
12% $HClO_4$
3 $N$ $K_2CO_3$ in 0.5 $M$ triethanolamine

Mitochondria incubated under standard conditions for 10 min are centrifuged at 10,000 $g$ for 10 min and the pellet is resuspended in MSH buffer to a protein concentration of 10 mg/ml. To determine the reduced intramitochondrial pyridine nucleotides the suspension is mixed with one-half of its volume 0.25 $N$ KOH in ethanol and incubated at 37° for 1 min. The pH is then adjusted to 8.5 with 1 $M$ triethanolamine and the precipitated proteins are seperated by centrifugation. Oxidized intramitochondrial pyridine nucleotides are determined after acid extraction. For this, the concentrated mitochondrial suspension is thoroughly mixed with an equal volume of 12% $HClO_4$, the pH is adjusted to 6.5 with 3 $N$ $K_2CO_3$ in 0.5 $M$ triethanolamine, and the precipitated proteins are separated by centrifugation. The extracted pyridine nucleotides are determined fluorimetrically in coupled enzymic assays.[8] To improve the signal-to-noise ratio all buffers and solutions should be filtered before use. The fluorescence signal can be calibrated with known amounts of pyridine nucleotides as standards. In mitochondria not loaded with $Ca^{2+}$ the hydroperoxide-induced decrease of the level of reduced pyridine nucleotides is balanced by a corresponding increase of the level of oxidized pyridine nucleotides. In $Ca^{2+}$-loaded mitochondria a complete loss of pyridine nucleotides can be observed.[5,7,9]

## $Ca^{2+}$ Movements Across the Inner Mitochondrial Membrane in the Presence of Hydroperoxides

Uptake and release of $Ca^{2+}$ may be measured by techniques described previously.[10] In the following experiments arsenazo III is used as $Ca^{2+}$ indicator. $Ca^{2+}$ movements are measured under standard conditions in the presence of 50 $\mu M$ arsenazo III in a dual-wavelength spectrophotometer at 675–685 nm.

[8] J. R. Williamson and B. E. Corkey, this series, Vol. 13 [65].
[9] H. R. Lötscher, K. H. Winterhalter, E. Carafoli, and C. Richter, *J. Biol. Chem.* **255**, 9325 (1980).
[10] M. Crompton and E. Carafoli, this series, Vol. 56 [28].

*Reagents*

Arsenazo III, purified as described[11]

Ruthenium red ($[(NH_3)_5Ru-O-Ru-(NH_3)_4-O-Ru(NH_3)_5]Cl_6$)

*Uptake of $Ca^{2+}$ by Rat Liver Mitochondria Is Not Influenced by the Redox State of Intramitochondrial Pyridine Nucleotides.* This can be demonstrated by comparing the initial rate of $Ca^{2+}$ uptake by mitochondria preincubated under standard conditions with or without hydroperoxide before the addition of $Ca^{2+}$. To oxidize intramitochondrial pyridine nucleotides 170 $\mu M$ $t$-butyl hydroperoxide is added 45 sec before $Ca^{2+}$ (100 $\mu M$). The initial $Ca^{2+}$ uptake rate is the same as that observed when the pyridine nucleotides are not oxidized by hydroperoxide.

*Hydroperoxides Induce Release of $Ca^{2+}$ from Rat Liver Mitochondria.* The amount and rate of $Ca^{2+}$ released from rat liver mitochondria upon addition of hydroperoxides are influenced by several parameters. Increasing $Ca^{2+}$ loads and increasing amounts of hydroperoxide both enhance extent and rate of $Ca^{2+}$ release.[9,12] A lowered intramitochondrial ATP content likewise facilitates release while oligomycin together with 20 $\mu M$ ATP completely prevents the hydroperoxide-induced release.[7] In the presence of ruthenium red, an inhibitor of the $Ca^{2+}$ uptake route, the release rate is faster than in its absence up to a load of about 100 nmol $Ca^{2+}$/mg protein.[12]

In the following experiment the hydroperoxide-induced release of $Ca^{2+}$ from mitochondria and its stimulation by ruthenium red can be observed. Mitochondria incubated under standard conditions are exposed to such an amount of $Ca^{2+}$ as to result in a total load of 30 nmol $Ca^{2+}$/mg protein when uptake is complete. [The endogenous $Ca^{2+}$ content of isolated mitochondria ranges from 10 to 12 nmol/mg protein. It can be determined with arsenazo III after uncoupling of mitochondria with carbonylcyanide $p$-trifluoromethoxyphenylhydrazone (FCCP).] Three minutes after addition of $Ca^{2+}$, efflux is initiated with 100 $\mu M$ $t$-butyl hydroperoxide and 2 nmol ruthenium red/mg protein, or with ruthenium red alone (Fig. 1).

*Hydroperoxide-Induced $Ca^{2+}$ Release Is Accompanied by Intramitochondrial Hydrolysis of Pyridine Nucleotides*

The inner membrane of isolated intact mitochondria is not permeable to pyridine nucleotides. Radioactive labeling of intramitochondrial pyridine nucleotides must therefore be performed *in vivo*.

---

[11] A. Scarpa, this series, Vol. 56 [27].

[12] S. Baumhüter and C. Richter, *FEBS Lett.* **148,** 271 (1982).

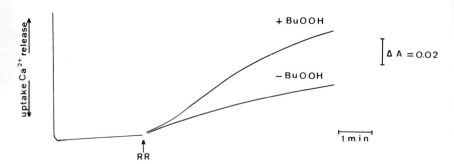

FIG. 1. Hydroperoxide-induced release of $Ca^{2+}$ from rat liver mitochondria. Mitochondria are incubated under standard conditions in the presence of the $Ca^{2+}$ indicator arsenazo III in a total volume of 3 ml. They are preloaded with 30 nmol $Ca^{2+}$/mg protein, and after 3 min $Ca^{2+}$ efflux is initiated with ruthenium red (RR) (2 nmol/mg protein) ± 100 $\mu M$ *t*-butyl hydroperoxide (BuOOH).

### Reagents

[*carboxyl*-$^{14}$C]Nicotinic acid (12.5 $\mu$Ci, 0.223 $\mu$mol)
Millipore filters (0.45 $\mu$m pore size)
*Labeling Procedure.* A rat of 150–200 g body weight is fasted overnight and then injected intravenously with [$^{14}$C]nicotinic acid. After 3 hr the animal is killed and liver mitochondria are isolated. The specific radioactivity of these mitochondria is 2500–3000 cpm/mg protein. Over 90% of the mitochondrial radioactivity is localized in the nicotinamide moiety of pyridine nucleotides.[9]

*Hydrolysis of Pyridine Nucleotides.* Radioactively labeled mitochondria are incubated under standard conditions. Then 100 nmol $Ca^{2+}$/mg protein is added, and after 2 more min 85 nmol *t*-butyl hydroperoxide is added to allow oxidation and hydrolysis of intramitochondrial pyridine nucleotides. At the desired times, 250-$\mu$l aliquots are withdrawn from the mitochondrial suspension, filtered through Millipore filters, and rinsed twice with 400 $\mu$l cold MSH buffer. The radioactivity in the filtrate is determined in a liquid scintillation counter. Intramitochondrial pyridine nucleotide hydrolysis and release of nicotinamide from mitochondria is complete within 5–10 min after hydroperoxide addition.[7,9] The other product besides nicotinamide of intramitochondrial pyridine nucleotide hydrolysis is ADPribose.[13] In the presence of 20 m$M$ nicotinamide the reaction is reversible[9] according to Eq. (1).

$$NAD^+ + H_2O \rightleftharpoons ADPribose + nicotinamide + H^+ \tag{1}$$

[13] S. Baumhüter and C. Richter, unpublished observation (1982).

*Comments*

A large number of compounds can evoke $Ca^{2+}$ release from rat liver mitochondria.[14,15] The physiological significance of these compounds is questionable since they act mostly by a reversal of the uptake route and by a collapse of the mitochondrial membrane potential. In contrast, the hydroperoxide-induced release of $Ca^{2+}$ from rat liver mitochondria occurs is high [12,16] and may therefore be of physiological importance. Pyridine nucleotide hydrolysis and protein ADPribosylation[17] may be responsible for the observed $Ca^{2+}$ release.

[14] E. Carafoli, *FEBS Lett.* **104**, 1 (1979).
[15] G. Fiskum and A. L. Lehninger, *Fed. Proc. Fed. Am. Soc. Exp. Biol.* **39**, 2432 (1980).
[16] H. R. Lötscher, K. H. Winterhalter, E. Carafoli, and C. Richter, *Eur. J. Biochem.* **110**, 211 (1980).
[17] C. Richter, K. H. Winterhalter, S. Baumhüter, H. R. Lötscher, and B. Moser, *Proc. Natl. Acad. Sci. U.S.A.* **80**, 3188 (1983).

# Section IV

# Pathology, Cancer, Aging

# [59] Glutathione Disulfide (GSSG) Efflux from Cells and Tissues

*By* HELMUT SIES and THEO P. M. AKERBOOM

Cells such as erythrocytes[1] or tissues such as eye lens[2] or liver[3] were found to respond to oxidizing conditions with a release of glutathione disulfide (GSSG) into the extracellular space. Other tissues have been studied as well (see the table). Under steady-state conditions, the rate of metabolism of an externally added model hydroperoxide, *t*-butyl hydroperoxide, was observed to be linearly related to the rate of efflux of GSSG from the isolated perfused rat liver[4]; in this organ about 3% of the flux through GSH peroxidase (see Ref. 5) is represented by an extracellular appearance of GSSG. Thus, even though most cell types contain an appreciable activity of GSSG reductase, a rise in the intracellular production of GSSG is accompanied by an export of the disulfide; apparently, there are multiple ways of disposing of GSSG that may accumulate.

Based on the relation between hydroperoxide metabolism and GSSG efflux,[4] it was proposed that the latter might be a useful extracellular parameter to assess hydroperoxide turnover in cells and tissues. Due to the multiple factors which may influence the rate of GSSG efflux, appropriate caution is required in each particular case in which the parameter is used (see below). Recent work has established that when oxygen radicals are generated in cells, there is an accompanying efflux of GSSG. Thus, not only hydroperoxides that are added or generated directly by oxidases, but also $H_2O_2$ formed by two successive one-electron steps involving the intermediate formation of the superoxide anion, $O_2^{\cdot-}$, is of interest here.

## Sample Handling

*General.* A major problem in determining GSSG in biological samples is the simultaneous presence of GSH. The generally very high GSH/GSSG ratio and the proneness of GSH for autoxidation to GSSG may lead to erroneously high GSSG data. Autoxidation is dependent on temperature, metal ions, and pH, so that sampling is carried out at low tempera-

---

[1] S. K. Srivastava and E. Beutler, *J. Biol. Chem.* **244**, 9 (1969).
[2] S. K. Srivastava and E. Beutler, *Biochem. J.* **112**, 421 (1969).
[3] H. Sies, C. Gerstenecker, H. Menzel, and L. Flohé, *FEBS Lett.* **27**, 171 (1972).
[4] H. Sies and K. H. Summer, *Eur. J. Biochem.* **57**, 503 (1975).
[5] L. Flohé and W. Günzler, this volume [12].

METHODS IN ENZYMOLOGY, VOL. 105
Copyright © 1984 by Academic Press, Inc.
All rights of reproduction in any form reserved.
ISBN 0-12-182005-X

GSSG Efflux from Cells and Tissues[a]

| Biological material | Addition | GSSG | GSH | GSH + GSSG (GSH-eq) |
|---|---|---|---|---|
| **Cells** | | | | |
| Erythrocytes[b] | | | | |
| Human[c,d] | None | 0 | | |
| | t-Butyl hydroperoxide | 0.8 | | |
| | Hydrogen peroxide | 1.1 | | |
| Rat[d] | Hydrogen peroxide | 1.3 | | |
| Isolated hepatocytes[e] | | | | |
| Rat[f] | None | 0.5 | 4.1 | |
| **Tissues**[e] | | | | |
| Liver, bile plus caval perfusate | | | | |
| Rat | | | | |
| Perfused, normal[g] | None | 1.0 | 12 | 14 |
| | t-Butyl hydroperoxide | 19 | 11 | 49 |
| Phenobarbital pretreated[h] | None | 1.0 | 15 | |
| | Ethylmorphine | 2.7 | 16 | |
| Liver, bile | | | | |
| Rat | | | | |
| Anesthetized[i] | None | 0.6 | 2.1 | |
| | Chronic ethanol | 1.4 | 2.3 | |
| Perfused in situ | None[j] | 0.4 | 1.1 | |
| | Menadione | 6.5 | 2.9 | |
| | Benzylamine[j] | 11.2 | 5.5 | |
| Isolated perfused,[j,k] | None | 0.3 | 0.1–1.6 | |
| Heart | | | | |
| Rat, perfused[l] | None | | | 0.5 |
| | t-Butyl hydroperoxide | 7.8 | | 15 |
| Lung | | | | |
| Rat | | | | |
| Perfused, normal[m] | None | | | 0.9 |
| | Hyperbaric oxygen | | | 1.5 |
| | t-Butyl hydroperoxide | | | 14 |
| Tocopherol deficient[m] | None | | | 1.0 |
| | Hyperbaric oxygen | | | 9.0 |

[a] Selected literature data on GSSG efflux and, where available, also GSH efflux.

[b] Values expressed in nanomoles per minute per milliliter of cells.

[c] S. K. Srivastava, Y. C. Awasthi, and E. Beutler, *Biochem. J.* **139**, 289 (1974).

[d] S. K. Srivastava and E. Beutler, *Biochem. J.* **114**, 833 (1969).

[e] Values expressed in nanomoles per minute per gram wet weight of cells or tissue.

[f] M. W. Farris and D. J. Reed, *in* "Isolation, Characterization, and Use of Hepatocytes" (R. A. Harris and N. W. Cornell, eds.), pp. 349–355, Elsevier, New York, 1983; assuming $10^8$ cells/g wet wt.

[g] G. M. Bartoli and H. Sies, *FEBS Lett.* **86**, 89 (1978).

[h] H. Sies, G. M. Bartoli, R. F. Burk, and C. Waydhas, *Eur. J. Biochem.* **89**, 113 (1978).

[i] H. Sies, O. R. Koch, E. Martino, and A. Boveris, *FEBS Lett.* **103**, 287 (1979).

[j] T. P. M. Akerboom, M. Bilzer, and H. Sies, *J. Biol. Chem.* **257**, 4248 (1982).

[k] K. E. Hill and R. F. Burk, *J. Biol. Chem.* **257**, 10668 (1982).

[l] T. Ishikawa, unpublished work.

[m] K. Nishiki, D. Jamieson, N. Oshino, and B. Chance, *Biochem. J.* **160**, 343 (1976); GSSG was not measured separately.

ture, in the presence of EDTA and at low pH. Acidification also leads to the precipitation of proteins which can be removed by centrifugation.

Chemical trapping of GSH is the most effective way of preventing autoxidation, e.g., by $N$-ethylmaleimide (NEM),[6] vinylpyridine,[7] or iodoacetic acid.[8,9] While NEM is very reactive and excess NEM has to be removed before enzymic assay, other compounds like vinylpyridine which are less reactive may not prevent autoxidation sufficiently.

*Cells in Suspension.* A sample from the suspension of cells (erythrocytes, hepatocytes) is centrifuged at 0° for 2 min at 3000 $g$ to separate the cells from the surrounding medium. An aliquot of the supernatant is treated with an equal volume of 1 $M$ $HClO_4$ containing 50 m$M$ NEM and 2 m$M$ EDTA; protein precipitated is removed by centrifugation. The supernatant is neutralized slowly with 2 $M$ KOH containing 0.3 $M$ $N$-morpholinopropanesulfonic acid (MOPS). Excess NEM is removed from the sample by ether extraction[10] or by chromatography,[11] and the samples are analyzed for GSSG by the enzymic assays described below or by high-performance liquid chromatography (HPLC).[9] When it is known that glutathione is released from cells only in the form of GSSG (e.g., in erythrocytes[1]) the NEM treatment is not necessary.

*Organ Perfusate.* Samples are collected into Eppendorf cups at 0° and assayed for GSSG immediately without sample treatment. Under normal perfusion conditions the liver releases no detectable amounts of GSSG into the caval perfusate, all glutathione appearing representing the reduced form.[12] Under conditions of oxidative stress and when there is membrane damage, however, significant amounts of GSSG may be found in the perfusate. If immediate assay is inconvenient, the method of masking GSH by NEM can be applied.

*Bile.* The bile duct should be cannulated with a tube of minimal length in order to obtain the sample at an early time point. Bile samples can be assayed for GSSG without further sample treatment if the bile is collected directly into Eppendorf cups at 0° (on ice) during short time intervals (2–5 min) followed by immediate assay.[12] Autoxidation of GSH in bile has been shown to occur[13]; therefore, if samples are to be collected or stored for longer times, bile should be collected into cups containing an equal vol-

[6] H. Güntherberg and J. Rost, *Anal. Biochem.* **15**, 205 (1966).
[7] O. W. Griffith, *Anal. Biochem.* **106**, 207 (1980).
[8] C. W. Tabor and H. Tabor, *Anal. Biochem.* **78**, 543 (1977).
[9] D. J. Reed, J. R. Babson, P. W. Beatty, A. E. Brodie, W. W. Ellis, and D. W. Potter, *Anal. Biochem.* **106**, 55 (1980).
[10] F. Tietze, *Anal. Biochem.* **27**, 502 (1969).
[11] T. P. M. Akerboom and H. Sies, this series, Vol. 77, p. 373.
[12] T. P. M. Akerboom, M. Bilzer, and H. Sies, *J. Biol. Chem.* **257**, 4248 (1982).
[13] D. Eberle, R. Clarke, and N. Kaplowitz, *J. Biol. Chem.* **256**, 2115 (1981).

ume of 5% metaphosphoric acid. Usually, 10-$\mu$l aliquots are sufficient for the assay of GSSG and total glutathione so that the samples can be used without further neutralization.

*Plasma.* Blood samples are rapidly collected in EDTA, to a final concentration of 5 m$M$, and immediately centrifuged for 1.5 min at 10,000 $g$. The plasma is rapidly acidified with an equal volume of 1 $M$ HClO$_4$ containing 50 m$M$ NEM, and processed as described above, or the plasma is acidified with an equal volume of 10% sulfosalicylic acid and treated with vinylpyridine (0.02 volumes) and triethanolamine (0.05 volumes).[14]

### GSSG Assay

*Principle.* GSSG is determined by its reaction with NADPH catalyzed by glutathione reductase.[15] For low concentrations, the reaction is monitored at the wavelength pair 340–400 nm using a dual-wavelength spectrophotometer or fluorometrically at 366 nm and 400- to 3000-nm excitation and emission wavelength, respectively.

#### Reagents

Potassium phosphate buffer, 0.1 $M$, pH 7.0, containing 1 m$M$ EDTA
NADPH, 5 m$M$, solution in 0.5% NaHCO$_3$
Glutathione reductase (yeast); dilute the commercial enzyme to 20 U/ml in the phosphate buffer containing additionally 50 $\mu M$ NADPH, prepare daily

*Procedure.* Pipet into cuvette 1 ml of buffer, 10 $\mu$l NADPH, and sample containing up to 25 nmol GSSG. After temperature equilibration at 25°, the reaction is started by the addition of 5 $\mu$l glutathione reductase and the change in absorbance is recorded.

*Comments.* Enzyme blanks may disturb assay at high-sensitivity determinations; they are minimized when the enzyme solution contains NADPH at a concentration similar to the mixture present in the cuvette.

### Enzymic Assay of GSSG Plus GSH

This kinetic assay which determines the sum of GSH and GSSG[10] can be useful as an assay for GSSG either if no GSH is present in the sample (for instance after trapping with NEM) or when the amount of GSH can be determined accurately in a separate assay, so that GSSG can be calculated by difference.[16]

*Principle.* The continuous reduction of 5,5'-dithiobis(2-nitrobenzoic

[14] M. E. Anderson and A. Meister, *J. Biol. Chem.* **255**, 9530 (1980).
[15] T. W. Rall and A. L. Lehninger, *J. Biol. Chem.* **194**, 119 (1952).
[16] G. M. Bartoli and H. Sies, *FEBS Lett.* **86**, 89 (1978).

acid) (DTNB) by NADPH is catalyzed by GSH or GSSG and glutathione reductase (see Refs. 10, 11, 17, 18). The reaction rate is linearly dependent on the glutathione concentration up to 2 $\mu M$, and is followed spectrophotometrically at 405 or 412 nm.

### Reagents

Phosphate buffer, 0.1 $M$, pH 7.0, containing 1 m$M$ EDTA
5,5′-Dithiobis(2-nitrobenzoic acid) (DTNB), 1.5 mg/ml, solution in 0.5% NaHCO$_3$; prepare daily
NADPH, 4 mg/ml, solution in 0.5% NaHCO$_3$
GSSG, 10 $\mu M$; prepare daily from stock solution (1 m$M$)
Glutathione reductase (yeast), dilute commercial enzyme to 6 U/ml in the phosphate buffer; prepare daily.

*Procedure.* Pipet into a thermostated cuvette (25°): 1 ml of buffer, 50 $\mu$l NADPH, 20 $\mu$l DTNB, 20 $\mu$l glutathione reductase, and 100 $\mu$l of sample containing 0.5–2 nmol glutathione. After mixing the contents of the cuvette, the reaction rate is measured as an increase in absorbance at 405 nm (Hg-line filter) or 412 nm. A reference assay without GSSG and a standard assay with a known amount of GSSG (usually 100 $\mu$l GSSG 10 $\mu M$, analyzed by the method described above) are run separately.

*Comments.* In some cases the reaction rate is influenced by physiological effectors present in the sample.[18,19] The addition of standard amounts of GSSG to the sample cuvette (internal standard) may circumvent this problem. If the internal and external standard amounts show different rates, the value for the internal standard should be used in the calculation, and the reference rate should be corrected accordingly by multiplying with the ratio of the internal standard rate to external standard rate.

Although the assay is calibrated with a known amount of the disulfide, the data are usually expressed in GSH-equivalents: GSH + 2 GSSG.

### General Remarks

*Critique.* GSSG release from cells can be used as an indicator for intracellular hydroperoxide metabolism. In fact, the addition of hydroperoxides to cells, or the intracellular generation of hydrogen peroxide, leads to elevated efflux of GSSG; in addition, chronic ethanol treatment was associated with an increased biliary GSSG release, leaving GSH release unaltered (see the table).

On the other hand, an increase in GSSG release must be interpreted with caution, since it also depends on several factors not directly related

[17] C. W. I. Owens and R. V. Belcher, *Biochem. J.* **94,** 705 (1965).
[18] N. Oshino and B. Chance, *Biochem. J.* **162,** 509 (1977).
[19] D. Häberle, A. Wahlländer, and H. Sies, *FEBS Lett.* **108,** 335 (1979).

to hydroperoxide metabolism. These include influences on activity of GSH peroxidase (selenium status)[20] and GSSG reductase (NADPH supply, inhibition by endogenous or exogenous inhibitors). Furthermore, the total non-protein-bound glutathione level, which varies under different diurnal[21] and nutritional[22] states, obviously is important; GSSG release is depressed in GSH-depleted livers. Finally, translocation of GSSG across the plasma membrane was shown to be dependent on energy in erythrocytes,[1] and translocase activity can be influenced by glutathione conjugates as shown with liver[23] and erythrocyte inside-out vesicles.[24]

Nonenzymic GSSG formation in cells may occur, as shown for example with nitrosobenzene[25]; thus, GSSG release observed upon addition of certain compounds will not indicate hydroperoxide formation.

GSSG release was found to be relatively variable (see the table); reasons for this include (1) methodological problems addressed above, and also (2) the experimental system employed (cells vs intact tissue) as well as biological variation.

*Compartmentation.* The transport of GSSG across the plasma membrane reflects localized cytosolic GSSG concentrations.[12] Therefore, GSSG efflux does not indicate total cellular hydroperoxide metabolism. For example, mitochondrial GSSG formation is normally not associated with efflux across the mitochondrial inner membrane.[26] Peroxisomal $H_2O_2$ metabolism is efficiently catalyzed by catalase, so that no peroxisomal $H_2O_2$ may leak into the cytosolic when catalase is functional.[18,27] Also, when catalase is present in the same (cytosolic) compartment as is GSH peroxidase, it can introduce an indicator error.[28]

Recent evidence that a GSH peroxidase activity is also involved in the reduction of microsomal and mitochondrial membrane-bound hydroperoxides has been reported for liver[29] and heart.[30] Therefore, cytosolic GSSG formation from membrane-bound hydroperoxides could contribute to the total cellular GSSG release.

[20] R. A. Lawrence and R. F. Burk, *J. Nutr.* **108,** 211 (1978).

[21] J. Isaacs and F. Binkley, *Biochim. Biophys. Acta* **497,** 192 (1977).

[22] N. Tateishi, T. Higashi, S. Shinya, A. Naruse, and Y. Sakamoto, *J. Biochem.* **75,** 93 (1974).

[23] T. P. M. Akerboom, M. Bilzer, and H. Sies, *FEBS Lett.* **140,** 73 (1982).

[24] T. Kondo, M. Murao, and N. Taniguchi, *Eur. J. Biochem.* **125,** 551 (1982).

[25] P. Eyer, *Chem. Biol. Interact.* **24,** 227 (1979).

[26] P. C. Jocelyn, *Biochim. Biophys. Acta* **396,** 427 (1975).

[27] B. Chance, H. Sies, and A. Boveris, *Physiol. Rev.* **59,** 527 (1979).

[28] G. Cohen and P. Hochstein, *Biochemistry* **2,** 1420 (1963).

[29] F. Ursini, M. Maiorino, M. Valente, L. Ferri, and C. Gregolin, *Biochim. Biophys. Acta* **710,** 197 (1982).

[30] M. Maiorino, F. Ursini, M. Leonelli, N. Finato, and C. Gregolin, *Biochem. Int.* **5,** 575 (1982).

General membrane damage as may be caused by aggressive compounds can alter the partitioning between bile and plasma; at high concentrations of *t*-butyl hydroperoxide significant amounts of GSSG are observed in the liver perfusate, whereas at low concentrations all GSSG is released into the biliary compartment. On the other hand, GSH release is unaltered when *t*-butyl hydroperoxide is metabolized[16]; however, the amount of GSH appearing in bile may slightly increase as shown in the table for menadione and benzylamine.

# [60] Stable Tissue Free Radicals

*By* W. Lohmann and H. Neubacher

Animal and plant tissues contain a great number of paramagnetic species consisting of free radicals and complexes of transition metal ions. The concentrations of these species vary, in general, between the different tissues. The purpose of this chapter is to provide a brief review of free radicals existing in animal tissues and some applications to cancer. Since the information that can be obtained from electron spin resonance (ESR) measurements of wet or frozen samples is rather limited, results obtained from lyophilized samples are also discussed. Nothing will be said about the role of free radicals in complex biological systems despite the fact that they influence these systems considerably.

## Free Radicals in Wet and Frozen Tissues

The very early ESR studies of wet tissues demonstrated the existence of different free radicals at $g \approx 2$ in the various organs and subcellular structures.[1] Since the signals were localized in the mitochondria it was suggested that they are associated with enzymic redox activity. Subsequent work, performed with frozen samples, also supported this idea.[2] Another essential source of free radicals was found to be present in microsomes.[3,4] In brain microsomes of rats, an ESR signal at $g = 2.004$ with a linewidth of 1.1 mT could be detected. These ESR signals have been attributed to flavin or other semiquinone radicals.[5] Identification of indi-

[1] B. Commoner and J. L. Ternberg, *Proc. Natl. Acad. Sci. U.S.A.* **47,** 1374 (1961).
[2] M. Waldschmidt, H. Mönig, and J. Schole, *Z. Naturforsch.* **23B,** 798 (1968).
[3] Y. Hashimoto, T. Yamano, and H. S. Mason, *J. Biol. Chem.* **237,** PC 3843 (1962).
[4] Z. Kometani and R. H. Cagan, *Biochim. Biophys. Acta* **135,** 1083 (1967).
[5] H. Beinert and G. Palmer, *Adv. Enzymol.* **27,** 105 (1965).

METHODS IN ENZYMOLOGY, VOL. 105
Copyright © 1984 by Academic Press, Inc.
All rights of reproduction in any form reserved.
ISBN 0-12-182005-X

vidual radical species from the envelope of the signals located in the $g \approx 2$ region has not been achieved yet.

Ruuge et al.[6,7] described an additional triplet located at $g \approx 2$ with a hyperfine (hf) splitting of about 2 mT in frozen rat and mouse tissues. It was also measured by Benedetto[8] in normal and tumorous tissues of human cervix and uterus in frozen powdered form. It was suggested that this triplet is a peroxy-type signal mainly of artifactual origin which might give some information about the paramagnetic behavior of normal and abnormal cervix. Additional signals have been observed at $g = 2.11$ to 2.15 which also change their intensity with development of malignancy. These signals are believed to reflect prostaglandin metabolism and lipid peroxidation.

It seems to be rather difficult to obtain detailed information on these radicals, because there may be quite a number of different radicals.[7] Moreover, there are several paramagnetic metal complexes in tissues with spectral components at $g \approx 2$.[9] Fortunately, their saturation behavior is different from that of the radicals, because metal ions usually have relatively short relaxation times due to their spin–orbit coupling. Therefore, the radical contribution to an ESR spectrum dominates at low microwave power, whereas at high microwave power and low temperature the contributions of the metals are predominant.[10] Often, however, it is not possible to separate the two contributions completely, and measurements on radical concentrations contain some contributions of metal complexes.[11,12]

Another important factor is the spin concentration. It seems to be the highest in liver, followed by kidney and heart.[13] Much less was found in spleen, liver, and muscle. The spin concentrations determined ranged from $5 \times 10^{14}$ to $5 \times 10^{15}$ radicals/g.

Usually fewer free radicals were found in tumor tissues than in comparable normal tissues.[1,14–19] This phenomenon is not universal, however, as was shown by Gutierrez et al.[20] Breast tumors of rats induced by 7,12-dimethylbenzanthracene had about twice the ESR signal intensity of nor-

[6] E. K. Ruuge and I. A. Kornienko, Biofizika 14, 752 (1969).

[7] E. K. Ruuge and A. G. Chetverikov, Biofizika 15, 478 (1970).

[8] C. Benedetto, in "Free Radicals, Lipid Peroxidation and Cancer" (D. C. H. McBrien and T. F. Slater, eds.), p. 27. Academic Press, New York, 1982.

[9] N. J. F. Dodd, in "Metal Ions in Biological Systems" (H. Siegel, ed.), Vol. 10, p. 95. Dekker, New York, 1980.

[10] H. M. Swartz and R. P. Molenda, Science 148, 94 (1965).

[11] M. Kent and J. R. Mallard, Phys. Med. Biol. 14, 431 (1969).

[12] H. M. Swartz, in "Biological Applications of Electron Spin Resonance" (H. M. Swartz, J. R. Bolton, and D. C. Borg, eds.), p. 155. Wiley (Interscience), New York, 1972.

[13] J. R. Mallard and M. Kent, Nature (London) 210, 588 (1966).

[14] J. R. Mallard and M. Kent, Nature (London) 204, 1192 (1964).

mal breast tissue. Tumor tissue of melanotic melanomas of mice also showed an increase in radical concentration.[21] In this case, radical activity may partly result from melanin. Considerable differences between the ESR transition metal spectra of normal and tumor tissue were obtained.[9,22]

A doublet at $g = 2.0052$ with an hf splitting of 0.18 mT was detected in homogenates of mouse melanoma and several rat tumor tissues.[23] It was identified as the ascorbyl radical[23,24] since it resembles very closely the ascorbyl radical determined by Lagercrantz[25] and Ruf and Weiss.[26] It is, however, not unique to tumors since it was also detected in fresh whole blood from normal and tumor-bearing animals as well as in normal human blood.[24] Furthermore, it is not present in all types of tissue, e.g., in adrenals, despite the fact that they have the highest ascorbic acid concentration of any tissue. Changes in the concentration of this radical were detected also in blood throughout the development of mouse myeloid leukemia which had been produced by an iv injection of $10^6$ leukemic spleen cells. It should be emphasized that this radical is not stable; it decays by about 50% within the first 2 hr.[27]

In general, it has been suggested that the ascorbyl radical reflects the oxidation rate of ascorbic acid rather than the ascorbic acid content of the tissue. Thus, an increase in concentration of this radical during disease may reflect a change in the relative concentration of ascorbic acid and oxidant caused by altered metabolism or cell lysis. These changes may be associated with inflammation or immune reactions.[28]

Melanin radicals are found in melanin-containing tissues and in melanomas.[29,30] Oxygen radicals may also be involved in melanin reactions since the latter consume oxygen and produce superoxide anions and

[15] A. J. Vithayathil, J. L. Ternberg, and B. Commoner, *Nature (London)* **207**, 1246 (1965).
[16] I. Kolomitseva, K. L'Vov, and L. Kayushin, *Biofizika* **5**, 636 (1960).
[17] J. Duchesne and A. van de Vorst, *Bull. Cl. Sci. Acad. R. Belg.* **56**, 433 (1970).
[18] M. Sentjurc and M. Schara, *Naturwissenschaften* **57**, 459 (1970).
[19] H. M. Swartz and P. L. Gutierrez, *Science* **198**, 936 (1977).
[20] P. L. Gutierrez, H. M. Swartz, and E. J. Wilkinson, *Br. J. Cancer* **39**, 330 (1979).
[21] I. L. Mulay and L. N. Mulay, *J. Natl. Cancer Inst.* **39**, 735 (1967).
[22] D. W. Nebert and H. S. Mason, *Cancer Res.* **23**, 833 (1963).
[23] P. S. Duke, *Exp. Mol. Pathol.* **8**, 112 (1968).
[24] N. J. F. Dodd, *Br. J. Cancer* **28**, 257 (1973).
[25] C. Lagercrantz, *Acta Chem. Scand.* **18**, 562 (1964).
[26] H. H. Ruf and W. Weis, *Biochim. Biophys. Acta* **261**, 339 (1972).
[27] N. J. F. Dodd and J. M. Giron-Conland, *Br. J. Cancer* **32**, 451 (1975).
[28] J. M. Silcock and N. J. F. Dodd, *Br. J. Cancer* **34**, 550 (1976).
[29] M. S. Blois, A. B. Zahlan, and J. E. Mailing, *Biophys. J.* **4**, 471 (1964).
[30] R. C. Sealy, C. C. Felix, J. S. Hyde, and H. M. Swartz, in "Free Radicals in Biology" (W. A. Pryor, ed.), Vol. IV, p. 210. Academic Press, New York, 1980.

hydrogen peroxide. These oxygen-related radicals located at $g \approx 2$ have attracted considerable interest during the last few years because of their involvement in a number of biological systems and reactions.[31] The importance of paramagnetic metal ions for free radical studies is based on the fact that they can generate free radicals by participating in redox reactions and can interact with free radicals, forming new species. These subjects are discussed elsewhere in this volume.

### Free Radicals in Lyophilized Tissues

Lyophilization was used for the early ESR investigations of biological material because of the sensitivity limitations of the spectrometers. Water-containing samples cause dielectric losses in the cavity and, therefore, reduce the instrumental sensitivity. In 1958, Truby and Goldzieher,[32] by comparing linewidth and signal intensities of lyophilized and frozen liver samples, found that the radicals seen in freeze-dried tissues are mainly artifacts, created during lyophilization. Hence, these signals were not considered to be identical with the free radicals present in functioning tissue. Furthermore, the paramagnetic behavior of the material might not be retained during the process of lyophilization. The copper spectrum of plasma ceruloplasmin could no longer be detected[33] and the oxidation state of manganese was changed considerably after lyophilization.[20,34] A similar effect was measured for the nonheme iron-sulfur protein signal at $g = 1.94$,[35] from which it might be concluded that changes in the oxidation state and the ligand sphere will also occur for other transition metal complexes in tissue samples.

During the last few years, however, it was demonstrated that there is a correlation between these radicals and the metabolic state. Especially during the investigation of tumor tissue, reproducible changes in the content of these radicals demonstrated the usefulness of this method. Therefore, this preparation method has been developed further by several groups, and the ESR signals obtained have been investigated in more detail. It should be emphasized, however, that special care must be taken with regard to oxygen tension, humidity, and storage temperature, e.g., in order to obtain reproducible results.

The general lineshape and $g$ factor of the ESR spectra of most lyophi-

---

[31] T. F. Slater, in "Free Radicals, Lipid Peroxidation and Cancer" (D. C. H. McBrien and T. F. Slater, eds.), p. 243. Academic Press, New York, 1982.

[32] F. K. Truby and J. W. Goldzieher, *Nature* (*London*) **182**, 1371 (1958).

[33] C. Mailer, H. M. Swartz, M. Konieczny, S. Ambegaonkar, and V. L. Moore, *Cancer Res.* **36**, 637 (1974).

[34] P. L. Gutierrez and H. M. Swartz, *Br. J. Cancer* **39**, 24 (1979).

[35] A. F. Vanin, A. G. Chetverikov, and L. A. Blyumenfeld, *Biofizika* **13**, 66 (1968).

lized biological material of animal and plant tissue are similar. The asymmetrical singlet at $g \approx 2$ with $\Delta H_m \approx 0.8$ mT exhibits a line anisotropy without any hyperfine structure which is typical for an axially symmetrical radical in nonmonocrystalline solid state. At 23 GHz (K band) and 35 GHz (Q band) measurements, the $g$ value anisotropy was somewhat more apparent.[20,34,36,37] From this, it was concluded that the asymmetry is not formed by superposition of two or more lines with different $g$ values but should be due to a $g$ value anisotropy of one absorption line only.

There are several investigations on the origin of this ESR signal.[38] In the early studies it was believed that semiquinones might be a precursor of this radical which will be stabilized by their adsorption on a protein matrix. An increase in radical content after lyophilization could be observed if tissue samples were treated with solutions of, e.g., gallates or ascorbic acid. The same results were obtained with a number of substances, which form semiquinone states, on their interaction with other materials, e.g., $Ba(OH)_2$,[39] filter paper,[40] quartz powder,[37] or anion exchange material.[41] The concentration of the semiquinone radicals was given by the hydroquinone/quinone equilibrium which depends on pH, temperature, and the presence of oxygen or other electron acceptors.

Until recently there was no consensus among investigators with regard to the origin of the ESR signal observed in tissues at $g \approx 2$. There were certain indications that it probably arises from an ascorbic acid–protein system.[42]

A more detailed explanation was given more recently.[43] The authors pointed out that the asymmetric signal in tissues really consists of at least two components. The first one was assigned to the ascorbyl radical bound to an organic matrix. It is an intermediate in the naturally occurring redox system ascorbic acid/dehydroascorbic acid and plays an especially predominant role in ESR studies on cancer.[43,44] No definitive assignment

[36] I. Miyagawa, W. Gordy, N. Watabe, and K. M. Wilbur, *Proc. Natl. Acad. Sci. U.S.A.* **44**, 613 (1958).

[37] I. G. Kharitonenkov, *Biofizika* **12**, 736 (1961).

[38] I. I. Naktinis and L. C. Cerniauskiene, *Biofizika* **19**, 1039 (1974).

[39] A. E. Kalmanson, L. P. Lipchina, and A. G. Chetverikov, *Biofizika* **6**, 410 (1961).

[40] A. G. Chetverikov, L. A. Blyumenfeld, and G. V. Fomin, *Biofizika* **10**, 476 (1965).

[41] K. G. Bensch, O. Körner, and W. Lohmann, *Biochem. Biophys. Res. Commun.* **101**, 312 (1981).

[42] E. K. Ruuge, T. M. Kerimov, and A. V. Panemanglov, *Biofizika* **21**, 124 (1976).

[43] W. Lohmann, J. Schreiber, and W. Greulich, *Z. Naturforsch.* **34**, 550 (1979).

[44] W. Lohmann, K. G. Bensch, H. Sapper, A. Pleyer, J. Schreiber, S. O. Kang, H. Löffler, H. Pralle, K. Schwemmle, and R. D. Filler, *in* "Free Radicals, Lipid Peroxidation and Cancer" (D. C. H. McBrien and T. F. Slater, eds.), p. 55. Academic Press, New York, 1982.

could be given yet for the other, broader signal located at a slightly lower $g$ value. It was believed that it was associated with protein since the two components were obtained by the authors originally by an ascorbic acid–copper protein interaction.

Dodd and Swartz[45] came to a similar conclusion. They describe the asymmetric signal in tissues consisting of one signal with $\Delta H = 0.6$ mT located at $g = 2.005$ which decays rapidly in moist air, and another one located at $g = 2.004$ with $\Delta H = 0.9$ mT which is much more stable.

The asymmetrical signal has been observed in several types of healthy and cancerous tissues.[44,46] Of special interest is the fact that only patients with acute lymphatic leukemia exhibit this signal in their erythrocytes. Depending on the stage of the disease, the concentration of this free radical is increased accordingly. It should be pointed out that the free radical concentration is independent of the leukocyte count. Kinetic studies have shown that the spin concentration decreases during treatment of the patients and returns to normal values concomitantly with an apparent curing of the patients.[47]

There are fewer free radicals in tumors than in comparable normal tissues according to most studies on this subject, as pointed out earlier. However, concerning lyophilized tissues a few investigators reported an increase in the level of free radical concentration above that of healthy tissue in an early stage of cancer followed by a decrease.[47,48] Since a characteristic change in spin concentration with increasing concentrations of ascorbic acid has been observed,[47,49] the early rise as well as the subsequent decrease in spin concentration in neoplastic tissues were suggested to be the result of alterations in the interaction of tumor cells with ascorbic acid. A determination of the free radical concentration of lyophilized samples of healthy and tumorous pulmonary tissues exhibited great differences: the spin concentration was the largest at the rim of the tumor and decreased toward the healthy tissue and center of the tumor as well. In the latter case, it was less than the concentration in healthy tissue.[50]

This brief review could focus only on a few applications of stable free radicals in tissue. The results presented may eventually provide insights into biochemical and pathological changes that occur in certain types of diseases.

[45] N. J. F. Dodd and H. M. Swartz, *Br. J. Cancer* **42**, 349 (1980).
[46] N. M. Emanuel, *Q. Rev. Biophys.* **9**, 283 (1976).
[47] W. Lohmann, J. Schreiber, W. Strobelt, and Ch. Müller-Eckhardt, *Blut* **39**, 317 (1979).
[48] A. N. Saprin, F. V. Klochko, K. E. Kruglyakova, V. M. Chibrikin, and N. M. Emanuel, *Dokl. Akad. Nauk SSSR* **169**, 222 (1966).
[49] E. K. Ruuge and L. A. Blyumenfeld, *Biofizika* **10**, 689 (1965).
[50] W. Lohmann, K. G. Bensch, J. Schreiber, and E. Müller, *Z. Naturforsch.* **36**, 5 (1981).

## [61] Assay of Superoxide Dismutase Activity in Tumor Tissue

*By* LARRY W. OBERLEY and DOUGLAS R. SPITZ

The assay of superoxide dismutase (SOD) activity in tumor cells is difficult for three main reasons.

(1) Tumor cells are generally low in SOD activity.[1] The degree of loss of SOD activity is directly proportional to the degree of differentiation of the tumors, with the poorly differentiated tumors having low SOD activity and the well-differentiated tumors having high SOD activity. Thus, the most undifferentiated tumors will have very low SOD activities. This low SOD activity means that any SOD assay that is used must have very high sensitivity. This requirement for sensitivity immediately rules out the most commonly used assay for SOD: the xanthine oxidase–cytochrome $c$ method.[2] This method has a very poor sensitivity with 1 unit of activity equal to 200 ng of pure bovine Cu,Zn-SOD.

Another point of confusion here is that SOD activity is not low in some tumors in absolute terms. However, the activity will be low in comparison to the differentiated cell counterpart from which the tumor arose.[1] In tumor therapy, one is not concerned with the absolute value of SOD activity, but rather that some difference exists between susceptible normal tissue and tumor tissue so that a therapeutic ratio can be achieved. Thus, it is very important when comparing SOD activity in tumor tissue to that in normal tissue to compare it to the activity in a proper normal tissue. A proper normal tissue is usually not a whole organ such as the lung which comprises at least 40 cell types. Rather, a proper control is the differentiated cell type from which the tumor arose. This means that some cell isolation procedure must be used to obtain a single cell population in order to obtain a proper control.

(2) Tumor cell whole homogenates must be used, because one should not discard any portion of the cell homogenate without verifying the amount of SOD activity that is in the discarded fraction. Since tumor cells are lower in SOD, one must be careful not to remove any cell fraction which might contain SOD activity. Tumor cells and normal cells might lose activity at different rates during fractionation. This could lead to spurious interpretations of the losses of SOD activity in tumor and normal tissue.

[1] L. W. Oberley, *in* "Superoxide Dismutase" (L. W. Oberley, ed.), Vol. 2, p. 127. CRC Press, Boca Raton, Florida, 1982.
[2] J. M. McCord and I. Fridovich, *J. Biol. Chem.* **244,** 6049 (1969).

METHODS IN ENZYMOLOGY, VOL. 105
Copyright © 1984 by Academic Press, Inc.
All rights of reproduction in any form reserved.
ISBN 0-12-182005-X

The fact that whole homogenates should be used is another reason why the cytochrome *c* assay should be avoided. As a general rule, it is dangerous to measure absorbance with protein concentrations about 2 mg/ml because higher protein concentrations cause light scatter and Beer's law is then not applicable. In order to measure SOD activities in the more malignant tumors with the cytochrome *c* assay, protein concentrations above 2 mg/ml are often needed to obtain 1 unit of activity.

(3) Tumor cells have abnormal biochemistry. In particular, they are much more resistant to lipid peroxidation than corresponding normal cells.[3] This altered reactivity to lipid peroxidation affects any assay based on superoxide-propagated chain reactions such as the epinephrine method[4] or the pyrogallol assay.[5] For these reasons they should not be used in assay of whole-tumor homogenates, because tumor cells may contain higher levels of radical reaction-terminating substances than normal cells. Indeed, the few studies using these assays have shown inconsistent results.[6]

For the above reasons, only two assays have been used with any success. The first of these is the direct assay of Marklund.[7] This assay uses elevated pH to slow the spontaneous dismutation of superoxide sufficiently to allow its observation on a time scale of minutes. This assay uses commercially available $KO_2$ as the source of superoxide and is extremely sensitive. It is also the only direct assay for SOD (i.e., it measures directly the dismutation of $O_2^-$). However, this assay has not been used much by free radical researchers, for two reasons: (1) the assay is difficult to run, apparently because of the effects of impurities and interferences; and (2) it is necessary to run the assay at a nonphysiological pH (10.2), at which Mn-SOD is known to be inhibited. This inhibition means an estimate must be made of the degree of inhibition in any particular tissue.

We have found that the best method for assaying SOD activity in tumor tissue is the xanthine oxidase–nitro blue tetrazolium (NBT) assay, first described by Beauchamp and Fridovich[8] and modified by our research group.[9] The assay has the necessary high sensitivity (1 unit = 15 ± 3 ng of pure bovine Cu,Zn-SOD activity) to measure SOD activity in crude homogenates of tumor cells with low SOD activity. The NBT assay

[3] R. M. Arneson, V. J. Aloyo, G. S. Germain, and J. E. Chenevey, *Lipids* **13**, 383 (1978).
[4] H. P. Misra and I. Fridovich, *J. Biol. Chem.* **247**, 3170 (1972).
[5] S. Marklund and G. Marklund, *Eur. J. Biochem.* **47**, 469 (1974).
[6] G. Dinescu-Romalo and C. Mihai, *Cell. Mol. Biol.* **25**, 101 (1979).
[7] S. Marklund, *J. Biol. Chem.* **251**, 7504 (1976).
[8] C. Beauchamp and I. Fridovich, *Anal. Biochem.* **44**, 276 (1971).
[9] G. R. Buettner, L. W. Oberley, and S. W. H. C. Leuthauser, *Photochem. Photobiol.* **28**, 693 (1978).

does not have as high a specificity as the cytochrome $c$ assay. However, as indicated earlier, even though the cytochrome $c$ assay has high specificity, it does not have the sensitivity necessary to measure some tumor tissue. The NBT assay also has the ability to distinguish between Cu,Zn-SOD activity and Mn-SOD activity. This is important because tumor cells have been found to be generally (but not always) low in Cu,Zn-SOD activity, but always low in Mn-SOD activity when compared to a differentiated normal cell counterpart.[1] Thus, measurement of total SOD activity is not enough; Mn-SOD activity must also be measured.

Inability to accurately quantitate Mn-SOD is another reason why the cytochrome $c$ assay cannot be used. We have found that 5 m$M$ cyanide is necessary to totally inhibit Cu,Zn-SOD activity in liver crude homogenates. This amount of cyanide cannot be used in the cytochrome $c$ assay because it inhibits cytochrome $c$ reduction (but not NBT reduction). Two millimolar cyanide is the maximum allowable in the cytochrome $c$ assay; this amount of cyanide will lead to overestimation of Mn-SOD activity.

A last argument for the use of the NBT assay is that we have verified our conclusions with an immunoassay.[10] Our immunoassay results indicated the amount of SOD protein in normal liver and rat hepatomas is very similar to that seen with our SOD assay. Thus, not only is SOD activity low in these tumors, but so is the amount of SOD protein. This correlation between immunoassay and our enzymic assay is perhaps the most compelling reason for using our assay.

Our assay differs from that of the commonly used NBT/cytochrome $c$ SOD assays in the following ways. First of all, we use diethylenetriaminepentaacetic acid (DETAPAC) instead of ethylenediaminetetraacetic acid (EDTA). We have found that DETAPAC makes the SOD assay more sensitive, probably because Fe-DETAPAC does not react with $O_2^-$ whereas Fe-EDTA does.[9] Second, we always include catalase in our assay mixture, whereas most commonly used assays do not. We include catalase for two reasons: (1) $H_2O_2$ inhibits and even inactivates Cu,Zn-SOD, so removal of $H_2O_2$ is necessary to prevent this inhibition, and (2) for kinetic reasons, the assay should be run with the product ($H_2O_2$) as low as possible, so that the equilibrium will not be shifted in favor of $O_2^-$ production.

With these changes in mind, our procedure is essentially that of Beauchamp and Fridovich.[8] Xanthine–xanthine oxidase is used to generate a reproducible flux of $O_2^-$ and nitro blue tetrazolium is used as an indicator of superoxide production. SOD will then compete with NBT for $O_2^-$. The

[10] S. W. H. C. Leuthauser, A. P. Autor, and L. W. Oberley, unpublished observations (1982).

percentage inhibition of NBT reduction is a measure of the amount of SOD present. In the assay procedure, varying concentrations of SOD activity are added until maximum inhibition is obtained. One unit of activity is that amount of protein which gives half-maximal inhibition. This assay differs from that of the cytochrome $c$ method in that 95–100% inhibition is usually not achieved. With pure enzyme, 80–90% inhibition is obtained, whereas with most tissue homogenates 70–80% inhibition is observed. This is why half-maximal inhibition is used rather than 50% inhibition as in the cytochrome $c$ assay. It is important to realize that different maximal inhibitions will be observed in different tissues; thus, the point where 1 unit is reached will change. Because of this variation of maximal inhibition, a full inhibition curve must be obtained for every sample.

Enzyme activities are usually expressed as units per milligram of protein, with protein measured by the method of Lowry.[11] Studies done on tissue culture cells where the number of cells can be quantitated are usually expressed as units/$10^6$ cells.

*Tissue Preparation*

As mentioned earlier, we recommend the use of whole homogenates for tumor cell studies. For many tissues, we have found that homogenization on ice in a motor-driven Teflon pestle homogenizer in 0.05 $M$ potassium phosphate buffer, pH 7.8, will allow the expression of maximal SOD activity.[12] Tris buffer should not be used because we have found that it appears to have SOD activity.[13] For some tissues, sonication is also necessary. We sonicate at the maximum output of a Biosonik IV sonicator for 2 to 3 min in 15-sec bursts, while on ice. This tissue preparation is also in marked contrast to the cytochrome $c$ assay, where sonication is usually necessary to measure full SOD activity. We think the difference is that NBT can penetrate into cellular compartments, whereas cytochrome $c$ cannot. Thus, in the cytochrome $c$ method, sonication is necessary to release the enzyme so that it is accessible to the indicator.

We have also had satisfactory results preparing the tissue with a Tekmar homogenizer (which both homogenizes and sonicates). Samples are homogenized using 5-sec bursts of full power for 1 min, with the sample on ice. We have found that for tissue culture cells, neither a Teflon pestle homogenizer nor the Tekmar can be used because cells adhere to

[11] O. H. Lowry, *J. Biol. Chem.* **193**, 265 (1951).
[12] S. W. H. C. Leuthauser, L. W. Oberley, and T. D. Oberley, unpublished observations (1982).
[13] D. P. Loven and L. W. Oberley, unpublished observations (1982).

the instruments. We have used the Biosonik sonicator alone for this work with very satisfactory results. A last point is that, in spite of our previous cautions, it may sometimes be necessary to use a low-speed centrifugation to clean up debris, fat, etc. Checks must be made to ascertain that the supernatant obtained contains the same SOD activity as the homogenate.

*Reagents*

Potassium phosphate buffer (0.05 $M$), pH 7.8; xanthine, 1.8 m$M$ in above phosphate buffer (prepare fresh every week); NaCN, 0.33 $M$ in above phosphate buffer (prepare fresh every 2 days); NBT, 2.24 m$M$ in above phosphate buffer (keep in brown bottle); DETAPAC, 1.33 m$M$ in above phosphate buffer; catalase, 40 units/ml, in above phosphate buffer; xanthine oxidase, about $10^{-2}$ units/ml, in 0.05 $M$ phosphate buffer with 1.33 m$M$ DETAPAC (diluted only at time of assay). All reagents can be obtained from Sigma and used without further purification. The reagents will give the following final concentrations in each tube: 1 m$M$ DETA-PAC, 1 unit of catalase, $5.6 \times 10^{-5}$ $M$ NBT, $10^{-4}$ $M$ xanthine, and 5 m$M$ NaCN.

*Procedure*

1. For a 20 tube assay, the following reagents are added to an Erlenmeyer flask: 13.0 ml of 0.05 $M$ phosphate buffer, pH 7.8, with 1.333 m$M$ DETAPAC, 0.50 ml catalase in 0.05 $M$ phosphate buffer, 0.50 ml of NBT in phosphate buffer, 1.70 ml of xanthine in 0.05 $M$ phosphate buffer, and 0.30 ml of phosphate buffer (to measure total SOD) or 0.33 $M$ NaCN in phosphate buffer (to measure Mn-SOD). This solution is mixed well.

2. Of this solution 0.8 ml is added to each of 20 tubes.

3. Of pure SOD (1 to 1000 ng) or tissue (1 to 500 $\mu$g) 100 $\mu$l is added to each tube in varying amounts; 100 $\mu$l of 0.05 $M$ phosphate buffer is added to at least five tubes to serve as blanks.

4. To assay Mn-SOD, after the CN$^-$ is added, the solution is incubated for at least 30 min and no more than 2 hr.

5. Xanthine oxidase (100 $\mu$l) is added to the blanks. Xanthine oxidase is diluted in phosphate buffer with DETAPAC. Dilutions are made until the tubes used as blanks (with no SOD) give an absorbance rate between 0.015 to 0.025/min at 560 nm. The rate of change of absorbance is recorded for at least 2 min starting when a good straight line is first obtained (usually 30 sec after addition of xanthine oxidase).

6. Xanthine oxidase is added to each succeeding tube with SOD activity and the change in absorbance is measured as a function of time.

*Precautions*

1. In all tissues, NBT reductase activity must be checked. Some tissues will reduce NBT without xanthine oxidase, although this occurrence is very rare. A tube with 500 $\mu$g of tissue and the standard assay mixture should be followed at 560 nm for 10 min to check for NBT reductase activity. If NBT reductase activity should be found, the value of NBT reduction in absorbance units per minute without xanthine oxidase present should be subtracted from that seen with xanthine oxidase.

2. It is possible for a tissue to have xanthine oxidase inhibitory activity. This activity would make the tissue appear to have extra SOD activity, because there would be less NBT reduction. We have never seen such activity, but it should be checked nevertheless. The production of uric acid can be followed at 290 mn to check the xanthine oxidase inhibitory activity. The rate of uric acid formation should be followed in the presence and absence of tissue. The standard assay mixture can be used in such studies, but it is probably better to replace the NBT with phosphate buffer, so that no spectrophotometric interference occurs.

3. During the course of study of SOD activities in breast, intestine, salivary glands, and some tumors, biochemical interference in the NBT–xanthine–xanthine oxidase assay as well as in the cytochrome $c$ assay was noted. This interference manifested itself in four observable ways. First, the formation of a spontaneous bluish-purple color occurred in the NBT assay tubes containing NaCN before addition of xanthine oxidase. This interference was noted in all the above-mentioned organ homogenates. Second, the NBT assay tubes containing these tissue homogenates and NaCN had higher slopes (rates of blue formazan production) than the blanks containing assay solution and NaCN, but no tissue. Third, an increased amount of curvature in the slope of the absorbance change per minute was noted in the cytochrome c assay tubes containing NaCN and tissue. Finally, a low maximum percentage inhibition of blue formazan production was also observed in both NaCN containing and non-NaCN containing tubes using the NBT assay.

We were puzzled when we originally looked at these observations and felt that this interference was affecting our ability to detect Mn-SOD in the presence of NaCN. We began to develop a hypothesis of why this interference was mainly seen in the presence of NaCN. Since $CN^-$ is a powerful blocker of site 3 in the electron transport chain and electron transport chain blockers are known to enhance $O_2^-$ production, we felt that this could be the source of the interference. We tried substituting diethyldithiocarbamate (DDC) (a Cu,Zn-SOD inhibitor as well as an inhibitor of cytochrome oxidase) for CN in the NBT assay. This was expected to also cause the same interference. When we did the experiment, we found that DDC did indeed enhance the interfering reactions. We tried another blocker (rotenone) to see what effect it had. We found that again the interference was enhanced. We then began to think that wherever we blocked within the chain we would see an increase in $O_2^-$ leakage off the chain. We believed that once the chain was blocked, the electrons that entered the chain above the block would leak off and combine with $O_2$ to form $O_2^-$. We then thought that the only way to inhibit this interference was to block the electrons before they could enter the chain. Therefore, we tried to block the entry to the second site with thenoyltrifluoroacetone (TTA). TTA blocks the entry of electrons into the chain from succinate. Since no electrons can enter the chain from substrates that feed into the chain above the block, we expected to see a reduction in the interference seen in the NBT assay system. Our hypotheses were strenghtened when it was observed by Nohl et al. that TTA can prevent ·OH formation from mitochondria.[14]

We utilized a final concentration of 2 mM TTA to block this electron entry point. We initially encountered solubility problems with the assay mixture. TTA was soluble in 95% ethanol (EtOH). When this solution of EtOH and TTA was added to the reaction tubes without NBT, no precipitate formed. However, when NBT was added to the tubes, a precipitate did form which inhibited our ability to measure SOD. We noticed that increasing amounts of protein in the reaction tubes lowered the amount of this precipitate. Therefore, we tried using 1 mg/ml bovine serum albumin (BSA) in the reaction mixture to see if we could stop the precipitation. When BSA was added, the precipitate did not form and the solution was easy to read on a spectrophotometer. All the interference seen in the NBT assay without TTA was inhibited by the presence of TTA. The "new" assay solution containing 2 mM TTA, 1 mg/ml BSA, and 0.163 M EtOH was then used to measure pure bovine Cu,Zn-SOD to see if the new reagents had any effect on the ability to detect pure enzyme. We found that the "new" NBT assay had the same sensitivity as the "old" system. The interference effects seen in tissue measurements (purple color formation, higher sample slopes than blanks, and low maximum percentage inhibition) were all nearly totally inhibited in the NBT assay system containing TTA, EtOH, and BSA. All of these effects are apparently caused by $O_2^-$ and ·OH production in the presence of CN. For instance, the low percentage inhibition is probably due to the fact that we are adding $O_2^-$ to our assays at the same time we are adding SOD. Using this "new" system, we also began to detect Mn-SOD in samples in which we had previ-

[14] H. Nohl, W. Jordan, and D. Hegner, Hoppe-Seylers Z. Physiol. Chem. 363, 599 (1982).

ously *not* been able to measure enzymic activity due to strong interference. We are now in the process of testing the concentration of TTA that is most effective at inhibiting interference from competing tissue reactions. We also plan to characterize our observations more completely in the hope of developing a more accurate and efficient indirect NBT–SOD assay system free of biochemical interference from spontaneous $O_2^-$ production. We also hope that when tested these observations will hold true for interference in the cytochrome *c*–SOD assay system. This modified assay should be used whenever the observed interferences are seen.

# [62] *In Vitro* Cell Cultures as Tools in the Study of Free Radicals and Free Radical Modifiers in Carcinogenesis

## By CARMIA BOREK

Cell cultures offer powerful tools in carcinogenesis research. In these *in vitro* systems cells are grown under defined conditions free from complex homeostatic mechanisms that prevail *in vivo*. These systems afford us the opportunity to assess at a cellular level the carcinogenic potential of physical and/or chemical agents.[1,2]

We expose single cells to the agent(s) and can then determine the ensuing short-term toxic actions on the cells such as cell killing, and damage to DNA and chromosomes and cellular membranes as well as their long-term effects in causing malignant transformation.

Having these defined models we can identify modulating factors which enhance transformation rates (carcinogens, promotors) or inhibit and at best eliminate the induction and development of malignant transformation.[1,3]

### Cell Transformation *in Vitro*

An important contribution which served as a foundation for studies on transformation *in vitro* was the development of the clonal assay by Puck and Marcus[4] showing the dose-related effect of radiation on the survival

---

[1] C. Borek, *Adv. Cancer Res.* **37**, 159 (1982).
[2] C. Heidelberger, *Annu. Rev. Biochem.* **44**, 79 (1975).
[3] C. Borek, *Proc. MD Anderson Symp. Mol. Interr. Nutr. Cancer*, 337 (1982).
[4] T. T. Puck and P. I. Marcus, *J. Exp. Med.* **103**, 653 (1956).

Copyright © 1984 by Academic Press, Inc.
All rights of reproduction in any form reserved.
ISBN 0-12-182005-X

of single cells. This quantitative assessment of cellular toxicity and reproductive death following exposure to a toxic agent made it possible in later
years to evaluate which of these surviving cells has been transformed into
a neoplastic state following exposure to radiation[1] and in the similar manner which of the surviving cells exposed to chemical carcinogens would
undergo neoplastic transformation.[2]

*Culture Systems Currently Used in Transformation Studies*

Cells systems currently used in most transformation studies are composed of fibroblast-like cells where morphological criteria serve well in
the quantitative assay of transformation.[1,2] Because in humans there exists a preponderance of carcinomas over sarcomas there is a need to
develop epithelial cultures to study transformation. A number of epithelial cell systems have been developed and used in quantitative studies on
chemically induced transformation[5-8] but so far not applied in radiogenic
radiation carcinogenesis. There is a uniformity in fibroblast-like cells that
does not exist in epithelial cells whose susceptibility to transformation
may be related to specific differentiated qualities as well as the source and
age of the tissue from which they are derived. A further difficulty with
epithelial cells arises from the fact that criteria for the neoplastic state of
epithelial cells are phenotypically expressed in a less constant manner
than in fibroblasts especially during early stages of transformation.[1,6,9]

Among the fibroblast cell cultures used in transformation studies there
are two main cell systems, primary culture cell strains and established cell
lines.

*Primary Cultures.* Primary cultures and the cell strains propagated
from them are freshly derived from animal or human tissue. These are
diploid cells which constitute direct descendants of the cells *in situ*. These
cultures have a finite life span that differs from one cell source to another
and is related to the longevity of the species from which they originate.
The life span of these cell strains is sensitive to the mode of handling and
conditions in which they are maintained in culture. Thus an optimum

[5] C. Borek, *Radiat. Res.* **79**, 209 (1979).

[6] C. Borek, *in* "Advances in Modern Environmental Toxicology, I. Mammalian Cell Transformation by Chemical Carcinogens." 297–318. (V. Dunkel, N. K. Mishra, and M.
Mehlman, eds.), pp. 297–318. Senate Press, Princeton, New Jersey, 1980.

[7] C. Borek, *in* "Radioprotectors and Anticarcinogens" (A. Nygaard and M. Simic, eds.),
pp. 495–513. Academic Press, New York, 1983.

[8] C. Borek and G. M. Williams, *Ann. N.Y. Acad. Sci.* **349**, 1 (1980).

[9] I. B. Weinstein, L. S. Lee, P. B. Fisher, A. Mufson, and H. Yamasaki, *in* "Environmental
Carcinogenesis: Occurence, Risk Evaluation and Mechanisms" (P. Emmelot and E.
Kriek, eds.), pp. 265–285. Elsevier, Amsterdam, 1979.

seeding density, high-quality serum, optimal pH (7.2), and minimal exposure to trypsin upon subculture all serve to lengthen the number of passages a cell strain will undergo before showing signs of senescence (lower plating efficiency, cell spreading, and longer cell cycle).

The primary and short-term cultures most commonly used in carcinogenesis studies *in vitro* are those derived from hamster embryo cells.[1,2,10–18] These are mixed asynchronous cell populations whose main advantage lies in the fact that they are composed of normal diploid cells. Since they undergo senescence upon progressive culture *in vitro* they allow "immortal" transformants to emerge against a background of dying cells thus ascertaining *in vitro* their phenotypic transformed state.

In this culture system cell survival and cell transformation are scored simultaneously in the same dishes and the rate of spontaneous transformation is less than $10^{-6}$.[12,13] Because of their low plating efficiency freshly cultured hamster cells are usually cloned by seeding at low cell density on feeder cells;[4] these may be syngeneic[12] or allogeneic" and consist of cells which have been irradiated with 4000 rad and have undergone reproductive death.[19] If serum is of exceptionally high quality allowing for high plating efficiency of the hamster cells feeder cells may not be necessary. The protocol is illustrated in Fig. 1.

The number of single cells seeded prior to treatment with physical or chemical carcinogens must be ascertained experimentally following dose–response relationship for cell survival after treatment by the agent being studied;[12,13] since the quality of serum determines the plating efficiency of control untreated cultures quantitative evaluation must take this into account.[12]

In the hamster cell system expression time for transformation is 8–10 days after treatment, a relatively short period, and the cells can be cryopreserved.[20]

[10] C. Borek and L. Sachs, *Nature (London)* **210,** 276 (1966).
[11] C. Borek and L. Sachs, *Proc. Natl. Acad. Sci. U.S.A.* **57,** 1522 (1967).
[12] C. Borek and E. J. Hall, *Nature (London)* **243,** 450 (1973).
[13] C. Borek, E. J. Hall, and H. H. Rossi, *Can. Res.* **38,** 2997 (1978).
[14] J. A. DiPaolo, P. J. Donovan, and N. C. Popesen, *Radiat. Res.* **66,** 310 (1976).
[15] Y. Berwald and L. Sachs, *Nature (London)* **200,** 1182 (1963).
[16] E. Huberman, R. Mager, and L. Sachs, *Nature (London)* **264,** 360 (1976).
[17] J. A. DiPaolo and P. J. Donovan, *Cancer Res.* **33,** 3250 (1973).
[18] J. C. Barrett and P. O. P. Ts'o, *Proc. Natl. Acad. Sci. U.S.A.* **75,** 3297 (1978).
[19] P. F. Kruse and J. R. Patterson (eds.), "Tissue Culture, Methods and Applications." Academic Press, New York, 1973.
[20] R. J. Pienta, *in* "Carcinogens: Identification and Mechanisms of Action" (A. C. Griffin and C. R. Shan, eds.), pp. 121–141. Raven, New York, 1979.

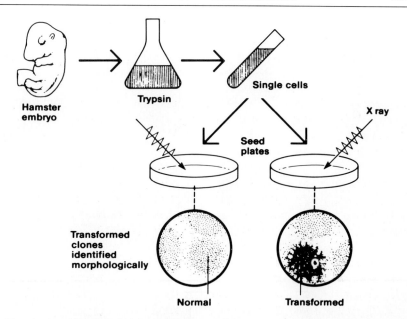

FIG. 1. Assay for hamster embryo cells transformed *in vitro* by radiation. Hamster embryos in midterm gestation are removed, minced, and trypsinized progressively with 0.25% trypsin. After removal of the trypsin by centrifugation, cells are suspended in complete medium and seeded as single cells on feeder layers.[4] They are exposed to radiation 24 hr later. After 8–10 days in incubation, cultures are fixed and stained. Transformed colonies are distinguished morphologically from controls. Chemically induced transformation follows the same procedure. The chemicals are added to the medium for 1 to 24 hr depending on the substance used (see text) and then removed.

Treatment of cells by the carcinogens is usually 24 hr after seeding.[12,16,17] While acute exposure to X rays is short[13] treatment time with chemical carcinogens depends on the nature of the compound being used and can last from 1 hr for exposure by an alkylating agent or 24 hr by hydrocarbons.[2,21]

Transformed colonies are identifiable by dense multilayered cells, random cellular arrangement, and haphazard cell–cell orientation accentuated at the colony edge (Fig. 2A and B); normal counterparts are usually flat with an organized cell–cell orientation. Because of the mixed population in this cell system one often finds untransformed colonies which may possess a higher cell density than normally found in the flat colonies.

[21] C. Heidelberger, *Adv. Cancer Res.* **18,** 317 (1973).

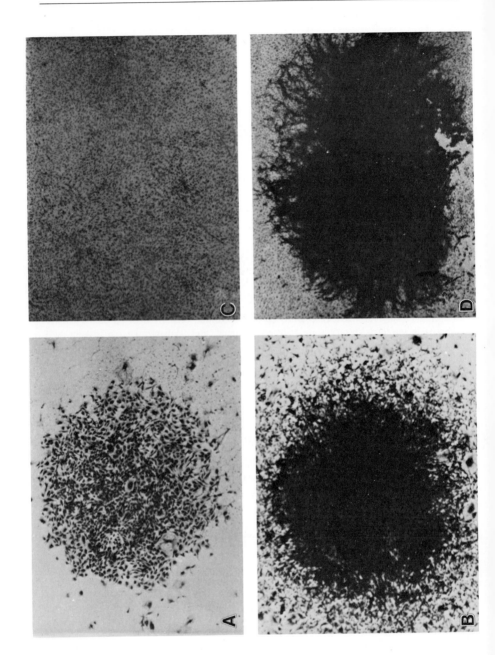

These however do not exhibit the randomness at the colony edge, and do not exhibit malignant potential.[13]

Human primary cultures used in transformation studies are derived from adult human skin,[22,23] human embryos,[24] or foreskin.[25–27] The assay for early transformation is a focus assay in culture or the ability to lose anchorage dependence and proliferate in agar.[1] The field of human cell transformation is young and reproducible transformation assays are still in developmental stages.

*Established Cell Lines.* Established cell lines which originated from primary cultures have the advantage that they possess an unlimited life span allowing the use of the cells over long periods of time without the cells undergoing senescence. On the other hand, cell lines are heteroploid and are subject to a higher spontaneous transformation rate than primary cultures. This frequency of spontaneous transformation is directly proportional to cell passage in culture, a fact which limits the use of these cells in transformation studies to early passage of 5–12.[1,2,28]

FIBROBLASTS. Examples of fibroblast cell lines that have been used extensively in transformation studies are BALB 3T3 cell line developed[29] and further cloned[30] to establish susceptible and nonsusceptible lines[31] and the C3H/10T-1/2 cells developed in Heidelberger's laboratory by Reznikoff *et al.*[32,33] (Fig. 2C and D). The transformation assay in these cell

---

[22] T. Kakunaga, *Proc. Natl. Acad. Sci. U.S.A.* **75**, 1334 (1978).

[23] C. Borek, *Nature (London)* **283**, 776 (1980).

[24] B. M. Sutherland, J. S. Cimino, N. Delihas, A. G. Shih, and R. P. Oliver, *Cancer Res.* **40**, 1934 (1980).

[25] K. C. Silinkas, S. A. Kateley, J. E. Tower, V. M. Maher, and J. J. McCormick, *Cancer Res.* **41**, 1334 (1981).

[26] R. J. Zimmerman and J. B. Little, *Carcinogenesis* **2**, 1303 (1981).

[27] G. E. Milo and J. A. DiPaolo, *Nature (London)* **275**, 130 (1973).

[28] N. Mishra, V. Dunkel, and M. Mehlman (eds.), "Advances in Modern Environmental Toxicology." Senate Press, Princeton, New Jersey, 1980.

[29] G. J. Todoro and H. Green, *J. Cell Biol.* **17**, 229 (1963).

[30] T. Kakunaga, *Int. J. Cancer* **12**, 465 (1973).

[31] T. Kakunaga and J. D. Crow, *Science* **209**, 505 (1980).

[32] C. A. Reznikoff, D. W. Brankow, and C. Heidelberger, *Cancer Res.* **33**, 3231 (1973).

[33] C. A. Reznikoff, J. S. Bertram, D. L. Branigan, and C. Heidelberger, *Cancer Res.* **33**, 3239 (1973).

---

FIG. 2. Morphological criteria for transformation. (A and B) Clonal assay in hamster embryo cell cultures; (C and D) focus in mouse embryo 10T-1/2 cells. (A) Normal hamster embryo cell clone, 14 days old. (B) A 14-day-old clone of hamster cells transformed *in vitro* by 300 rad of X rays. (C) A monolayer of untransformed 10T-1/2 cells 6 weeks in culture. (D) A type III focus of 10T-1/2 cells transformed by 300 rad of X rays, 6 weeks in culture. Note the flatness of the normal cells and their regular orientation. In contrast, the transformed cells form multilayers and display a lack of orientation that is clearly distinguishable at the periphery. From Borek.[1]

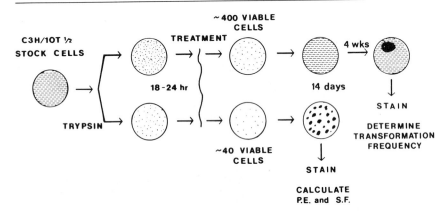

Fíg. 3. Protocol for experiments with 10T-1/2 cells. Cells are seeded at two cell concentrations into 50-cm² petri dishes. To assess the plating efficiency (P.E.) and cell surviving fraction (S.F.) a sufficient number of cells are seeded so that an estimated number of about 50 cells survive the subsequent treatment; these are incubated for 2 weeks, after which they are fixed and stained and the number of discreet colonies per dish is counted. For the assessment of transformation, cells are seeded so that an estimated 400 reproductively viable cells survive the subsequent treatment. Cells are allowed to attach by overnight incubation at 37° before being irradiated, following which they are incubated for 6 weeks, with the growth medium changed weekly. Reproduced from Hall and Miller.[47]

lines is a focus assay which scores pleomorphic dense foci of cells growing upon a sheet of confluent untransformed cells (Fig. 2C and D). In the C3H/10T-1/2 cells three types of foci are identifiable—I, II, and III. Their morphology has been related to their malignant potential in animals with type III being the most malignant.[33,34] Expression time of transformation is 6 weeks and cell survival must be evaluated in separate sets of dishes (Fig. 3).

EPITHELIAL CELL LINES. The sources of tissues from which epithelial lines have been cultured for use in transformation studies are limited.[5,6,28] Of these liver cell lines have been most useful.[6,8,35–38] While early stages of transformation may go undetected if morphological criteria are used, at later stages following many generations in culture the cells acquire anchorage independence and can grow in agar,[36] a criterion which has been used as a transformation assay in these cells.[38]

[34] M. Terzaghi and J. B. Little, *Cancer Res.* **36,** 1367 (1976).
[35] G. M. Williams, J. M. Elliot, and J. J. Weisburger, *Cancer Res.* **33,** 606 (1973).
[36] C. Borek, *Proc. Natl. Acad. Sci. U.S.A.* **69,** 956 (1972).
[37] C. Borek, *in* "Gene Expression and Carcinogenesis in Cultured Liver Cells" (L. E. Gersherson and E. B. Thompson, eds.), pp. 62–93. Academic Press, New York, 1975.
[38] R. H. C. San, M. F. Lapsia, A. I. Soiefer, C. J. Maslansky, J. M. Rice, and G. M. Williams, *Cancer Res.* **39,** 1026 (1979).

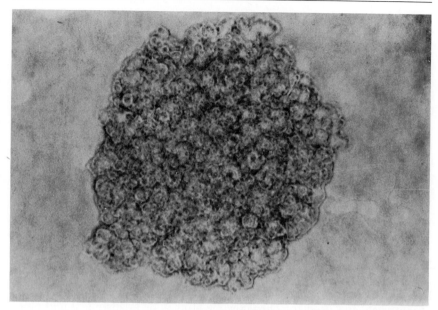

FIG. 4. A colony of transformed hamster fibroblasts growing in agar.

## Methods and Criteria Used in Transformation Studies

*Initiation and Expression.* One can roughly delineate stages in neo-plastic events as follows.

1. Initiation of transformation associated with genetic damage and requiring cell replication soon after treatment to fix transformation, a hereditary property.[11,33,39,40]

2. Expression of transformation requiring additional cell replications to express the transformed phenotype[1,11,33] (Fig. 4). This expression may be in the form of altered morphology in cell culture, or an acquisition of the ability to grow in semisolid medium, a characteristic which in cells of solid tissue is associated with a neoplastic phenotype.[1] In rodent fibro-blasts, growth in agar appears at later stages after morphological transfor-mation,[13,18] though in human cell transformation the loss of anchorage independence is acquired at the same time as altered morphology.[1,41]

[39] C. Borek and L. Sachs, *Proc. Natl. Acad. Sci. U.S.A.* **59**, 83 (1968).
[40] T. Kakunaga, *Cancer Res.* **35**, 1637 (1975).
[41] T. Kakunaga, J. D. Crow, and C. Augl, *in* "Radiation Research" (S. Okada, M. Imamura, T. Terashima, and H. Yamaguchi, eds.), pp. 589–595. Japanese Assoc. for Radiation Research, Tokyo, 1979.

3. The ultimate expression of transformation and the uniequivocal demonstration of the neoplastic state are proof of malignancy by isolating transformed foci or colonies, propagating them, and injecting them into the appropriate host for tumor growth; usually $10^5$ to $10^6$ are injected.[1]

*Enhancement of Transformation.* The evaluation of cocarcinogenesis is carried out by exposing cells to an additional carcinogen at the time of initiation, and assessing transformation frequency compared to that observed with the single agent.

PROMOTION. The events in promotion, by tumor promotors, can be assessed by enhancing transformation rates at the time of expression.[3,7,42,43] A promotor which has been used extensively is the phorbol ester derivative TPA (12-*O*-tetradecanoylphorbol 13-acetate).[44]

*Practical Aspects*

In transformation studies described above cells can be exposed in the following manner to potential carcinogens and/or modulations.

Exposure as single cells allowing proliferation. Transformation is detected by characteristic colony or foci formation, which can be isolated, propagated, and used for chromosome studies or biochemical and morphological changes.[1]

Exposure of large populations of proliferating cells (mass cultures) with continuous propagation in mass culture (epithelial cells or fibroblasts until morphological changes appear[6,10,15] or until an ability to grow in semisolid medium is expressed by seeding $10^5$ cells into 0.33% agarose over a bottom layer of 0.5% agar[45] (Fig. 4). Periodically, the cells at any stage of propagation can be cloned out for quantitative evaluation of transformation. Modulators (inhibitors or promotors) can be added at initiation or during expression and transformation assayed subsequently as above, along with evaluating other endpoints.

Most criteria for quantitative assays in transformation are based on the loss of cell–cell contact and anchorage independence *in vitro* (growth in agar) or *in vivo* (tumor formation). Chromosomal analysis in early transformation has not been satisfactory though sister chromatid exchanges have often served to detect initiating agents in transformation but not modulators.[1,3]

[42] A. R. Kennedy, S. Mondal, C. Heidelberger, and J. B. Little, *Cancer Res.* **38,** 439 (1978).
[43] C. Borek and W. Troll, *Proc. Natl. Acad. Sci. U.S.A.* **80,** 1304 (1983).
[44] E. Hecker, *Meth. Cancer Res.* **6,** 439 (1971).
[45] I. McPherson, *in* "Tissue Culture, Methods and Application," (P. F. Kruse and M. K. Patterson, eds.), pp. 276–280. Academic Press, New York, 1973.

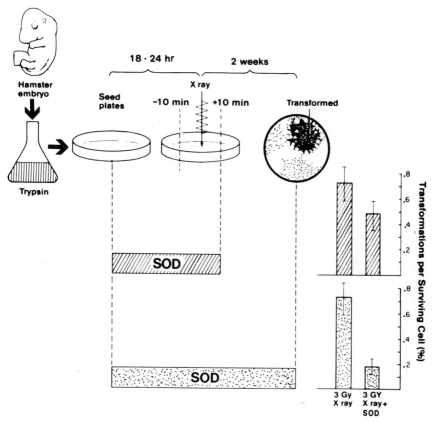

Fig. 5. The effect of SOD on the incidence of oncogenic transformation in hamster embryo cells produced by 3-Gy X rays. The SOD (10 units/ml) was added in two different protocols as illustrated. From Borek and Troll.[43]

## Applying Transformation Systems to Study Free Radical Action in Carcinogenesis

The role of free radicals in carcinogensis at a cellular level can be evaluated in the transformation systems described above.[3,43,46]

Radiation and specific chemicals which produce free radicals can serve as initiators or as promotors.[3,43] While in principle one can assess the production of free radicals following exposure to the carcinogens using methods described in this volume, their fleeting persistence and the

[46] C. Borek and E. J. Hall, *Ann. N.Y. Acad. Sci.* **397**, 193 (1982).

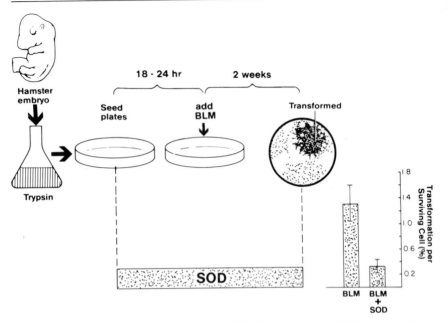

FIG. 6. The effect of SOD (10 units/ml) on the incidence of oncogenic transformation in hamster embryo cells produced by the chemotherapy agent bleomycin (BLM; 1.0 μg/ml).

low frequency of transformation may result in a questionable quantitative relationship between the free radicals and transformation.

An alternate method is to use single or combined protectors which reduce directly or indirectly the availability of free radicals to the cell and to score transformation in their presence and absence.[3,43,46]

In practical terms the following studies can be carried out some of which are illustrated in Figs. 5–8 for the two cell systems described above (hamster embryo and C3H/10T-1/2). In these studies radiation and the drugs bleomycin and mizonidazole[43,46,47] were used to initiate transformation. TPA served as a promotor, and SOD was used as a free radical modifier. Other modifying agents such as catalase,[43] selenium, and some retinoids[3] have been assessed in vitro. The role of vitamins E and C as inhibitors[48] is still not clearly established in vitro. It is important to stress that inherent cellular protective systems vary with cell type and probably with age and tissue from which they are derived.[7,43,49] Thus, the effective-

[47] E. J. Hall and R. C. Miller, Radiat. Res. 87, 203 (1981).
[48] M. S. Arnott, J. Van Eys, and Y. M. Wang (eds.), "Molecular Interactions of Nutrition and Cancer." Raven, New York, 1982.
[49] C. Borek, Ann. N.Y. Acad. Sci. 407, 284 (1983).

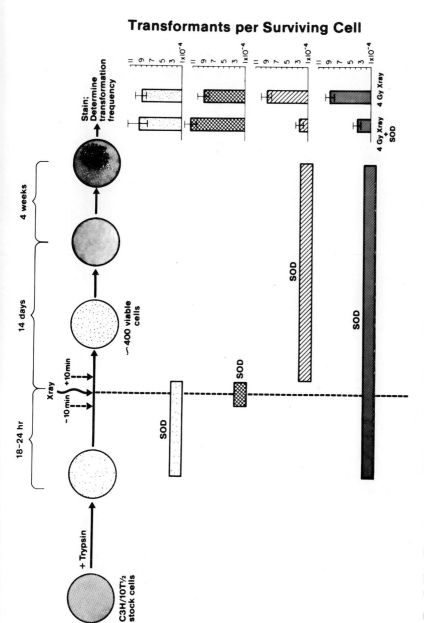

FIG. 7. The effect of superoxide dismutase (SOD) on the incidence of transformation produced by 4 Gy of X rays. The SOD at a concentration of 10 units/ml was applied in various protocols, as shown. Borek and Hall[46] and R. C. Miller, R. S. Osmak, M. Zimmerman, and E. J. Hall, *Int. J. Radiat. Oncol. Biol. Phys.* **8**, 771 (1982).

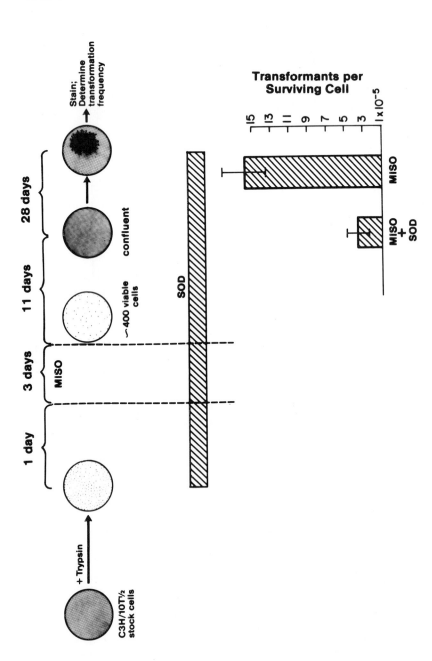

FIG. 8. The effect of superoxide dismutase (10 units/ml) on the incidence of oncogenic transformation produced by misonidazole (mM for 3 days). From Borek and Hall[46] and R. C. Miller, R. S. Osmak, M. Zimmerman, and E. J. Hall, *Int. J. Radiat. Oncol. Biol.*

ness of a scavenging agent may be dose related and vary quantitatively with the cell system studied.[43] It is therefore important to determine the cellular content of scavenging and protective agents using methods described in this volume, in order to understand the events taking place during modulation of transformation. The assessment of free radical modifiers in transformation would be based in the protocols and concepts described earlier utilizing various criteria for transformation.

Preincubation with a modifier, e.g., selenium[1] or SOD[43] prior to treatment with the carcinogen and resulting in an inhibitory action would illuminate the role of radicals in initiation and promotion. However, the addition of the agent at some time postexposure to the initiator would evaluate effectiveness of the modifier on late effects in carcinogenesis.

The assessment of various scavengers or protectors (vitamin E, vitamin A analogs, SOD catalase) on promotion would be carried out effectively either by adding the protector simultaneously with the promotor

TABLE I

INHIBITION OF CELL TRANSFORMATION BY SOD AND CATALASE[a] FOLLOWING EXPOSURE TO X RAYS AND TPA

| Treatment | Surviving fraction[b] $\pm$ SE (plating efficiency[c] $\pm$ SE) | Percentage transformation $\pm$ SE |
|---|---|---|
| Control | (4.6 $\pm$ 0.12) | 0 |
| SOD | 0.95 $\pm$ 0.04 | 0 |
| Catalase | 0.97 $\pm$ 0.03 | 0 |
| TPA | 0.92 $\pm$ 0.05 | 0 |
| DMSO | 0.95 $\pm$ 0.04 | 0 |
| 3 Gy | 0.75 $\pm$ 0.12 | 0.74 $\pm$ 0.14 |
| 3 Gy + SOD | 0.91 $\pm$ 0.09 | 0.17 $\pm$ 0.06 |
| 3 Gy + catalase | 0.80 $\pm$ 0.11 | 0.66 $\pm$ 0.09 |
| 3 Gy + SOD + catalase | 0.87 $\pm$ 0.12 | 0.13 $\pm$ 0.10 |
| 3 Gy + TPA | 0.69 $\pm$ 0.14 | 1.62 $\pm$ 0.12 |
| 3 Gy + SOD + TPA | 0.89 $\pm$ 0.10 | 0.55 $\pm$ 0.08 |
| 3 Gy + catalase + TPA | 0.82 $\pm$ 0.13 | 1.32 $\pm$ 0.15 |
| 3 Gy + catalase + SOD + TPA | 0.93 $\pm$ 0.12 | 0.41 $\pm$ 0.09 |
| 3 Gy + SOD[d] | 0.80 $\pm$ 0.10 | 0.21 $\pm$ 0.07 |
| 3 Gy + TPA[d] | 0.64 $\pm$ 0.12 | 1.56 $\pm$ 0.08 |
| 3 Gy + SOD[d] + TPA[d] | 0.82 $\pm$ 0.13 | 0.61 $\pm$ 0.11 |

[a] SOD and/or catalase where present on the cells throughout the experiment.
[b] Surviving fraction = number of colonies counted/(number of cells plated $\times$ plating efficiency).
[c] Plating efficiency = (number of colonies counted/number of cells plated) $\times$ 1000.
[d] SOD and/or TPA where added 24 hr after irradiation, and left until the end of the experiment.

TABLE II
CELLULAR CONTENT OF CATALASE AND SOD$^a$ IN HAMSTER
EMBRYO (HE) CELLS$^b$

| Cells | Catalase | SOD |
|---|---|---|
| HE primary cultures | 6 units/96 $\mu$g $\pm$ 2% | 36 ng/72 $\mu$g $\pm$ 2% |
| HE secondary cultures | 7.5 units/130 $\mu$g $\pm$ 2% | 37 ng/97.5 $\mu$g $\pm$ 2% |

$^a$ No SOD or catalase was detected in fetal calf serum or calf serum, within the sensitivity of the assay (100 ng SOD/1 unit catalase).
$^b$ HE/$\mu$g protein.

sometime after exposure to the carcinogen[43] or by exposing the cells to the initiator in the presence of the protector, removing the protector at the time or shortly after the addition of a promotor,[3,50] and scoring for transformation with data pertaining to toxicity and adquate controls safely on hand.

Using these methods we have so far found that in the hamster embryo cell system SOD (10 units/ml) inhibits initiation but predominantly acts on promotion by radiation alone or following treatment with TPA (Table I).[43] Catalase (10 $\mu$g/ml) is not effective. Since the cells are relatively rich in catalase (Table II) their content may serve as adequate protection against the oncogenic action of X rays and bleomycin.

In the C3H/10T-1/2 cells SOD had no effect on initiation but rather appeared to inhibit (at 10 units/ml) later stages in radiogenic transformation[46] (Figs. 7 and 8).

Further work is required in fibroblast and epithelial systems which lend themselves to transformation assays. The mechanistic role of free radicals in malignant transformation can be assessed, while evaluating cellular enzymes, thiols, and pathways and modes in cellular toxicity such as lipid peroxidation.[51,52] In addition, from a pragmatic point of view since many drugs and modalities used in cancer therapy are producers of free radicals[46,52] knowledge of the inhibitory action of modifiers would be important. Many environmental pollutants either produce free radicals or interact with them (e.g., ozone) in polluted atmospheres[49] or may have oncogenic potential themselves. Thus, the experimental models to evaluate modulation of toxicity and oncogenesis caused by these free radical-interacting agents would contribute to our knowledge on the control of environmental carcinogenesis.

[50] R. C. Miller, C. R. Geard, R. S. Osmak, M. Rutledge-Freeman, A. Ong, H. Mason, A. Napholz, N. Perez, L. Harisiadis, and C. Borek, Cancer Res. 41, 655 (1981).
[51] M. T. Smith, H. Thor, P. Hartzell, and S. Orrenius, Biochem. Pharmacol. 31, 19 (1982).
[52] H. Thor, M. T. Smith, P. Hartzell, G. Bellomo, S. A. Jewell, and S. Orrenius, J. Biol. Chem. 257, 12419 (1982).

Acknowledgment

This work was supported by Contract CA 12536-11 to the Radiological Research Laboratory/Department of Radiology, awarded by the National Cancer Institute, DHHS.

# [63] Radicals in Melanin Biochemistry

By ROGER C. SEALY

Melanin pigments are complex heterogeneous polymers from the enzymic oxidation of 3,4-dihydroxyphenylalanine (dopa) and/or its metabolite 5-$S$-cysteinyldopa.[1] Their major function is photoprotection. They are classified either as *eumelanins* or *pheomelanins:* eumelanins are derived predominantly from dopa, pheomelanins from 5-$S$-cysteinyldopa. Most natural melanins are copolymers formed from a mixture of the two precursor molecules.

Melanins contain free radicals under all known experimental conditions. The free radical content is sensitive to light, reducing and oxidizing equivalents, and other changes in the polymer environment.[2] The reliable characterization and quantitation of radicals in melanins are important in gaining an understanding of the role these radicals play in biochemical and photochemical systems of biological importance. Much work is required to more precisely identify the free radical species and to define mechanisms of radical formation and decay. Here we consider the experimental measurements that should be made and some of the possible pitfalls. Radicals that are localized on the melanin polymer and which are detectable by electron spin resonance (ESR) spectroscopy are emphasized. Note, however, that other radical species (e.g., $O_2^-$, $e_{aq}^-$, and H·) appear to be involved in melanin photochemistry and potentially could participate in other melanin reactions.

## Magnetic Parameters and Spin Concentrations

ESR spectra of free radicals in melanins from dopa and from cysteinyldopa are readily distinguishable (Fig. 1).[3] The dopa melanin has a single line, whereas the cysteinyldopa melanin has three features characteristic

[1] G. Prota, *J. Invest. Dermatol.* **75,** 122 (1980).
[2] R. C. Sealy, C. C. Felix, J. S. Hyde, and H. M. Swartz, *Free Radicals Biol.* **4,** 209 (1980).
[3] R. C. Sealy, J. S. Hyde, C. C. Felix, I. A. Menon, G. Prota, H. M. Swartz, S. Persad, and H. F. Haberman, *Proc. Natl. Acad. Sci. U.S.A.* **75,** 5395 (1978).

METHODS IN ENZYMOLOGY, VOL. 105
Copyright © 1984 by Academic Press, Inc.
All rights of reproduction in any form reserved.
ISBN 0-12-182005-X

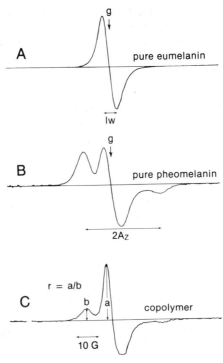

FIG. 1. Spectra and characterization of radicals in melanins from dopa ("pure" eumelanin), from cysteinyldopa ("pure" pheomelanin), and from a copolymer. (Spectra were obtained at pH 4.5 and −196°; 3 mM $Zn^{2+}$ was used to enhance radical concentrations.) (A) Dopa melanin; lw = linewidth. (B) Cysteinyldopa melanin; $2A_z$ = separation of the outer extrema. (C) A melanin copolymer; the ratio $r = (a/b)$ is related to the relative amounts of dopa and cysteinyldopa incorporated into the polymer.[4]

of an immobilized radical that is partly nitrogen centered. Many melanins have ESR spectra that approximate one or the other of these two basic types. In such cases, minimal characterization of a dopa melanin-type spectrum is by $g$ value and linewidth, whereas a cysteinyldopa melanin-type spectrum should be characterized by $g$ value and the separation of the outer extrema ($2A_z$). Spectra of natural melanins that are copolymeric to the extent that they contain appreciable amounts of both precursors show features of both kinds of radical (Fig. 1). Such spectra are conveniently parameterized in terms of an $r$ value,[4] which may be used to estimate the relative amounts of precursors incorporated into the polymer.

[4] R. C. Sealy, J. S. Hyde, C. C. Felix, I. A. Menon, and G. Prota, *Science* **217**, 545 (1982).

Standard procedures[5] for the precise measurement of magnetic parameters should be followed: $g$ values are obtained either from direct measurements of magnetic field and microwave frequency or by use of a reference material; linewidths or other magnetic field separations simply require a field calibration. In measuring linewidths, care should be taken to avoid overmodulation and saturation with microwave power. Note that magnetic parameters can show small or even substantial changes with temperature,[6] pH,[7,8] complexing metal ions,[9,10] etc.

Spin concentrations usually are expressed in terms of spins/g of melanin. It is important that the measurement of spins be made on an aliquot of hydrated material *before* the material is dried to constant weight, because drying tends to introduce irreversible changes in the melanin, including its free radical content. Measurements of spin concentration[5] require an appropriate reference standard. Overmodulation may be used, but microwave saturation should be avoided unless an appropriate correction is applied to the data. If paramagnetic metal ions are present (as seems common in melanins from commercial sources) the *apparent* spin concentration may be greatly reduced (by a factor of 10 or more) because of magnetic interaction between the radicals and the metal ions.[2]

An investigator inexperienced with melanin systems might wish to confirm the following data that have been obtained in this laboratory for synthetic melanin (from dopa), which is readily obtained in a soluble form by the following procedure. Dopa is dissolved, with stirring, in deionized or distilled water to give a concentration of 5 g/liter. Sodium hydroxide solution (1–10 $M$) is subsequently added to increase the pH of the dopa solution to 8. Air is then bubbled through the stirred solution (an aquarium pump provides effective aeration) for 72 hr. After removal of any sediment by paper filtration, the reaction mixture is passed through an Amicon membrane filter (XM-50) to remove material of low molecular weight. The concentrated solution is then diluted and the procedure repeated several times. Alternatively, dialysis can be used to remove low-molecular-weight material. The characteristics of the material prepared in

[5] J. R. Bolton, D. C. Borg, and H. M. Swartz, in "Biological Applications of Electron Spin Resonance" (H. M. Swartz, J. R. Bolton, and D. C. Borg, eds.), p. 63. Wiley, New York, 1972.
[6] S.-S. Chio, J. S. Hyde, and R. C. Sealy, *Arch. Biochem. Biophys.* **199,** 133 (1980).
[7] F. J. Grady and D. C. Borg, *J. Am. Chem. Soc.* **90,** 2949 (1968).
[8] S.-S. Chio, J. S. Hyde, and R. C. Sealy, *Arch. Biochem. Biophys.* **215,** 100 (1982).
[9] C. C. Felix, J. S. Hyde, T. Sarna, and R. C. Sealy, *J. Am. Chem. Soc.* **100,** 3922 (1978).
[10] T. Sarna, W. Korytowski, M. Pasenkiewicz-Gierula, and E. Gudowska, in "Pigment Cell 1981" (M. Seiji, ed.), p. 23. Univ. of Tokyo Press, Tokyo, 1981.

this way are[8,11] C 58.0, H 2.93, N 8.75, O 30.4; $\varepsilon_{300\,nm}$ = 39.7 (mg/ml)$^{-1}$ cm$^{-1}$; $g$ = 2.0036 ± 0.0001; linewidth = 3.8 ± 0.1 G; spin concentration 2 × 10$^{18}$ spins/g. (Data given are for pH 7 and ambient temperature.)

*Radical Structure: Information from the Effects of pH (pD) and Complexing Metal Ions*

Polymerization of dopa and 5-*S*-cysteinyldopa involves oxidative cyclization to give structures related to 5,6-dihydroxyindole and hydroxy-1,4-benzothiazine, respectively.[1] Radicals in eumelanins and pheomelanins are thus expected to be indolesemiquinones and semiquinonimines of types 1 and 2. For any melanin, the following procedures may prove

1                          2

useful in identifying radical structure: (1) variation of pH (radicals such as 1 and 2 should participate in acid–base equilibria), (2) exchange with D$_2$O (to test for exchangeable protons), and (3) treatment with diamagnetic metal ions (*o*-semiquinones and related species form chelate complexes).

Both eumelanins and pheomelanins show changes in ESR spectra with pH.[3,7,8] To an approximation these changes can, in the systems studied to date, be accounted for in terms of two spectral species. (Note that multifrequency ESR experiments[3,7] can facilitate the separation of spectra.) However, because of changes in radical concentration that also occur (see below), it is not clear at the present time whether these *spectral* species are acid–base forms of a single *chemical* species, or two different species where the concentration of one of them changes in a pH-dependent manner. A comparison of solutions in H$_2$O and D$_2$O is particularly important for pheomelanins. Experiments at high hydrogen ion concentration have revealed the presence of an exchangeable proton in radicals from pheomelanin[3]: hyperfine splitting from this proton collapses in D$_2$O.

A chelating structure for the free radicals can be confirmed by experiments with either 2+ or 3+ diamagnetic metal ions. Millimolar concentrations of metal ion in weakly acidic solution are usually sufficient for complexation to be observed without problems from metal ion precipitation. The most useful ions are those with a nuclear moment, which results in an additional hyperfine splitting in the complex (generally observed as a

[11] T. Sarna and R. C. Sealy, *Photochem. Photobiol.* **38**, in press (1983).

line broadening).[9] An example of such an ion is $^{113}Cd^{2+}$ which has a nuclear spin $I = 1/2$ and gives complexes with ESR spectra from which the hyperfine splitting to $^{113}Cd$ can be extracted by spectral reconstruction.[9] Further information on radical structure can be obtained from the magnitude of this splitting: whereas simple $o$-semiquinones (e.g., from catechol,[12] or a melanin from oxidation of catechol) have a cadmium splitting of ~7 G, radicals in eumelanins (presumably indolesemiquinones) and pheomelanins (presumably $o$-semiquinonimines) have splittings of 3–4 and ~15 G, respectively.[3,9]

## Kinetics and Mechanisms of Radical Reactions

The ESR signals that melanins invariably show must reflect the presence of either chemically inert, persistent radicals and/or more transient radicals that exist in steady-state concentration as a result of an equilibrium between (probably) diamagnetic and paramagnetic sites on the polymer. A significant fraction of the radicals detected in fluid systems participate in such an equilibrium.[6]

Changes in radical concentration that occur during visible or UV irradiation or other chemical reactions for the most part appear to be changes in steady-state levels. In order to determine radical yields in a system one therefore needs to determine changes in *rates* of radical formation and/or decay. These rates can then be related to measured rates of photon fluence, disappearance of reagents, etc.

In a time-resolved experiment the rate of radical formation can be obtained either from the initial buildup to the steady state *or* from the steady-state radical concentration and the radical lifetime or rate constant for decay. Thus, for a steady-state radical concentration $R_{ss}$ and second-order decay with rate constant $2k$, the rate of radical formation $R$ is given by:

$$R = 2k[R_{ss}]^2$$

Under such conditions, other things being equal, the rate of radical formation is proportional to $[R_{ss}]$.[2] Alternatively, if radical termination were first order, the rate would be $\propto [R_{ss}]$.

In the absence of kinetic profiles it may not be possible to distinguish between changes in steady-state concentration arising from a change in the rate of radical formation and from a change in the radical termination rate. One should be aware of this possibility, particularly in systems where induced radicals are spectroscopically different from those initially present in the system.

[12] C. C. Felix and R. C. Sealy, *J. Am. Chem. Soc.* **104**, 1555 (1982).

## [64] Assay of Lipofuscin/Ceroid Pigment *in Vivo* during Aging

*By* R. S. SOHAL

*Terminology*

A consistent morphological feature of cellular aging is the progressive accumulation of autofluorescent, lipoidal, pigmented granules called lipofuscin within the cytoplasm of nondividing, long-lived cells such as neurons and cardiac myocytes. Structures exhibiting considerable morphological and cytochemical similarity to lipofuscin are also formed in response to certain experimental and pathological conditions, e.g., vitamin E deficiency. Such lipopigments, induced by disease or experimental manipulation, are termed ceroid. The genesis of both lipofuscin and ceroid has been associated with free radical and lipid peroxidative reactions.[1] Despite certain chemical differences, ceroid has been suggested to be an early stage of lipofuscin.[2] In this discussion lipofuscin and ceroid will be considered together as lipopigments. The term "fluorescent pigment" denotes the fluorescent material present in chloroform extracts of tissues which exhibits Schiff base-like fluorescent characteristics. Quantitative and homologous relationships between lipopigments and fluorescent pigment have not as yet been firmly established; therefore they will be treated separately.

*Identification of Lipopigments*

In unstained histological sections, lipopigments appear as yellowish-brown granules, rounded or oblong in contour, usually 1–5 $\mu$m in diameter. In electron micrographs, lipopigments are bounded by a limiting membrane and contain variable amounts of materials exhibiting vacuolar, granular, and lamellar organization. Due to the variations in chemical composition of the lipopigments, at different stages of their formation, there is no single histochemical method available which can be universally used for their identification. Lipopigments have been histochemically characterized as follows: basophilic; periodic acid–Schiff positive; resist complete extraction by polar and nonpolar solvents; stain positively

---

[1] A. L. Tappel, *Free Radicals Biol.* **4,** 1 (1980).
[2] A. G. E. Pearse, "Histochemistry," 3rd Ed. Vol. 2. Williams & Wilkins, Baltimore, Maryland, 1972.

Copyright © 1984 by Academic Press, Inc.
All rights of reproduction in any form reserved.
ISBN 0-12-182005-X

for neutral lipids with oil red O, Sudan black B, and osmium tetroxide; and exhibit positive localization of lysosomal hydrolytic enzymes.[3,4]

The single most universal characteristic of lipopigments, which can be used for their identification, is the emittance of intense orange-yellow autofluorescence, when excited with UV light. The excitation maxima of the *in situ* and isolated granules have been reported to range from 360 to 395 nm and the emission maxima range from 430 to 460 nm. Additional emission bands between 540 and 580 nm and at 610 nm have also been reported.[4]

## Quantification of Lipopigments

The lipopigment content of tissues can be ascertained by light or electron microscopy using standard morphometric point counting techniques.[5] Relative levels of lipopigments can also be estimated in histological sections on the basis of the intensity of emission spectra by microspectrofluorometry.[6] The validity and reliability of point counting morphometry for the measurement of lipofuscin has been discussed by Reichel and co-workers.[5] For more detailed information, readers are referred to a recent volume on stereology by Weibel.[7] Briefly, for light microscopic morphometry, tissue samples are processed for paraffin embedding and randomly selected tissue blocks are sectioned at a thickness of 5–6 $\mu$m. Sections are deparaffinized in xylol, rehydrated in an ascending series of water concentrations, and mounted in glycerol. Unstained sections are examined with a fluorescence microscope using an HBO 50-W mercury vapor lamp or a xenon lamp, a dark field condenser, a BG 12 excitor filter, and an OG 515 barrier filter. Under these optical conditions lipopigments appear orange to yellowish. For quantification of lipopigments, a 5-mm reticulate grid, ruled to 0.5-mm squares, with a total of 121 intersections, is superimposed on the microscopic field by introducing the grid in the left ocular of the microscope. Tissue sections are randomly moved and the percentage of grid intersections overlying the lipopigments is counted. Due to the thinness of the section, this percentage is believed to be a valid approximation of the tissue volume occupied by the lipopigment granules. The intracellular volume of the lipopigments can also be ascertained by this method provided that after each measurement the

[3] B. L. Strehler, *Adv. Gerontol. Res.* **1**, 343 (1964).
[4] M. Elleder, in "Age Pigments" (R. S. Sohal, ed.), p. 204. Elsevier, Amsterdam, 1981.
[5] W. Reichel, J. Hollander, J. H. Clark, and B. L. Strehler, *J. Gerontol.* **23**, 71 (1968).
[6] G. L. Wing, G. C. Blanchard, and J. J. Weiter, *Invest. Opthalmol. Visual Sci.* **17**, 601 (1978).
[7] E. R. Weibel, "Stereological Methods," Vol. 1. Academic Press, New York, 1979.

same field is examined by bright field or phase contrast optics and the number of grid intersections overlying the cell bodies (rather than extracellular space) is counted. Results can be computed as follows:

$$\frac{\text{percentage section volume occupied by lipopigments}}{\text{percentage section volume occupied by cell bodies}}$$

$$= \text{percentage lipopigments volume in cells}$$

### Fluorescent Pigment

On the basis of spectral similarities, Tappel and co-workers[1,8] have identified the fluorophores, extracted with chloroform from purified lipopigments and from tissues, as disubstituted conjugated Schiff bases. Schiff base substances exhibit a characteristic fluorescent pattern, having an excitation maximum of about 365 nm and emission maximum of about 435 nm. Although high concentrations of fluorescent material, exhibiting these spectral characteristics, have been noted in tissues with large accumulations of morphologically detectable lipopigment,[1,8] a quantitative correlation between the two has not as yet been clearly demonstrated. In fact, in some tissues, differences in the volume of lipopigments are not reflected in the centration of chloroform-extractable fluorescent material.[9,10] Furthermore, chloroform extracts of tissues contain a mixture of fluorescent substances exhibiting Schiff base-like fluorescent characteristics.[9,10] It has not as yet been determined whether these substances are derived from lipopigments or other precursors. It is suggested that these reservations and limitations should be considered in the evaluation of experimental results.

### Procedure for the Measurement of Fluorescent Pigment

The procedure outlined below is a modification of the method developed by Fletcher and co-workers.[8] Tissues are weight and homogenized, at room temperature, in 20 vol (v/wet weight of tissue) of 2 : 1 chloroform : methanol (spectral grade), using a Potter-type glass homogenizer. Although time and speed of homogenization may vary depending upon tissue type, it is imperative that the tissues be thoroughly homogenized to maximize extraction. The resulting brei is mixed with an equal amount of distilled water and centrifuged at 1500 $g$ for 10 min. The bottom, chloroform, layer is removed for the measurement of fluorescent intensity, with

[8] B. L. Fletcher, C. J. Dillard, and A. L. Tappel, *Anal. Biochem.* **52,** 1 (1978).

[9] H. Shimasaki, T. Nozawa, O. S. Privett, and W. R. Anderson, *Arch. Biochem. Biophys.* **183,** 443 (1977).

[10] J. G. Bieri, T. J. Tolliver, and W. G. Robinson, *Lipids* **15,** 10 (1980).

a spectrofluorometer, at excitation and emission maxima of 365 and 435 nm, respectively. Cloudiness of the chloroform extract, if present, can be eliminated by adding 0.1 ml methanol per milliliter of chloroform. The spectrofluorometer is standardized to give a deflection of 50 or 100 at the above wavelengths with a 1 $\mu$g/ml solution of quinine bisulfate in 0.1 $M$ $H_2SO_4$. The results are expressed as relative fluorescent units/ml chloroform/g wet tissue weight. Retinol, a fluorescent contaminant which is sometimes encountered in chloroform extracts of tissues, can be removed by exposure to high-intensity UV light for 30 sec. Tappel and co-workers[1,8] have devised several tests which are indicative of the presence of Schiff-base components in chloroform extracts, e.g., a decrease in the fluorescence intensity following the addition of 10 m$M$ NaOH or 0.1–0.2 m$M$ europium(III) 2,2,6,6-tetramethylheptane-1,3-dione.

Finally, it should be emphasized that the terms lipofuscin and ceroid should be used only in the morphological context. The fluorescent material present in chloroform extracts of tissues should be referred to by a distinctly separate term, e.g., "fluorescent age pigment" or "extractable fluorescent substances."

## Section V

Enzymes, Viral Activity, and Cell Viability as End Points
for Study of Free Radical Damage

## [65] The Use of Selective Free Radical Probes to Study Active Sites in Enzymes and Viruses

*By* J. Leslie Redpath

The use of oxidative free radicals as probes for active sites and critical residues in biologically active protein structures requires that such free radicals be selective in their reactions with the constituents of proteins, i.e., amino acids and peptides.

In aqueous systems, the hydroxyl radical ($\cdot OH$) is frequently generated under certain conditions, e.g., in the presence of metal ions and peroxide (Fenton's reagent), or by radiolysis. The $\cdot OH$ radical is a powerful oxidizing agent, and while it will react with some amino acids at a more rapid rate than with others[1] it cannot be considered as selective in nature in its reactions with these compounds. However, it is possible to generate secondary radicals, which are analogous to $\cdot OH$ radicals in that they are oxidative in nature, but which are selective in their reactions with amino acids, by reacting $\cdot OH$ with certain inorganic anions. Such anions include $SCN^-$, $Br^-$, $CO_3^{2-}$, $I^-$, and corresponding resulting radical anions are $\cdot(SCN)_2^-$, $\cdot Br_2^-$, $\cdot CO_3^-$, and $\cdot I_2^-$.

A general schema for the production of these radicals is given below.

$$\cdot OH + X^- \rightarrow \cdot X + OH^- \qquad (1)$$
$$\cdot X + X \rightleftharpoons X_2^- \qquad (2)$$

### The Radiolytic Generation of Selective Oxidative Free Radicals

In most studies to date on the use of selective free radicals to study active sites in enzymes the radicals have been generated radiolytically. Since their production is dependent upon the production of $\cdot OH$ radicals, conditions are generally used which are optimal for the production of such radicals. In radiation chemical experiments in aqueous solution this translates to conditions of $N_2O$ saturation since $N_2O$ reacts with the radiation-produced hydrated electron to produce $\cdot OH$, and thus doubles the yield of radiolytically generated $\cdot OH$, viz.,

$$H_2O \rightarrow e_{aq}^- (2.7), \cdot OH (2.7), \cdot H (0.6) \qquad (3)$$

The numbers in parentheses represent $g$ values (radicals/100 eV deposited energy)

$$e_{aq}^- + N_2O \rightarrow N_2 + OH + \cdot OH \qquad (4)$$

[1] L. M. Dorfman and G. E. Adams, *N.S.R.D.S.—NBS* **46** (1973).

METHODS IN ENZYMOLOGY, VOL. 105

Copyright © 1984 by Academic Press, Inc.
All rights of reproduction in any form reserved.
ISBN 0-12-182005-X

In order to ensure that the desired secondary radicals are produced the added anion $X^-$ must be present at a concentration which will effectively scavenge most of the $\cdot OH$ radicals in competition with any additional solutes, e.g., the enzyme or protein under study. Since most of the above-mentioned anions $(X^-)$, and most enzymes and proteins, react with $\cdot OH$ at diffusion controlled rates,[1] $X^-$ is usually present in at least 100 times the concentration of the enzyme or protein.

In addition, the concentration of enzyme or protein used in these experiments is limited by the concentration of $N_2O$ in a saturated aqueous solution (20 m$M$), since the $N_2O$ has to effectively scavenge all of the hydrated electrons in competition with added enzyme. Again, since $e_{aq}^-$ reacts with $N_2O$ and enzymes at diffusion-controlled rates[2] the concentration of enzyme should not exceed 0.2 m$M$. Thus the relative rates of reaction of $e_{aq}^-$ with $N_2O$ and the enzyme, together with the solubility of $N_2O$ in water, dictate the maximum concentration of enzyme that should be used in these experiments. Furthermore the relative rates of reaction of $\cdot OH$ with the enzyme and solute $X^-$ dictate the concentration of $X^-$ to be used. With these conditions established it is feasible to begin experiments.

### Determination of the Selectivity of Reaction of Secondary Oxidative Free Radicals

The technique that has been most widely used to study the reactivity of the secondary free radicals $(\cdot X_2^-)$ is pulse radiolysis. Detailed descriptions of this technique can be found in various references.[3-5] Briefly, the technique involves exposing the sample to a brief ($<10^{-6}$ sec) pulse of ionizing radiation (usually an electron beam) and then following chemical changes in the solution as a function of time. The resolution in time that can be obtained using this technique is dictated by the pulse width and by the detection technique. The most common detection technique used is absorption spectrophotometry; however, others that can be used are emission spectrophotometry (e.g., if a radiolysis product emits light), conductivity (e.g., if the reactant or product is charged), and electron spin resonance (if the reactant or product is a free radical).

Studies of the reactivity of these radicals $(\cdot X_2^-)$ have been done with absorption spectrophotometry since these species have strong optical absorptions in the visible region of the spectrum and are also relatively long

[2] M. Anbar, M. Bambenek, and A. B. Ross, N.S.R.D.S.—NBS **43** (1973).

[3] J. P. Keene, Nature (London) **188**, 843 (1960).

[4] M. S. Matheson and L. M. Dorfman, J. Chem. Phys. **32**, 1870 (1960).

[5] L. M. Dorfman, J. Chem. Educ. **58**, 84 (1981).

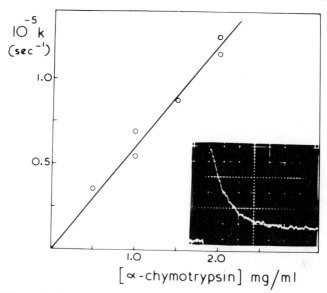

FIG. 1. First-order decay rate of $\cdot Br_2^-$ absorption as a function of $\alpha$-chymotrypsin concentration. Inset: oscillogram showing decay of $\cdot Br_2^-$ absorption in the presence of 1.0 mg/ml $\alpha$-chymotrypsin (1 large division = 10 $\mu$sec).

lived (Ref. 6 and references therein). By measuring the first-order rate of decay of the $\cdot X_2^-$ absorption in the presence of different added concentrations of reactant it is possible to determine the bimolecular rate constant for reaction of $\cdot X_2^-$ with the reactant (see Fig. 1).

A survey of the reactivity of various radical anions of the $\cdot X_2^-$ type with individual amino acids has been made.[6] It was found that the aromatic and sulfur-containing amino acids were the most reactive (see Table I). It can be seen from Table I that at neutral pH $\cdot(SCN)_2^-$ is selective for tryptophan while $\cdot I_2^-$ is selective for cysteine. It should be stressed that these rate constants are pH dependent (see Ref. 6 for details). Therefore it is feasible, for example, to change the distribution of sites of attack of these radicals on an enzyme by altering the pH (see next section).

## Transient Absorption Spectra as Indicators of Sites of Attack

Apart from the determination of rate constants, the pulse radiolysis technique can be used to determine absorption spectra of transient reac-

[6] G. E. Adams, J. E. Aldrich, R. H. Bisby, R. B. Cundall, J. L. Redpath, and R. L. Willson, *Radiat. Res.* **49**, 298 (1972).

TABLE I

REACTIVITY OF SOME AMINO ACIDS WITH RADICAL ANIONS[a,b]

| Radical anion | Trypto-phan | Tyrosine | Histidine | φ-Ala-nine | Cysteine | Methio-nine |
|---|---|---|---|---|---|---|
| ·$Br_2^-$ | 77 | 2.0 | 1.5 | 0.1 | 18 | 1.1 |
| | | pH 7.5 | pH 7.6 | | pH 6.6 | pH 7.3 |
| ·$(CNS)_2^-$ | 27 | 0.5 | 0.1 | 0.1 | 5 | 0.2 |
| | | | | | pH 6.6 | |
| ·$I_2^-$ | 0.1 | 0.1 | 0.1 | 0.1 | 11 | 0.1 |
| | | | | | pH 6.8 | |
| ·$CO_3^-$ pH 11.2 ± 0.3 | 44 | 29 | 0.7 | 0.1 | 27 | 12 |
| ·$Cl_2^-$ pH 1.8 ± 0.2 | 260 | 27 | 1.4 | 0.6 | 85 | 0.7 |

[a] Reproduced from Adams et al.[6] with permission.
[b] pH 7.0 ± 0.2 except where stated; temperature, 22°; rate constants, units of 1.0 ± 0.1 × $10^7$ $M^{-1}$ $sec^{-1}$; inorganic salt concentration, $10^{-1}$ $M$, excepting iodide ($5 \times 10^{-1}$ $M$).

tion products. For example, examination of the oscillogram in Fig. 1 reveals that the absorption does not return to baseline following completion of the initial reaction. The residual absorption is due to a transient reaction product. By monitoring this residual absorption at a suitable time after the pulse, corresponding to completion of the initial reaction, as a function of wavelength, the absorption spectrum of the transient product can be measured.

Such transient spectra have been determined for the reactions of ·$X_2^-$ with several amino acids and are shown in Figs. 2 and 3 for two of the most reactive amino acids, tyrosine and tryptophan. It can be seen that these spectra are quite different and thus can be used as indicators of sites of attack. This is illustrated in Fig. 4 where at neutral pH tryptophan and cysteine are major sites of attack of ·$(SCN)_2^-$ on papain, whereas in alkaline solution tyrosine is the major site of attack.

### The Use of Selective Oxidative Free Radical Probes to Study Active Sites in Enzymes and Viruses

Once the baseline information on free radical reactivity and transient spectra has been obtained using pulse radiolysis techniques, this information can be applied to studies of enzyme and viral inactivation under steady-state irradiation conditions. In the sections below the results of some such studies are summarized.

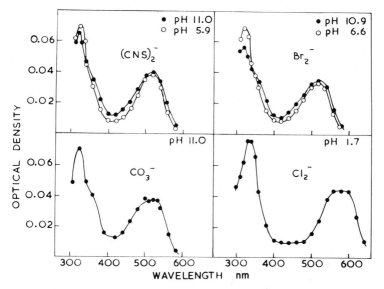

Fig. 2. Transient spectra from pulse radiolysis of $N_2O$-saturated solutions of tryptophan ($10^{-3}$ $M$) and the respective anions ($10^{-1}$ $M$). Spectra measured 50 $\mu$sec after the pulse. Dose: 2400 rads. (Reproduced from Adams et al.[6] with permission.)

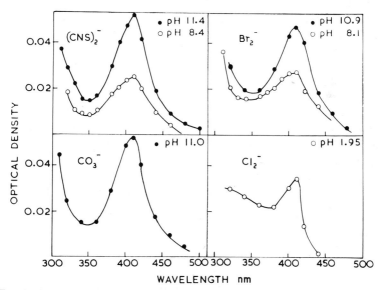

Fig. 3. Transient spectra from pulse radiolysis of $N_2O$-saturated solutions of tyrosine ($10^{-3}$ $M$) and the respective anions ($10^{-1}$ $M$). Spectra measured 50 $\mu$sec after the pulse. Dose: 2400 rads. (Reproduced from Adams et al.[6] with permission.)

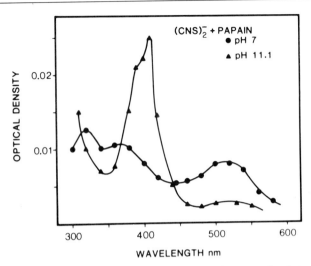

FIG. 4. Transient spectra from pulse radiolysis of $N_2O$-saturated solutions of papain (0.75–1.0 mg/ml) containing KCNS ($5 \times 10^{-2}$ $M$). Spectra measured 50 $\mu$sec after a 1000 rad pulse. [Reproduced from J. L. Redpath, *J. Chem. Educ.* **58**, 131 (1981) with permission.]

*Enzyme Systems*

The first enzyme to be studied using the selective free radical technique was lysozyme.[7] This enzyme was inactivated at neutral pH by ·$(SCN)_2{}^-$ at the same rate as by ·OH. Since the amino acid reactivity data indicate ·$(SCN)_2{}^-$ to be selective for tryptophan, the inactivation results imply that tryptophan residues are critical to the activity of lysozyme, a fact which is well established from classical enzymology. Measurement of the transient absorption spectrum of the product of reaction of ·$(SCN)_2{}^-$ with lysozyme indicates that all the radicals react with tryptophan residues in the enzyme (Fig. 5).

Since this technique was introduced it has been used to probe active sites in many different enzymes some of which are listed in Table II, together with literature citations to which readers are referred for details. As can be seen from Table II, the technique has had success in many different enzyme systems, including both metal-containing and thiol-containing enzymes.

[7] G. E. Adams, R. L. Willson, J. E. Aldrich, and R. B. Cundall, *Int. J. Radiat. Biol.* **16**, 33 (1969).

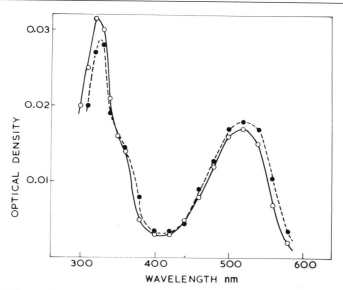

FIG. 5. Comparison of product spectrum from reaction of $\cdot(CNS)_2^-$ with tryptophan with that from reaction of $\cdot(CNS)_2^-$ with lysozyme. Solid line: $10^{-3}\,M$ tryptophan, $10^{-1}\,M$ KCNS, pH 5.9, $N_2O$ saturated. Dashed line: lysozyme 2 mg/ml, $10^{-1}\,M$ KCNS, pH 5.4, $N_2O$ saturated. Spectra measured 50 $\mu$sec after pulse. Dose in each case, 1050 rads. (Reproduced from Adams *et al.*[6] with permission.)

## Viral Systems

This technique of inactivation by selective free radicals has been applied to bacteriophage systems which can be inactivated by free radical attack on the protein coats of such organisms, with the tail being a particularly sensitive target.[8] Some striking differences between *Escherichia coli* phages $T_7$, $T_6$, and $T_2$ have been demonstrated[9-11] (Table III). From Table III it can be seen that all three of these phage are most sensitive to inactivation by $\cdot Br_2^-$, a radical which will react with Trp, Tyr, and His at neutral pH. However, it can also be seen that $T_7$ and $T_2$ are relatively insensitive to inactivation by $\cdot(SCN)_2^-$ compared to $T_6$. All three phage are more sensitive to $\cdot Br_2^-$ than $\cdot(SCN)_2^-$, implying that oxidation of Tyr and/or His is more important with respect to inactivation of the phage protein than oxidation of Trp residues. However, comparison of the rela-

[8] C. E. Clarkson and D. L. Dewey, *Radiat. Res.* **54,** 531 (1973).
[9] J. L. Redpath and D. L. Dewey, *Int. J. Radiat. Biol.* **25,** 189 (1974).
[10] J. L. Redpath, *Int. J. Radiat. Biol.* **27,** 493 (1975).
[11] K. O. Hiller and R. L. Willson, personal communication.

TABLE II

Some of the Enzymes Investigated by the Selective Free
Radical Technique

| Enzyme | Crucial residues identified | Reference |
|---|---|---|
| Lysozyme | Trp | 7 |
| Ribonuclease | His | a |
| Chymotrypsin | His | b |
| Trypsin | His | c |
| Papain | Trp, Cys | d |
| Carboxypeptidase | Tyr (Trp?) | e |
| Subtilisins, Carlsberg and Novo | His | f |
| Bovine carbonic anhydrase | Trp, Tyr, His | g |
| Superoxide dismutase | His | h |
| Yeast alcohol dehydrogenase | His, Cys | i |

a G. E. Adams, R. H. Bisby, R. B. Cundall, J. L. Redpath, and R. L. Willson, *Radiat. Res.* **49**, 290 (1972).
b K. Baverstock, R. B. Cundall, G. E. Adams, and J. L. Redpath, *Int. J. Radiat. Biol.* **16**, 39 (1974).
c G. E. Adams, J. L. Redpath, R. H. Bisby, and R. B. Cundall, *J. Chem. Soc. Faraday Trans. 1* **69**, 1068 (1973).
d G. E. Adams and J. L. Redpath, *Int. J. Radiat. Biol.* **25**, 129 (1974).
e P. B. Roberts, *Int. J. Radiat. Biol.* **24**, 143 (1973).
f R. H. Bisby, R. B. Cundall, G. E. Adams, and J. L. Redpath, *J. Chem. Soc. Faraday Trans. 1* **70**, 2210 (1974).
g J. L. Redpath, R. Santus, J. Ovadia, and I. I. Grossweiner, *Int. J. Radiat. Biol.* **28**, 243 (1975).
h P. B. Roberts, E. M. Fielden, G. Rotilio, L. Calabrese, J. V. Bannister, and W. H. Bannister, *Radiat. Res.* **60**, 441 (1974).
i R. Badiello, M. Tamba, and M. Quintilliani, *Int. J. Radiat. Biol.* **26**, 311 (1974).

TABLE III

A Comparison of the Sensitivities of *E. coli*
Phages T$_2$, T$_6$, and T$_7$ to Inactivation by
·(SCN)$_2^-$ and ·Br$_2^-$

| Inactivating radical | Sensitivity S $(1/D_{37}$ krad$^{-1})$ | | |
|---|---|---|---|
| | T$_2$[11] | T$_6$[10] | T$_7$[9] |
| ·(SCN)$_2^-$ | 0.56 | 1.8 | 0.2 |
| ·Br$_2^-$ | 10 | 5.0 | 3.33 |

tive sensitivities to $\cdot(SCN)_2^-$ indicates that Trp residues are more essential to the biological function of $T_6$ than $T_7$ or $T_2$.

In a more detailed analysis of this type of data[12] it has been shown that direct comparison of $D_{37}$ values (as in Table III) can be misleading. For example, while the phage are more sensitive when irradiated in the presence of $Br^-$, it can be shown that $\cdot OH$ is more lethal than $\cdot Br_2^-$ at the $T_7$ phage surface by a factor of 2. It is the longer lifetime of $\cdot Br_2^-$ (which is linked to its lower, and more selective, reactivity) that is a key factor. Furthermore, it has been shown that the sensitivity of phage irradiated in the presence of $Br^-$ is dependent upon $Br^-$ concentration, even at concentrations above that required for complete scavenging of $\cdot OH$ radicals. This effect may be due to the production of a tertiary radical such as $\cdot Br_3^{2-}$ which is more lethal than $\cdot Br_2^-$. Alternatively, $Br^-$ binding to the phage coat may promote enhanced lethality.

### The Use of Selective Reductive Free Radical Probes to Study Structural Aspects of Enzymes and Proteins

The hydrated electron is a reducing radical which reacts rapidly with many amino acids, in particular histidine and oxidized cysteine, i.e., cystine. The formate radical ion, $\cdot CO_2^-$, formed by reaction of hydroxyl radicals with formate, viz.

$$\cdot OH + HCOO^- \rightarrow \cdot CO_2^- + H_2O \tag{5}$$

is a more selective reducing species than the hydrated electron. At neutral pH this radical reacts with disulfide bridges with a rate constant of $\sim 10^{-9}$ $M^{-1}$ $sec^{-1}$ and with all other amino acids at $< 10^7$ $M^{-1}$ $sec^{-1}$.

$$\cdot CO_2^- + RSSR \rightarrow \cdot RSSR^- + CO_2 \tag{6}$$

In acid solution (pH $< 4.5$) a measurable reactivity with histidine ($\sim 10^7$ $M^{-1}$ $sec^{-1}$) can be seen by using pulse radiolysis.

An interesting application of the selectivity of reaction of $\cdot CO_2^-$ with disulfide bridges is to probe enzymes for the relative accessibility of these structures. Unlike the hydrated electron, $\cdot CO_2^-$ can only react with accessible disulfide bridges since its size does not permit diffusion into the inner regions of the enzyme structure (Fig. 6). Pulse radiolysis studies have shown that the disulfide bridges in lysozyme are more accessible than those in ribonuclease or $\alpha$-chymotrypsin.[13] Pulse radiolysis of an

[12] D. Becker, J. L. Redpath, and L. I. Grossweiner, *Radiat. Res.* **73,** 51 (1978).
[13] R. H. Bisby, J. L. Redpath, G. E. Adams, and R. B. Cundall, *J. Chem. Soc. Faraday Trans. I* **72,** 51 (1976).

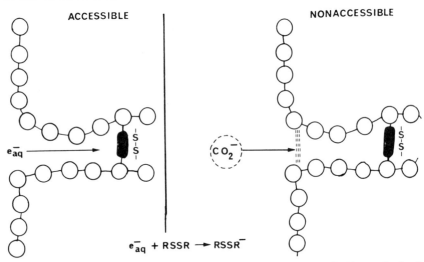

FIG. 6. Structural inhibition of electron transfer from the formate radical ion to the S—S bridges in enzymes (see text). [Reproduced from J. L. Redpath, *J. Chem. Educ.* **58,** 131 (1981), with permission.]

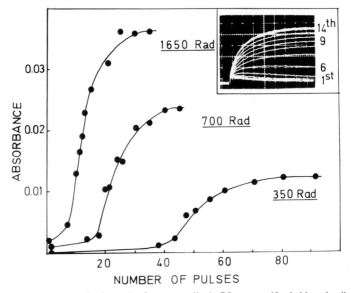

FIG. 7. Electron transfer from the formate radical $\cdot CO_2^-$, to sulfur bridges in ribonuclease. Main figure: effect of number of pulses on the absorption of RSSR$^-$ at 410 nm. Ribonuclease concentration 2.0 mg/ml, formate concentration $= 4 \times 10^{-2} M$, pH 8.0. Inset: oscillogram showing rate of formation of RSSR$^-$ at 410 nm. Each curve recorded after indicated number of prepulses. Dose per pulse $= 2$ krad. Sweep speed 20 $\mu$sec/cm. [Reproduced from G. E. Adams, J. L. Redpath, R. H. Bisby, and R. B. Cundall, *Isr. J. Chem.* **10,** 1079 (1972), with permission.]

$N_2O$-saturated solution of ribonuclease containing formate shows no reaction of the $\cdot CO_2^-$ radical with the disulfide bridges in RNase. However, by repeatedly pulsing the solution the enzyme structure can be broken up to expose the —S—S— bonds and allow for full reaction of the $\cdot CO_2^-$ radical at these sites (Fig. 7). An alternative way of doing the same thing is to heat the enzyme up to 55° whereupon it loses its structural integrity.

# Section VI

## Drugs: Environmental Induction of Radical Formation and Radical Species

# [66] Detection and Measurement of Drug-Induced Oxygen Radical Formation

*By* MARTYN T. SMITH, HJÖRDIS THOR, and STEN ORRENIUS

A wide variety of drugs and related compounds are known to induce the formation of active oxygen species in mammalian cells.[1,2] For brevity, we restrict ourselves in this chapter to describing methods for the detection and quantification of superoxide anion radicals ($O_2^-$) and hydrogen peroxide ($H_2O_2$) only. The reader is referred elsewhere for information concerning the detection of hydroxyl radicals ($HO\cdot$)[3,4] and singlet oxygen ($^1\Delta_g O_2$).[5,6]

## Detection and Measurement of Drug-Induced Superoxide Radical Formation

The cooxidation of adrenaline to adrenochrome has been widely used for the spectrophotometric measurement of $O_2^-$ formation in biological systems.[7,8] This assay is, however, unreliable[9] and insensitive and has been replaced, in recent years, by the use of ferricytochrome $c$, either in its native form[10] or in a partially acetylated[11] or succinoylated state.[1] Modification of the native cytochrome drastically lowers the ability of enzymes such as NADPH–cytochrome $c$ ($P$-450) reductase to reduce it and therefore greatly increases its sensitivity to $O_2^-$. Specificity is achieved by subtracting the rate of cytochrome $c$ reduction in the presence of superoxide dismutase (SOD) from that obtained in its absence. The rate of native and modified cytochrome $c$ reduction is monitored by dual-wavelength spectrophotometry using either the wavelength pair 550–

[1] H. Kuthan, H. Tsuji, H. Graf, V. Ullrich, J. Werringloer, and R. W. Estabrook, *FEBS Lett.* **91,** 343 (1978).

[2] D. P. Jones, H. Thor, B. Andersson, and S. Orrenius, *J. Biol. Chem.* **253,** 6031 (1978).

[3] G. Czapski, this volume [24].

[4] G. Cohen, *Photochem. Photobiol.* **28,** 669 (1978).

[5] C. S. Foote, this volume [3].

[6] E. Cadenas, R. P. Daniele, and B. Chance, *FEBS Lett.* **123,** 225 (1981).

[7] K. Asada, K. Kiso, and K. Yoshikawa, *J. Biol. Chem.* **249,** 2175 (1974).

[8] G. M. Bartoli, T. Galeotti, G. Palombini, G. Parisi, and A. Azzi, *Arch. Biochem. Biophys.* **184,** 276 (1977).

[9] W. Bors, M. Saran, E. Lengfelder, C. Michel, C. Fuchs, and C. Frenzel, *Photochem. Photobiol.* **28,** 629 (1978).

[10] J. M. McCord and I. Fridovich, *J. Biol. Chem.* **244,** 6049 (1969).

[11] A. Azzi, C. Montecucco, and C. Richter, *Biochem. Biophys. Res. Commun.* **65,** 597 (1975).

Copyright © 1984 by Academic Press, Inc.
All rights of reproduction in any form reserved.
ISBN 0-12-182005-X

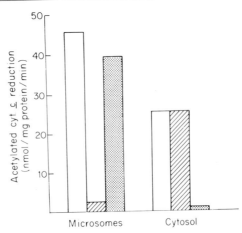

FIG. 1. Menadione-stimulated acetylated cytochrome $c$ reduction by liver subcellular fractions. Liver microsomes and cytosol from phenobarbital-treated rats were incubated (1 mg protein/ml) at 25° with 10 $\mu M$ menadione and 1 m$M$ NADPH. SOD (0.2 mg/ml) (striped bar) or dicoumarol (30 $\mu M$) (stippled bar) were added separately to test the specificity of the reactions. Acetylated cytochrome $c$ reduction was measured using an Aminco DW-2 spectrophotometer and the wavelength pair 550–540 nm [cf. H. Thor, M. T. Smith, P. Hartzell, G. Bellomo, S. A. Jewell, and S. Orrenius, *J. Biol. Chem.* **257,** 12419 (1982)].

540 or 550–557 nm. The absorption coefficients used are 19.6[11] and 21.0[12] m$M^{-1}$ cm$^{-1}$, respectively.

Recently, we have used the above technique with acetylated cytochrome $c$ to measure the rate of $O_2^-$ formation induced by menadione (2-methyl-1,4-naphthoquinone) in both liver subcellular fractions and isolated hepatocytes.[13] The incubation conditions used were as follows.

*Subcellular fractions* (0.01–1 mg protein/ml) were incubated with menadione in 0.1 $M$ Tris–HCl, pH 7.4, containing 50 m$M$ KCl at 25° in the presence of 0.2 mg/ml acetylated cytochrome $c$, and either 1 m$M$ NADPH or NADH. Addition of 0.2 mg/ml SOD (Sigma) enabled the background reduction rate to be subtracted. Powis and co-workers[14] have used similar incubation conditions in conjunction with purified enzymes. Typical results obtained when isolated liver microsomes and cytosol are incubated in this manner with 10 $\mu M$ menadione and 1 m$M$ NADPH are shown in Fig. 1. The majority of the acetylated cytochrome $c$ reduction produced by the microsomal fraction could be attributed to $O_2^-$, since it

[12] B. F. Van Gelder and E. C. Slater, *Biochim. Biophys. Acta* **58,** 593 (1962).
[13] H. Thor, M. T. Smith, P. Hartzell, G. Bellomo, S. A. Jewell, and S. Orrenius, *J. Biol. Chem.* **257,** 12419 (1982).
[14] G. Powis, B. A. Svingen, and P. Appel, *Mol. Pharmacol.* **20,** 387 (1981).

was prevented by SOD addition (Fig. 1). In contrast, the menadione-dependent reduction of cytochrome $c$ by the cytosolic fraction was unaffected by the addition of SOD (Fig. 1). This result shows the importance of using SOD inhibition to specifically determine the rate of $O_2^-$ formation. It appears that menahydroquinone was responsible for the non-SOD inhibitable reduction of acetylated cytochrome $c$ in this system, because the addition of dicoumarol, a powerful inhibitor of the chiefly cytosolic flavoprotein NAD(P)H dehydrogenase (quinone) [NAD(P)H : (quinone-acceptor) oxidoreductase (DT-diaphorase)], almost totally prevented the cytochrome $c$ reduction (Fig. 1). DT-diaphorase catalyzes the reduction of menadione to menahydroquinone in the presence of NADPH or NADH without forming menasemiquinone or oxygen radicals.[13] These findings also illustrate the fact that native and modified cytochrome $c$ is readily reduced by a number of compounds.

*Freshly isolated hepatocytes* ($10^6$ cells/ml) were incubated in Krebs–Henseleit buffer, pH 7.4, containing 25 m$M$ HEPES ($N$-2-hydroxyethylpiperazine-$N'$-2-ethanesulfonic acid)[13] and 0.2 mg/ml acetylated cytochrome $c$ in a stirred 3 ml cuvette. Figure 2 shows typical spectrophotometric traces of the reduction of extracellular acetylated cytochrome $c$ before and during menadione metabolism in hepatocytes isolated from phenobarbital-treated rats and incubated by this procedure. The subsequent addition of 0.2 mg/ml SOD caused an approximately 70% decrease in the rate of reduction of the extracellular acetylated cytochrome $c$ (Fig. 2). This result indicates that $O_2^-$ and possibly also mena-

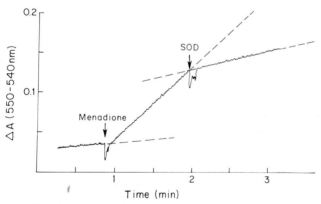

FIG. 2. Menadione-induced $O_2^-$ formation by isolated hepatocytes. Hepatocytes ($10^6$ cells/ml) isolated from phenobarbital-treated rats were incubated at 37° in Krebs–Henseleit buffer, pH 7.4, containing 25 m$M$ HEPES and 0.2 mg/ml acetylated cytochrome $c$, prepared as described by A. Azzi, C. Montecucco, and C. Richter, *Biochem. Biophys. Res. Commun.* **65,** 597 (1975). Menadione (100 $\mu M$) and SOD (0.2 mg/ml) were added as indicated.

hydroquinone traverse the plasma membrane of hepatocytes. It is unlikely, however, that $O_2^-$ does this via specific anion channels since certain sulfonate stilbenes, which block $O_2^-$ transport in erythrocytes and
granulocytes, have no marked effect on menadione-induced $O_2^-$ formation by hepatocytes.[14] It is possible that $O_2^-$ passes through the membrane
directly,[15] but it is much more likely that it is converted to its protonated
form, the perhydroxyl radical (HOO·), which then passes through the
lipid bilayer and reappears as $O_2^-$ in the aqueous incubation medium. The
$O_2^-$ appearing in the incubate probably represents only a small fraction of
the $O_2^-$ formed intracellularly. It has been calculated, for example, that
only 2–4% of the $O_2^-$ formed during menadione metabolism in hepatocytes is released.[14] At the present time, no suitable method exists for the
quantification of intracellular $O_2^-$ formation. One should also consider the
fact that the partial modification of ferricytochrome $c$ by acetylation or
succinoylation can greatly modify its reactivity toward $O_2^-$. This point
has recently been considered in depth by Kuthan *et al.*[16] and equations for
the calculation of maximal rates of $O_2^-$ production in complex biological
systems have been presented.

An alternative approach to the detection and quantification of $O_2^-$
formation by spectrophotometry is the use of certain spin traps in combination with electron paramagnetic resonance spectroscopy, which is adequately discussed elsewhere in this volume. One of the drawbacks to this
approach is, however, that up to 100 mmol of the spin trap/liter may be
present.

### Detection and Measurement of Drug-Induced $H_2O_2$ Formation

The drug-induced rate of formation of $H_2O_2$ in subcellular fractions
can be determined by a variety of different methods.[17] In intact cells two
more complex procedures can be used,[18] based either (1) on the determination of the steady-state level of the catalase-$H_2O_2$ intermediate (compound I), or (2) on measuring the flux through a peroxidatic reaction
catalyzed either by catalase itself or a peroxidase such as glutathione
peroxidase.

*Determination of the Catalase Compound I Level.* The measurement
of drug-induced $H_2O_2$ production rates by direct optical measurement of
the steady-state level of compound I has been described in detail by

[15] S. Mabrey, G. Powis, J. B. Schenkman, and T. R. Tritton, *J. Biol. Chem.* **252**, 2929 (1977).
[16] H. Kuthan, V. Ullrich, and R. W. Estabrook, *Biochem. J.* **203**, 551 (1982).
[17] A. G. Hildebrandt, I. Roots, M. Tjoe, and G. Heinemeyer, this series, Vol. 52, p. 342.
[18] H. Sies, this series, Vol. 77, p. 15.

Sies.[18] Briefly the rate of $H_2O_2$ production, $d[H_2O_2]/dt$, is estimated from methanol titrations of compound I using the equation

$$d[H_2O_2]/dt = k \times [Cat] \times [A]_{1/2}$$

where [Cat] is catalase heme concentration, $k$ is the constant for methanol as hydrogen donor (31.5 $min^{-1}$), and $[A]_{1/2}$ is the methanol concentration required for 50% titration of compound I.

Experimentally, $[A]_{1/2}$ has been determined in perfused organs,[19,20] in the exposed organ *in situ*,[21] and in isolated cells.[2,22] Drug-stimulated $H_2O_2$ production in isolated hepatocytes is, however, observable only in hepatocytes which have been depleted of glutathione by pretreatment of the animals with diethyl maleate.[2] This is due to the major role glutathione peroxidase plays in the decomposition of $H_2O_2$ formed during drug metabolism at the endoplasmic reticulum.[22]

An alternative method for measuring $H_2O_2$ production in the presence of a drug which is actively demethylated, and thus forms formaldehyde and formic acid, has been proposed by Jones *et al.*[22] This method relies upon the maintenance of a steady-state formic acid concentration in the cells. The rate of $H_2O_2$ production is calculated according to the equation

$$d[H_2O_2]/dt = \frac{2n}{[Cat]/p_m - (n + 1)} (k_4 \cdot p_m \cdot [HCOOH])$$

where $n = 1.5$, [Cat] is the catalase heme concentration, $p_m$ is compound I concentration, and $k_4$ is the rate constant for the peroxidation reaction of catalase with formic acid. Intracellular [HCOOH] is calculated from formate measurements. For further details see Jones *et al.*[22]

*GSSG Efflux from Glutathione Peroxidase Mediated $H_2O_2$ Metabolism.* The release of oxidized glutathione (GSSG) from tissues and cells is a commonly used indicator of oxidative stress. For example, in the perfused liver the GSSG efflux is dependent on the rate of $H_2O_2$ infusion.[23] However, glutathione peroxidase catalyzes the breakdown of organic and lipid hydroperoxides as well as $H_2O_2$.[24] The interpretation of GSSG release in terms of $H_2O_2$ metabolism, rather than of lipid peroxides, is therefore best suited for use with drugs which also inhibit lipid peroxida-

---

[19] N. Oshino, B. Chance, H. Sies, and T. Bucher, *Arch. Biochem. Biophys.* **154**, 117 (1973).
[20] H. Sies, T. Bucher, N. Oshino, and B. Chance, *Arch. Biochem. Biophys.* **154**, 106 (1973).
[21] N. Oshino, D. Jamieson, T. Sugano, and B. Chance, *Biochem. J.* **146**, 67 (1975).
[22] D. P. Jones, L. Eklow, H. Thor, and S. Orrenius, *Arch. Biochem. Biophys.* **210**, 505 (1981).
[23] H. Sies and K. H. Summer, *Eur. J. Biochem.* **57**, 503 (1975).
[24] A. Wendel, this series, Vol. 77, p. 325.

tion, such as ethylmorphine, menadione, and aminopyrine.[25] It should also be borne in mind that the release of GSSG from cells does not correspond stoichiometrically to the rate of glutathione peroxidase reaction or the rate of $H_2O_2$ production, because the majority of the GSSG formed will be reduced back to GSH through the action of GSSG reductase.[22]

### Acknowledgment

This research was supported by grants from the Swedish Medical Research Council, Karolinska Institutet, and Northern California Occupational Health Center.

[25] M. T. Smith, H. Thor, P. Hartzell, and S. Orrenius, *Biochem. Pharmacol.* **31**, 19 (1982).

## [67] Alloxan and 6-Hydroxydopamine: Cellular Toxins

### By GERALD COHEN and RICHARD E. HEIKKILA

6-Hydroxydopamine and alloxan are cellular toxins with similar modes of action, but different tissue specificity. Alloxan, injected *in vivo*, destroys the insulin-producing $\beta$ cells of the pancreas to produce an experimental form of diabetes.[1] 6-Hydroxydopamine, on the other hand, destroys catecholamine nerve terminals and cell bodies.[2]

The polyphenolic (reduced) forms of both agents react rapidly with molecular oxygen to yield quinoidal products and several forms of reduced oxygen, namely, superoxide and hydroxyl free radicals, and hydrogen peroxide.[3] Although the quinones and oxygen metabolites can all contribute to tissue damage, current evidence[4] indicates that it is the reduced forms of oxygen that predominate in the destruction of target tissues. Intermediate semiquinone radicals have not been evaluated for their toxic actions.

Why do these toxic agents exhibit such prominent tissue selectivity? Why are other cell types, such as red blood cells or liver not affected?

[1] C. C. Rerup, *Pharmacol. Rev.* **22**, 485 (1970).
[2] H. Thoenen and J. P. Tranzer, *Naunyn-Schmeideberg's Arch. Pharmakol.* **261**, 271 (1968).
[3] G. Cohen, R. E. Heikkila, and D. MacNamee, *J. Biol. Chem.* **249**, 2447 (1974).
[4] D. G. Graham, S. M. Tiffany, W. R. Bell, Jr., and W. F. Gutknecht, *Mol. Pharmacol.* **14**, 644 (1978).

Copyright © 1984 by Academic Press, Inc.
All rights of reproduction in any form reserved.
ISBN 0-12-182005-X

Why does alloxan not lesion catecholamine neurons, and why does 6-hydroxydopamine not induce experimental diabetes? The answers, in large measure, can be found in the specific accumulation of the cell toxins within their respective target cells. The toxins destroy those cells in which they accumulate.

In general, tissue reducing agents, such as ascorbate,[5] amplify the cellular toxicity by reducing the quinoidal forms of the cellular toxins. The reduced forms then recycle through oxidative reactions with molecular oxygen to increase a flux of evanescent and destructive free radical intermediates. Thus, an apparently paradoxical situation arises in which reducing agents amplify oxidative damage.

## 6-Hydroxydopamine

When 6-hydroxydopamine is injected iv or ip, it specifically lesions the norepinephrine-secreting nerve terminals of the peripheral sympathetic nervous system, i.e., it produces a peripheral sympathectomy.[6] It does not cross the blood–brain barrier to any great extent. The same is true for 6-aminodopamine and several other neurotoxic, polyphenolic phenylethylamines. To lesion catecholamine cell clusters (or nuclei) of the brain, such as the noradrenergic locus ceruleus or the dopaminergic substantia nigra, it is necessary to inject the neurotoxin, by stereotaxic methodology, directly into or close to the cell clusters. Injections into the ventricular systems of brain (e.g., the cerebral ventricles) will damage catecholamine nerve terminals, but not the cell bodies.

On the other hand, the amino acid analog, 6-hydroxydopa, does penetrate freely into the brain. After ip injection, both peripheral and central catecholamine nerve terminals are destroyed. There is evidence to suggest that the mechanism of action involves transformation to 6-hydroxydopamine.[6]

For experiments conducted *in vitro* with, for example, brain tissue slices, the addition of 6-hydroxydopamine to the bathing medium will induce internal damage from the 6-hydroxydopamine that penetrates into the cells, and external damage from the autoxidation of 6-hydroxydopamine in the medium. The latter can be seen by adding, for example, catalase, which protects the nerve terminals from damage by hydrogen peroxide in the external environment.[7]

For *in vivo* experiments, particularly those concerned with peripheral sympathetic nerves, the nerve terminals can be protected by drugs, such

[5] R. E. Heikkila and G. Cohen, *Mol. Pharmacol.* **8**, 241 (1972).
[6] R. M. Kostrzewa and D. M. Jacobowitz, *Pharmacol. Rev.* **26**, 199 (1974).
[7] R. E. Heikkila and G. Cohen, *Science* **172**, 1257 (1971).

as desmethylimipramine or cocaine, which prevent the "uptake" of 6-hydroxydopamine into the nerve terminals.[6] Hence, damage *in vivo* occurs almost completely intracellularly. The accumulated 6-hydroxydopamine within nerve terminals in experiments conducted *in vivo*[8] is in the range of 25–50 m$M$.

### Technical Notes Common to Both 6-Hydroxydopamine[9] and Alloxan[10]

Details are provided for experiments with Swiss–Webster mice, although other animal species, such as rats, are also commonly used.

1. We prefer iv injections into a tail vein; a typical injected volume is 0.2 ml, made with a glass 0.25 or 0.50 ml syringe. If the iv injection does not proceed smoothly, the animal is discarded. When ip injections are used, the dosage should be increased by a factor of 3- to 5-fold.

2. Alloxan and 6-hydroxydopamine are unstable in aqueous solution. The solutions for injection are prepared in ice-cold isotonic (0.9% w/v) saline just prior to use. For 6-hydroxydopamine, the saline is sparged thoroughly with nitrogen gas prior to dissolving the 6-hydroxydopamine, in order to prevent autoxidation. Solutions are kept on ice.

3. Mice are typically divided into experimental groups in which individual animals do not differ by more than ±1.0 g from the mean body weight. Different shipments of mice may be in the range of 20–32 g.

4. Injections are carried out as rapidly as possible. With practice, a 1-min interval between injections can be achieved. Injections are rotated among the experimental groups, i.e., an animal from the first group, followed by an animal from the second group, and so on.

5. Protective agents (e.g., free radical scavengers) are generally given ip 1 hr beforehand.

### 6-Hydroxydopamine Methodology

1. Convenient doses of 6-hydroxydopamine (HBr salt) for the study of the sympathetic innervation of the left atrium of the heart are in the range 4–10 mg/kg, given iv to Swiss–Webster mice. Doses may need adjustment for the study of other innervated organs (e.g., the iris or vasculature).

[8] C. Sachs and G. Jonsson, *Biochem. Pharmacol.* **24,** 1 (1975).
[9] G. Cohen, R. E. Heikkila, B. Allis, F. Cabbat, D. Dembiec, D. MacNamee, C. Mytilineou, and B. Winston, *J. Pharmacol. Exp. Ther.* **199,** 336 (1976).
[10] R. E. Heikkila, B. Winston, G. Cohen, and H. Barden, *Biochem. Pharmacol.* **25,** 1085 (1976).

2. The sympathetic (noradrenergic) innervation of an organ is evaluated by measuring the high-affinity uptake and accumulation of [³H]norepinephrine (³H-NE). Uptake can be measured either *in vivo* (after iv injection of ³H-NE), or *in vitro* with isolated atria.[9] We have tended to work *in vitro*.

3. The destruction of peripheral sympathetic nerves is rapid. Measurements can be made as early as 4 hr after 6-hydroxydopamine. For convenience, we routinely use 24 hr.

4. Mice are killed rapidly by cervical dislocation (or decapitation). The left atria are immediately removed and immersed in ice-cold Krebs–Ringer buffer. After all atria are collected, they are opened by two cuts with a pair of microdissecting scissors, and rinsed in fresh cold medium to remove entrapped blood.

5. Groups of three to four atria are placed into 10 ml of Krebs–Ringer solution contained in 25-ml Erlenmeyer flasks. The buffer composition is 118 mM NaCl, 32 mM sodium phosphate, 4.7 mM KCl, 1.8 mM CaCl₂, 1.2 mM MgSO₄, 5.6 mM glucose, 1.7 mM ascorbic acid, and 1.3 mM disodium EDTA, adjusted to pH 7.4.

6. Atria are temperature-equilibrated by shaking in a water bath at 37° for 15 min.

7. Then, 2.5 µCi of ³H-NE (in 0.1 ml) is added (specific activity 10–40 Ci/mmol; final concentration 6.25–25 nM). After 10 min at 37° with shaking, the contents of the flasks are poured through a Gooch crucible, which retains the atria.

8. The atria in the crucible are quickly rinsed by pouring through 20 ml of fresh buffer.

9. The crucible with its contents is immersed in 20 ml of fresh buffer contained in a 50-ml beaker. Reincubation at 37° with shaking for 15 min rinses the atria free from nonneuronal ³H-NE.

10. The crucible is removed from the wash fluid and the atria are quickly rinsed with 20 ml fresh buffer.

11. Individual atria are removed with a pair of forceps, briefly blotted, and then added to 3 ml absolute ethanol contained in a vial that will subsequently be used for scintillation spectrometry. Incubation with shaking at 37° for 20 min extracts the ³H from the tissue.

12. Scintillation fluid (e.g., Liquiscint) is then added to each vial and the radioactivity is measured in a scintillation spectrometer. The radioactivity retained is an index of the sympathetic innervation. The decrease in retained radioactivity compared to control animals assesses the degree of damage induced by 6-hydroxydopamine.

13. Sample data from an experiment with 6-hydroxydopamine (6-OHDA HBr salt, 10 mg/kg, iv) and phenylthiazolylthiourea (PTTU, hy-

droxyl radical scavenger, 200 mg/kp, ip, 1 hr beforehand) are presented in the tabulation below. The $^3$H-NE used to measure neuronal "uptake" had a specific activity of 9.8 Ci/mmol.

| Group | $^3$H-NE uptake (pmol/Atrium + SEM) |
|---|---|
| Control | 0.28 ± 0.02 ($N$ = 7) |
| 6-OHDA | 0.09 ± 0.01 ($N$ = 8) (68% destruction) |
| PTTU + 6-OHDA | 0.29 ± 0.02 ($N$ = 7) |
| PTTU alone | 0.30 ± 0.02 ($N$ = 7) |

14. Assessment of neuronal damage may also be made by direct visualization of the nerve plexus under the fluorescence microscope,[9] or by measuring tyrosine hydroxylase, which serves as a marker for catecholamine neurons. Changes in endogenous norepinephrine can also serve as a guide, providing that sufficient time has elapsed (e.g., 48 hr) for resynthesis of released stores of norepinephrine in intact nerve terminals.

15. Other notes: (1) Flasks containing groups of three to four atria are staggered onto the water bath (step 6) at 1-min intervals, in order to provide sufficient time for processing each flask at later steps. Rotation is made among the experimental groups. (2) It is essential that agents that protect *in vivo* be tested for their ability to simply block the entry of 6-hydroxydopamine into nerve terminals. This is readily accomplished[9] by measuring the accumulation in the left atrium of a tracer dose of $^3$H-NE (100 $\mu$Ci/kg in 0.2 ml) injected in place of 6-hydroxydopamine. The atria are removed 10 min later.

## Alloxan

Alloxan and streptozotocin are widely used as experimental tools to induce a diabetic state in laboratory animals.[1] Alloxan has been in use for approximately 40 years, while streptozotocin was introduced in the early 1960s. Both agents destroy the beta cells of the pancreas. Although dissimilar in structure and certain of their toxicological properties, recent studies have suggested that these two agents may act via a common mechanism.[11]

Alloxan is a quinoidal compound. The reduced form is dialuric acid.

[11] H. Yamamoto, Y. Uchigata, and H. Okamoto, *Biochem. Biophys. Res. Commun.* **103**, 1014 (1981).

Either alloxan or dialuric acid may be administered to induce the diabetic state; but, in practice, alloxan is the agent that is used predominantly.

Experiments with alloxan can be carried out either *in vivo* (ip or iv injection) or *in vitro* (studies with isolated pancreatic islets). The ability of injected superoxide dismutase to protect islet cells *in vivo*[12] leads to the inference that major damage, mediated by superoxide, occurs directly on or in the cellular membrane.

### Alloxan Methodology

1. Doses of alloxan for studies with Swiss-Webster mice are in the range of 25–100 mg/kg given intravenously.

2. High circulating levels of blood glucose protect against the diabetogenic action of alloxan. Hence, we routinely withhold food from animals for 3–4 hr prior to alloxan injection. Even longer periods of food deprivation, e.g., 24–48 hr, are used if other agents (e.g., free radical scavengers), given prior to alloxan, tend to elevate the blood glucose. The food is returned to the animals at 1 hr after the alloxan injection.

3. The diabetic state is conveniently and simply assessed by measuring blood glucose.[10] We generally do this at 3 days after alloxan injection; other investigators use 1–4 days.

4. Mice are decapitated and blood from the neck is collected in a convenient container (e.g., small beaker). Other techniques include sampling from the tail, heart puncture, or orbital sinus puncture.

5. To prepare samples for assay, 0.1 ml of blood is lysed in 3.9 ml water. The blood is deproteinized by the addition of 0.5 ml of 1.8% barium hydroxide, followed by 0.5 ml of 2% zinc sulfate. Samples are filtered.

6. Glucose is assayed with a glucose oxidase kit (e.g., Boehringer–Mannheim or Worthington). In, for example, the Boehringer–Mannheim method, an aliquot of filtrate (0.5 ml) is added to 4.0 ml of buffered reagent and samples are incubated at room temperature. The hydrogen peroxide formed by glucose oxidase is coupled to the oxidation of an indicator dye by horseradish peroxidase. The reaction is complete in 20 min. The absorbance of the samples is read at 520 nm and compared to glucose standards. Blank values obtained from reagents are subtracted.

7. Sample data from an experiment with alloxan (50 mg/kg, iv) and ethanol (hydroxyl radical scavenger, 4 g/kg, ip 30 min beforehand) are presented in the tabulation below. Blood glucose was measured 72 hr after alloxan.

[12] K. Grankvist, S. Marklund, and I. B. Taljedal, *Nature (London)* **294**, 158 (1981).

| Group | Blood glucose (mg/100 ml + SEM) |
|---|---|
| Control | 153 ± 4 (N = 7) |
| Alloxan | 439 ± 43 (N = 11) |
| Ethanol and Alloxan | 180 ± 21 (N = 8) |

In this experiment, food was withheld for 48 hr prior to alloxan in order to suppress a rise in blood glucose induced by ethanol. Blood glucose at the time of injection of alloxan was 137 ± 7 mg/100 ml.

8. Several other methods are available for assessing the diabetic state. (1) Insulin levels in serum or plasma can be measured by radioimmunoassay.[13,14] Commercial kits are available from a number of sources. In a representative study,[13] control mice showed plasma levels of 120 uU immunoreactive insulin per ml, while levels in diabetic mice were virtually nondetectable. (2) Histologic examination can evaluate directly the damage to pancreatic $\beta$ cells. Tissues are fixed in Zenker Formol and stained with aldehyde fuchsin. Representative specimens can be seen in published experiments.[10,15] Normal $\beta$ cells are deeply stained; diabetic mice show only a few scattered stained cells. (3) Diabetic mice respond abnormally to a glucose tolerance test.[16] They also show high water intake,[17] which can be measured, along with correspondingly increased urine volume. Marked weight loss also occurs.

[13] A. A. Like, M. C. Appel, R. M. Williams, and A. A. Rossini, *Lab. Invest.* **38**, 470 (1978).
[14] P. H. Wright, D. R. Makulu, D. Vichick, and K. E. Sussman, *Diabetes* **20**, 33 (1971).
[15] R. E. Heikkila, H. Barden, and G. Cohen, *J. Pharmacol. Exp. Ther.* **190**, 501 (1974).
[16] W. J. Riley, T. J. McConnell, N. K. Maclaren, J. V. McLaughlin, and G. Taylor, *Diabetes* **30**, 718 (1981).
[17] J. A. Cohn and A. Cerami, *Diabetologia* **17**, 187 (1979).

# [68] Microsomal Oxidant Radical Production and Ethanol Oxidation

*By* ARTHUR I. CEDERBAUM and GERALD COHEN

Rat liver microsomes generate a potent oxidant during electron transfer reactions initiated by NADPH. The oxidant species exhibits the oxidizing power of the hydroxyl radical ($\cdot$OH) and appears to be formed from $H_2O_2$ as precursor. Our interest was stimulated by the known oxidation of

Copyright © 1984 by Academic Press, Inc.
All rights of reproduction in any form reserved.
ISBN 0-12-182005-X

ethanol, a good ·OH scavenger, by liver microsomes in a NADPH-requiring reaction.[1] Although a minor pathway of ethanol oxidation compared to alcohol dehydrogenase, the microsomal system assumes greater significance at elevated concentrations of ethanol, and is inducible by chronic consumption of ethanol.[2-4] Some characteristics and the significance of the microsomal ethanol-oxidizing pathway have been described in a previous volume of this series.[5] The mechanism at the molecular level for the oxidation of ethanol by liver microsomes is not clear. Ethanol can react with ·OH to produce acetaldehyde.[6] It was previously noted that ethanol was oxidized by ·OH generated from model systems.[7] Ethanol oxidation by liver microsomes was inhibited by a series of competitive ·OH scavengers.[8-10] Typical ·OH scavengers were oxidized during NADPH-dependent electron transfer.[11,12] In the sections below, techniques to assay the generation of ·OH-like oxidant species by microsomes, and to evaluate its role in microsomal ethanol oxidation are described.

## Scavengers of ·OH and Their Reaction Products

Dimethyl sulfoxide (DMSO) reacts with ·OH to produce methyl radicals (·CH$_3$),[13,14] which can abstract hydrogen to produce methane, or dimerize to yield ethane, or react with O$_2$ to ultimately produce formaldehyde.[12,15-17]

$$(CH_3)_2SO + \cdot OH \rightarrow CH_3SOOH + \cdot CH_3$$
$$\cdot CH_3 \rightarrow CH_4, CH_3CH_3, CH_2O$$

[1] W. H. Orme-Johnson and D. M. Ziegler, *Biochem. Biophys. Res. Commun.* **21,** 78 (1965).
[2] N. Grunnet, B. Quistorff, and H. I. D. Thieden, *Eur. J. Biochem.* **40,** 275 (1973).
[3] C. S. Lieber and L. M. DeCarli, *J. Biol. Chem.* **245,** 2505 (1970).
[4] C. S. Lieber and L. M. DeCarli, *J. Pharmacol. Exp. Ther.* **181,** 279 (1972).
[5] C. S. Lieber, L. M. DeCarli, S. Matsuzaki, K. Ohnishi, and R. Teschke, this series, Vol. LII, p. 355.
[6] M. Anbar and P. Neta, *Int. J. Appl. Radiat. Isot.* **18,** 483 (1967).
[7] G. Cohen, in "Alcohol and Aldehyde Metabolizing Systems" (R. G. Thurman, H. Drott, J. R. Williamson, and B. Chance, eds.), p. 403. Academic Press, New York, 1977.
[8] A. I. Cederbaum, E. Dicker, and G. Cohen, *Biochemistry* **17,** 3058 (1978).
[9] A. I. Cederbaum, E. Dicker, E. Rubin, and G. Cohen, *Biochemistry* **18,** 1187 (1979).
[10] A. I. Cederbaum and G. Cohen, *Arch. Biochem. Biophys.* **204,** 397 (1980).
[11] G. Cohen and A. I. Cederbaum, *Science* **204,** 66 (1979).
[12] G. Cohen and A. I. Cederbaum, *Arch. Biochem. Biophys.* **199,** 438 (1980).
[13] W. J. Dixon, R. O. C. Norman, and A. L. Buley, *J. Chem. Soc.* 3625 (1964).
[14] C. Lagercrantz and S. Forshult, *Acta Chem. Scand.* **23,** 811 (1969).
[15] N. R. Brownlee, J. J. Huttner, R. V. Panganamalla, and D. G. Cornwell, *J. Lipid Res.* **18,** 635 (1977).
[16] J. E. Repine, J. W. Eaton, M. W. Anders, J. R. Hoidal, and R. B. Fox, *J. Clin. Invest.* **64,** 1642 (1979).
[17] S. M. Klein, G. Cohen, and A. I. Cederbaum, *Biochemistry* **20,** 6006 (1981).

2-Keto-4-thiomethylbutyrate (KTBA) (or an analog, methional) reacts with ·OH to produce an easily measurable product, ethylene gas.[18-20]

$$CH_3S-CH_2CH_2-CO-COOH + \cdot OH \rightarrow CH_2=CH_2 + 2CO_2 + \tfrac{1}{2}(CH_3S)_2$$

Benzoic acid can be decarboxylated upon interaction with ·OH, and can also form hydroxylated benzoic acid derivatives.[21,22] The production of $^{14}CO_2$ from [7-$^{14}C$]benzoate during decarboxylation makes this scavenger useful for the detection of ·OH.[23,24]

$$C_6H_5COOH + \cdot OH \rightarrow CO_2 + C_6H_5OH + \text{other products}$$

Primary aliphatic alcohols react with ·OH mainly at the α position to produce hydroxylalkyl radicals,[25] which subsequently give rise to aldehydes by dismutation or loss of an electron, e.g., to a ferric chelate.[26]

$$CH_3CH_2OH + \cdot OH \rightarrow CH_3 \cdot CHOH \rightarrow CH_3CHO$$

t-Butanol does not have an α hydrogen and produces either the β-hydroxylalkyl radical or the alkoxy radical upon reaction with ·OH.[25,27]

$$(CH_3)_3-C-OH + \cdot OH \rightarrow \cdot CH_2-C-(CH_3)_2-OH \quad \text{or} \quad (CH_3)_3-C-O\cdot$$

The alkoxy radical can undergo spontaneous fission to produce acetone and ·$CH_3$[28]

$$(CH_3)_3-C-O\cdot \rightarrow (CH_3)_2-CO + \cdot CH_3$$

with ·$CH_3$ ultimately producing formaldehyde.[29]

## Basic Reaction Conditions for Microsomal Radical Production and Substrate Oxidation

Rat liver microsomes are prepared by differential centrifugation, washed twice, and suspended in 125 mM KCl. Reactions are carried out at 37° in sealed 25-ml Erlenmeyer flasks in a shaking water bath. The

[18] S. F. Yang, J. Biol. Chem. **244**, 4360 (1969).
[19] C. Beauchamp and I. Fridovich, J. Biol. Chem. **245**, 4641 (1970).
[20] R. E. Heikkila, B. Winston, G. Cohen, and H. Barden, Biochem. Pharmacol. **25**, 1085 (1976).
[21] R. W. Matthews and D. F. Sangster, J. Phys. Chem. **69**, 1930 (1965).
[22] J. Hoigne and H. Bader, Science **190**, 782 (1975).
[23] A. L. Sagone, M. A. Decker, R. M. Wells, and C. DeMocko, Biochim. Biophys. Acta **628**, 90 (1980).
[24] G. W. Winston and A. I. Cederbaum, Biochemistry **21**, 4265 (1982).
[25] L. M. Dorfman and G. E. Adams, NSRDS, Natl. Bur. Standards, **46** (1973).
[26] A. Shafferman and G. Stein, Science **183**, 428 (1974).
[27] C. Walling, Acc. Chem. Res. **8**, 125 (1975).
[28] J. H. Raley, F. F. Rust, and W. E. Vaughan, J. Am. Chem. Soc. **70**, 88 (1948).
[29] A. I. Cederbaum and G. Cohen, Biochem. Biophys. Res. Commun. **97**, 730 (1980).

phosphate and pyrophosphate solutions and the water used to make all other solutions are passed through columns of Chelex 100 resin to remove contaminating metals. The standard assay medium consists of 83 m$M$ potassium phosphate, pH 7.4, 10 m$M$ sodium pyrophosphate, 10 m$M$ MgCl$_2$, 0.3 m$M$ NADP$^+$, 10 m$M$ glucose 6-phosphate, 7 units of glucose-6-phosphate dehydrogenase, and about 3 to 7 mg microsomal protein in a final volume of 3 ml. The addition of 0.1 m$M$ EDTA augments the oxidation of substrates. Convenient substrates include ethanol, 1-butanol, KTBA, DMSO, $t$-butanol, or benzoate, in the concentration range of 10 to 50 m$M$. When benzoate is used, [7-$^{14}$C]benzoate is added to a final specific activity of 5.5 $\mu$Ci/mmol. Reactions are initiated by the addition of the NADPH-generating system. At suitable time points, reactions are terminated with acid and products analyzed as described below. All values are corrected for zero-time controls, which consist of the acid added before the NADPH-generating system.

*Specific Assays*

  *Production of Acetaldehyde or 1-Butyraldehyde from Ethanol or 1-Butanol.*[8] Two different procedures can be used to measure the production of the aldehydes. Reactions can be carried out in center-well flasks containing 0.6 ml of 15 m$M$ semicarbazide HCl in 180 m$M$ potassium phosphate, pH 7.4, in the center well. Reactions are terminated by the addition of trichloroacetic acid to a final concentration of 4.5% w/v. The sealed flasks are incubated overnight at room temperature. Aliquots of the center-well contents, e.g., 0.2 ml, are diluted to 3 ml with H$_2$O and the absorbance of the aldehyde–semicarbazone complex determined at 224 nm. Standard curves can be prepared by adding known amounts of acetaldehyde or 1-butyraldehyde to zero-time controls, or an extinction coefficient of 9.4 m$M^{-1}$ cm$^{-1}$ can be used. A second procedure involving head space gas chromatography can also be used. Reactions are carried out in flasks sealed with tight-fitting serum caps. After terminating reactions by injecting perchloric acid (final concentration of 7%) through the cap, the flasks are incubated at 60° for about 30 min. A 1-ml sample of the head space is removed with a gas-tight 1-ml syringe. The syringe plunger is drawn back and forth 10 times to ensure thorough mixing of the gas phase. The sample is injected into a gas chromatograph equipped with a flame ionization detector, e.g., a Hewlett–Packard Model 5840A. A 6-ft column of Poropak N, 50–80 mesh, or of Carbowax 20M-Haloport F, 30–60 mesh, can be used to detect the aldehydes. For the Carbowax column, operating conditions are column, 50°; inlet, 100°; detector, 150°; nitrogen (carrier) flow, 35 ml/min. Under these conditions, the following retention times (minutes) are obtained: acetaldehyde, 0.40; 1-butyraldehyde, 0.80; ethanol, 1.1; and 1-butanol, 3.7. Relative peak areas are quantitated by using standard curves prepared by adding appropriate external standards to zero-time controls.

*Ethylene Production from KTBA.*[12] Reactions are carried out and terminated as described for the gas chromatographic determination of aldehydes. Samples can be left at room temperature. One-milliliter samples of head space are injected onto a 6-ft column of Poropak N, 50–80 mesh. Operating conditions are column, 60°; inlet, 190°; detector, 190°; nitrogen flow, 35 ml/min. The retention time for ethylene is 1.6 min. Relative peak areas are quantitated by comparison with suitably prepared ethylene gas standards or commercial gas mixtures (e.g., olefins in helium, Supelco Co., Bellefonte, Pa.).

*Production of Formaldehyde, Methane, and Ethane from DMSO.*[12,17] Reactions are terminated by the addition of trichloroacetic acid through the sealed serum cap. The production of methane and ethane is determined by gas chromatographic analysis of a 1-ml aliquot of the head space. Reaction conditions are identical to those described for detecting ethylene. Retention times are 0.50 min for methane and 2.2 min for ethane. Subsequently, the flasks are opened and the contents centrifuged for 10 min to remove precipitated protein. The concentration of formaldehyde in the supernatant is determined by reacting 1.5 ml of sample with 1.5 ml of Nash reagent.[30] An extinction coefficient of 8 m$M^{-1}$ cm$^{-1}$ is used to quantitate the amount of formaldehyde produced. With the microsomal system, considerably greater amounts of formaldehyde are produced compared to methane and ethane.[17] For greater sensitivity or for scaled-down procedures, a fluorometric modification of the method of Nash may be used.[17,31]

*Production of Formaldehyde*[29] *and Acetone from t-Butanol.* Reactions are carried out in center-well flasks containing semicarbazide HCl as described above for ethanol. Reactions are terminated with trichloroacetic acid and the sealed flasks incubated overnight. An aliquot of the center-well contents is removed, diluted with $H_2O$ to 3 ml, and the absorbance of the acetone–semicarbazone complex determined at 224 nm. A standard curve is prepared by adding known amounts of acetone to zero-time controls. Formaldehyde is not sufficiently volatile to be measured in this procedure. After completing the acetone measurements, the contents of the main compartment are centrifuged and the formaldehyde in the supernatant fluid is determined by reacting a 1.5-ml aliquot with 1.5 ml Nash reagent.

[30] T. C. Nash, *Biochem. J.* **55,** 416 (1953).
[31] C. Steffen and K. J. Netter, *Toxicol. Appl. Pharmacol.* **47,** 593 (1979).

MICROSOMAL OXIDATION OF ALCOHOLS AND HYDROXYL
RADICAL SCAVENGERS[a]

| Substrate | Product | Rate of product formation[b] | |
| | | − Azide | + Azide[c] |
| --- | --- | --- | --- |
| Ethanol | Acetaldehyde | 7.65 ± 0.66 | 6.44 ± 1.03 |
| 1-Butanol | 1-Butyraldehyde | 2.06 ± 0.26 | 5.24 ± 0.29 |
| KTBA | Ethylene | 0.52 ± 0.13 | 1.36 ± 0.09 |
| DMSO | Formaldehyde | 0.55 ± 0.20 | 2.50 ± 0.35 |
| t-Butanol | Formaldehyde | 0.60 ± 0.10 | 2.15 ± 0.25 |
| | Acetone | 0.35 ± 0.10 | 1.75 ± 0.20 |
| Benzoate | $^{14}CO_2$ | 0.25 ± 0.03 | 1.35 ± 0.20 |

[a] Results are from four to eight experiments in the presence
of 0.1 m$M$ EDTA.
[b] Rates are expressed as nanomoles of product per minute
per milligram microsomal protein.
[c] Final concentration of azide was 1.0 m$M$.

$^{14}CO_2$ *Production from [7-$^{14}$C]Benzoate.*[24] Reactions are carried out in
flasks fitted with gas-sealing rubber caps containing hanging plastic center
wells. After terminating the reaction with perchloric acid, 0.3 ml of hy-
amine hydroxide is added to the center well with a syringe directly
through the rubber caps. The flasks are allowed to incubate for 60 min to
absorb the $CO_2$ into the center well. Fitting the center wells with a piece of
fluted filter paper increases the surface area available to absorb the $CO_2$.
The center wells are removed, placed into vials containing 10 ml of
Econofluor, and shaken vigorously. Radioactivity is counted in a liquid
scintillation counter with automatic quench control.

*Sample Data*

The table shows that in the presence of an NADPH-generating system, rat liver microsomes catalyze the oxidation of primary aliphatic
alcohols such as ethanol and 1-butanol, as well as a variety of ·OH scavengers. The reactions are linear for about 15 min with ethanol, 1-butanol,
or KTBA as substrates, and for at least 30 min with the other substrates.
With the exception of ethanol, azide increases the rate of formation of
product by 2.5- to 5.5-fold. This augmentation may reflect an accumulation of $H_2O_2$ in the system when catalase (which is present as a contaminant) is inhibited by azide. Ethanol, unlike 1-butanol, is a substrate for the
peroxidatic activity of catalase; therefore, the addition of azide may lower
the catalase pathway of ethanol oxidation, while augmenting the alternate

pathway (·OH-dependent). There is little or no product formation when microsomes, the substrates, or any component of the NADPH-generating system are omitted from the system. $H_2O_2$, added in the presence of azide, cannot replace the NADPH-generating system in supporting the oxidation of alcohols and of ·OH scavengers. Studies have shown that there is cross-competition among ethanol and the ·OH scavengers for oxidation by the microsomes, e.g., benzoate inhibits the oxidation of ethanol and ·OH scavengers, and in turn, the oxidation of benzoate is inhibited by ethanol and ·OH scavengers.

Endogenous iron in the microsomes may be necessary to catalyze the production of ·OH. The addition of EDTA augments the oxidative activity. The addition of iron-EDTA produces a further increase in oxidative activity.[32] The iron chelating agent, desferrioxamine, inhibits the stimulation induced by the addition of EDTA or iron-EDTA.[33] By contrast, the activity of aminopyrine demethylase, a typical mixed-function oxidase activity, is not affected by EDTA, iron-EDTA, or desferrioxamine.

The microsomal metabolism of ethanol shares many similarities with that of the ·OH scavengers. At least part of the mechanism underlying the oxidation of ethanol by microsomes can be ascribed to the interaction with ·OH generated from microsomal electron transfer. However, some differences between the metabolism of ethanol and ·OH scavengers exist, e.g., sensitivity to desferrioxamine,[33–35] effect of organic hydroperoxide.[24,35] Results described elsewhere[33–35] may indicate the existence of two pathways for the NADPH-dependent microsomal oxidation of ethanol, one pathway which involves ·OH and another pathway which appears to be independent of oxygen radicals.

Current evidence indicates that the oxidation of various scavengers by liver microsomes is mediated by a reactive species with the oxidizing power of the hydroxyl radical. $H_2O_2$ appears to be a precursor and iron is also required. Microsomal electron transfer serves to reduce iron to the ferrous state, as well as to generate $H_2O_2$. As described elsewhere, the mixed-function oxidase pathway can be dissociated from the pathway that oxidizes ·OH scavengers. The nature of the iron catalyst in microsomes or in intact cells remains to be determined.

[32] A. I. Cederbaum, E. Dicker, and G. Cohen, *Biochemistry* **19**, 3695 (1980).
[33] A. I. Cederbaum and E. Dicker, *Biochem. J.* **210**, 107 (1983).
[34] G. W. Winston and A. I. Cederbaum, *J. Biol. Chem.* **258**, 1508 (1983).
[35] G. W. Winston and A. I. Cederbaum, *J. Biol. Chem.* **258**, 1514 (1983).

# [69] Exacerbation of Superoxide Radical Formation by Paraquat*

*By* HOSNI M. HASSAN

Paraquat ($PQ^{2+}$; 1,1'-dimethyl-4,4'-bipyridinium dichloride), also known as methyl viologen, is the active ingredient of many commercially available broad-spectrum herbicides. Paraquat seems to be universally toxic both in prokaryotes and eukaryotes and incidents of fatal paraquat poisonings have been reported in man and animals.

Paraquat is readily reduced by a single electron to a stable but dioxygen-sensitive monocation radical ($PQ^{\ddot{+}}$). The reaction between the paraquat radical and dioxygen ($O_2$) generates the true toxic species, the superoxide radical ($O_2^-$) and subsequently hydrogen peroxide ($H_2O_2$).[1] Hydroxyl radicals (OH·) may also be generated as a consequence of secondary interactions between $O_2^-$ and $H_2O_2$. In this chapter methods for measuring and identifying the partially reduced oxygen species generated during the reaction of $PQ^{\ddot{+}}$ and $O_2$ are presented.

## Methods for Generation of $PQ^{\ddot{+}}$

### Pulse Radiolysis

Irradiation of a $10^{-4}$ $M$ aqueous solution of paraquat dichloride with < 0.1 sec, 600-rad pulses yields $PQ^{\ddot{+}}$. The second-order rate constant for the reaction of $e_{aq}^-$ with paraquat[2] is $8.4 \times 10^{10}$ $M^{-1}$ $sec^{-1}$. Irradiation of aqueous solutions usually generates OH· which can react with $PQ^{\ddot{+}}$; therefore, it is customary to perform the irradiation in 0.1 $M$ sodium formate to eliminate this side reaction. The yield of $PQ^{\ddot{+}}$ is doubled,[2] moreover, because formate reacts with OH· to generate $CO_2^{\ddot{-}}$ which is a reducing species. The $g$ value for $PQ^{\ddot{+}}$ formed in aqueous formate is 5.9,[2] and $\Delta\varepsilon_{650\,nm} = 1.33 \times 10^4$.

* Paper Number 8792 of the Journal Series of the North Carolina Agricultural Research Service, Raleigh, NC 27650. The use of trade names in this publication does not imply endorsement by the North Carolina Agricultural Research Service of the products named, nor criticism of similar ones not mentioned.

[1] A. Calderbank, *Adv. Pest Control. Res.* **8**, 127 (1968).
[2] J. A. Farrington, M. Ebert, E. J. Land, and K. Fletcher, *Biochim. Biophys. Acta* **314**, 372 (1973).

Copyright © 1984 by Academic Press, Inc.
All rights of reproduction in any form reserved.
ISBN 0-12-182005-X

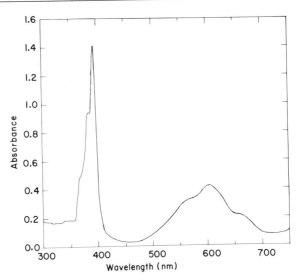

FIG. 1. Optical spectrum of $PQ^{+}$ produced by the reduction of $3 \times 10^{-5}$ $M$ $PQ^{+}$, dissolved in 0.05 $N$ NaOH, with sodium dithionite in the absence of air.

## Chemical Reduction

Paraquat radical may be formed by chemical reduction with zinc or alkaline sodium dithionite.[3]

*Procedure.* Dissolve $PQ^{2+}$ (30 $\mu M$) in 0.05 $N$ NaOH and place in a stoppered cuvette. Add a small amount of sodium dithionite and quickly stopper to protect from reoxidation by $O_2$. Immediately read at 605 nm or scan between 300 and 750 nm (Fig. 1).

## Enzymic Reduction

Crude cell-free extracts from plants, animals, and bacterial origins have been shown to reduce $PQ^{2+}$ to $PQ^{+}$ in the presence of NADPH or an NADPH-generating system (i.e., glucose 6-phosphate, glucose-6-phosphate dehydrogenase, and $NADP^{+}$). The enzyme that reduces $PQ^{2+}$ is a diaphorase-like enzyme usually present in the cytoplasm and specific for NADPH.[4,5]

*Procedure.* Prepare cell-free extracts from *Escherichia coli* B (or other source) growing in the logarithmic phase. In a glass-stoppered cuvette add

[3] J. G. Carey, J. F. Cairns, and J. E. Colchester, *J. Am. Chem. Soc. D,* 1280 (1969).
[4] R. C. Baldwin, A. Pasi, J. T. MacGregor, and C. H. Hine, *Toxicol. Appl. Pharmacol.* **32,** 298 (1975).
[5] H. M. Hassan and I. Fridovich, *J. Biol. Chem.* **253,** 8143 (1978).

0.1 m$M$ PQ$^{2+}$, 0.4 m$M$ NADPH in 2.9 ml of 50 m$M$ potassium phosphate pH 7.8. Start the reaction by adding 0.1 ml (~200 $\mu$g protein) of the cell-free extract and seal the cuvette with the glass stopper to exclude O$_2$. Follow the increase in absorbance at 605 nm. Admit air into the cuvette and observe the rapid regeneration of the colorless parent compound, PQ$^{2+}$.

### Reactions of PQ$^+$ with Dioxygen

PQ$^+$ reacts with dioxygen to generate O$_2^-$ at a rate constant equal to 7.7 × 10$^8$ $M^{-1}$ sec$^{-1}$ [Eq. (1)].[2] PQ$^+$ can further react with O$_2^-$ to generate H$_2$O$_2$ at a rate constant equal to 6.5 × 10$^8$ $M^{-1}$ sec$^{-1}$ [Eq. (2)].[2] It can also react with H$_2$O$_2$ to generate OH· [Eq. (3)], which in turn may oxidize PQ$^+$ to regenerate PQ$^{2+}$ [Eq. (4)].

$$PQ^+ + O_2 \rightarrow PQ^{2+} + O_2^- \tag{1}$$
$$PQ^+ + O_2^- + 2H^+ \rightarrow PQ^{2+} + H_2O_2 \tag{2}$$
$$PQ^+ + H_2O_2 \rightarrow PQ^{2+} + OH^- + OH· \tag{3}$$
$$PQ^+ + OH· \rightarrow PQ^{2+} + OH^- \tag{4}$$

The extent of these reactions depends on the concentrations of PQ$^+$ and dioxygen. Oxygen limitation has been shown[2] to favor the reaction between PQ$^+$ and O$_2^-$.

### Methods for Detection and Measurement of PQ$^{2+}$-Mediated Oxy-Radical Formation

#### Cyanide-Insensitive Respiration

The generation of O$_2^-$ in living cells during aerobic metabolism has been demonstrated by EPR using spin traps.[6–8] However, measurements of the amounts of O$_2^-$ made *in vivo* have been hampered by the ubiquity of superoxide dismutases. This problem has been circumvented in *Streptococcus faecalis* by measuring O$_2^-$ production using cell-free extracts in the presence of an SOD inhibiting antibody.[9] Recent studies in *E. coli*[10,11] and *Salmonella typhimurium* (unpublished data) have indicated the utility of cyanide-insensitive respiration as an approximate index of O$_2^-$ plus H$_2$O$_2$ production. This practical index is based on the assumption that

[6] R. Schmid, *FEBS Lett.* **60**, 98 (1975).
[7] A. N. Saprin and L. H. Piette, *Arch. Biochem. Biophys.* **180**, 480 (1977).
[8] E. J. Rauckman, G. M. Rosen, and B. B. Kitchell, *Mol. Pharmacol.* **15**, 131 (1979).
[9] L. Britton, D. P. Malinowski, and I. Fridovich, *J. Bacteriol.* **134**, 229 (1978).
[10] H. M. Hassan and I. Fridovich, *J. Biol. Chem.* **252**, 7667 (1977).
[11] H. M. Hassan and I. Fridovich, *Arch. Biochem. Biophys.* **196**, 385 (1979).

oxygen consumption by cells can be divided into two categories. One, which is due to the reduction of $O_2$ to $H_2O$ by the action of a cytochrome oxidase which is inhibited by cyanide and the second, which is due to all other oxygen-consuming reactions that may involve the generation of $O_2^-$ and $H_2O_2$ which are insensitive to cyanide. This arbitrary division has been tested and found to be useful.[10-12]

In determining the rate of cyanide-insensitive respiration in response to a given concentration of paraquat, it should be realized that $O_2^-$ will rapidly dismute to $H_2O_2$ [Eq. (5)] either enzymically or spontaneously. Therefore, the rate of cyanide-insensitive respiration is one-half the rate of $O_2^-$ actually generated by paraquat.

$$O_2^- + O_2^- + 2H^+ \rightarrow H_2O_2 + O_2 \tag{5}$$

Furthermore, the addition of cyanide inhibits the ubiquitous heme catalases, normally present in aerobic cells, that would prevent the accumulation of $H_2O_2$ by converting it to $\frac{1}{2}O_2$ plus $H_2O$.

## Materials

Paraquat (1–100 m$M$)

E. coli B, $B_{12}^-$ (ATCC 29682), or any similar test organism

Glucose minimal medium[13] containing, per liter, $MgSO_4 \cdot 7H_2O$, 0.2 g; citric acid $\cdot H_2O$, 2.0 g; $K_2HPO_4$, 10 g; $NaNH_4PO_4 \cdot 4H_2O$, 3.5 g; vitamin $B_{12}$, 1.0 mg; and glucose, 5.0 g adjusted to pH 7.0

Trypticase soy-yeast extract (TSY) medium containing 3% trypticase soy broth and 0.5% yeast extract (both from Baltimore Biological Laboratories)

50 m$M$ potassium phosphate containing 0.1 m$M$ $MgSO_4$, pH 7.0

Rotary shaker water bath at 37°

Gilson 5 ml water-jacketed cell fitted with a Clark-type polarographic electrode and a strip chart recorder

Procedure. Grow the test organism overnight in the glucose minimal medium. Dilute the culture to an initial $OD_{600}$ of 0.2 in the TSY medium. Allow growth at 37° and 200 rpm to the midlogarithmic phase ($OD_{600} \simeq$ 3.0). Collect the cells by centrifugation at 4° for 15 min at 10,000 $g$. Resuspend the cells in the phosphate-$Mg^{2+}$ buffer to an $OD_{600} \simeq$ 7–9 and keep on ice until used (use the same day). Under these conditions 1 mg dry weight of E. coli (ATCC 29682) cells per ml is equal to 3.57 $OD_{600}$ units.[10] For measuring oxygen uptake and paraquat-mediated cyanide-insensitive respiration, prepare a reaction mixture containing glucose or

[12] C. S. Moody and H. M. Hassan, Proc. Natl. Acad. Sci. U.S.A. **79**, 2855 (1982).

[13] H. J. Vogel and D. M. Bonner, J. Biol. Chem. **218**, 97 (1956).

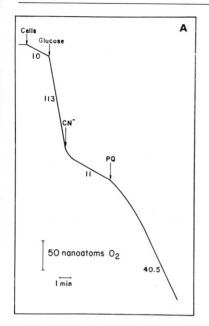

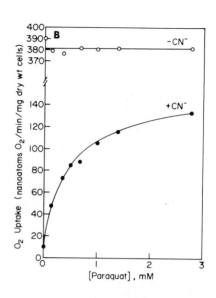

FIG. 2. Effects of $PQ^{2+}$ on the rate of cyanide-resistant respiration in *E. coli* B. (A) Trace showing the changes in oxygen concentration upon adding potassium phosphate/$Mg^{2+}$, 50 m$M$/0.1 m$M$; cells, 170 μg/ml; glucose, 0.5%; NaCN, 1 m$M$; $PQ^{2+}$, 6 m$M$ in a 5 ml total volume. Numbers shown on the tracing indicate the slopes expressed as nanoatoms of $O_2$/min. From H. M. Hassan and I. Fridovich, *J. Biol. Chem.* **254**, 10846 (1979). (B) Effects of $PQ^{2+}$ concentrations on total oxygen uptake and on the rate of cyanide-insensitive respirations. Conditions as in (A). From H. M. Hassan and I. Fridovich, *J. Biol. Chem.* **252**, 7667 (1977).

any suitable carbon source (0.5%), phosphate-$Mg^{2+}$ buffer to a total volume of 4.9 ml. Equilibrate the Gilson oxygraph cell using a circulating water bath set at 36–37°. Start the reaction (oxygen uptake) by adding 0.05 ml of the cell suspension (final concentration ~ 100–200 μg/ml). Record the rate of oxygen consumption for 1–2 min, then add cyanide (1.0 m$M$) and observe the inhibition of $O_2$ uptake. After a steady rate is established add the desired concentration of $PQ^{2+}$ (0.1–5.0 m$M$) and observe the increase in $O_2$ uptake (Fig. 2A). Calculate the rate of oxygen consumption as nanoatoms of $O_2$ per minute per milligram dry weight of cells. Figure 2B demonstrates the effects of different concentrations of $PQ^{2+}$ on the rates of total oxygen consumption and cyanide-insensitive respiration in *E. coli* B.[10,14]

[14] H. M. Hassan and I. Fridovich, *J. Biol. Chem.* **254**, 10846 (1979).

## Oxygen-Dependent $PQ^{2+}$ Toxicity

If the toxic effect of $PQ^{2+}$ is due to its ability to mediate the generation of $O_2^-$, $H_2O_2$, and $OH\cdot$, then the presence of dioxygen should be an obligatory requirement. Indeed, this is the case in bacteria,[5,12] plants,[2] and animals.[4]

*Procedure.* Grow the test organism overnight in the glucose minimal medium as indicated under Cyanide-Insensitive Respiration. Dilute the culture about 20-fold (an initial $OD_{600} \simeq 0.05-0.1$) in a prewarmed fresh glucose minimal medium containing different concentrations of $PQ^{2+}$ (0–1.0 m$M$). Incubate at 37° and 200 rpm. Follow the growth for ~6 hr at 20- to 30-min intervals by measuring $OD_{600}$. Determine the effect of anaerobiosis by repeating the same experiment using anaerobic cuvettes with a special side arm. Sterilize the cuvettes by using 90% ethanol or any suitable sterilant. To the cuvettes, aseptically add ~3 ml of sterile glucose minimal medium containing the desired concentrations of paraquat. Add enough inocula (initial $OD_{600} \simeq 0.1$) in the side arms. Flush the cuvettes for 2 hr with a stream of $N_2$ that has been passed over a column of hot colloidal copper[15] to remove traces of $O_2$. Seal the cuvettes tightly, making sure not to introduce any air. Tip the inocula from the side arms, incubate the cuvettes at 37°, and follow the growth at 600 nm. Plot the data as in Fig. 3. Notice that 1 m$M$ $PQ^{2+}$ inhibits growth aerobically while it has no effect anaerobically.[5,12]

## Induction of SOD by $PQ^{2+}$

*Escherichia coli* and *S. typhimurium* contain two superoxide dismutases. One contains iron (Fe-SOD) and appears to be constitutive[16] and the second contains manganese (Mn-SOD) and is inducible by the flux of $O_2^-$ inside the cells.[10–12,16,17] A positive correlation is seen[10] between the rate of $PQ^{2+}$-mediated cyanide-insensitive respiration and the concentration of Mn-SOD synthesized by the cells in response to the same concentration of $PQ^{2+}$. This fact, that Mn-SOD is induced by its substrate, $O_2^-$, supports the superoxide theory of oxygen toxicity.[18]

*Procedure.* Grow the test organism (*E. coli* B or other) aerobically overnight in the glucose minimal medium as indicated under Cyanide-Insensitive Respiration. Use this culture to inoculate TSY media contain-

[15] F. B. Meyer and G. Ronge, *Z. Angew. Chem.* **52,** 637 (1939).

[16] H. M. Hassan and I. Fridovich, *J. Bacteriol.* **129,** 1574 (1977).

[17] H. M. Hassan and I. Fridovich, *in* "Enzymatic Basis of Detoxication" (W. B. Jakoby, ed.), Vol. 1, p. 311. Academic Press, New York, 1980.

[18] J. M. McCord, B. B. Keele, Jr., and I. Fridovich, *Proc. Natl. Acad. Sci. U.S.A.* **86,** 1024 (1971).

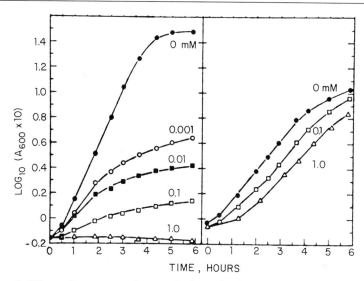

FIG. 3. Effects of paraquat on the growth of *E. coli* B in glucose minimal medium in the presence (left) and absence (right) of oxygen. From H. M. Hassan and I. Fridovich, *J. Biol. Chem.* **253**, 8143 (1978).

ing paraquat (0–5.0 m*M*) to an initial $OD_{600}$ of 0.2. Incubate the cultures aerobically at 37° and 200 rpm for 3 to 4 hr. At the end of the growth period, measure the final $OD_{600}$. This will provide an indication of the extent of growth inhibition by $PQ^{2+}$. Harvest the cells by centrifugation at 10,000 *g* for 20 min. Resuspend the cells in 50 m*M* potassium phosphate, 0.1 m*M* EDTA, pH 7.8 ($KP_i$/EDTA buffer) and disrupt by sonication[16] at 4°. Remove unbroken cells and particulates by centrifugation at 35,000 *g* for 1 hr. Dialyze the supernatant fraction against the $KP_i$/EDTA buffer. Determine the total SOD activity using the cytochrome *c* assay.[19,20] For determining the relative amounts of the SOD isozymes, aliquots of the soluble extracts estimated to contain 3.0 units of total SOD should be applied to 10% polyacrylamide gels.[10,16] After electrophoresis, stain the gels for SOD activities[21] and subject them to linear scanning densitometry. Estimate the relative amount of the SOD isozymes from the areas under the densitogram troughs[16] and calculate the absolute amounts of each isozyme from the total activity determined in the extract by the cytochrome *c* method. Plot the data as in Fig. 4.

[19] J. M. McCord and I. Fridovich, *J. Biol. Chem.* **244**, 6079 (1969).
[20] J. D. Crapo, J. M. McCord, and I. Fridovich, this series, Vol. 53, p. 283.
[21] C. Beauchamp and I. Fridovich, *Anal. Biochem.* **44**, 276 (1971).

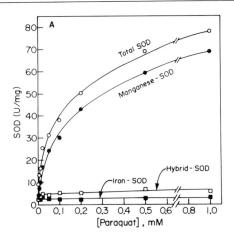

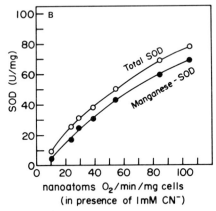

Fig. 4. (A) Effects of varying the concentration of $PQ^{2+}$ on the synthesis of the different SOD isozymes. (B) Correlation between the rate of cyanide-insensitive respiration and the synthesis of Mn-SOD as both influenced by paraquat concentration. From H. M. Hassan and I. Fridovich, *J. Biol. Chem.* **252**, 7667 (1977).

## Measurements of $PQ^{2+}$-Mediated $O_2^-$ Production in Cell-Free Extracts

*E. coli* contains a soluble diaphorase, which catalyzes the reduction of paraquat to $PQ^{+}$ by NADPH.[14] The generation of $O_2^-$ by $PQ^{2+}$ is dependent on NADPH as the electron donor and on the diaphorase as the catalyst. The rate of $O_2^-$ generated by a given concentration of $PQ^{2+}$ can be estimated by measuring the rate of SOD-inhibitable ferricytochrome *c* reduction as indicated in this volume [53]. Since the crude cell-free ex-

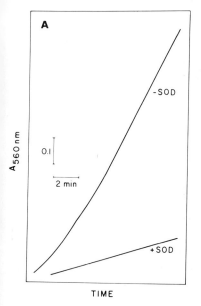
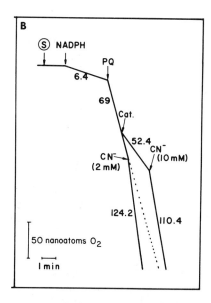

TIME

Fig. 5. Paraquat and the generation of $O_2^-$. (A) Measurements of $O_2^-$ generation in terms of SOD-inhibitable NBT reduction. Conditions are listed in the text. (B) Paraquat-dependent oxygen uptake in the presence of cell-free extracts and NADP. From H. M. Hassan and I. Fridovich, *J. Biol. Chem.* **254,** 10846 (1979).

tracts, which are used as a source of the diaphorase enzyme also contain SOD, the rate of $O_2^-$ measured is an underestimate. It is, therefore, recommended to use cell-free extracts that have a very low concentration of SOD, or to use partially purified extracts where the SOD activities are removed.[20] Alternately, the rate of $O_2^-$ generation can be determined by measuring the rate of oxygen consumption by the cell-free extracts in the presence of NADPH and $PQ^{2+}$. Under these conditions SOD has no effect on the rate. It should be realized, however, that the rate of observed $O_2$ uptake is only half the rate of $O_2^-$ generated ($2O_2^- + 2H^+ \rightarrow H_2O_2 + O_2$). The presence of catalases will break down $H_2O_2$ to $\frac{1}{2}O_2$ and $H_2O$, but catalases may be inhibited by adding 1 m$M$ NaCN.

*Procedure.* The ability of paraquat to generate $O_2^-$ may be estimated in terms of SOD-inhibitable reduction of cytochrome $c$ or of nitro blue tetrazolium (NBT) (Fig. 5A). Prepare dialyzed cell-free extracts (as indicated under Induction of SOD by $PQ^{2+}$) from cells that have been grown under conditions that result in a minimal SOD content (i.e., anaerobic conditions or in a glucose minimal medium). Add 50 $\mu$l of the cell-free extracts to a reaction mixture containing: NADPH, 0.4 m$M$; ferricytochrome $c$, $2.5 \times 10^{-5}$ $M$ (or NBT, $10^{-4}$ $M$); paraquat, 50 $\mu M$; ± bovine Cu,Zn-SOD, 20 $\mu$g/ml; and potassium phosphate, 50 m$M$ pH 7.8 containing 0.1 m$M$ EDTA in a total volume of 3 ml. Follow the reduction of cytochrome $c$ at 550 nm (NBT at 560 nm) at 25°.

The alternate procedure is to measure the rate of paraquat-dependent oxygen uptake. Prepare dialyzed cell-free extracts (as indicated under Induction of SOD by PQ$^{2+}$). Using the Clark-type polarographic electrode for measuring O$_2$ consumption add the following reactants: cell-free extracts, 440 $\mu$g protein/ml; NADPH, 0.25 m$M$; paraquat, 0–5.0 m$M$; potassium phosphate, 50 m$M$ (pH 7.0) in a total volume of 2–5 ml. Record the rate of oxygen consumption after each addition (Fig. 5B). Notice that adding NaCN (2 m$M$) resulted in almost double the rate of oxygen-uptake measured in its absence (i.e., 69 vs 124 nanoatoms O$_2$/min). This is due to the inhibition of catalase. Furthermore, adding extra catalase (7 $\mu$g/ml) decreased the rate from 69 to 52.4 nanoatoms of O$_2$/min. This effect of catalase is removed by adding excess cyanide (10 m$M$) (Fig. 5B). Notice that the actual rate of O$_2^-$ generation is twice the rate of oxygen consumption observed in the presence of cyanide (i.e., 124 × 2 = 248 nanoatoms of O$_2^-$/min/the amount of protein added).

### Acknowledgments

This work was supported, in part, by the NIH Biomedical Research Support Grant 88204, and NSF Grant PCM-8213853. I wish to thank Irwin Fridovich and Carmella S. Moody for critically reading this manuscript, and Ann Farmer for expert typing.

# [70] Ethidium Binding Assay for Reactive Oxygen Species Generated from Reductively Activated Adriamycin (Doxorubicin)

By J. WILLIAM LOWN

The anthracycline glycoside antibiotics and antitumor agents are characterized by the presence of a four-ring quinonoid chromophore to which is attached one or more amino sugar moieties (Fig. 1).[1] The anthracyclines having the greatest current clinical interest include daunorubicin, isolated

[1] F. Arcamone, in "Topics in Antibiotic Chemistry" (P. G. Sammes, ed.), Vol. 2, p. 99. Wiley, New York, 1978.

Copyright © 1984 by Academic Press, Inc.
All rights of reproduction in any form reserved.
ISBN 0-12-182005-X

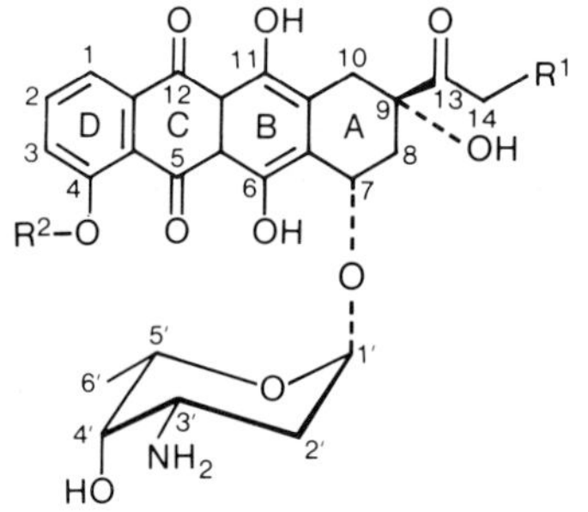

$R^1 = H$, $R^2 = Me$, Daunorubicin
$R^1 = OH$, $R^2 = Me$, Adriamycin

Fig. 1. Structural formulas of daunorubicin and adriamycin (doxorubicin).

from *Streptomyces peucetius*[2] and adriamycin (doxorubicin) isolated from *Streptomyces peucetius* var. *caesius*.[3] The antitumor properties of the anthracyclines including adriamycin, which shows the broadest activity clinically,[1,4] correlate with their capacity to bind intercalatively to duplex DNA, although other cellular effects appear to be significant.[1,4] The administration of adriamycin and daunorubicin is accompanied by severe risk of cardiotoxicity.[5,6] Growing evidence[7–12] indicates that this side effect may be related to factors including (1) the accumulation of the drug in cardiac tissue,[1,4] (2) reductive enzymatic activation[13] at the chromophore which in the presence of oxygen produces reactive oxygen species,[14,15]

[2] A. Di Marco, M. Gaetani, P. Orezzi, B. M. Scarpinato, R. Silvestrini, M. Soldati, T. Dasdia, and L. Valentini, *Nature (London)* **201**, 706 (1964).
[3] A. Di Marco, M. Gaetani, and B. M. Scarpinato, *Cancer Chemother. Rep.* **53**, 33 (1969).
[4] F. Arcamone, "Doxorubicin Anticancer Antibiotics." Academic Press, New York, 1981.
[5] G. Bonadonna and S. Monfardini, *Lancet* **1**, 837 (1969).
[6] B. Smith, *Br. Heart J.* **31**, 607 (1969).
[7] E. Bachmann, E. Weber, and G. Zbinden, *Agents Actions* **5**, 383 (1975).
[8] G. Zbinden, E. Bachmann, and C. Holdnegger, *Antibiot. Chemother.* **23**, 255 (1978).
[9] N. R. Bachur, S. L. Gordon, and M. V. Gee, *Mol. Pharmacol.* **13**, 901 (1977).
[10] J. W. Lown, H. H. Chen, J. A. Plambeck, and E. M. Acton, *Biochem. Pharmacol.* **31**, 575 (1982).
[11] J. W. Lown, H. H. Chen, J. A. Plambeck, and E. M. Acton, *Biochem. Pharmacol.* **28**, 2563 (1979).
[12] T. Kishi, T. Watanabe, and K. Folkers, *Proc. Natl. Acad. Sci. U.S.A.* **73**, 4653 (1976).
[13] T. Komiyama, T. Oki, and T. Inui, *J. Antibiot.* **32**, 1219 (1979).
[14] J. W. Lown, S. K. Sim, K. C. Majumdar, and R. Y. Chang, *Biochem. Biophys. Res. Commun.* **76**, 705 (1976).
[15] V. Berlin and W. A. Haseltine, *J. Biol. Chem.* **256**, 4747 (1981).

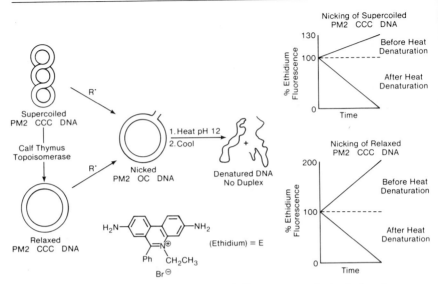

Fig. 2. Schematic description of ethidium binding assays employing both native super-coiled and relaxed PM2 covalently closed circular DNAs for the detection and comparison of free radical-induced DNA scission produced by anthracyclines reduced *in situ*.

(3) concomitant oxidative damage of membrane lipids,[16] and (4) the diminished levels in the heart of enzymes which effect protection against cellular oxidative lesions.[17]

Assays designed to detect and quantify the production of individual reactive oxygen species from activated anthracyclines may therefore be useful in comparing drugs designed to alleviate this serious clinical limitation of cardiotoxicity.[10,11]

*Assay Method*

*Principle.* Advantage is taken of the extreme sensitivity of supercoiled covalently closed circular (CCC) DNAs to oxidative lesions to detect the generation of reactive free radicals especially when used in conjunction with certain enzymes involved in cellular protection, growth and repair. A planar chromophore having dimensions compatible with the Watson–Crick duplex may insert between adjacent base pairs and be held there by van der Waals, stacking, and electrostatic forces. Ethidium bromide (Fig. 2) binds intercalatively with little apparent sequence selectivity (binding constant $K_b \approx 10^5–10^7 \ M^{-1}$) to duplex DNA and in the process shows an

[16] J. Goodman and P. Hochstein, *Biochem. Biophys. Res. Commun.* **77**, 797 (1977).
[17] J. H. Doroshow, G. Y. Locker, and C. E. Myers, *J. Clin. Invest.* **65**, 128 (1980).

enhancement in its fluorescence intensity (excitation 525 nm, emission 600 nm) of ~25-fold[18] as a result of reduced proton exchange leading to a longer lifetime for the excited state.[19] Native negatively supercoiled PM2 CCC DNA binds a certain amount of ethidium and shows a fluorescence reading arbitrarily assigned a value of 100 (Fig. 2). The action of reactive free radicals, generated from an activated antibiotic such as adriamycin, causes nicking of the strands of the CCC DNA leading to the open circular (OC) form.[14,15] Since the latter is not subject to topological constraints (unlike the initial CCC form) it can consequently accept more ethidium. This results in a 30–38% increase in the fluorescence reading of the DNA–ethidium complex depending on the superhelix density of the preparation of PM2 CCC DNA.[20] Heating of the PM2 OC DNA sample at 96° for 3–4 min in an assay medium at pH 11.8 followed by rapid cooling to 25° results in denaturation and the high pH ensures removal of small regions of accidental self-complementarity in the resulting single-stranded DNA.[20] Consequently when each of the original CCC DNA molecules has been nicked at least once all potential duplex intercalation sites for ethidium are removed and the fluorescence reading in the assay will be zero. In contrast in the control experiment on PM2 CCC DNA, or in the case of unreacted CCC DNA, heating at 96° for 3 to 4 min and cooling to 25° results in renaturation owing to the aforementioned topological constraints. Ideally the fluorescence reading of the CCC DNA–ethidium complex should return to 100% on the arbitrary scale (Fig. 2). In practice the small percentage of the OC form (typically 5 to 16% depending on the preparation) will be revealed by a proportionate decrease in fluorescence reading (Fig. 3). Therefore the structural integrity of the initial PM2 CCC DNA may be quantified and the determined proportion of OC DNA taken into account when assaying the reactivity of the anthracycline (q.v.). The topological relationships between the CCC and OC DNAs imply that at a given time point in the assay the rise in fluorescence before heat denaturation is approximately one-third of the corresponding decrease in fluorescence reading after heat denaturation. This redundant information serves as an internal control and deviations from this behavior may be used to detect concomitant molecular events involving the DNA.[21,22]

The sensitivity of the assay may be increased by first relaxing the supercoils of the DNA with a topoisomerase enzyme.[20] During this pro-

[18] J. B. Le Pecq and C. Paoletti, *J. Mol. Biol.* **27,** 87 (1967).
[19] J. Olmsted and D. R. Kearns, *Biochemistry* **16,** 3647 (1977).
[20] A. R. Morgan, J. S. Lee, D. E. Pulleyblank, N. L. Murray, and D. H. Evans, *Nucleic Acids Res.* **7,** 547, 571 (1979).
[21] J. W. Lown and L. W. McLaughlin, *Biochem. Pharmacol.* **28,** 1631 (1979).
[22] J. W. Lown, L. W. McLaughlin, and Y. M. Chang, *Bioorg. Chem.* **7,** 97 (1978).

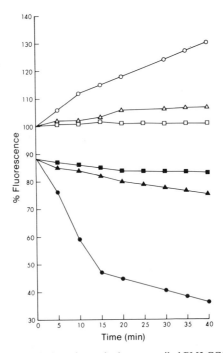

FIG. 3. Single-stranded scission of negatively supercoiled PM2 CCC DNA by adriamycin reduced *in situ* and its selective inhibition. Reactions were performed at 37° in 40 m$M$ potassium phosphate buffer, pH 7.0, and contained 1.02 $A_{260}$ units/ml of PM2 CCC DNA (91% CCC), $1 \times 10^{-4}$ $M$ adriamycin, and $3.3 \times 10^{-3}$ $M$ sodium borohydride. The before-heat denaturation fluorescence readings are shown as open symbols and the closed symbols are fluorescence readings after the denaturation at 96° and rapid cooling to 25°. Additional components were ($\triangle$) $5 \times 10^{-5}$ g/ml catalase or superoxide dismutase; ($\square$) 0.5 $M$ sodium benzoate or mannitol; ($\bigcirc$) none.

cess the number of intercalation sites for ethidium is decreased by 30–38% (again depending on the average superhelix density of the initial preparation of PM2 DNA). The relaxation process can be monitored by the change in fluorescence with time. The topological relationship between relaxed PM2 CCC DNA and OC DNA requires that nicking of each DNA molecule at least once now corresponds to a relative 100% increase in fluorescence reading of the DNA–ethidium complex in the pH 11.8 assay medium.[20,21] When using this relaxed PM2 CCC DNA the sensitivity of the assay is such that one chemical event may be detected in a DNA of the equivalent of $6 \times 10^7$ daltons.

A particular advantage of this assay is that by adding selective enzymes and chemical inhibitors one may identify individual reactive oxygen species generated from the anthracycline by observing their inhibitory effect on the DNA scission process thereby permitting the establishment of a chemical mechanism for the process.[14] These enzymes include superoxide dismutase (for $O_2^-$) and catalase (for $H_2O_2$) or glutathione peroxidase. It is necessary when employing these enzymes in conjunction with the assay to perform controls with the heat-denatured enzymes to ensure any inhibition is not due to nonspecific protein effects.[14,20] Similarly the accelerating effect due to specific metal ion catalysis (or its suppression by chelating agents) may be readily monitored by this assay.[14,20] The action of the selective inhibitors provides the following chemical basis for the assay.[14]

$$A + 2[H] \rightarrow AH_2$$
$$AH_2 + O_2 \rightarrow AH\cdot + HO_2^-$$
$$HO_2^- \rightleftharpoons H^+ + O_2^-$$
$$2HO_2^- \rightarrow H_2O_2 + O_2$$
$$H_2O_2 + O_2^- \rightarrow OH\cdot + OH^- + H_2O$$
$$OH\cdot + DNA \rightarrow \text{strand scission}$$

It should be emphasized that although this assay for the detection of individual oxygen species generated from activated anthracyclines employs nucleic acids it is not thereby restricted to the study of the action of anthracyclines on nucleic acids. The results of the assay may therefore be applied to the general problem of macromolecular oxidative lesions including lipid peroxidation.[15] It may similarly be readily adapted to the study of the enzymic reduction of the anthracycline chromophore[13] involved in such processes as reductive glycosidation.

## Materials

### Reagents

Potassium phosphate buffer 40 m$M$, pH 7.0
Potassium phosphate buffer 20 m$M$, pH 11.8
EDTA, 0.5 m$M$
Ethidium bromide (2,7-diamino-10-ethyl-9-phenylphenanthridinium bromide) 0.5 $\mu$g/ml
Sodium borohydride
Mannitol A.R.
Sodium benzoate A. R.
Acetonitrile, spectral grade

### Enzymes and DNAs

Catalase (beef liver) (EC 1.11.1.6)
Calf thymus topoisomerase
Superoxide dismutase (EC 1.15.1.1)
PM2 CCC DNA at ~1 $A_{260}$
Calf thymus DNA, both native and heat denatured at 1 $A_{260}$

*Procedure*

The fluorescence assay solution is made up freshly in 1 liter quantities containing 0.5 $\mu$g/ml of ethidium bromide, 20 m$M$ tripotassium phosphate, pH 11.8, and 0.5 m$M$ EDTA and stored in a light-proof Repipet Ltd. dispensing container at room temperature.[19] This assay medium upon treatment with heat-denatured calf thymus DNA should give rise to 2–5% of the fluorescence of native calf thymus DNA. Fresh blanks and standards are prepared on a daily basis. The standard contains 10 $\mu$l of calf thymus DNA at 1 $A_{260}$ (0.5 $\mu$g DNA) added to 2 ml of the assay solution. Using the Turner and Associates spectrofluorimeter Model 430, excitation is at 525 nm and emission at 600 nm with the slit widths at their maximum settings. The fluorescence scale is arbitrary, the scale being set at $\times 100$ normally and the sensitivity adjusted so that having set the pH 11.8 ethidium assay solution at 0 for the blank, a standard containing 10 $\mu$l of calf thymus DNA at 1 $A_{260}$ (0.5 $\mu$g) reads 50. The $10 \times 75$-mm tubes used for the assay are first heated with the pH 11.8 assay solution for $>10$ min to ensure no fluorescing material contaminates them. Handling of the tubes should similarly be minimized. Samples are added to ethidium solution in the tubes and may be read immediately or left until a convenient number has accumulated. If samples containing CCC DNA are set aside they must be kept in the dark, otherwise the ethidium gradually nicks the DNA in a light-catalyzed reaction.[23] The fluorimeter cell compartment and circulating water bath (for cooling heated samples) are thermostated to 25°. After the "before heat" fluorescence readings are taken the solutions are heated in the assay tubes at 96° for 3–4 min. The tubes are then cooled in ice, then in a circulating water bath at 25° for a further 2 min.

Reactions are performed at 37° in a volume of 200 $\mu$l containing 20 m$M$ potassium phosphate, pH 7.0, ~1.0 $A_{260}$ units of PM2 CCC DNA[24] (91% CCC), $2 \times 10^{-4}$ $M$ anthracycline, $3.3 \times 10^{-3}$ $M$ of sodium borohydride, and 10% acetonitrile and other components as indicated in the legends to Fig. 3. If relaxed PM2 CCC DNA is to be used this may be prepared readily by prior treatment of the native negatively supercoiled PM2 CCC DNA with calf thymus topoisomerase.[25] At intervals 20-$\mu$l aliquots of the reaction mixture are added to 2 ml of the ethidium assay solution at pH 11.8 and readings of fluorescence taken both before and after heat denaturation and cooling as described above. The effects of selective inhibitors both enzymic and chemical may be monitored in a similar fashion.[14]

From the data of Fig. 3 the before- and after-heat denaturation read-

[23] I. A. Deniss and A. R. Morgan, *Nucleic Acids Res.* **3,** 315 (1976).
[24] A. R. Morgan and D. E. Pulleyblank, *Biochem. Biophys. Res. Commun.* **61,** 396 (1974).
[25] G. Herrick and B. Alberts, *J. Biol. Chem.* **251,** 2124 (1976).

ings are 100 and 88, respectively, so 12 units are due to OC DNA. After single strand scission the before heat reading reaches a maximum of 130 units, so $130 - 12 = 118$ units are due to DNA which was originally CCC. Therefore the percentage increase in fluorescence is $(118 - 88)/88 = 34\%$. In addition the percentage OC DNA in the original sample is $(12 \times 100)/[12 + (88 \times 1.34)] = 9.2\%$. A similar analysis is possible when topoisomerase-relaxed DNA is employed. In this way the extent of oxygen radical-induced DNA scission produced by different anthracyclines under standard conditions may be compared by making allowance for the differences in OC content of the preparation of PM2 DNA used.[10,11]

### Acknowledgment

This work was supported by the National Cancer Institute of Canada.

## [71] Chemical Carcinogenesis: Benzopyrene System

*By* Stephen A. Lesko

The importance of metabolic activation in the mechanism of action of many carcinogenic substances has been recognized for a considerable time. The pioneering work of the Millers[1-3] led to the hypothesis that metabolic activation of carcinogens produces electrophilic intermediates which react with nucleophilic sites of vital macromolecules in cells, thereby initiating neoplastic transformation. More recently it has been realized that metabolic activation of chemical carcinogens may lead to free radical intermediates which may be related to the action of chemical carcinogens.[4,5] There is considerable interest in evaluating the importance of such radical intermediates on the toxic, mutagenic, tumor-initiating, and tumor-promoting capacities of the relevant carcinogens.

The metabolic activation of benzo[*a*]pyrene (BP),[6] an important environmental carcinogen, has been extensively investigated.[7] Several major

[1] J. A. Miller, *Cancer Res.* **30**, 559 (1970).
[2] E. C. Miller, *Cancer Res.* **38**, 1479 (1978).
[3] J. A. Miller and F. C. Miller, *in* "Chemical Carcinogenesis" (P. O. P. Ts'o and J. A. DiPaolo, eds.), Part A, p. 61. Dekker, New York, 1974.
[4] D. C. H. McBrien and T. F. Slater, eds., "Free Radicals, Lipid Peroxidation and Cancer." Academic Press, New York, 1982.
[5] P. O. P. Ts'o, W. J. Caspary, and R. J. Lorentzen, *in* "Free Radicals in Biology" (W. H. Pryor, ed.), Vol. 3, p. 251. Academic Press, New York, 1977.
[6] Abbreviation: BP, benzo[*a*]pyrene.
[7] H. V. Gelboin, *Physiol. Rev.* **60**, 1107 (1980).

Copyright © 1984 by Academic Press, Inc.
All rights of reproduction in any form reserved.
ISBN 0-12-182005-X

BP metabolites, viz. 6-hydroxy-BP, BP-1,6-dione, BP-3,6-dione, and BP-6,12-dione give rise to free radical intermediates and reactive reduced oxygen species as the result of autoxidative processes and redox cycles.[8,9] These metabolites induce DNA strand scission *in vitro*[5,9] and are cytotoxic to mammalian cells in culture.[10,11] More importantly, 6-hydroxy-BP shows moderate mutagenicity in strain TA98 of *Salmonella typhimurium*[12] and Chinese hamster V79 cells[12] and induces morphological transformation of early passage Syrian hamster embryo fibroblasts.[11] This chapter reviews the preparation, purification, and chemical properties of 6-hydroxy-BP and the three isomeric BP diones.

## Methods

### Preparation of 6-Hydroxy-BP

The procedure described is that of Lorentzen *et al.*[8] which is a modification of that originally described by Fieser and Hershberg.[13] 6-Acetoxy-BP is prepared from BP according to the procedure of Fieser and Hershberg.[14] BP (485 mg) is dissolved in 35 ml of benzene. To this is added 1.2 g of lead tetraacetate in 40 ml of glacial acetic acid. The solution is allowed to stand at room temperature for 30 min and the solvent is then removed by rotary evaporation. The 6-acetoxy-BP was dissolved in hot benzene–chloroform (1 : 1) and purified by silica gel chromatography by elution with hexane–benzene (1 : 1) and then benzene. The purified product, mp 209–209.5°, is recrystallized from benzene after removal of the majority of the solvent by rotary evaporation. This 6-acetoxy-BP is then added to a dry ether slurry of lithium aluminum hydride (5- to 10-fold excess). The mixture is stirred under an atmosphere of nitrogen in the dark for 4 hr at room temperature. Careful neutralization with 1 $M$ HCl is carried out at 0° as rapidly as conditions permit. Do not allow the mixture to become acidic. The mixture is extracted with cold ether, dried with anhydrous $Na_2SO_4$, and flask evaporated at 0°. The yellow solid is sublimed *in vacuo*

[8] R. J. Lorentzen, W. J. Caspary, S. A. Lesko, and P. O. P. Ts'o, *Biochemistry* **14**, 3970 (1975).
[9] R. J. Lorentzen and P. O. P. Ts'o, *Biochemistry* **16**, 1467 (1977).
[10] R. J. Lorentzen, S. A. Lesko, K. McDonald, and P. O. P. Ts'o, *Cancer Res.* **39**, 3194 (1979).
[11] L. M. Schechtman, S. A. Lesko, R. J. Lorentzen, and P. O. P. Ts'o, *Proc. Am. Assoc. Cancer Res.* **15**, 66 (1974).
[12] P. G. Wislocki, A. W. Wood, R. L. Chang, W. Levin, H. Yagi, O. Hernandez, P. M. Dansette, D. M. Jerina, and A. H. Conney, *Cancer Res.* **36**, 3350 (1976).
[13] L. F. Fieser and E. B. Hershberg, *J. Am. Chem. Soc.* **61**, 1565 (1939).
[14] L. F. Fieser and E. B. Hershberg, *J. Am. Chem. Soc.* **60**, 2542 (1938).

at 150–160° giving a yellow, microcrystalline product: mp 180°; $\lambda_{max}$ in benzene is 388 nm ($\varepsilon = 20,300$); mass spectrum 268 (molecular ion), 252 (M–O), 240 (M–CO), and 239 (M–CHO). 6-Hydroxy-BP is not stable in solution; purification by solution methods did not give satisfactory results. Air exposure gradually turns the solid compound to a dark yellow-green color and results in increasing amounts of BP-dione impurities. It should be stored under nitrogen or under vacuum desiccation.

### Preparation of 6-Oxo-BP Radical[8]

6-Hydroxy-BP is dissolved in nitrogen-sparged benzene at a concentration of $10^{-3}$ to $10^{-4}$ M with the aid of glove bag filled with nitrogen. An equal volume of a nitrogen-sparged, aqueous solution of $K_3Fe(CN)_6$ at at $10^{-2}$ M is added to the benzene solution. The mixture is shaken and the yellow benzene layer turns almost immediately to a yellowish green color, indicating conversion to the 6-oxo-BP radical. The $K_3Fe(CN)_6$ is required in large excess. The benzene layer is removed and dried with anhydrous $N_2SO_4$. The 6-oxo-BP radical is stable under nitrogen in dried benzene.

### Preparation of BP-Diones

BP-1,6-dione, BP-3,6-dione, and BP-6,12-dione are prepared by autoxidation of 6-OH-BP and purified by extensive column and thick-layer chromatography on alumina.[8] 6-Hydroxy-BP is dissolved in 95% ethanol and an equal volume of $10^{-2}$ M sodium phosphate buffer, pH 7, is added. Concentration of 6-hydroxy-BP is $5 \times 10^{-4}$ M. The solution is allowed to stand at 22° in the dark for at least 2 days. Some of the ethanol is evaporated and the colored products are extracted into benzene. The extract is applied to an alumina (activity grade III) column and first eluted with distilled benzene. The first compound to elute from the column is the BP-6,12-dione which is further purified by thick-layer chromatography (benzene–EtOH, 99 : 1) and recrystallization from toluene. This dione forms rust-colored, fine needles: mp 320–321°, mass spectrum 282 (molecular ion). Next to be eluted is BP-1,6-dione. The first half of this fraction is further purified by thick-layer alumina chromatography (benzene–EtOH, 99 : 1) and recrystallization from toluene. This dione forms orange plates: mp 287–288°; mass spectrum 282 (molecular ion). The third compound to elute is BP-3,6-dione. Complete separation of the 1,6- and 3,6-diones is not easily accomplished because a small amount of 1,6-dione tails into the 3,6-dione fraction. It is necessary to do five successive thick-layer chromatographic runs to obtain a purified product which is then recrystallized from toluene. The 3,6-dione forms red, fine needles: mp 288–289°; mass spectrum 282 (molecular ion). Elution of the column is then continued with benzene–EtOH (98 : 2). A small amount of a violet-colored (absorption $\lambda_{max}$ 527, 568, and 602 nm in benzene), orange fluorescing ($\lambda_{max}$ emission 577 nm in benzene) material is obtained. This material displays an ESR spectrum (singlet, $g = 2.005$, linewidth 7G in benzene). No hyperfine structure is observed even upon degassing of the solution.

The amount of BP-diones obtained is 90–92% of the theoretical yield. This determination is made by absorption spectroscopy using the extinction coefficient data obtained from the highly purified products shown in Fig. 2. When the product mixture is not eluted quickly from the column, the yields are lower.

## Measurement of Oxygen Uptake by 6-Hydroxy-BP

The purified compound is dissolved in 95% ethanol and an equal volume of 10 mM sodium phosphate buffer, pH 7, is added. Hydrocarbon concentration is 0.52 mM. The solution is quickly transferred to completely fill a closed, thermostated chamber (3 ml) fitted with an oxygen electrode. The electrode potential is measured at 22° for 24 hr by a potentiometer and recorded automatically.

## Assay for Hydrogen Peroxide

The assay described is that of Lorentzen et al.[8] which is a modification of the method of Gregory.[15] Prepare a 0.52 mM solution of 6-hydroxy-BP in 1 : 1 (v/v) 95% ethanol–10 mM sodium phosphate, pH 7. Add 50 $\mu$l of 1% dianisidine in methanol and 50 $\mu$l of horseradish peroxidase in water (1 mg/ml). Mix and quickly measure the absorbance at 600 nm against a reference which does not contain peroxidase. It is necessary to use this wavelength rather than the absorbance maximum (460 nm) of the oxidized form of o-dianisidine because of the high extinction of 6-hydroxy-BP and BP-dione products in this region. The extinction coefficient of the oxidized form at 600 nm is $1.38 \times 10^3 \ M^{-1} \ cm^{-1}$.

## Reduction of BP-Diones to BP-Diols[9]

BP-dione is dissolved in distilled, nitrogen-sparged benzene at a concentration of $10^{-3}$ to $10^{-5}$ M. Sodium dithionite (20 mM) is prepared in nitrogen-sparged water. These two immiscible solvents are shaken vigorously, resulting in immediate production of the BP-diol in the benzene layer accompanied by a dramatic color change. The benzene layer is removed in a nitrogen-filled glove bag and dried with sodium sulfate.

[15] R. Gregory, Biochem. J. 101, 582 (1966).

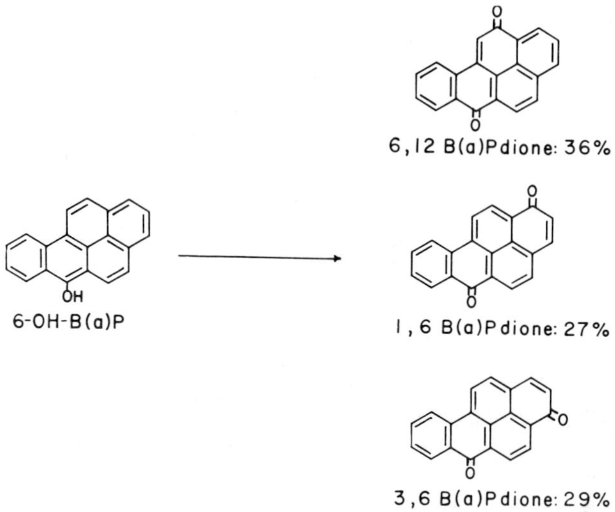

6,12 B(a)Pdione: 36%

6-OH-B(a)P

1,6 B(a)Pdione: 27%

3,6 B(a)Pdione: 29%

FIG. 1. The three major products and their yields obtained from the autoxidation of 6-hydroxy-BP in 1:1 (v/v) 95% ethanol–10 m$M$ sodium phosphate, pH 7. Reprinted with permission from Lorentzen et al.[8] Copyright 1975 American Chemical Society.

When BP-diol is desired in a different solvent, the benzene is evaporated under vacuum, the vacuum broken with nitrogen, and the solid BP-diol redissolved in the appropriate nitrogen-sparged solvent. Any exposure to the atmosphere results in some autoxidation of the BP-diol to BP-dione which is easily detected by absorption spectroscopy. It is desirable to prepare BP-diols immediately before use.

Autoxidation of 6-Hydroxy-BP and the Generation of Reduced
    Oxygen Species

6-Hydroxy-BP is a labile compound being autoxidized in aqueous buffer-ethanol (1 : 1) solution to three stable BP-diones (6,12; 1,6; 3,6) plus a small amount of an unidentified paramagnetic, violet-colored material. Figure 1 shows the percent yields of these diones after incubation for 2 days at room temperature. The absorption spectra and extinction coefficients are shown in Fig. 2. Molecular oxygen is consumed during the autoxidation. Some of this oxygen is reduced by electron transfer which is evident by the accumulation of hydrogen peroxide as the reaction proceeds. Hydrogen peroxide is a very stable species and easy to assay. Due to the paramagnetic nature of ground-state oxygen, spin restriction dictates that, whenever energetically feasible, autoxidations proceed by one-

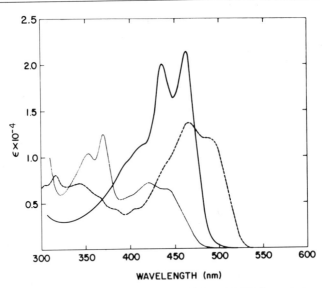

FIG. 2. Absorption spectra and extinction coefficients of BP-diones in benzene; 6,12-dione, $\cdots$; 1,6-dione, ——; 3,6-dione, ----. Reprinted with permission from Ref. 8. Copyright 1975 American Chemical Society.

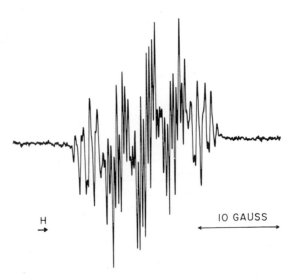

FIG. 3. Hyperfine ESR spectrum of 6-oxo-BP radical in 1:1 (v/v) 95% ethanol–10 m$M$ sodium phosphate, pH 7; modulation amplitude 0.1 G; power 6 mW; $g = 2.0038$. Reprinted with permission from Lorentzen et al.[8] Copyright 1975 American Chemical Society.

electron steps, a notion supported in this specific case by the formation of the 6-oxo-BP radical. Therefore, it is virtually certain that the superoxide radical anion is transiently formed. Hydrogen peroxide is a product of the dismutation of superoxide.

Figure 3 shows the characteristic ESR spectrum of the 6-oxo-BP radical. It produces a five-line spectrum at high modulation (>1 G) indicating that the electron interacts with four protons with approximately equal coupling constants. At lower modulations, additional hyperfine couplings from other protons are observed. Molecular orbital analysis of the 6-oxo-BP radical structure by Inomata and Nagata[16] revealed that four positions in the molecule should have high and approximately equal spin densities leading to a basic five-line spectrum.

The 6-oxo-BP radical is relatively more stable in benzene than in polar solvents. The radical can be isolated and prevented from further oxidation by removing oxygen. Like 6-hydroxy-BP, this isolated free radical is autoxidized in aqueous buffer ethanol solutions to yield the identical BP-dione products. Studies with $^{18}O$ show that molecular oxygen couples directly with the 6-oxo-BP radical. The subsequent conversion of the peroxyl radical to dione could conceivably take place by several different mechanisms.

Nagata et al.[17,18] demonstrated that 6-hydroxy-BP can be detected during the metabolism of BP in rat liver homogenates, as well as in mouse and rat skin homogenates, via the 6-oxo-BP radical. This observation has been confirmed and extended by Lesko et al.[19] The only products detected after incubation of 6-hydroxy-BP in rat liver homogenates were the three isomeric BP-diones. The percentage of BP metabolism proceeding through the 6-hydroxy-BP pathway is about 20% as determined in rat liver homogenates.[19]

## BP-Dione/BP-Diol Redox Cycles: Generation of Reduced Oxygen Species

BP-diones are electron deficient and particularly prone to participate in reversible, one-electron redox reactions involving BP-diols and semi-quinone radical forms. The oxidation potentials of BP-diones are very low at neutral pH[9,20] and they readily accept electrons even from mild biologi-

[16] M. Inomata and C. Nagata, *Gann* **63**, 119 (1972).
[17] C. Nagata, M. Inomata, M. Kodama, and Y. Tagashira, *Gann* **59**, 289 (1968).
[18] C. Nagata, Y. Tagashira, and M. Kodama, *in* "Chemical Carcinogenesis" (P. O. P. Ts'o and J. A. Paolo, eds.), Part A, p. 87. Dekker, New York, 1974.
[19] S. A. Lesko, W. J. Caspary, R. J. Lorentzen, and P. O. P. Ts'o, *Biochemistry* **14**, 3978 (1975).
[20] E. Moriconi, B. Rakoczy, and W. O'Connor, *J. Org. Chem.* **27**, 2772 (1962).

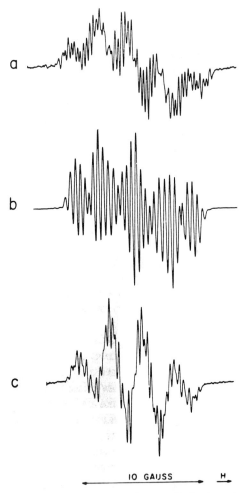

FIG. 4. Hyperfine ESR spectra of BP-semiquinone radical anions obtained from BP-diols $(5 \times 10^{-4} M)$ in 1 : 1 (v/v) 95% ethanol–1 $M$ NaOH by careful exposure to air; (A) BP-6,12-semiquinone radical, modulation amplitude 0.1 G, power 3 mW, $g = 2.0043$; (B) BP-1,6-semiquinone radical, modulation amplitude 0.2 G, power 1 mW, $g = 2.0041$; (C) BP-3,6-semiquinone radical, modulation amplitude 0.05 G, power 2 mW, $g = 2.0043$. Reprinted with permission from Lorentzen and Ts'o.[9] Copyright 1977 American Chemical Society.

cal reducing agents such as NAD(P)H, cysteamine, and glutathione in the absence of oxygen. The BP-diols are extremely sensitive to oxygen. When exposed to the atmosphere, they are immediately and quantitatively oxidized to diones producing an equivalent of hydrogen peroxide.

$$BP\text{-diol} + O_2 \rightarrow [BP\text{-semiquinone radical}] \rightarrow BP\text{-dione} + H_2O_2$$

$$\uparrow \hspace{2cm} \text{Reduction} \hspace{3cm}$$

These autoxidations of BP-diols occur by one-electron steps and the ESR spectra of the three semiquinone radical anions obtained by careful exposure of synthetic BP-diols in alkaline solution to small amounts of air are presented in Fig. 4. At neutral pH, the un-ionized forms of the semiquinone radicals are not sufficiently stable to be observed by ESR spectroscopy. One-electron transfer from BP-diols to molecular oxygen produces transient superoxide radical anions and semiquinone radicals; the superoxide rapidly dismutates to molecular oxygen and hydrogen peroxide, the latter being a detectable product of the autoxidations. The one-electron transfer is further supported by the fact that solutions of BP-diols exposed to air readily reduce nitro blue tetrazolium, a common reagent used for detection of the superoxide radical.

The extreme lability of BP-diols makes their metabolic detection, as such, unlikely. However, Falk et al.[21] identified glucuronide conjugates of BP-diols in the bile of rats treated with BP, and this indicates that BP-dione reduction had taken place in vivo. Other, more general evidence exists indicating that a wide variety of quinone compounds are substrates for cellular reductases.[22-25] Figure 5 demonstrates the ability of small amounts of BP-dione to induce the oxidation of much larger amounts of NADH in the presence of a reductase. This has also been observed using purified rat liver microsomes in place of reductase. A nearly equivalent amount of $H_2O_2$ is produced in this BP-dione-induced oxidation of NADH, indicating oxygen is the ultimate acceptor.

Single-strand breaks are introduced into $T_7$ DNA by BP-diones as measured by alkaline sedimentation.[9] Strand scission is substantially enhanced by the presence of $Cu^{2+}$ or NADPH and suppressed by the presence of EDTA or by the removal of oxygen. These data suggest a free radical mechanism of DNA strand scission involving BP-diones and oxygen. BP-diones are highly toxic to cells in culture.[10] The cellular toxicity is dependent upon molecular oxygen. The replacement of oxygen by nitrogen in both the culture medium and the atmosphere around the cells substantially reduces the capacity of BP-diones to inhibit cellular DNA

[21] H. L. Falk, P. Kotin, S. Lee, and A. Nathan, J. Natl. Cancer Inst. 28, 699 (1962).
[22] N. Bachur, S. Gordon, and M. Gee, Med. Pharmacol. 13, 901 (1977).
[23] N. Bachur, S. Gordon, and M. Gee, Cancer Res. 38, 1745 (1978).
[24] N. Bachur, S. Gordon, M. Gee, and H. Kon, Proc. Natl. Acad. Sci. U.S.A. 76, 954 (1979).
[25] T. Iyanagi and T. Yamazaki, Biochim. Biophys. Acta 172, 370 (1969).

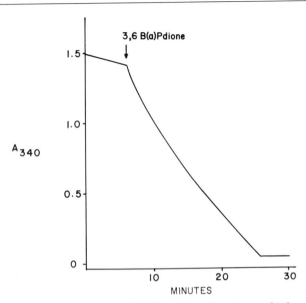

FIG. 5. Oxidation of NADH ($2.4 \times 10^{-4}\ M$), measured by decrease in absorbance at 340 nm, with NADH dehydrogenase from *Clostridium kluyverii* (2 units) using a catalytic amount of BP-3,6-dione ($5 \times 10^{-6}\ M$) as an electron acceptor. After 30 min, $H_2O_2$ concentration was determined to be $1.9 \times 10^{-4}\ M$. Reprinted with permission from Lorentzen and Ts'o.[9] Copyright 1977 American Chemical Society.

synthesis and to cause cell death. Figure 6 presents a model for explaining the cytotoxic effects of BP-diones on mammalian cells in culture.

## Concluding Remarks

BP-diones and 6-hydroxy-BP induce damage to DNA *in vitro*[5,9] and exhibit biological activity when administered to mammalian cells in culture.[10,11] Conditions which modify the biological and biochemical effects of BP-diones and 6-hydroxy-BP indicate that reduced oxygen species propagate the free radical reactions responsible for these effects. Reduced oxygen species have been shown to induce DNA strand scission,[26] DNA interstrand cross-links,[27] DNA–protein crosslinks,[27] and saturate the 5,6 double bond of thymine.[26,28] These various types of DNA damage ob-

[26] S. A. Lesko, R. J. Lorentzen, and P. O. P. Ts'o, *Biochemistry* **19**, 3023 (1980).

[27] S. A. Lesko, J.-L. Drocourt, and S.-U. Yang, *Biochemistry* (1982).

[28] K. A. Schellenberg, *Fed. Proc., Fed. Am. Soc. Exp. Biol.* **21**, 5010 (1979).

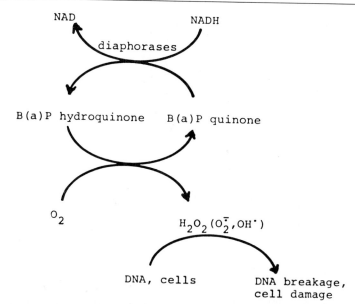

FIG. 6. A cyclic scheme showing the involvement of BP-dione/BP-diol redox couples in the production of reactive reduced oxygen species at the expense of cellular reducing power. Reprinted with permission from Lorentzen and Ts'o.[9] Copyright 1977 American Chemical Society.

served *in vitro* are caused by hydroxyl radicals generated in Fenton-type reactions which involve chelated transition metal ions and hydrogen peroxide.[26,27] Superoxide plays an important role by supplying the reducing power to recycle the oxidized metal ion. Oxygen radical-induced DNA damage has not received much consideration in the initiation of BP carcinogenesis mainly because of the emphasis placed on the covalent binding of electrophilic metabolites to critical cellular nucleophiles. However, the role of free radical damage in radiation-induced carcinogenesis is widely accepted.

Slaga *et al.*[29] have reported that benzoyl peroxide and other free radical-generating compounds are effective skin tumor promoters. Phorbol ester tumor promoters stimulate the production of reduced oxygen species in polymorphonuclear leukocytes[30] and this has been associated with

[29] T. J. Slaga, A. J. P. Klein-Szanto, L. L. Triplett, and L. P. Yotti, *Science* **213**, 1024 (1981).

[30] B. D. Goldstein, G. Witz, M. Amoruso, D. S. Stone, and W. Troll, *Cancer Lett.* **11**, 257 (1981).

extensive DNA strand scission[31] and chromosomal aberrations.[32] This has led Birnboim[31] to postulate that DNA damage may be related to the action of phorbol 12-myristate 13-acetate as a skin tumor promoter in animals. The metabolic conversion of BP to reactive electrophiles as well as intermediates which generate reactive oxygen species may account for its ability to serve as a complete carcinogen, i.e., having both initiating and promoting activities. The role of reactive reduced oxygen species in the tumor-promoting aspects of chemical carcinogenesis is an area that has recently come under active investigation.

[31] H. C. Birnboim, *Science* **215,** 1247 (1982).
[32] I. Emerit and P. A. Cerutti, *Nature (London)* **293,** 144 (1981).

# Author Index

Numbers in parentheses are footnote reference numbers and indicate that an author's work is referred to although the name is not cited in the text.

# Subject Index

## A

relative to hydrogen peroxide, 357
*t*-Butylperoxymaleic acid
mutagenicity testing, 250, 251
reactivity with dichlorofluorescein,
relative to hydrogen peroxide, 357
*N-tert*-Butyl-α-phenylnitrone, spin trap,
19–22
1-Butyraldehyde, production, in microsomal system, 519, 520, 521

# C

Calcium
in neutrophil activation, 389, 390–392
in superoxide production, 371
Calmodulin
activation of specific enzymes, 392, 393
antagonists, 390, 392
in neutrophil activation, 389, 390
calcium dependence, 391, 392
drug inhibition studies, 390, 392
in neutrophils, assay, 390, 391
Carbon dioxide, radiolabeled
production, from [7-$^{14}$C]benzoate, in
microsomal system, 520, 521
release from carboxylated compounds,
as assay of hydrogen peroxide,
397
Carbonic anhydrase, bovine, investigation
by selective free radical technique,
498
Carbon tetrachloride, 387
Carbonyl compounds, excited
generation, 228
photoemission, 221
Carboxypeptidase, investigation by selective free radical technique, 498
Carcinogen
chemical, metabolic activation, 539
*Salmonella* mutagenicity test, 249
Carcinogenesis, *see also* Cell, transformation
benzypyrene, radicals in, 539–550
free radical action in, cell transformation studies of, 473–478
induction, in cell cultures, 467
study, cell cultures in, 464–479
β-Carotene, 39, 160, 162
Carotenoids
assay, 155–162

extraction, 156, 157
high-performance liquid chromatography, 157–160
identification, 160, 161
plasma concentration, 165
quantitation, 161, 162
thin-layer chromatography, 157
CAT, 218, 219
CAT$_{12}$, 218, 219
Catalase, 231, 450
assay, 122–125
in blood, 123
in tissues, 123, 124
in assay for oxygen species generated
from activated anthracyclines, 537
compound I, measurement, in determination of drug-induced hydrogen
peroxide production, 508, 509
cell transformation modifier, 474–478
defense against oxygen toxicity, 47–49
definition of units, 124
determination
automated procedure, 126
by H$_2$O$_2$ removal, 125, 126
by O$_2$ production, 125, 126
titrimetric methods, 125, 126
by UV spectrophotometry, 125, 126
effects, on mutagenicity of test compounds, 263
function, 121
immunoprecipitation, 125, 126
inhibition of endothelial cell injury by
neutrophils, 383
*in vitro*, 121–126
kinetic properties, 121, 122
release of O$_2$, in hydrogen peroxide
assay, 397, 398
screening techniques, 125, 126
specific activity, 124, 125
calculation, 122
in human erythrocytes, 122
stability, 124
subcellular distribution, in rat liver,
107–111
substrate, 26
in superoxide dismutase assay, 459
Catecholamine, 59
Cell
damage, by oxy-radicals, assessment,
410–412